Stable Isotopes and Plant Carbon–Water Relations

Physiological Ecology
A Series of Monographs, Texts, and Treatises

Series Editor
Harold A. Mooney
Stanford University, Stanford, California

Editorial Board
Fakhri Bazzaz   F. Stuart Chapin   James R. Ehleringer
Robert W. Pearcy   Martyn M. Caldwell

List continues at end of this volume

Stable Isotopes and Plant Carbon–Water Relations

Edited by

James R. Ehleringer

Stable Isotope Ratio Facility
for Environmental Research
Department of Biology
University of Utah
Salt Lake City, Utah

Anthony E. Hall

Department of Botany and Plant Science
University of California
Riverside, California

Graham D. Farquhar

Plant Environmental Biology Group
Research School of Biological Sciences
Australian National University
Canberra, Australia

Academic Press, Inc.

A Division of Harcourt Brace & Company

San Diego   Boston   New York
London   Sydney   Tokyo   Toronto

Cover photograph: Fagus Forest, Bayreuth, Germany.

This book is printed on acid-free paper. ∞

Academic Press, Inc.
1250 Sixth Avenue, San Diego, California 92101-4311

United Kingdom Edition published by
Academic Press Limited
24–28 Oval Road, London NW1 7DX

Library of Congress Cataloging-in-Publication Data

Stable isotopes and plant carbon/water relations / edited by James R.
  Ehleringer, Anthony E. Hall, Graham D. Farquhar.
      p.   cm. -- (physiological ecology series)
    Includes bibliographical references and index.
    ISBN 0-12-233380-2
    1. Plant-water relationships.   2. Plants--Metabolism.   3. Carbon-
  -Isotopes.   4. Stable isotopes in plant physiology research.
  I. Ehleringer, J. R.   II. Hall, A. E. (Anthony Elmit),   Date.
  III. Farquhar, Graham D.   IV. Series: Physiological ecology.
  QK870.S73   1993
  581. 1' 33--dc20   93-1090
                                                            CIP

PRINTED IN THE UNITED STATES OF AMERICA
93 94 95 96 97 98   MM   9 8 7 6 5 4 3 2 1

7206641

Contents

Contributors   xv
Preface   xix

Part I
History and Theoretical Considerations

1. Introduction: Water Use in Relation to Productivity   3
    I. Introduction   3
    II. A Long-Standing Interest in Water-Use Efficiency   3
    III. Concepts of Water Use in Relation to Productivity   4
    References   7

2. Historical Aspects of Stable Isotopes in Plant Carbon and Water Relations   9
James R. Ehleringer and John C. Vogel
    I. First Observations of Carbon Isotope Fractionation in Plants   10
    II. Discovery of C_4 Photosynthesis   11
    III. Carbon Isotopes and Photosynthetic Pathway Distribution   12
    IV. Understanding the Basis of ^{13}C Variation in Plants   12
    V. Water and Evaporative Enrichment   13
    VI. Environmental Signals in D/H and $^{18}O/^{16}O$ of Plant Tissues   14
    VII. Water Movement into Plants   15
    References   15

3. Biochemical Basis of Carbon Isotope Fractionation   19
Marion H. O'Leary
    I. Introduction   19
    II. Definition of Terms   19
    III. Components of Isotope Fractionation in Plants   20
    IV. Theory of Isotope Fractionation in Plants   24
    V. Summary   26
    References   26

4. Variability of Carbon Isotope Fractionation during Photosynthesis   29
John C. Vogel
    I. Introduction   29
    II. Two-Stage Isotope Fractionation   31

    III. Separation of the Diffusional Steps   34
    IV. Daylight Respiration of C_3 Plants   35
    V. Discrimination in C_4 Plants   37
    VI. Experimental Verification   40
    VII. Discussion and Conclusions   43
    References   45

5. Carbon and Oxygen Isotope Effects in the Exchange of Carbon Dioxide between Terrestrial Plants and the Atmosphere   47
Graham D. Farquhar and Jon Lloyd
    I. Carbon Isotope Effects   47
    II. Oxygen Isotope Effects   58
    References   66

6. Environmental and Biological Influences on the Stable Oxygen and Hydrogen Isotopic Composition of Leaf Water   71
Lawrence B. Flanagan
    I. Introduction   71
    II. Isotopic Fractionation during Transpiration   72
    III. Environmental and Biological Effects on Leaf-Water Isotopic Composition   76
    IV. Leaf-Water $^{18}O/^{16}O$ Ratio and $C^{18}O^{16}O$ Discrimination during Photosynthesis   86
    V. Conclusions   88
    References   88

Part II
Ecological Aspects of Carbon Isotope Variation

7. Carbon Isotope Discrimination in Leaf Soluble Sugars and in Whole-Plant Dry Matter in *Helianthus annuus* L. Grown under Different Water Conditions   93
Marco Lauteri, Enrico Brugnoli, and Luciano Spaccino
    I. Introduction   93
    II. Materials and Methods   94
    III. Results and Discussion   96
    IV. Summary   106
    References   107

8. Carbon Isotope Discrimination and the Coupling of CO_2 Fluxes within Forest Canopies   109
Mark S. J. Broadmeadow and Howard Griffiths
    I. Introduction   109
    II. Carbon Isotope Composition of Organic Material and Source CO_2 within Forest Canopies   110
    III. Carbon Isotope Discrimination within Tropical Forest Formations in Trinidad   115

IV. Carbon Isotope Discrimination within a Temperate Coniferous Canopy   119
V. Variations in δ_p and δ_a: Implications for the Coupling of CO_2 and H_2O Fluxes within Canopies   125
VI. Summary   127
References   127

9. Environmental and Physiological Influences on Carbon Isotope Composition of Gap and Understory Plants in a Lowland Tropical Forest   131
Paula C. Jackson, Frederick C. Meinzer, Guillermo Goldstein,
Noel M. Holbrook, Jaime Cavelier, and Fermin Rada
I. Introduction   131
II. Materials and Methods   133
III. Results   134
IV. Discussion   137
V. Summary   139
References   139

10. Carbon Isotope Fractionation in Tree Rings in Relation to the Growing Season Water Balance   141
N. J. Livingston and D. L. Spittlehouse
I. Introduction   141
II. Methods   142
III. Results   146
IV. Discussion   150
V. Conclusions   152
References   152

11. Carbon and Water Relations in Desert Plants: An Isotopic Perspective   155
James R. Ehleringer
I. Desert Climate and Life Form Variation   155
II. Carbon Isotope Composition as a Measure of the Set Point for Gas Exchange Activity   157
III. Δ as a Reliable Indicator of Intercellular CO_2 Concentration and Water-Use Efficiency   159
IV. Desert Environments Are Characterized by Low Δ Values   160
V. Interpopulation-Level Variation in Carbon Isotope Discrimination   163
VI. Carbon Isotope Discrimination and Competition for Water   164
VII. Selection for Variation in Carbon Isotope Discrimination Values   166
VIII. Summary   170
References   171

12. Carbon Isotope Discrimination and Resource Availability in the Desert Shrub *Larrea tridentata*   173
Philip W. Rundel and M. Rasoul Sharifi
   I. Introduction   173
  II. Physiological Ecology of *Larrea tridentata*   174
 III. Field Studies and Laboratory Experiments   175
  IV. Components of Variation in Gas Exchange and $\delta^{13}C$ 176
   V. Summary   182
      References   184

13. Altitudinal Variation in Carbon Isotope Discrimination by Conifers   187
John D. Marshall and Jianwei Zhang
   I. Introduction   187
  II. Carbon Isotope Discrimination and Water-Use Efficiency in Conifers   188
 III. Altitudinal Variation in Carbon Isotope Discrimination   191
  IV. Genetic Variation in Carbon Isotope Discrimination in Conifers   191
   V. Plasticity of Δ across Altitudinal Gradients   194
  VI. Potential Mechanisms for the Plastic Response   195
 VII. Summary   197
      References   197

14. Characterization of Photobiont Associations in Lichens Using Carbon Isotope Discrimination Techniques   201
C. Máguas, H. Griffiths, J. Ehleringer, and J. Serôdio
   I. Introduction   201
  II. Carbon Isotope Discrimination and Gas Exchange Methodology   202
 III. Characterization of Three Lichen Groups as Determined by Photobiont Association   204
  IV. Evaluation of Lichen Physiology Using Carbon Isotope Discrimination and Gas Exchange Techniques: Discussion   208
   V. Summary   210
      References   211

15. Carbon Isotope Discrimination and Gas Exchange in Ozone-Sensitive and -Resistant Populations of Jeffrey Pine   213
Mark T. Patterson and Philip W. Rundel
   I. Introduction   213
  II. Materials and Methods   215
 III. Results   217
  IV. Discussion   221
   V. Summary   223
      References   224

16. Carbon Isotope Composition and Gas Exchange of Loblolly and Shortleaf Pine as Affected by Ozone and Water Stress   227
Christine G. Elsik, Richard B. Flagler, and Thomas W. Boutton
    I. Introduction   227
    II. Experimental Methods   228
    III. Results   231
    IV. Discussion   236
    V. Summary   241
    References   242

Part III
Agricultural Aspects of Carbon Isotope Variation

17. Genetic and Environmental Variation in Transpiration Efficiency and Its Correlation with Carbon Isotope Discrimination and Specific Leaf Area in Peanut   247
Graeme C. Wright, Kerry T. Hubick, Graham D. Farquhar, and R. C. Nageswara Rao
    I. Introduction   247
    II. Peanut Cultivar Variation in Transpiration Efficiency and Correlation with Δ at the Whole-Plant Level   248
    III. Peanut Cultivar Variation in Transpiration Efficiency and Correlation with Δ in Field Canopies   252
    IV. Genotype X Environment Interaction and Heritability for W and Δ 257
    V. Relationships between Specific Leaf Area, W, and Δ 258
    VI. Negative Association between W and Harvest Index   261
    VII. Selection for W in Peanut-Breeding Programs   264
    VIII. Summary   265
    References   266

18. Genotypic and Environmental Variation for Carbon Isotope Discrimination in Crested Wheatgrass, a Perennial Forage Grass   269
Douglas A. Johnson, Kay H. Asay, and John J. Read
    I. Introduction   269
    II. Association of Δ with W 271
    III. Effect of Environment on Δ 273
    IV. Genotypic Variation in Δ 276
    V. Stability of Δ Response   276
    VI. Relationship between Δ and Yield   277
    VII. Summary   279
    References   279

19. Carbon Isotope Discrimination, Water Relations, and Gas Exchange in Temperate Grass Species and Accessions   281
Richard C. Johnson and Larry L. Tieszen
    I. Introduction   281
    II. Field Evaluation of Δ in Temperate Grass Species and Accessions   282
    III. Water-Use Efficiency in Whole Plants and Δ 286
    IV. Gas Exchange and Δ in Tall Fescue Genotypes   288
    V. Germplasm Enhancement in Tall Fescue   293
    VI. Summary   295
    References   296

20. Environmental and Developmental Effects on Carbon Isotope Discrimination by Two Species of *Phaseolus*   297
Qingnong A. Fu, Thomas W. Boutton, James R. Ehleringer, and Richard B. Flagler
    I. Introduction   297
    II. Methods and Materials   298
    III. Results and Discussion   299
    IV. Summary   307
    References   308

21. Diversity in the Relationship between Carbon Isotope Discrimination and Transpiration Efficiency when Water Is Limited   311
Kerry T. Hubick and Ann Gibson
    I. Introduction   311
    II. Factors Affecting W and Δ 312
    III. The Use of Δ to Predict W in Plant Breeding   316
    IV. Relationships between W and Δ in Diverse C_3 Species   318
    V. Conclusion   323
    References   324

22. Carbon Isotope Discrimination and Gas Exchange in Coffee during Adjustment to Different Soil Moisture Regimes   327
Frederick C. Meinzer, Guillermo Goldstein, and David A. Grantz
    I. Introduction   327
    II. Responses to Progressive Soil Drying in the Field   329
    III. Dynamics of Adjustment to Altered Soil Moisture Regimes   333
    IV. Sensing of Soil Water Availability   335
    V. Importance of Spatial and Temporal Scales   339
    VI. Discrimination as a Predictor of Performance   341
    VII. Summary   343
    References   344

Part IV
Genetics and Isotopic Variation

23. Implications for Plant Breeding of Genotypic and Drought-Induced Differences in Water-Use Efficiency, Carbon Isotope Discrimination, and Gas Exchange   349
Anthony E. Hall, Abdelbagi M. Ismail, and Cristina M. Menendez
    I. Introduction   349
   II. Choosing Tissue to Be Sampled for C-Isotope Composition and Breeding Nursery Environments   350
  III. Consistency of Genotypic Ranking for C-Isotope Discrimination   352
  IV. Physiological Basis of Plant Differences in C-Isotope Discrimination   353
    V. Inheritance of C-Isotope Discrimination and Water-Use Efficiency   355
  VI. Heritability of C-Isotope Discrimination and Genetic Correlations with Other Traits   360
 VII. Breeding Strategies for C-Isotope Discrimination and Progress in Breeding   361
VIII. Summary   367
     References   368

24. Analysis of Restriction Fragment Length Polymorphisms Associated with Variation of Carbon Isotope Discrimination among Ecotypes of *Arabidopsis thaliana*   371
Josette Masle, Jeong Sheop Shin, and Graham D. Farquhar
    I. Introduction   371
   II. Analysis of Naturally Occurring DNA Polymorphisms versus Analysis by Mutagenesis   372
  III. Principle of DNA Polymorphism Analysis   373
  IV. Analysis of DNA Polymorphisms Associated with Variation of Carbon Isotope Discrimination in *Arabidopsis thaliana*   375
    V. Prospects   383
     References   385

25. Implications of Carbon Isotope Discrimination Studies for Breeding Common Bean under Water Deficits   387
Jeffrey W. White
    I. Introduction and Review   387
   II. Variation in Carbon Isotope Discrimination in Bean Germplasm   388
  III. Inheritance of Carbon Isotope Discrimination   391
  IV. Associations of Δ with Physiological Traits, Crop Growth, and Seed Yield   393
    V. Implications for Breeding Common Bean   396
  VI. Summary   397
     References   397

26. Potential of Carbon Isotope Discrimination as a Selection Criterion in Barley Breeding   399
E. Acevedo
　　I. Introduction   399
　　II. Carbon Isotope Discrimination   400
　　III. Screening for Carbon Isotope Discrimination in Early
　　　　Generations   410
　　IV. Summary   414
　　　　References   415

27. Genetic Analyses of Transpiration Efficiency, Carbon Isotope Discrimination, and Growth Characters in Bread Wheat   419
Bahman Ehdaie, David Barnhart, and J. G. Waines
　　I. Introduction   419
　　II. Materials and Methods   420
　　III. Results and Discussion   423
　　IV. Breeding Implications   431
　　V. Summary   432
　　　　References   433

28. Exploiting Genetic Variation in Transpiration Efficiency in Wheat: An Agronomic View   435
A. G. Condon and R. A. Richards
　　I. Introduction   435
　　II. Genetic Variation in Transpiration Efficiency in Wheat   436
　　III. Relationships between Dry Matter Production or Grain Yield and
　　　　Carbon Isotope Discrimination in Field Experiments   437
　　IV. Sources of Variation in Carbon Isotope Discrimination
　　　　in Wheat   439
　　V. Testing the Effect of "Scale" on Season-Long Transpiration
　　　　Efficiency of Wheat   441
　　VI. The Effect of Changing Photosynthetic Capacity   447
　　VII. Summary   448
　　　　References   449

29. Challenges Ahead in Using Carbon Isotope Discrimination in Plant-Breeding Programs   451
R. A. Richards and A. G. Condon
　　I. Improving Water-Use Efficiency   451
　　II. Measurement of Carbon Isotope Discrimination   453
　　III. Genetic Challenges   455
　　IV. The Relationship between Δ and W in Field-Grown Plants   456
　　V. Use of Δ in Breeding Programs   460
　　VI. Summary and Conclusions   461
　　　　References   461

Part V
Water Relations and Isotopic Composition

30. Water Sources of Plants as Determined from Xylem-Water Isotopic Composition: Perspectives on Plant Competition, Distribution, and Water Relations   465
Todd E. Dawson
  I. Introduction   465
  II. Water Sources: Using δD as a "Tracer"   467
  III. Source- and Plant-Water Determinations and Their Use in Studying Plant Distributions and Interactions   469
  IV. Long-Term Studies of δD Variation in Source and Plant Waters   486
  V. Conclusions   491
     References   492

31. Hydrogen Isotopic Fractionation by Plant Roots during Water Uptake in Coastal Wetland Plants   497
Guanghui Lin and Leonel da S. L. Sternberg
  I. Introduction   497
  II. Materials and Methods   498
  III. Results   500
  IV. Discussion   505
  V. Summary   508
     References   509

32. The Source of Water Transpired by *Eucalyptus camaldulensis*: Soil, Groundwater, or Streams?   511
Peter J. Thorburn and Glen R. Walker
  I. Introduction   511
  II. Water Relations of *E. camaldulensis* on the River Murray Floodplain   512
  III. Methods   512
  IV. Results   517
  V. Discussion   520
  VI. Summary   525
     Appendix: Calculating the Proportion of Groundwater in Tree Sap from δD and δ^{18}O Data   525
     References   527

33. The ^{18}O of Water in the Metabolic Compartment of Transpiring Leaves   529
Dan Yakir, Joseph A. Berry, Larry Giles, and C. Barry Osmond
  I. Introduction   529
  II. Material and Methods   531
  III. Results   535
  IV. Discussion   537
     References   539

Index   541

Contributors

Numbers in parentheses indicate the pages on which the authors' contributions begin.

Edmundo Acevedo (399), International Maize and Wheat Improvement Center, (CIMMYT), Wheat Program, 06600 Mexico, D.F., Mexico

Kay H. Asay (269), U.S. Department of Agriculture, Agricultural Research Service, Forage and Range Research Laboratory, Utah State University, Logan, Utah 84322-6300

David Barnhart (419), Department of Botany and Plant Sciences, University of California, Riverside, Riverside, California 92521

Joseph A. Berry (529), Department of Plant Biology, Carnegie Institution of Washington, Stanford, California 94305

Thomas W. Boutton (297), Department of Rangeland Ecology and Management, Texas A&M University, College Station, Texas 77843

Mark S. J. Broadmeadow (109), Department of Agricultural and Environmental Science, University of Newcastle upon Tyne, Newcastle NE1 7RU, United Kingdom

Enrico Brugnoli (93), CNR, Istituto per l'Agroselvicoltura, via Marconi 2, Porano (TR), 05010 Italy

Jaime Cavelier (131), Departamento de Biologia, Universidad de Los Andes, Bogota, Apdo Aereo 4976, Colombia

Anthony G. Condon (435, 451), Commonwealth Scientific Industrial Research Organization, Division of Plant Industry, Canberra, ACT 2601, Australia

Todd E. Dawson (465), Section of Ecology and Systematics, Cornell University, Ithaca, New York 14853

Bahman Ehdaie (419), Department of Botany and Plant Sciences, University of California, Riverside, Riverside, California 92521

James R. Ehleringer (9, 155, 201, 297), Stable Isotope Ratio Facility for Environmental Research and Department of Biology, University of Utah, Salt Lake City, Utah 84112

Christine G. Elsik (227), Department of Forest Science, Texas Agricultural Experiment Station, Texas A&M University, College Station, Texas 77843

Graham D. Farquhar (47, 247, 371), Research School of Biological Sciences, Institute of Advanced Studies, Australian National University, Canberra, ACT 2601, Australia

Richard B. Flagler (227, 297), Department of Forest Science, Texas A&M University, College Station, Texas 77843

Lawrence B. Flanagan (71), Department of Biology, Carleton University, Ottawa, Ontario, Canada K1S 5B6

Qingnong A. Fu[1] (297), Department of Rangeland Ecology and Management, Texas A&M University, College Station, Texas 77843

Ann Gibson (311), Department of Forestry, Australian National University, Canberra, ACT 2601, Australia

Larry Giles (529), Botany Department, Duke University, Durham, North Carolina 27706

Guillermo Goldstein (131, 327), Department of Botany, University of Hawaii, Honolulu, Hawaii 96822

David A. Grantz (327), University of California, Kearney Agricultural Center, Parlier, California 93648

Howard Griffiths (109), Department of Agricultural and Environmental Science, University of Newcastle upon Tyne, Newcastle NE1 7RU, United Kingdom

Anthony E. Hall (349), Department of Botany and Plant Sciences, University of California, Riverside, Riverside, California 92521

Noel M. Holbrook (131), Department of Biological Sciences, Stanford University, Stanford, California 94305

Kerry T. Hubick (247, 311), Exploitable Science and Technology, Department of Industry, Technology, and Commerce, Canberra, ACT 2601, Australia

Abdelbagi M. Ismail (349), Department of Botany and Plant Sciences, University of California, Riverside, Riverside, California 92521

Paula C. Jackson (131), Department of Biology and Laboratory of Biomedical and Environmental Sciences, University of California, Los Angeles, Los Angeles, California 90024

Douglas A. Johnson (269), U.S. Department of Agriculture, Agricultural Research Service, Forage and Range Research Laboratory, Utah State University, Logan, Utah 84322-6300

Richard C. Johnson (281), U.S. Department of Agriculture, Agricultural Research Service, Western Regional Plant Introduction Station, Pullman, Washington 99164

Marco Lauteri (93), CNR, Istituto per l'Agroselvicoltura, via Marconi 2, Porano (TR), 05010 Italy

Guanghui Lin (497), Department of Biology, University of Miami, Coral Gables, Florida 33124

Nigel J. Livingston (141), Department of Biology, University of Victoria, Victoria, British Columbia, Canada V8W 2Y2

[1]Present address: Department of Horticultural Sciences, Texas A&M University, College Station, Texas 77843.

Jon Lloyd (47), Plant Environmental Biology Group, Research School of Biological Sciences, Institute of Advanced Studies, Australian National University, Canberra, ACT 2601, Australia

Christina Máguas (201), Botanical Department, Ecology Section, Lisbon University, 1700 Lisbon, Portugal

John D. Marshall (187), Department of Forest Resources, University of Idaho, Moscow, Idaho 83843

Josette Masle (371), Research School of Biological Sciences, Institute of Advanced Studies, Australian National University, GPO Box 475, Canberra, ACT 2601, Australia

Frederick C. Meinzer (131, 327), Hawaiian Sugar Planters' Association, Aiea, Hawaii 96701

Cristina M. Menendez (349), Department of Botany and Plant Sciences, University of California, Riverside, Riverside, California 92521

Marion H. O'Leary (19), Department of Biochemistry, University of Nebraska at Lincoln, Lincoln, Nebraska 68583

C. Barry Osmond (529), Botany Department, Duke University, Durham, North Carolina 27706

Mark T. Patterson (213), Laboratory of Biomedical and Environmental Sciences, and Department of Biology, University of California, Los Angeles, Los Angeles, California 90024

Fermin Rada (131), Departamento de Biologia, Universidad de Los Andes, Merida 5101 Venezuela

R. C. Nageswara Rao (247), Legumes Program, International Crop Research Institute for the Semi-Arid Tropics, Hyderabad, India

John J. Read (269), U.S. Department of Agriculture, Agricultural Research Service, Crops Research Laboratory, Fort Collins, Colorado 80526

Richard A. Richards (435, 451), Commonwealth Scientific Industrial Research Organization, Division of Plant Industry, Canberra, ACT 2601, Australia

Philip W. Rundel (173, 213), Laboratory of Biomedical and Environmental Sciences, and Department of Biology, University of California, Los Angeles, Los Angeles, California 90024

João Serôdio (201), Botanical Department, Ecology Section, Lisbon University, 1700 Lisbon, Portugal

M. Rasoul Sharifi (173), Department of Biology and Laboratory of Biomedical and Environmental Sciences, University of California, Los Angeles, Los Angeles, California 90024

Jeong Sheop Shin (371), Department of Agronomy, College of Natural Resources, Korea University, Seoul 136-075, Korea

Luciano Spaccino (93), CNR, Istituto per l'Agroselvicoltura, via Marconi 2, Porano (TR), 05010 Italy

David L. Spittlehouse (141), British Columbia Ministry of Forests, Victoria, British Columbia, Canada V8W 3E7

Leonel da S. L. Sternberg (497), Department of Biology, University of Miami, Coral Gables, Florida 33124

Peter J. Thorburn (511), Centre for Goundwater Studies, Commonwealth Scientific Industrial Research Organization, Division of Water Resources, Glen Osmond SA 5064, Australia

Larry L. Tieszen (281), Department of Biology, Augustana College, Sioux Falls, South Dakota 57197

John C. Vogel (9, 29), Quaternary Dating Research Unit, Commonwealth Scientific Industrial Research Organization, CSIR, Pretoria 0001, South Africa

J. Giles Waines (419), Department of Botany and Plant Sciences, University of California, Riverside, Riverside, California 92521

Glen R. Walker (511), Centre for Goundwater Studies, Commonwealth Scientific Industrial Research Organization, Division of Water Resources, Glen Osmond SA 5064, Australia

Jeffrey W. White (387), Centro International Agricultura Tropicale, Apartado Aereo 6713, Cali, Colombia

Graeme C. Wright (247), Queensland Department of Primary Industries, J. Bjelke-Peterson Research Station, Kingaron, Qld 4610, Australia

Dan Yakir (529), Botany Department, Duke University, Durham, North Carolina 27736

Jianwei Zhang (187), Department of Forest Resources, University of Idaho, Moscow, Idaho 83843

Preface

Stable isotope analyses are receiving an increasing emphasis in plant biology, ranging from agricultural through ecological studies. This attention stems from an improved theoretical understanding of the basis for isotopic fractionation in plants, an awareness of the utility of stable isotopes for scaling physiological processes across temporal and spatial scales, and the prospects for using stable isotopes as a tool for plant-breeding programs. The contributions to this volume address these three topics, providing a unique opportunity to compare common themes in ecological and agricultural research.

The purpose of this book was to bring together investigators to examine the usefulness of stable isotopes in addressing a broad range of questions in agricultural and ecological sciences. Attention to this topic had originated earlier with a conference held nearby at Lake Arrowhead, California, in 1986. As will be seen in the contributions that follow, a number of new and productive directions have emerged since the Lake Arrowhead conference.

One focus of a subsequent meeting in Riverside, California, was to examine new developments over the intervening six years. Another major focus was a critical evaluation of the prospects that stable isotopic analyses will directly contribute to agricultural breeding programs, in which the interest is in applying basic information to the improvement of crop performance under limited soil moisture environments.

We are grateful for the sponsorship and the joint cooperation of the Australian Department of Industry, Technology, and Commerce; the U.S. National Science Foundation; the U.S. Department of Agriculture; and the University of California, Riverside. We hope that this volume will serve two purposes: first, to summarize the progress that has been achieved in the application of stable isotopes to the understanding of plant carbon and water relations and, second, to provide a framework and to stimulate future work in this area.

James R. Ehleringer
Anthony E. Hall
Graham D. Farquhar

I

History and Theoretical Considerations

The contributions to this section focus on the fundamentals of plant carbon–water relations, the historical development of stable isotopes in plant biology, and the biochemical basis of fractionation of stable isotopes of carbon, hydrogen, and oxygen. The introductory chapter sets the stage with a summary of basic aspects of gas exchange as they relate to carbon dioxide and water vapor fluxes between plants and the atmosphere and also establishes a common set of terms relating water loss and carbon gain. That chapter is followed by a historical view of stable isotopes in plants by Ehleringer and Vogel, who point out that the initial basic observations on this topic are all derived from geochemistry. O'Leary follows with a discussion of the basic isotope fractionation processes and establishes the linkages between carbon isotopes and plant gas exchange. In the subsequent chapter, Farquhar and Lloyd derive a theory linking plant gas exchange and atmospheric gas composition, illustrating that both carbon and oxygen isotopes in carbon dioxide provide insightful information that can be used to understand physiological processes at the leaf level and then can be used to scale for global considerations. In the last chapter of this section, Flanagan discusses the theory behind how both hydrogen and oxygen stable isotope analyses of leaf water reveal information on biological and atmospheric processes.

1

Introduction: Water Use in Relation to Productivity

I. Introduction

Water loss with respect to photosynthesis has been the topic of numerous books and reviews and has attracted widespread interest from diverse groups, ranging from those interested in basic mechanistic aspects of plant physiology to those concerned with constraints on agricultural production (Fischer and Turner, 1978; Taylor *et al.*, 1983; Sinclair *et al.*, 1984; Stanhill, 1986; Turner, 1986). The intent of this section is to briefly introduce the roots of this interest and to provide a set of terms relating to plant-water use that will be useful when reading the remaining chapters in this volume.

II. A Long-Standing Interest in Water-Use Efficiency

For over a century there has been interest in understanding the relationships between water consumption by plants and overall productivity. Ecologists have been interested in how leaves of different species varied in these parameters, especially in response to seasonal and geographical changes in moisture availability and how these responses influenced both structural and physiological features of natural vegetation. This interest was pioneered by natural scientists and plant geographers (Haberlandt, 1884; Schimper, 1898; Warming, 1909) and later with an ecological emphasis (Livingston and Shreve, 1921). At the same time, agronomists, whose primary goal was increasing productivity, were especially interested in the water requirements for growth of a broad range of crop species (Briggs and Shantz, 1913b).

The pioneering experiments in this area involved growing plants in pots and other containers, where precise (but very time-consuming) information was gained by measuring plant growth and water loss over extended time periods (Briggs and Shantz, 1913a; Shantz and Piemiesel, 1927). From these studies of crop and native plants, it became clear that there was substantial variation in the relationships between water consumption and biomass production. While the mechanistic bases were unknown at the time, plants could be separated into two distinct groups on the basis of their water requirements for growth. We now know that these two groupings represent the C_3 and C_4 photosynthetic pathways. Furthermore, these early studies established that there could be significant differences among cultivars of a crop in water consumption requirements in relation to biomass production.

These observations showed that atmospheric conditions were of overriding importance in determining absolute relations between water consumption and biomass productivity of a crop. The large site-to-site and year-to-year variations in absolute water consumption characteristics (driven by temperature and humidity conditions) of the same crop were confounding and made it difficult to accept the usefulness of information derived from pot studies (Stanhill, 1986). In part because of the necessity to better understand the climatic factors influencing transpiration and in part because of the need to extend from pot to field studies, micrometeorological aspects predominated from the 1930s through the 1960s (Penman, 1948; Penman and Schofield, 1951; Lemon *et al.*, 1961).

A shift toward appreciating the significance of plants (particularly at the leaf level) in affecting water and carbon fluxes with the atmosphere reemerged with the pioneering efforts of de Wit (1958). Shortly thereafter, Bierhuizen and Slatyer (1965) published their studies of the interactions between carbon gain through photosynthesis and water loss through transpiration. An analogous situation occurred in the ecological sciences, where the emphasis shifted from understanding community dynamics and structure, which predominated until the early 1960s, to ecophysiological studies of natural systems (Mooney *et al.*, 1987). Today field micrometeorological and physiological approaches are routinely combined in field investigations, and carbon isotope analysis has emerged as a means of spatially and temporally integrating these carbon and water relations parameters (Farquhar *et al.*, 1989).

III. Concepts of Water Use in Relation to Productivity

In evaluating relationships between water loss and growth, a diffuse, sometimes overlapping, and frequently confusing terminology has emerged, which often impedes progress and understanding. In part the differences in terms are driven by different objectives: the interests of a physiologist in understanding mechanisms at the leaf level are not the same as those of an agricultural engineer interested in the marketable crop yield relative to the amount of irrigation applied to a field. Also along the way, terms such as "efficiency" have emerged, which only add further confusion since they typically do not refer to an efficiency in the same sense as it is understood in engineering. This baggage may not be avoidable in all cases. Nevertheless, useful concepts and terms have emerged to help us better understand the interactions between water and carbon fluxes in evaluating plant physiological responses, growth, and survival.

At the leaf level, carbon and water fluxes involve *net photosynthesis* (*A*), measured as CO_2 uptake, and *transpiration* (*E*), measured as water loss. At steady state these fluxes can be quantified by Ohm's law analogy relationships:

$$E = \nu g, \tag{1}$$

and

$$A = (c_a - c_i)g/1.6, \tag{2}$$

where ν is the gradient in water-vapor pressure between the leaf and atmosphere divided by total atmospheric pressure, g is the leaf conductance to water vapor, and 1.6 is the ratio of gaseous diffusivities of CO_2 and water vapor in air. c_a and c_i are the external and internal CO_2 concentrations. The term ν is equal to

$$\nu = \frac{e_i - e_a}{P}, \tag{3}$$

where e_i and e_a are the water-vapor pressures inside the leaf (assumed to be at saturation) and in the ambient air outside the boundary layer, respectively, and P is the total atmospheric pressure. By expressing the driving gradient for transpiration in this manner, the effects of elevation changes (*P*) and temperature on diffusion do not have an impact on the estimates of the stomatal conductance (*g*), which should only be influenced by stomatal pore characteristics (Cowan, 1977). In Eq. (3), the numerator is similar to a term many previous studies have presented as vapor pressure deficit (vpd). The term ν has been presented as "Δw" in many previous studies, but such an expression leads to confusion with respect to carbon isotope discrimination (Δ) (*sensu* Farquhar *et al.*, 1989).

Instantaneous water-use efficiency is defined as the ratio of the fluxes of net photosynthesis (*A*) and transpiration (*E*). Since *A* and *E* share a diffusion pathway, the stomata, instantaneous water-use efficiency can be determined without an estimate of g:

$$\frac{A}{E} = \frac{c_a(1 - c_i/c_a)}{1.6\nu}. \tag{4}$$

Considering the strong role that atmospheric conditions play in determining water-use efficiency through its influence on determining ν, many investigators chose to evaluate only the role that biological components play in determining water–carbon exchange relationships. The *intrinsic water-use efficiency* is defined as A/g, which is really the instantaneous water-use efficiency multiplied by v. However, the advantage of intrinsic water-use efficiency as a term is that it allows direct comparison of intrinsic physiological considerations, factoring out the confounding effects of temperature and humidity gradient differences between plants.

While instantaneous and intrinsic water-use efficiencies might be useful in evaluating water and carbon exchange on a short-term basis, they do not necessarily scale to long-term considerations, such as might be of interest to those studying canopy productivity or growth. Yet the basic notion of assimilation and transpiration can be extended to longer time periods. Here the basic measure is one of the actual biomass produced to the total amount of water consumed in producing that biomass.

Transpiration efficiency (W) is the expression for long-term measures of biomass or yield with respect to water loss at the whole plant level. It can be related to Eq. (4) as

$$W = \frac{c_a(1 - c_i/c_a)(1 - \phi_c)}{1.6v(1 + \phi_w)}. \tag{5}$$

A fraction of the fixed carbon will be lost through respiration (ϕ_c) and a fraction of the water (ϕ_w) may be lost at night if stomata do not completely close. The consequence is that W is not equivalent to A/E. W is defined in terms of molar abundance ratio, yet can just as easily be expressed on a grams-carbon-per-grams-water or grams-biomass-per-grams-water basis. As such, it is particularly important to make clear just exactly which units are being used in the expression.

As will be seen in later chapters in this volume, the carbon isotope composition ($^{13}C/^{12}C$) is often used in developing a time-integrated surrogate estimate of either A/E or W. In the strict sense, such an extrapolation is only possible provided the following assumptions are met: (1) the evaporative gradient between leaf and environment (v) is equivalent and known for the species and environments compared and (2) plants and/or species compared do not differ in the fraction of the carbon gain that is lost through respiratory processes (Farquhar and Richards, 1984). Neither of these conditions is likely to be entirely met in a field study of many species. Although carbon isotopic composition values of leaf tissue are unlikely to provide a direct, quantitative comparison of any measure of water-use efficiency, they should provide a relative index for ranking A/E or W among species. Furthermore, calculations can be made to determine the extent to which errors arising from differences in leaf temperatures among plants being compared affect interpretation of carbon isotope composition data.

Frequently, agronomists are not so much interested in W but in the crop yield (biomass of harvestable product such as seed) per unit of water applied to the crop. This expression is not equal to W but is exactly equal to the product of W and the harvest index.

The expression W also lacks consideration of evaporative moisture loss occurring from the soil surface. Such losses can occur early in the growing cycle of a crop or because inadequate moisture levels limit growth and prevent complete canopy cover of the soil surface. Of importance to the agronomists will be a measure that incorporates the transpiration efficiency to the fraction of water actually consumed by the crop. *Crop water-use efficiency* (C) is defined as

$$C = W \cdot \int E / \int E_{\text{total}}, \tag{5}$$

where E is the transpirational water loss through the crop and E_{total} is the total water loss from the soil (Passioura, 1977). Both water-loss rates are integrated over the life of the crop and expressed on a unit ground area basis.

None of the above "efficiency" terms are clearly measures of efficiency as typically considered in engineering, although the last definition does come close since it measures the extent to which the plant is able to process the moisture loss from the soil.

Alternatively, we could express relationships between carbon and water fluxes in an inverse manner. That is, instead of A/E, we could express the patterns through E/A. Such was actually the approach used in the pioneering efforts at the beginning of this century. The "water requirement" was defined as the ratio of the amount of water transpired through the plant relative to the amount of biomass produced. Mathematically, the water requirement would be equivalent to $1/W$. Today the expression "water requirement" is not used. However, in its place, it is common in the literature to find the term *transpiration ratio*.

References

Bierhuizen, J. F., and R. O. Slatyer. 1965. Effects of atmospheric concentration of water vapour and CO_2 in determining transpiration-photosynthesis relationships of cotton leaves. *Agric. Meteor.* **2:** 259–270.

Briggs, L. J., and H. L. Shantz. 1913a. The water requirements of plants. I. Investigations in the Great Plains in 1910 and 1911. USDA Bureau of Plant Industry, Bulletin 284.

Briggs, L. J., and H. L. Shantz. 1913b. The water requirement of plants. II. A review of the literature. USDA Bureau of Plant Industry, Bulletin 285.

Cowan, I. R. 1977. Stomatal behavior and environment. *Adv. Bot. Res.* **4:** 117–227.

de Wit, C. T. 1958. "Transpiration and Crop Yields. Institute of Biological and Chemical Research on Field Crops and Herbage." Wageningen, The Netherlands.

Farquhar, G. D., and R. A. Richards. 1984. Isotopic composition of plant carbon correlates with water-use efficiency of wheat genotypes. *Aust. J. Plant Physiol.* **11:** 539–552.

Farquhar, G. D., J. R. Ehleringer, and K. T. Hubick. 1989. Carbon isotope discrimination and photosynthesis. *Annu. Rev. Plant Physiol. Mol. Biol.* **40:** 503–537.

Fischer, R. A., and N. C. Turner. 1978. Plant productivity in the arid and semi-arid zones. *Annu. Rev. Plant Physiol.* **29:** 277–317.

Haberlandt, G. 1884. "Physiologische Pflanzenanatomie." Verlag von Wilhelm Englemann, Leipzig.

Lemon, E. R., D. W. Stewart, and R. W. Shawcroft. 1961. The sun's work in a cornfield. *Science* **174:** 371–378.

Livingston, B. E., and F. Shreve. 1921. "The Distribution of Vegetation in the United States, as Related to Climatic Conditions." Carnegie Institution of Washington, Washington, D.C.

Mooney, H. A., R. W. Pearcy, and J. R. Ehleringer. 1987. Plant physiological ecology today. *BioScience* **37:** 18–20.

Passioura, J. B. 1977. Grain, yield, harvest index, and water use of wheat. *J. Aust. Inst. Agri. Sci.* **43:** 117–120.

Penman, H. L. 1948. Natural evaporation from open water, bare soil, and grass. *Proc. R. Soc. London A* **193:** 120–146.

Penman, H. L., and R. K. Schofield. 1951. Some physical aspects of assimilation and transpiration. *Sym. Soc. Exp. Biol.* **5:** 115–129.

Schimper, A. F. W. 1898. "Pflanzen-Geographie auf Physiolgischer Grundlage." Fisher, Jena.

Shantz, H. L., and L. N. Piemeisel. 1927. The water requirement of plants at Akron, Colorado. *J. Agri. Res.* **34,** 1093–1190.

Sinclair, T. R., C. B. Tanner, and J. M. Bennett. 1984. Water-use efficiency in crop production. *BioScience* **34:** 36–40.

Stanhill, G. 1986. Water use efficiency. *Adv. Agron.* **39:** 53–85.

Taylor, H. M., W. R. Jordan, and T. R. Sinclair. 1983. "Limitations to Efficient Water Use in Crop Production." American Society of Agronomy, Madison, WI.

Turner, N. C. 1986. Crop water deficits: A decade of progress. *Adv. Agron.* **39:** 1–51.

Warming, E. 1909. "Oecology of Plants: An Introduction to the Study of Plant Communities." Clarendon Press, Oxford, U.K.

2

Historical Aspects of Stable Isotopes in Plant Carbon and Water Relations

James R. Ehleringer and John C. Vogel

Most of the history of stable isotopes in plant biological research began outside of biology. Interest in stable isotopes first developed in the physical sciences in the mid-1930s and then by the 1940s became a major component of geology. Only in the past 2 decades have plant biologists shown widespread interest in stable isotopes. The original botanical contributions in the area of stable isotopes are largely a history of contributions by geochemists with very broad interests who wanted to know more about natural variations in isotopic abundance levels. In this brief chapter, we attempt to reconstruct some of that history, specifically as it relates to plant water and carbon relations, and to describe the "how"s and "why"s of the development of botanical interests in stable isotopes.

Harold Urey and co-workers discovered a heavy form of hydrogen, deuterium, in the early 1930s. Once deuterium was known, it was only natural to investigate whether there was variation in the abundance of light and heavy isotopic forms of hydrogen in nature. Careful measurements of the density of water revealed considerable differences in the heavy isotope content of samples of different origin. While this approach was satisfactory for looking at heavy water, it was not practical for studying the abundance of isotopic forms of other elements that were being discovered during this period. A major breakthrough came when Nier and colleagues developed the modern isotope mass spectrometer (Nier, 1936, 1940, 1990), which allowed precise measurements of the relative abundances of light and heavy isotopic forms of hydrogen as well as other elements, such as carbon and oxygen. The development of mass spectrometers of sufficiently high precision in the late 1940s prompted systematic surveys of the variability in the isotopic composition of the light elements (H, C, N, O) in nature. These studies were notably undertaken in the laboratory of Harold Urey at the

University of Chicago by Harmon Craig (^{13}C), Sam Epstein (^{18}O), Irving Friedman (^{2}H), Cesare Emiliani (^{18}O thermometry), and other co-workers (see review by Nier, 1990).

I. First Observations of Carbon Isotope Fractionation in Plants

Using a mass spectrometer, Nier and Gulbransen (1939) were the first to observe that the heavy isotope of carbon, ^{13}C, was slightly depleted in plants with respect to inorganic carbonaceous materials such as limestone. Shortly thereafter, Murphy and Nier (1941) went on to show that there was variation in ^{13}C content in wood among different plant species.

The early surveys of the ^{13}C/^{12}C ratio of carbon in plants (Wickman, 1952; Craig, 1953) were accompanied by attempts to explain the observed ^{13}C depletion. Based on herbarium samples, Wickman (1952), working in Stockholm, observed that plants collected along railroad stops from central Asian deserts were more enriched in ^{13}C than plants from tropical rain forests. He suggested that there was a cyclic process in which the degree of depletion of ^{13}C in plants depended on the amount of recycling of soil-derived carbon dioxide. Craig (1954) pointed out, however, that this mechanism could not explain the observed fractionation satisfactorily. Harmon Craig favored an alternative hypothesis—that fractionation processes within leaves accounted for the variation in ^{13}C composition and that perhaps the environment played a role in influencing the magnitude of these fractionations. Meanwhile Baertschi (1953), working in Basel, showed in laboratory experiments that no fractionation occurred during the respiration of bean seedlings and that ^{13}C was depleted by about 26‰ during CO_2 assimilation/photosynthesis. The similarity of this fractionation factor with that found by Urey in 1948 (29‰) and by Calvin and Weigl in 1952 (27‰) for algae (cited by Baertschi, 1953) suggested to him that the fractionation process was basically the same for different plants and growth conditions. The values these authors obtained, incidentally, correspond closely to the maximum fractionation factor for photosynthesis observed in nature.

Craig's (1954) in-depth discussion of the biological fractionation process formed the basis for further experimental work by Park and Epstein (1960, 1961). Rod Park and Sam Epstein proposed a three-step model to account for the observed differences in ^{13}C composition of leaves from the CO_2 in the atmosphere. Diffusion and photosynthetic fractionation were important components of their model, with secondary plant metabolism accounting for only a small additional fractionation. They correctly reasoned that leaf-level fractionation must occur, since the isotopic composition of dissolved CO_2 was not the same as that of the carbohydrate formed. Park and Epstein suspected that translocation of enriched [13]CO_2 to the root system played an important part in determining the overall fractionation. At the time, there was supportive evidence from other investigators, showing that roots evolved CO_2 at high rates during the light. However, Park and Epstein (1960, 1961) were unable to confirm that the root CO_2 was enriched in [13]C. Here the quest to understand the fractionation process in plants rested for nearly 20 years.

II. Discovery of C₄ Photosynthesis

Without it yet being realized, C_4 plants had been represented in the first plant surveys: Wickman (1952) found four desert specimens with abnormally small discrimination against [13]C (*Calligonum acanthopterum* in the Polygonaceae and *Arthrophytum arborescens, Haloxylon aphyllum,* and *Salsola richteri* in the Chenopodiaceae). Since these plants derived from the central Asian deserts with high-carbonate soils, they formed an important aspect of Wickman's cyclic enrichment model. Wickman (1952) also found one emergent aquatic grass (*Paspalum distichum*) with high ^{13}C content. It is perhaps just one of these odd facts that this grass was collected at Lake Laugunita on the Stanford University campus, where 20 years later the Carnegie group on that campus would conduct their pioneering research on the genetics and physiology of C_4 photosynthesis.

Craig (1953) also had one unknown "grass from southwestern Kansas" that had high ^{13}C content, while all other plant samples collected from that pedocalcic site had "normal" values. He reexamined that grass sample after washing it with HCl to ensure that any carbonates would be removed. When he observed the same ^{13}C content as before, Harmon Craig concluded that "the sample has either utilized bicarbonate or carbonate ion directly in its photosynthesis, or that it has used carbon dioxide derived from the bicarbonate ion." Unfortunately, Craig was incorrectly referring to the possibility that CO_2 from calcium carbonate at the soil surface was serving as the photosynthetic CO_2 source for this grass (because either the air was enriched or there was direct uptake through the roots).

The connection between the C_4 syndrome and the ^{13}C content in terrestrial plants only became clear after the discovery of the C_4 pathway in tropical grasses by Kortshak et al. (1965) and Hatch and Slack (1966). Margaret Bender (1968) published the first ^{13}C data showing an association between enriched isotopic composition and the newly discovered photosynthetic pathway. Extension of this "anomalous" group of plants to other families soon followed (Bender, 1971; Smith and Epstein, 1971), and it was not long before isotopic studies led to the discovery that C_4 species were found to belong to at least 13 different families, suggesting parallel evolution rather than one common origin. The fact that certain succulents showed anomalous ^{13}C values was also noted at this time (Vogel and Lerman, 1969; Bender, 1971) and was soon linked to crassulacean acid metabolism (Bender et al., 1973).

III. Carbon Isotopes and Photosynthetic Pathway Distribution

The linkage between ^{13}C content and photosynthetic pathway type make it possible to more easily determine the extent of C_4 photosynthesis within the plant kingdom. The late 1960s and 1970s were an exciting time for tracing the phylogenetic distribution of C_3, C_4, and CAM photosynthesis among different plant taxa. The charge was lead by two distinct groups: established geochemists/physicists broadly interested in botany (such as Margaret Bender in Wisconsin and John Vogel in South Africa) and a group of plant physiologists who were either doing postdoctoral research in geochemical laboratories or had access to necessary instrumentation. Among the pioneering group of plant physiologists were Bruce Smith working in Texas; Bruce Tregunna, John Downton, and Joe Berry in Canada; John Troughton in New Zealand; Barry Osmond in Australia; Elien Deleens in France; and Klaus Winter and Hubert Ziegler in Germany. The 1969 U.S.–Australia meeting on photosynthesis and photorespiration in Canberra (Hatch *et al.*, 1971) served as a focal point for what was known at the time about carbon isotope composition and photosynthesis (Downton, 1971). The isotopic data base expanded rapidly during the 1970s and Downton (1975) and Raghavendra and Das (1976, 1978) produced summary lists of the families known to have C_4 taxa, pointing out that a number of the genera contained both C_3 and C_4 species.

Later studies and field surveys continued to fill gaps in our knowledge of the taxonomic distribution of C_3 and C_4 photosynthesis. One of the basic observations that emerged from these field studies was that there could be substantial variation in the ^{13}C content of C_3 photosynthesis, even within a species, while much less variation occurred among C_4 plants (O'Leary, 1988). While it was presumed that this isotopic variation encompassed both environmental and genetic components, the basis contribution of each component remained unclear until the basis of the fractionation events were elucidated.

IV. Understanding the Basis of ^{13}C Variation in Plants

Once detailed descriptions of the kinetic isotope fractionation process during photosynthesis were produced in the late 1970s (Vogel, 1980; O'Leary and Osmond, 1980; O'Leary, 1981; Farquhar *et al.*, 1982) the scene was set for the development of direct application of carbon isotope studies to physiological processes. It was especially Graham Farquhar, Marion O'Leary, and Joe Berry (1982) who showed that the carbon isotope ratio of an individual plant correlated with intercellular carbon dioxide levels and that this could be used in the selective breeding of high transpiration efficiency genotypes (Farquhar and Richards, 1984). Several chapters in this volume deal with further developments in this field.

Applications of leaf ^{13}C content to ecophysiology have also arisen out of the observation that the isotopic depletion in plants is influenced by environmental factors. One such application is related to the original cyclic depletion model proposed by Wickman (1952) and by atmospheric CO_2 observations of Keeling (1958); it can be observed in dense forests where recycling of respired carbon dioxide takes place—the so-called "canopy effect" (Vogel, 1978). While the phenomenon is somewhat more complicated than originally proposed, it seems clear that isotopic variation within canopies reflects both a recycling of respired carbon dioxide as well as increased intercellular carbon dioxide levels in the lower leaves of the canopy (Schleser and Jayasekera, 1985; Ehleringer *et al.*, 1986; Sternberg *et al.*, 1989). Several chapters in this volume deal with the current efforts to use ^{13}C composition of plants to address recycling of carbon within forest canopies.

V. Water and Evaporative Enrichment

Studies of hydrogen and oxygen isotope ratios are quickly assuming greater importance in plant biology, but again the history of research on these elements began outside of biology. Early workers in the 1950s in Europe and the United States measured the ^{18}O and ^{2}H content of different freshwater samples (Epstein and Mayeda, 1953; Friedman, 1953; Dansgaard, 1954), showing that fresh water typically contained less of the heavy isotopes than did seawater. However, it was the systematic investigations of Dansgaard (1961, 1964) in Denmark that provided the framework for a mechanistic understanding of the factors contributing to the geographic isotopic variation in meteoric waters.

Harold Urey (1947) calculated that atmospheric oxygen at equilibrium should be enriched in ^{18}O by 6‰ relative to seawater. Yet when Dole *et al.* (1954) measured atmospheric diatomic oxygen, they observed an ^{18}O excess of 23‰. Since this diatomic oxygen originates from photosynthesis and because evaporative processes were known to result in an enrichment of surface waters (Craig and Gordon, 1965), early investigations began to analyze the isotopic composition of leaf water which becomes enriched during transpiration (Gonfiantini *et al.*, 1965). The extent of the leaf-water evaporative enrichment could be modeled (Dongmann *et al.*, 1974; Zundel *et al.*, 1978), and one of the key parameters influencing enrichment of the heavier isotopes was relative humidity. Early in these efforts, Farris and Strain (1978) suggested that water stress could directly impact the extent of leaf-water enrichment. However, it now appears that a significant amount of the isotopic enrichment by water-stressed leaves may be the result of elevated leaf temperatures in stressed plants (Flanagan and Ehleringer, 1991).

There was limited attention by plant physiologists to leaf-water evaporative enrichment until recently when models linking carbon isotope fractionation and transpiration efficiency became better understood (see recent progress detailed by Farquhar *et al.*, Chapter 5; Flanagan, Chapter 6; and Yakir *et al.*, Chapter 33, in this volume). One ultimate goal of the renewed interest has been to determine whether or not isotopic evaporative enrichment of leaf water can be used as a direct measure of the leaf-to-air water vapor gradient, the driving force for transpiration. If so, this would make it possible to calculate absolute estimates of transpiration efficiency through combined studies of the hydrogen and oxygen isotopes of leaf water and the carbon isotope ratios of leaves. This effect should be especially useful for assessing the degree to which plants can limit water loss under hot, dry conditions in their natural environment.

VI. Environmental Signals in D/H and $^{18}O/^{16}O$ of Plant Tissues

The investigation of deuterium (^2H or D) in plant tissues was initiated by a general survey of deuterium in organic matter in the early 1970s (Schiegl, 1970, 1972; Schiegl and Vogel, 1970; Epstein *et al.*, 1976). These studies showed that the hydrogen in the tissue of plants is considerably depleted in deuterium with respect to the source water from which it was derived, viz the leaf sap or, in the case of hydrophytes, the aqueous substrate. Schiegel (1970) found that there was a correlation between D/H ratio and climate, with D/H ratio strongly correlated with temperature. Early in these studies, Sam Epstein and colleagues at Cal Tech focused on the D/H ratio of organic tissues as paleorecorders (Epstein *et al.*, 1976; Yapp and Epstein, 1982). In particular, they concentrated on cellulose in wood, since this material is laid down during a single season, clearly distinguishable as tree rings within the wood, and not modified thereafter. Recognizing that the hydrogen in hydroxyl groups of cellulose could continue to exchange with water within the plant, Epstein and colleagues restricted their analyses to only the D/H ratio of C—H groups in cellulose (Epstein *et al.*, 1976).

Physiological studies were soon to follow. Ziegler *et al.* (1976), Sternberg and De Niro (1983), and Leaney *et al.* (1985) showed that there were distinct differences in D/H ratio among C_3, C_4, and CAM plants, while Epstein and co-workers started to record the differences in isotopic composition of various chemical substances in the plant (Smith and Epstein, 1970; Epstein *et al.*, 1976, 1977; Lenhart, 1979; Ziegler, 1979).

Since techniques were developed for measuring the ^{18}O content of organic matter (e.g., Hardcastle and Friedman, 1974), this isotope has been analyzed together with deuterium in studies of the fractionation processes within the plant. Since oxygen is potentially derived not only from water but also from carbon dioxide, it was thought that the ^{18}O content of organic matter could conceivably show a different pattern from that of the D/H ratio. De Niro and Epstein (1979), however, established that equilibration with aqueous phase is established before ^{18}O incorporation in the tissue, so that the initial $^{18}O/^{16}O$ ratio is determined by the plant water alone.

The application of ^2H and ^{18}O analyses of plant tissue in ecological studies is actively being pursued (cf. Ziegler 1989; Sternberg 1989; White 1989) and will undoubtedly produce many interesting results in the near future. Some of these topics are discussed later by Farquhar and Lloyd (Chapter 5), by Flanagan (Chapter 6), and also by Yakir and colleagues (Chapter 33).

VII. Water Movement into Plants

At the same time as the leaf-water studies were being initiated, Gonfiantini *et al.* (1965) and Wershaw *et al.* (1966) showed that there was no hydrogen or oxygen isotope fractionation during water uptake through roots. These observations established that xylem sap analyses could be used for determining plant-water sources by comparing the nonenriched moisture in the stems of plants with that of the possible sources. White *et al.* (1985) and Sternberg and Swart (1987) were among the first to apply these principles to ecological studies; their results showed that adjacent plants could use different water sources and that plants could shift between sources over short time periods. This application is now gaining widespread interest because of its power in quantitatively assessing belowground activities. Later in this volume, Dawson (Chapter 30) and Thorburn and Walker (Chapter 32) examine the recent progress in this area.

References

Baertschi, P. 1953. Die Fraktionierung der natürlichen Kohlenstoffisotopen im Kohlendioxydstoffwechsel grüner Pflanzen. *Helv. Chim. Acta* **36:** 773–781.

Bender, M. M. 1968. Mass spectrometric studies of carbon-13 variations in corn and other grasses. *Radiocarbon* **10:** 468–472.

Bender, M. M. 1971. Variations in the $^{13}C/^{12}C$ rations of plants in relation to the pathway of photosynthetic carbon dioxide fixation. *Phytochemistry* **10:** 1239–1244.

Bender, M. M., I. Rouhani, H. M. Vines, and J. C. C. Black. 1973. $^{13}C/^{12}C$ ratio changes in crassulacean acid metabolism plants. *Plant Physiol.* **52:** 427–430.

Craig, H. 1953. The geochemistry of the stable carbon isotopes. *Geochim. Cosmochim. Acta* **3:** 53–92.

Craig, H. 1954. Carbon-13 in plants and the relationship between carbon-13 and carbon-14 variations in nature. *J. Geol.* **62:** 115–149.

Craig, H., and L. I. Gordon. 1965. Deuterium and oxygen-18 variations in the ocean and marine atmosphere. *In* Stable Isotopes in Oceanographic Studies and Paleotemperatures. Spoleto, Italy. Lischi & Figli, Pisa.

Dansgaard, W. 1954. The O^{18} abundance in fresh water. *Geochim. Cosmochim. Acta* **6:** 241–260.

Dansgaard, W. 1961. The isotopic composition of natural waters with special reference to the Greenland ice cap. Medd. Groenland **165(2):** 1–120.

Dansgaard, W. 1964. Stable isotopes in precipitation. *Tellus* **16:** 436–468.

De Niro, M. J., and S. Epstein. 1979. Relationship between oxygen isotope ratios of terrestrial plant cellulose, carbon dioxide and water. *Science* **204:** 51–53.

Dole, M., G. A. Lange, D. P. Rudd, and D. A. Zaukelies. 1954. Isotopic composition of atmospheric oxygen and nitrogen. *Geochim. Cosmochim. Acta* **6:** 65–78.

Dongmann, G., H. Förstel, and K. Wagner. 1972. ^{18}O-rich oxygen from land plants. *Nature* **240:** 127–128.

Dongmann, G., H. W. Nurnberg, H. Förstel, and K. Wagner. 1974. On the enrichment of H_2 ^{18}O in the leaves of transpiring plants. *Radiat. Environ. Biophys.* **11:** 41–52.

Downton, W. J. S. 1971. Check list of C_4 species, pp. 554–558. *In* M. D. Hatch, C. B. Osmond, and R. O. Slatyer (eds.), Photosynthesis and Photorespiration. Wiley, New York.

Downton, W. J. S. 1975. The occurrence of C_4 photosynthesis among plants. *Photosynthetica* **9:** 96–105.

Ehleringer, J. R., C. B. Field, Z. F. Lin, and C. Y. Kuo. 1986. Leaf carbon isotope and mineral composition in subtropical plants along an irradiance cline. *Oecologia* **70:** 520–526.

Epstein, S., and T. Mayeda. 1953. Variation of O^{18} content of waters from natural sources. *Geochim. Cosmochim. Acta* **4:** 213–224.

Epstein, S., C. J. Yapp, and J. H. Hall. 1976. The determination of the D/H ratio of nonexchangeable hydrogen in cellulose extracted from aquatic and land plants. *Earth Planet. Sci. Lett.* **30:** 241–251.

Epstein, S., P. Thompson, and C. J. Yapp. 1977. Oxygen and hydrogen isotopic ratios in plant cellulose. *Science* **198:** 1209–1215.

Farquhar, G. D., and R. A. Richards. 1984. Isotopic composition of plant carbon correlates with water-use efficiency of wheat genotypes. *Aust. J. Plant Physiol.* **11:** 539–552.

Farquhar, G. D., M. H. O'Leary, and J. A. Berry. 1982. On the relationship between carbon isotope discrimination and the intercellular carbon dioxide concentration in leaves. *Aust. J. Plant Physiol.* **9:** 121–137.

Farris, F., and B. R. Strain. 1978. The effects of water stress on leaf H_2 ^{18}O enrichment. *Radiat. Environ. Biophys.* **15:** 167–202.

Flanagan, L. B., and J. R. Ehleringer. 1991. Effects of mild water stress and diurnal changes in temperature and humidity on the stable oxygen and hydrogen isotopic composition of leaf water in *Cornus stolonifera* L. *Plant Physiol.* **97:** 298–305.

Friedman, I. 1953. Deuterium content of natural water and other substances. *Geochim. Cosmochim. Acta* **4:** 89–103.

Gonfiantini, R., S. Gratziu, and E. Tongiorgi. 1965. Oxygen isotopic composition of water in leaves, pp. 405–410. *In* Isotopes and Radiation in Soil-Plant Nutrition Studies. Intern. At. Energy Agency, Vienna.

Hardcastle, K. G., and I. Friedman. 1974. A method for oxygen isotope and analyses of organic material. *Geophys. Res. Lett.* **1:** 165–167.

Hatch, M. D., and C. R. Slack. 1966. Photosynthesis by sugarcane leaves: A new carboxylation reaction and the pathway of sugar formation. *Biochem. J.* **101:** 103–111.

Hatch, M. D., C. B. Osmond, and R. O. Slatyer (eds.). 1971. Photosynthesis and Photorespiration. Wiley, New York.

Keeling, C. D. 1958. The concentration and isotopic abundance of atmospheric carbon dioxide in rural areas. *Geochim. Cosmochim. Acta* **13:** 322–334.

Kortschak, H. P., C. E. Hartt, and G. O. Burr. 1965. Carbon dioxide fixation in sugarcane leaves. *Plant Physiol.* **40:** 209–213.

Leaney, F. W., C. B. Osmond, G. B. Allison, and H. Ziegler. 1985. Hydrogen-isotope composition of leaf water in C_3 and C_4 plants: Its relationship to the hydrogen-isotope composition of dry matter. *Planta* **164:** 215–220.

Lenhart, B. 1979. Untersuchungen zur Kohlenstoff- und Wasserstoffisotopen-Diskriminierung bei C_3-, C_4- und CAM-Pflanzen, Dissertation Technische Universität, München.

Murphy, B. F., and A. O. Nier. 1941. Variations in the relative abundance of the carbon isotopes. *Phys. Rev.* **59:** 771–772.

Nier, A. O. 1936. A mass-spectrographic study of the isotopes of argon, potassium, rubidium, zinc, and cadmium. *Phys. Rev.* **50:** 1041–1045.

Nier, A. O. 1940. A mass spectrometer for routine isotope abundance measurements. *Rev. Sci. Instrum.* **11:** 212–216.

Nier, A. O. 1990. Some reminiscences of isotopes, geochronology, and mass spectrometry, Vol. 3, parts 1 and 2, pp. 590–607. *In* J. Lederberg (ed.), Excitement and Fascination of Science: Reflections by Eminent Scientists. Annual Reviews, Inc., Palo Alto, CA.

Nier, A. O., and E. A. Gulbransen. 1939. Variations in the relative abundance of the carbon isotopes. *J. Am. Chem. Soc.* **61:** 697–698.

O'Leary, M. H. 1981. Carbon isotope fractionation in plants. *Phytochemistry* **20:** 553–567.

O'Leary, M. H. 1988. Carbon isotopes in photosynthesis. *BioScience* **38:** 328–336.

O'Leary, M. H., and C. B. Osmond. 1980. Diffusional contribution to carbon isotope fractionation during dark CO_2 fixation in CAM plants. *Plant Physiol.* **66:** 931–934.

Park, R., and S. Epstein. 1960. Carbon isotope fractionation during photosynthesis. *Geochim. Cosmochim. Acta* **21:** 110–126.

Park, R., and S. Epstein. 1961. Metabolic fractionation of ^{13}C and ^{12}C in plants. *Plant Physiol.* **36:** 133–138.

Raghavendra, A. S., and V. S. R. Das. 1976. Distribution of the C_4 dicarboxylic acid pathway of photosynthesis in local monocotyledonous plants and its taxonomic significance. *New Phytol.* **76:** 301–305.

Raghavendra, A. S., and V. S. R. Das. 1978. The occurrence of C_4-photosynthesis: A supplementary list of C_4 plants reported during late 1974–mid 1977. *Photosynthetica* **12:** 200–208.

Schiegl, W. E. 1970. Natural Deuterium in Biogenic Materials. Ph.D. thesis, University of South Africa, Pretoria.

Schiegl, W. E. 1972. Deuterium content of peat as a palaeoclimatic recorder. *Science* **175:** 512–513.

Schiegl, W. E., and J. C. Vogel. 1970. Deuterium content of organic matter. *Earth Planet. Sci. Lett.* **7:** 307–313.

Schleser, G. H., and R. Jayasekera. 1985. $\delta^{13}C$ variations of leaves in forests as an indication of reassimilated CO_2 from the soil. *Oecologia* **65:** 536–542.

Smith, B. N., and S. Epstein. 1970. Biogeochemistry of the stable isotopes of hydrogen and carbon in salt marsh biota. *Plant Physiol.* **46:** 738–742.

Smith, B. N., and S. Epstein. 1971. Two categories of $^{13}C/^{12}C$ ratios for higher plants. *Plant Physiol.* **47:** 380–384.

Sternberg, L. S. L. 1989. Oxygen and hydrogen isotope ratios in plant cellulose: Mechanisms and application, pp. 124–141. *In* P. W. Rundel, J. R. Ehleringer, and K. A. Nagy (eds.), Stable Isotopes in Ecological Research. Springer-Verlag, New York.

Sternberg, L. S. L., and M. J. DeNiro. 1983. Isotopic composition of cellulose from C_3, C_4, and CAM plants growing in the vicinity of one another. *Science* **220:** 947–948.

Sternberg, L. S. L., and P. K. Swart. 1987. Utilization of fresh water and ocean water by coastal plants of southern Florida. *Ecology* **68:** 1898–1905.

Sternberg, L. S. L., S. S. Mulkey, and S. J. Wright. 1989. Ecological interpretation of leaf carbon isotope ratios: Influence of respired carbon dioxide. *Ecology* **70:** 1317–1324.

Urey, H. C. 1947. The thermodynamic properties of isotopic substances. *J. Chem. Soc.* **1947:** 562–581.

Vogel, J. C. 1978. Recycling of carbon in a forest environment. *Oecol. Plant.* **13:** 89–94.

Vogel, J. C. 1980. Fractionation of the carbon isotopes during photosynthesis, pp. 111–135. *In* Sitzungsberichte der Heidelberger Akademie der Wissenschaften, mathematisch-naturwissenschaftliche Klasse Jahrgang 1980. 3. Abdandlung. Springer-Verlag, Berlin/Heidelberg/New York.

Vogel, J. C., and J. C. Lerman. 1969. Groningen Radiocarbon Dates. VIII. *Radiocarbon* **11:** 351–390.

Wershaw, R. L., I. Friedman, S. J. Heller, and P. A. Frank. 1966. Hydrogen isotope fractionation of water passing through trees, pp. 55–67. *In* G. D. Hobson (ed.), Advances in Organic Geochemistry. Pergamon, New York.

White, J. W. C. 1989. Stable hydrogen isotope ratios in plants: A review of current theory and some potential applications, pp. 142–162. *In* P. W. Rundel, J. R. Ehleringer, and K. A. Nagy (eds.), Stable Isotopes in Ecological Research. Springer-Verlag, New York.

White, J. W. C., E. R. Cook, J. R. Lawrence, and W. S. Broecker. 1985. The D/H ratios of sap in trees: Implications for water sources and tree ring D/H ratios. *Geochim. Cosmochim. Acta* **49:** 237–246.

Wickman, F. E. 1952. Variations in the relative abundance of the carbon isotopes in plants. *Geochim. Cosmochim. Acta* **2**: 243–252.

Yapp, C. J., and S. Epstein. 1982. Climatic significance of the hydrogen isotope ratios in tree cellulose. *Nature* **297**: 636–639.

Ziegler, H. 1979. Diskriminierung von Kohlenstoff- und Wasserstoffisotopen: Zusammenhänge mit dem Photosyntheses-Mechanismus und den Standortbedingungen. *Ber. Dtsch Bot. Ges.* **92**: 169–184.

Ziegler, H. 1989. Hydrogen isotope fractionation in plant tissues, pp. 105–123. *In* P. W. Rundel, J. R. Ehleringer, and K. A. Nagy (eds.), Stable Isotopes in Ecological Research. Springer-Verlag, New York.

Ziegler, H., C. B. Osmond, W. Stickler, and D. Trimborn. 1976. Hydrogen isotope discrimination in higher plants: Correlation with photosynthetic pathway and environment. *Planta* **128**: 85–92.

Zundel, G., W. Miekeley, B. M. Grisi, and H. Förstel. 1978. The $H_2^{18}O$ enrichment in the leaf water of tropic trees: Comparison of species from the tropical rain forest and the semi-arid region of Brazil. *Radiat. Environ. Biophys.* **15**: 203–212.

3

Biochemical Basis of Carbon Isotope Fractionation

Marion H. O'Leary

I. Introduction[1]

Plants fractionate carbon isotopes during photosynthesis. The magnitude of the fractionation varies with photosynthetic type, environment, genotype, and other factors, and this variation can be used to study a variety of issues in plant physiology (O'Leary, 1981, 1988; Troughton, 1979; Vogel, 1980; Farquhar *et al.*, 1989; Rundel *et al.*, 1989). The physical and biochemical phenomena underlying this fractionation are generally well understood, and this understanding permits us to consider applying isotopic methods to a variety of problems in plant physiology, ecology, and other areas.

The purpose of this article is first to summarize the physical and chemical components that give rise to isotope fractionation in plants. Then we show how these components are put together to form integrated theories of plant isotope fractionation.

II. Definition of Terms

About 1.1% of all carbon atoms in natural materials are the nonradioactive isotope carbon-13. Precise measurements of the ^{13}C content of CO_2 are carried out using an isotope ratio mass spectrometer specially designed for high-precision measurement of the ratio R, defined by

$$R = {}^{13}CO_2/{}^{12}CO_2. \tag{1}$$

[1]Abbreviations used: RuBP, ribulose bisphosphate; RuBisCO, ribulose bisphosphate carboxylase/oxygenase; PEP, phosphoenolpyruvate; PGA, 3-phosphoglyceric acid; p_i, internal gas-phase CO_2 partial pressure; p_a, atmospheric CO_2 partial pressure.

Other materials are converted to CO_2 prior to analysis, either by combustion (e.g., for leaves and other plant materials) or by chemical or enzymatic methods (for purified compounds). For convenience, R values are ordinarily converted to values of $\delta^{13}C$,

$$\delta^{13}C = [R(\text{sample})/R(\text{standard}) - 1] \times 1000\%o. \tag{2}$$

The standard is carbon dioxide obtained from "PDB" a limestone from the Pee Dee formation in South Carolina (Craig, 1957). The nondimensional units of $\delta^{13}C$ are "per mil," or $\%o$. Although there as been some tendency to move away from this formalism, which was originally introduced by Craig (1957), this still remains the most accessible and most universal approach to the description of isotopic contents.

In the absence of industrial activity, the carbon dioxide in air has a $\delta^{13}C$ value of $-8\%o$, and each year this value becomes a little more negative as a result of combustion of fossil fuels and deforestation (Keeling *et al.*, 1979). Since isotope fractionations should actually reflect the isotopic difference between source and product, it is preferable to use a fractionation scale based on the isotopic difference between source and product,

$$\text{discrimination} = \frac{\delta^{13}C_a - \delta^{13}C_p}{1 + \delta^{13}C_p/1000}, \tag{3}$$

where $\delta^{13}C_a$ is the $\delta^{13}C$ value of air and $\delta^{13}C_p$ is that of the plant. The second term in the denominator of Eq. (3) is quite small and is often neglected.

The reliability of individual $\delta^{13}C$ values bears some comment. Earlier measurements of $\delta^{13}C$ were carried out following manual combustion of organic material, in a high-pressure combustion bomb, in a sealed tube, or in a combustion train. The reproducibility of such values was typically only a little better than $\pm 1\%o$. Current analysis systems involve an automated combustion line interfaced to a CO_2 purification system that feeds directly into the mass spectrometer. Such systems can provide reproducibility near $\pm 0.1\%o$ (Wong *et al.*, 1992). However, the reader is still advised to be skeptical of isotopic differences less than $1\%o$, particularly where analyses from several laboratories are being compared. It should also be noted that there is some statistical variability associated with analysis of different members of a seemingly homogeneous plant population. In practice, it is usual to take a sample of several leaves to remove some of the inhomogeneity.

III. Components of Isotope Fractionation in Plants

Isotope fractionation in plants is best understood by beginning with the individual physical and chemical processes that contribute to the overall isotope fractionation. The numerical values of these fractionations are summarized in Table I, and the individual values are discussed below.

Table I Carbon Isotope Fractionations Associated with Photosynthesis

Process	Fractionation (‰)[a]	Reference
Solubility of CO_2 in water	1.1	O'Leary, 1984
Hydration of CO_2	−9.0	Mook *et al.*, 1974
CO_2 diffusion in air	4.4	O'Leary, 1981; Hersterberg and Siegenthaler, 1991
CO_2 diffusion in aqueous solution	0.7	O'Leary, 1984
Spontaneous hydration of CO_2	6.9	Marlier and O'Leary, 1984
Carbonic anhydrase-catalyzed hydration of CO_2	1.1	Paneth and O'Leary, 1985
Carboxylation of PEP	2.0	O'Leary *et al.*, 1981
Carboxylation of RuBP	29.0	Roeske and O'Leary, 1984

[a]Positive values in this table indicate that the product is depleted in ^{13}C compared to the starting state; negative values indicate enrichment.

Following this survey, we present the equations which are used to integrate these components into a model for whole-plant CO_2 uptake.

A. Thermodynamic versus Kinetic Fractionations

Isotope fractionations may be of two types: thermodynamic and kinetic. Thermodynamic fractionations reflect differences in equilibrium constants for isotopic species. Kinetic fractionations reflect differences in rate constants for isotopic species. Kinetic fractionations are usually larger than thermodynamic fractionations. For example, the kinetic fractionation associated with the carboxylation of ribulose bisphosphate is near 29‰, whereas the equilibrium fractionation for this processes is approximately zero (O'Leary and Yapp, 1978). In a multistep sequence, thermodynamic fractionations are additive, whereas kinetic ones are not.

B. Isotope Fractionations in Physical Processes

Diffusional processes, which involve mass transfer, often show small isotope discriminations. The heavier isotopic species invariably diffuses more slowly. The diffusion of $^{12}CO_2$ and $^{13}CO_2$ in air is subject to an isotope fractionation of 4.4‰ (O'Leary, 1981; Hesterberg and Siegenthaler, 1991). This diffusion has to be understood as a binary diffusion, in which a mythical "air" molecule and a CO_2 exchange places, thus preserving the center of mass of the system. The corresponding isotope fractionation for the diffusion of $H_2{}^{16}O$ and $H_2{}^{18}O$ in N_2 has been measured and is in accord with theory (Merlivat, 1978).

Diffusion of CO_2 dissolved in water shows a smaller isotope fractionation (0.7‰) (O'Leary, 1984) than diffusion of CO_2 in air (4.4‰). In liquid solution, much of the "work" of diffusion is connected with rearrangement of the highly hydrogen-bonded water structure, and this is independent of the mass of CO_2; thus, the isotope fractionation is small. The isotope fractionation associated with diffusion of HCO_3^- in solution has not been measured. It is safe to assume that this fractionation would be substantially smaller than that for dissolved CO_2 because of the higher mass and extensive solvation of HCO_3^-.

The dissolution of CO_2 in water shows an equilibrium isotope fractionation of $-1.1‰$ (Vogel *et al.*, 1970; O'Leary, 1984); that is, $^{13}CO_2$ is less soluble than $^{12}CO_2$ (i.e., taking gaseous CO_2 as the "source" and dissolved CO_2 as the "product"). This fractionation is in the opposite direction from what would ordinarily be predicted, presumably because of the solvation of dissolved CO_2. Because this fractionation is an equilibrium fractionation, it must be included in any treatment of CO_2 uptake. In treatments of plant isotope fractionation, this value is usually subsumed in the value used for RuBisCO or PEP carboxylase.

C. Isotope Fractionations in Enzymatic Processes

Isotope fractionations associated with enzyme-catalyzed reactions have now been under active study for about 20 years, and fractionations for a wide variety of enzymatic systems have been measured (O'Leary, 1989; Cleland, 1983, 1987; Cook, 1991).

Ribulose 1,5-bisphosphate carboxylase/oxygenase (RuBisCO) catalyzes the first step in CO_2 fixation in C_3 plants:

$$CO_2 + RuBP \longrightarrow 2 \quad 3\text{-PGA}. \tag{4}$$

It was recognized early that isotope fractionation by this enzyme was probably a source of the large isotope fractionation shown by C_3 plants, and in intervening years, this fractionation has been measured by a variety of investigators using a variety of methods (see Roeske and O'Leary, 1984, for a summary of earlier studies). The isotope fractionation can be measured by combustion analysis (Christeller *et al.*, 1976; Estep *et al.*, 1978; Wong *et al.*, 1979), by specific analysis of individual carbon atoms in substrate and product (Roeske and O'Leary, 1984), and by analysis of remaining CO_2 during the carboxylation (Schmidt *et al.*, 1978; Winkler *et al.*, 1982; Guy *et al.*, 1987). A variety of recent measurements give consistent results. The isotope fractionation is $29‰$ compared with dissolved CO_2, with an uncertainty of less than $1‰$ at 25°C, pH 8 (i.e., $30‰$ compared to gaseous CO_2).

The interconversion of CO_2 and HCO_3^- occurs readily in plants. In most organelles, this process is under control of carbonic anhydrase, which presumably keeps the reaction near equilibrium. At equilibrium, the fractionation is $-9.0‰$, with ^{13}C concentrating in HCO_3^- (Mook *et al.*, 1974). There is also a kinetic fractionation associated with the carbonic anhydrase-catalyzed hydration of CO_2 (Paneth and O'Leary, 1985), but this is probably not a factor under most conditions, and it is adequate to assume that the hydration reaction is at equilibrium (Holtum *et al.*, 1984; Hatch and Burnell, 1990). This hydration does not enter the C_3 fractionation model at all because RuBisCO takes up CO_2, rather than HCO_3^-.

The initial carboxylation in C_4 plants is carried out by PEP carboxylase, which catalyzes the reaction

$$PEP + HCO_3^- \longrightarrow OAA + P_i. \tag{5}$$

This fractionation has been measured by a number of investigators (Schmidt *et al.*, 1978; Winkler *et al.*, 1983; O'Leary, 1981; Whelan *et al.*, 1973) using both substrate depletion and product analysis methods. The isotope fractionation is 2‰. However, this is the fractionation starting from HCO_3^-, whereas in plants the reaction begins with CO_2. When referred to gaseous CO_2 the isotope fractionation becomes -5.7‰.

The activity of maize PEP carboxylase has recently been shown to depend on phosphorylation on serine-15 (in the maize enzyme numbering). In fact, the enzyme prepared by conventional means is often missing a polypeptide from the N-terminus, including this serine (Jiao *et al.*, 1991). Recent isotope effect studies have established that the isotope fractionation by PEP carboxylase is unaffected by phosphorylation state or proteolysis (Madhavan, Paneth, and O'Leary, unpublished).

D. Environmental Effects on Component Fractionations

The isotope fractionations reported above are well established, at least for the particular species and conditions that have been studied. However, there is still a question whether these same values are appropriate for all plants under all conditions. We must consider whether species effects, temperature effects, CO_2 concentration effects, or other natural variables might change the values of the individual isotope fractionations.

Isotope fractionations for physical and diffusive processes should not be affected by the range of environmental or concentration conditions likely to be encountered during normal plant growth. The isotope fractionation associated with the CO_2 hydration equilibrium has a small temperature effect (Mook *et al.*, 1974) that can be included when appropriate. The isotope fractionation due to RuBisCO is independent of CO_2 concentration (at least *in vitro*), almost independent of pH (Roeske and O'Leary, 1984), and independent of temperature (Christeller *et al.*, 1976; Fang and O'Leary, unpublished).

The possibility that there might be species effects on the fractionation due to RuBisCO is problematical. Most higher plant studies have been done with the enzyme from spinach, and studies with enzyme from other species are inadequate to establish whether there might be species variations. Further, enzymes from *Rhodospirillum rubrum* (Roeske and O'Leary, 1985) and *Anacystis nidulans* (Guy *et al.*, 1987) show isotope fractionations significantly different from that for the spinach enzyme.

Although it might seem that the same arguments could be made with regard to PEP carboxylase, the small size of the fractionation observed in that case and the relationship of that fractionation to the enzymatic mechanism make it less likely that the fractionation varies with species or conditions.

IV. Theory of Isotope Fractionation in Plants

Early workers (Craig, 1953) recognized that diffusion and carboxylation were likely to be the principal causes of isotope fractionation in plants. Following the demonstration that PEP carboxylase and RuBisCO have very different isotope fractionations, it was suggested that this difference might be responsible for the difference between C_3 and C_4 plants (Whelan *et al.*, 1973). Subsequent treatments have incorporated this idea and made it more quantitative; in many instances, quite specific predictions can be made about the relationship of isotope fractionation to physiology.

Work of Vogel (1980), O'Leary (1981), and Farquhar, O'Leary, and Berry (1982) successfully put these individual fractionations on a quantitative basis. The initial work of O'Leary (1981) using a model based on chemical kinetics has found less favor than the approach of Farquhar *et al.* (1982) based on gas exchange, and most subsequent work has been based on the Farquhar approach (See Farquhar *et al.*, 1989, for the most recent summary of this theory). A number of further developments of the theory have followed the initial papers (Farquhar *et al.*, 1982, 1989; Deleens *et al.*, 1983; Peisker, 1982, 1984, 1985; Farquhar, 1983). A key to all these theories has been the recognition that isotope fractionation is related to p_i, the internal gas-phase CO_2 concentration in the leaf.[2]

A. A Qualitative Approach to Isotope Fractionation

It is useful to give a simple physical picture of the isotope fractionation in a C_3 plant before proceeding to the more quantitative approaches. For this use, a simple, two-step CO_2 uptake scheme suffices:

$$CO_2 \text{ (external)} \xrightleftharpoons{\text{diffusion}} CO_2 \text{ (internal)} \xrightarrow{\text{carboxylation}} R{-}CO_2^-. \qquad (6)$$

In the first step, CO_2 diffuses through the stomata to the site of carboxylation; in the second step, this CO_2 is taken up irreversibly by the appropriate carboxylase. The isotope fractionation associated with diffusion is 4.4‰, whereas that associated with carboxylation is 29‰ in C_3 plants (30‰ when CO_2 dissolution is included).

We can consider two limiting cases. First, if the stomata are nearly closed, the overall CO_2 uptake rate is limited by the initial diffusion process and the internal CO_2 concentration is low. In these circumstances, the carboxylation process takes up virtually all carbon available, and the carboxylation isotope fractionation is not expressed. The isotope fractionation is small, approaching 4.4‰ at very small apertures. Thus, $\delta^{13}C$ for a C_3 plant should approach $-12‰$ ($-8\ -4.4$).

On the other hand, if the stomata are relatively open, the internal CO_2 concentration approaches the external CO_2 concentration, and there is a

---

[2]The original realization of this fact came in a long series of discussions during the Australian summer of 1978–1979, when O'Leary, Berry, and Farquhar were all at The Australian National University.

facile transfer of CO_2 between the external and internal pools. In this case, the diffusional fractionation is not expressed (diffusion approaches equilibrium) and the observed fractionation approaches the carboxylation fractionation. Leaf $\delta^{13}C$ would then approach $-38‰$ (-8 -30).

Real plants, of course, show behavior intermediate between these two extremes. The internal CO_2 concentration is perhaps half the external concentration, and the isotope fractionation is larger than the diffusional fractionation but smaller than the carboxylation fractionation. As stomatal aperture changes, the fractionation and the internal CO_2 concentration will change correspondingly.

B. Fractionation in C_3 Plants

For C_3 plants, isotope fractionation can be understood from

$$\text{fractionation} = [a + (b - a)p_i/p_a - d], \tag{7}$$

where a is the discrimination due to diffusion ($4.4‰$), b is the discrimination due to carboxylation ($30‰$ when corrected for the equilibrium effect on CO_2 dissolution), p_i is the internal gas-phase pressure of CO_2, and p_a is the external CO_2 pressure. The isotope fractionation varies as the internal CO_2 concentration varies, or, equivalently, as the relative resistivities of stomatal diffusion and carboxylation vary.

The term d involves contributions from respiration, liquid-phase diffusion, isotopic changes due to carbon export, CO_2 fixation in C_3 plants by PEP carboxylase, and a variety of other factors. Some of these issues have been discussed by Farquhar *et al.* (1989). Fortunately, the value of d is usually small, and variations in d within a single series of studies can usually be neglected.

Respiration represents an important part of the carbon budget of plants, and insofar as respired carbon is isotopically different from leaf carbon, d in Eq. (7) might be large and variable. The isotopic consequences of respiration are difficult to measure in the absence of significant contributions from other photosynthetic processes, but most available evidence suggests that the isotope fractionation associated with respiration is small (O'Leary, 1981; Farquhar *et al.*, 1989). Direct measurement of the isotope fractionation associated with photorespiration in soybean (*Glycine max*) provided an isotope fractionation of $7‰$ (Rooney, 1988).

C. Fractionation in C_4 Plants

Farquhar (1983) adapted the same kind of treatment used for C_3 photosynthesis to account for the features of C_4 photosynthesis,

$$\text{fractionation} = a + (b_4 + b_3\phi - a)p_i/p_a, \tag{8}$$

where a is the fractionation due to diffusion ($4.4‰$), b_4 is the fractionation from gaseous CO_2 through PEP carboxylase ($-5.7‰$), b_3 is the fractionation due to RuBisCO ($30‰$), and ϕ is the fraction of CO_2 released in the bundle sheath that leaks to the mesophyll, where it may be either taken up by PEP carboxylase or released to the atmosphere. Farquhar *et al.* (1989) suggest that this value is often near 0.37, but more recent studies (Henderson *et al.*, 1992) suggest that a value of 0.21 is more appropriate (see Farquhar, Chapter 17, this volume).

Comparison of Eq. (8) with Eq. (7) reveals that stomatal opening causes opposite effects in C_3 and C_4 plants. The results of this can be seen, for example, in studies of the effect of relative humidity on $\delta^{13}C$ in plants grown in controlled environments. Increasing relative humidity causes an increase in $\delta^{13}C$ in C_4 plants but a decrease in C_3 plants (Madhavan *et al.*, 1991).

V. Summary

The major processes contributing to carbon isotope fractionation in plants are CO_2 diffusion and carboxylation. The difference in isotopic fractionation between C_3 and C_4 plants is due primarily to differences in isotope fractionation between RuBisCO and PEP carboxylase and secondarily to different stomatal apertures in the two cases. Respiration, bundle-sheath leakage in C_4 plants, and other factors may also contribute. Carefully controlled studies can reveal changes in isotope fractionation that reflect changes in stomatal aperture, changes in water-use efficiency, changes in enzyme levels, and a variety of other factors. A number of these studies are described in the later chapters of this book.

References

Christeller, J. T., W. A. Laing, and J. H. Troughton. 1976. Isotope discrimination by ribulose 1,5-diphosphate carboxylase. *Plant Physiol.* **57**: 580–582.

Cleland, W. W. 1983. Use of isotope effects to elucidate enzyme mechanisms. *CRC Crit. Rev. Biochem.* **13**: 385–428.

Cleland, W. W. 1987. The use of isotope effects in the detailed analysis of catalytic mechanisms of enzymes. *Bioorg. Chem.* **15**: 283–302.

Cook, P. F. 1991. *Enzyme Mechanisms from Isotope Effects.* CRC Press, Boca Raton.

Craig, H. 1953. The geochemistry of the stable carbon isotopes. *Geochim. Cosmochim. Acta* **3**: 53–92.

Craig, H. 1957. Isotopic standards for carbon and oxygen and correction factors for mass-spectrometric analysis of carbon dioxide. *Geochim. Cosmochim. Acta* **12**: 133–149.

Deleens, E., A. Ferhi, and O. Queiroz. 1983. Carbon isotope fractionation by plants using the C_4 pathway. *Physiol. Veg.* **21**: 897–905.

Estep, M. F., F. R. Tabita, P. L. Parker, and C. Van Baalen. 1978. Carbon isotope fractionation by ribulose-1,5-bisphosphate carboxylase from various organisms. *Plant Physiol.* **61**: 680–687.

Farquhar, G. D. 1983. On the nature of carbon isotope discrimination in C_4 species. *Aust. J. Plant Physiol.* **19**: 205–226.

Farquhar, G. D., M. H. O'Leary, and J. A. Berry. 1982. On the relationship between carbon isotope discrimination and the intercellular carbon dioxide concentration in leaves. *Aust. J. Plant Physiol.* **9**: 121–137.

Farquhar, G. D., J. R. Ehleringer, and K. T. Hubick. 1989. Carbon isotope discrimination and photosynthesis. *Annu. Rev. Plant Physiol. Plant Mol. Biol.* **40:** 503–37.

Guy, R. D., M. F. Fogel, J. A. Berry, and T. C. Hoering. 1987. Isotope fractionation during oxygen production and consumption by plants. *Prog. Photosynth. Res.* **III:** 597–600.

Hatch, M. D., and J. N. Burnell. 1990. Carbonic anhydrase activity in leaves and its role in the first step of C_4 photosynthesis. *Plant Physiol.* **93:** 825–828.

Henderson, S. A., S. von Caemmerer, and G. D. Farquhar. 1992. Short-term measurements of carbon isotope discrimination in several C_4 species. *Aust. J. Plant Physiol.* **19:** 263–285.

Hesterberg, R., and U. Siegenthaler. 1991. Production and stable isotopic composition of CO_2 in a soil near Bern, Switzerland. *Tellus* **43B:** 197–205.

Holtum, J. A. M., R. Summons, C. A. Roeske, H. N. Comins, and M. H. O'Leary. 1984. Oxygen-18 incorporation into malic acid during nocturnal carbon dioxide fixation in Crassulacean acid metabolism plants. *J. Biol. Chem.* **259:** 6870–6881.

Jiao, J.-A., J. Vidal, C. Echevarria, and R. Chollet. 1991. *In vivo* regulatory phosphorylation site in C_4-leaf phosphoenolpyruvate carboxylase from maize and sorghum. *Plant Physiol.* **96:** 297–301.

Keeling, C. D., W. G. Mook, and P. P. Tans. 1979. Recent trends in the $^{13}C/^{12}C$ ratio of atmospheric carbon dioxide. *Nature* **277:** 121–123.

Madhavan, S., I. Treichel, and M. H. O'Leary. 1991. Effects of relative humidity on carbon isotope fractionation in plants. *Bot. Acta* **104:** 292–294.

Marlier, J. F., and M. H. O'Leary. 1984. Carbon kinetic isotope effects on the hydration of carbon dioxide and the dehydration of bicarbonate ion. *J. Am. Chem. Soc.* **106:** 5054–5057.

Merlivat, L. 1978. Molecular diffusivities of $H_2^{16}O$, $HD^{16}O$, and $H_2^{18}O$ in gases. *J. Chem. Phys.* **69:** 2864–2871.

Mook, W. G., J. C. Bommerson, and W. H. Staverman. 1974. Carbon isotope fractionation between dissolved bicarbonate and gaseous carbon dioxide. *Earth Planet Sci. Lett.* **22:** 169–175.

O'Leary, M. H. 1981. Carbon isotope fractionation in plants. *Phytochemistry* **20:** 553–567.

O'Leary, M. H. 1984. Measurement of the isotope fractionation associated with diffusion of carbon dioxide in aqueous solution. *J. Phys. Chem.* **88:** 823–825.

O'Leary, M. H. 1988. Carbon isotopes in photosynthesis. *BioScience* **38:** 328–336.

O'Leary, M. H. 1989. Multiple isotope effects on enzyme-catalyzed reactions. *Annu. Rev. Biochem.* **58:** 377–401.

O'Leary, M. H., and C. J. Yapp. 1978. Equilibrium isotope effect on a decarboxylation reaction. *Biochem. Biophys. Res. Commun.* **80:** 155–160.

O'Leary, M. H., J. E. Rife, and J. D. Slater. 1981. Kinetic and isotope effect studies of maize phosphoenolpyruvate carboxylase. *Biochem.* **20:** 7308–7314.

Paneth, P., and M. H. O'Leary. 1985. Carbon isotope effect on dehydration of bicarbonate ion catalyzed by carbonic anhydrase. *Biochemistry* **24:** 5143–5147.

Peisker, M. 1982. The effect of CO_2 leakage from bundle sheath cells on carbon isotope discrimination in C_4 plants. *Photosynthetica* **16:** 533–541.

Peisker, M. 1984. Modellvorstellungen zur Kohlenstoff-Isotopendiskriminierung bei der Photosyntheses von C_3- und C_4-Pflanzen. *Kulturpflanze* **32:** 35–65.

Peisker, M. 1985. Modelling carbon metabolism in C_3-C_4 intermediate species 2: Carbon isotope discrimination. *Photosynthetica* **19:** 300–311.

Roeske, C. A., and M. H. O'Leary. 1984. Carbon isotope effects on the enzyme-catalyzed carboxylation of ribulose bisphosphate. *Biochemistry* **23:** 6275–6284.

Roeske, C. A., and M. H. O'Leary. 1985. Carbon isotope effect on carboxylation of ribulose bisphosphate catalyzed by ribulosebisphosphate carboxylase from *Rhodospirillum rubrum*. *Biochemistry* **24:** 1603–1607.

Rooney, M. A. 1988. Short-term Carbon Isotope Fractionation in Plants. Ph.D. thesis, University of Wisconsin, Madison.

Rundel, P. W., J. R. Ehleringer, and K. A. Nagy. 1989. *Stable Isotopes in Ecological Research*. Springer-Verlag, New York.

Schmidt, H.-L., F. J. Winkler, E. Latzko, and E. Wirth. 1978. ^{13}C-Kinetic isotope effects in photosynthetic carboxylation reactions and δ^{13}C-values of plant material. *Isr. J. Chem.* **17**: 223–224.

Troughton, J. H. 1979. δ^{13}C as an indicator of carboxylation reactions. *Encycl. Plant Physiol., New Ser.* **6**: 140–149.

Vogel, J. C. 1980. Fractionation of the carbon isotopes during photosynthesis. *Sitzungsber. Heidelb. Akad. Wiss.* **3**: 111–135.

Vogel, J. C., P. M. Grootes, and W. G. Mook. 1970. Isotopic fractionation between gaseous and dissolved carbon dioxide. *Z. Phys.* **230**, 225–238.

Whelan T., W. M. Sackett, and C. R. Benedict. 1973. Enzymatic fractionation of carbon isotopes by phosphoenolpyruvate carboxylase from C_4 plants. *Plant Physiol.* **51**: 1051–1054.

Winkler F. J., H. Kexel, C. Kranz, and H.-L. Schmidt. 1982. Parameters affecting the $^{13}CO_2/$ $^{12}CO_2$ isotope discrimination of the ribulose-1,5-bisphosphate carboxylase reaction, pp. 83–89. *In* H.-L. Schmidt, H. Förstel, and K. Henzinger (eds.), *Stable Isotopes*. Elsevier, Amsterdam.

Winkler, F. J., H.-L. Schmidt, E. Wirth, E. Latzko, B. Lenhart, and H. Ziegler. 1983. Temperature, pH and enzyme-source dependence of the HCO_3^- carbon isotope effect on the phosphoenolpyruvate carboxylase reaction. *Physiol. Veg.* **21**: 889–895.

Wong, W. W., C. R. Benedict, and R. J. Kohel. 1979. Enzymic fractionation of the stable carbon isotopes of carbon dioxide by ribulose-1,5-bisphosphate carboxylase. *Plant Physiol.* **63**: 852–856.

Wong, W. W., L. L. Clarke, G. A. Johnson, M. Llaurador, and P. D. Klein. 1992. Comparison of two elemental-analyzer gas-isotope-ratio mass spectrometer systems in the simultaneous measurement of $^{13}C/^{12}C$ ratios and carbon content in organic samples. *Anal. Chem.* **64**: 354–358.

4

Variability of Carbon Isotope Fractionation during Photosynthesis

J. C. Vogel

I. Introduction

Although the basics of carbon isotope fractionation during photosynthesis are understood in broad terms, there are still various aspects of the process that need clarification. One may anticipate that the unravelling of these outstanding details will not only lead to a better description of the isotope ratio of plants, but also provide insights into the mechanisms of the physiological processes involved. The uncertainties that still remain need not, however, distract from the application of natural isotope analyses to practical problems such as their use in determining water-use efficiency in cultivars—a subject that will be addressed in later chapters.

Since this chapter is intended to be an introduction to the topic rather than a comprehensive review, it will be appropriate to present the basic concepts once again. In the presentation the mathematical terminology that was introduced by Harold Urey and his students in the early post-World War II period will be adopted: This includes the familiar δ notation where δ (in ‰ or parts per thousand) is the relative deviation of the $^{13}C/^{12}C$ ratio, R, from that in a reference standard. Defined in this way, δ is negative if the carbon sample contains less of the heavy isotope ^{13}C than the standard.

Early surveys showed that the carbon isotope ratios of C_3 and C_4 plants fall into two nonoverlapping categories (Bender, 1971; Smith and Epstein, 1971; Troughton, 1972). An example of this bimodal distribution for 351 species of the grass family is shown in Fig. 1 (Vogel, 1980). The C_3 grass species all have δ-values between −22‰ and −34‰, while the C_4 species fall between −9‰ and −16‰. As atmospheric CO_2 has a value of about −7.5‰, the average fractionation in the C_3 group is −19‰ compared with

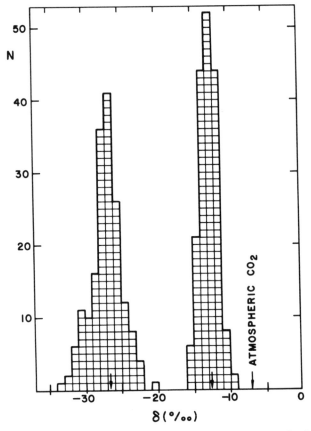

Figure 1. Histogram of the ^{13}C content of 351 species of the *Poaceae* family from Vogel (1980). The separation is unique: all the C_4 species have δ-values more positive than $-16‰$, (modal value $-12.6‰$), while only C_3 species have δ-values between -22 and $-34‰$ (modal value $-26.7‰$).

$-5‰$ for the C_4 group. This difference in isotope composition is clearly related to the two distinct metabolic pathways.

While the isotope ratios of the two groups of plants are related to function and, as is well known, also to structure, the variation within each category is due to the influence of environmental factors on the kinetics of photosynthesis. Isotope ratio measurements can thus be used for interpreting the relative magnitudes of competing processes that take place during metabolism. As will be seen later, this applies especially to C_3 plants, while the situation with regard to the C_4 pathway is still somewhat obscure.

As far as the carbon isotope composition is concerned, those succulents that have the ability to utilize the crassulacean acid metabolism (CAM) constitute a separate category. These plants do not specifically differ from C_3 plants in their structure, but they have the ability to fix CO_2 at night as malate along the C_4 pathway. The assimilated carbon is then further pro-

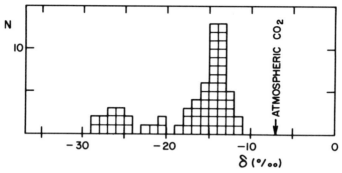

Figure 2.   Histogram of the ^{13}C content of 63 species of the genus *Aloe* from Vogel (1980). Those species with a strong tendency to utilize the crassulacean acid metabolism (CAM) have δ-values more positive than $-16‰$, while the group with δ-values less negative than $-22‰$ only utilize CAM under water stress.

cessed by ribulose bisphosphate (RuBP) carboxylase on the following day, when the stomata can remain closed and restrict transpiration. Those succulents that have a strong tendency to operate in the CAM mode and assimilate only small amounts of CO_2 during the day via the C_3 pathway have isotope ratios similar to those of C_4 plants. On the other extreme, such plants that have only a weak affinity for the CAM syndrome tend to have isotope ratios in the C_3 range. The latter group can usually be induced to fix carbon at night by subjecting the plants to water stress (cf., for instance, Osmond *et al.*, 1976), and the isotope ratios then become intermediate.

The genus *Aloe* may serve as an example of this category. In Fig. 2 a histogram of the isotope ratios of 63 species of aloe, grown mostly in their natural environment, is presented (Vogel, 1980). There are clearly two groups of C_3-like and C_4-like species with some intermediate ones. It is obvious that isotope ratio measurements can be used to establish the degree to which succulents utilize CAM and also to study the variability of this tendency.

II. Two-Stage Isotope Fractionation

Craig (1954) in his early discussion of the carbon isotopic composition of plants pointed out that isotope fractionation can only take place before the first irreversible step in the process of photosynthesis. Since the carboxylation reaction is essentially irreversible in both the C_3 and C_4 pathways, fractionation will occur in what is basically a two-stage process: The first stage comprises the diffusion of CO_2 to the site where carboxylation takes place, and the second is the carboxylation reaction itself. Actually the process is more complicated, but it is useful to consider this approximation first.

Several authors have presented mathematical expressions describing the fractionation (Vogel, 1980; O'Leary and Osmond, 1980; Farquhar, 1980, 1983; Farquhar *et al.*, 1982). These are all essentially the same, although expressed in different ways. Here we follow Urey (1947) and Craig (1954) who defined the *fractionation factor*, α, as the isotope ratio of the product divided by that of the source, viz,

$$\alpha = R_p/R_a,$$

and, since α is a quantity close to unity, the *fractionation*, ε, as the deviation of α from unity, viz,

$$\varepsilon = \alpha - 1.$$

In this notation ε will be negative if the product contains relatively less ^{13}C than the source material, in conformity with the δ notation.

The assimilation process described as diffusion followed by chemical reaction can be presented as

$$CO_2(atm) \rightarrow CO_2(int) + A \rightarrow ACO_2 \rightarrow$$

$[CO_2]$:	c_a	c_c	0
Isotope ratio:	R_a	R_c	R_p,

where A is the acceptor, viz, ribulose bisphosphate or phosphoenolpyruvate. The rate of CO_2 assimilation, or flux density of CO_2 through the system, is given by

$$j = (c_a - c_c)/r_d = c_c/r_c = c_a/r, \tag{1}$$

where c_a and c_c are the CO_2 concentrations in the environment and at the reaction site, respectively, r_d and r_c are the resistances (reciprocal conductances) with respect to diffusion and reaction, respectively, and r is the total resistance, $r = r_d + r_c$.

It is convenient for the mathematical deductions that are to follow to use the concept of resistances and to present the rate of photosynthesis in the form of Ohm's law whereby the flux density, j, is equivalent to current and the concentration of CO_2 to potential. The "resistance" of the carboxylation reaction, r_c, is merely the inverse of the rate constant, k. At low concentrations r_c is constant, and the reaction rate increases linearly with the concentration, but as the saturation concentration is approached, the rate becomes independent of concentration. The resistance r_c can thus be written (cf. Lommen *et al.*, 1971)

$$r_c = r_0(1 + \lambda c_c), \tag{2}$$

where r_0 and λ are constants. By using this formulation the general form of Ohm's law is preserved.

If Eq. (1) is taken to describe the flow of the abundant ^{12}C molecules, a similar expression denoted by a prime can apply to the heavy ^{13}C molecules. Dividing the two expressions gives

$$j'/j = c'_a/c_a \cdot r/r'.$$

But, since j'/j is the isotope ratio, R_p, of the ultimately absorbed carbon and c'_a/c_a is that of the free CO_2, R_a,

$$r/r' = R_p/R_a = \alpha.$$

Thus, the isotope fractionation,

$$\alpha = r/r' = (r_d + r_c)/(r'_d + r'_c),$$

and since the fractionation factor due to diffusion is $\alpha_d = r_d/r'_d$, and that of the reaction is $\alpha_c = r_c/r'_c$, the overall fractionation factor becomes

$$\alpha = \frac{\alpha_d + \alpha_c \cdot r'_c/r'_d}{1 + r'_c/r'_d}$$

and the fractionation becomes

$$\varepsilon = \frac{\varepsilon_d + \varepsilon_c \cdot r'_c/r'_d}{1 + r'_c/r'_d}. \tag{3a}$$

Because $r'_c/r'_d \approx r_c/r_d(1 + \varepsilon_d - \varepsilon_c)$, the resistance ratio for ^{13}C can be replaced, with good approximation, by that for ^{12}C, which is practically the same as that for the total carbon.

The equation can also be written in other forms, viz,

$$\varepsilon = \varepsilon_d \cdot r_d/r + \varepsilon_c \cdot r_c/r, \tag{3b}$$

or

$$\varepsilon = \varepsilon_d + (\varepsilon_c - \varepsilon_d) \cdot c_c/c_a, \tag{3c}$$

and, using Eq. (1), also as

$$\varepsilon = \varepsilon_d(c_a - c_c)/c_a + \varepsilon_c \cdot c_c/c_a. \tag{3d}$$

The expression shows that, when the diffusional resistance, r_d, is large compared to the resistance of the reaction, r_c (where $r = r_c + r_d$), the overall fractionation, ε, approaches ε_d, while when r_d is small, ε becomes equal to ε_c.

The fractionations ε_d and ε_c for the C_3 pathway are known by approximation: Diffusion theory predicts that, for the diffusion of CO_2 through a fixed path length of air, the fluxes will depend on the ratio of the reduced masses of the isotopic molecules so that $\varepsilon_d = -4.4‰$ (Craig, 1954; Vogel, 1980). Roeske and O'Leary (1984) measured the fractionation caused by the enzyme-catalyzed carboxylation of RuBP in aqueous solution to be $-29‰$ at pH 8 and 25°C. This value does not necessarily apply to the reaction *in vivo* in the chloroplast, and the experimentally deduced value of $\varepsilon_c = -27‰$ (Vogel, 1980; Farquhar *et al.*, 1982, and subsequent papers) is perhaps more appropriate for general use. ε_c in Eq. (3) actually includes the equilibrium fractionation between gaseous and dissolved CO_2, ε_{dg}, which is $-1.1‰$ (Vogel *et al.*, 1970) so that $\varepsilon_c = \varepsilon_{dg} + \varepsilon'_c$. Equation (3b) thus becomes

$$\varepsilon = \varepsilon_d \cdot r_d/r + (\varepsilon_{dg} + \varepsilon'_c)r_c/r. \tag{3e}$$

Equation (3d) is identical to that used by Farquhar (cf. Farquhar and Richards, 1984) viz,

$$\Delta = a + (b - a) \cdot p_i/p_a,$$

except that he uses the terms *discrimination*, Δ, a, and b, which are the same as the fractionation, ε, ε_d, and ε_c, as defined here, without the negative sign. He furthermore simplifies the expression by assuming that the resistance to the diffusion of CO_2 (and HCO_3^-) through the cell walls and cytoplasm to the chloroplasts is small. Thus $c_c \approx c_i$, the concentration of CO_2 in the intercellular spaces, and, finally, he expresses concentrations as partial pressure, p. The advantage of this approximation is that p_i/p_a is a quantity that can be obtained from the measurement of water loss by transpiration and the equation can be compared directly with experiment. In our terminology this equation is expressed as

$$\varepsilon \approx \varepsilon_d + (\varepsilon_c - \varepsilon_d) \cdot p_i/p_a. \tag{3f}$$

III. Separation of the Diffusional Steps

The boundary layer of air above the leaf surface sometimes acts as a rate-limiting step in fast growing plants and needs to be considered separately. The diffusion through such an instantaneous layer has a longer effective path length for the lighter, faster moving $^{12}CO_2$ molecules than for $^{13}CO_2$, with the result that the fractionation experienced differs from that through the stomata. Simple reasoning leads to the conclusion that the fractionation through the boundary layer, ε_b, should be only half that through still air (Vogel, 1980), while direct measurements (Kays, 1966) suggest a factor of ⅔ (Farquhar, 1983). Thus,

$$\varepsilon_b = \tfrac{2}{3}(-4.4‰) = -2.9‰.$$

To bring this boundary layer diffusion into account, the total diffusion step in the assimilation chain must be separated into two parts. Using the same reasoning as in Section II, we obtain an expression for the total diffusional fractionation, ε_d, viz,

$$\varepsilon_d = \varepsilon_b \cdot r_b/r_d + \varepsilon_l \cdot r_l/r_d,$$

where r_b and r_l are the resistances of the boundary layer and the leaf, respectively, and ε_l is the total fractionation inside the leaf.

The diffusion in the leaf can, in its turn, be separated into gaseous diffusion through the stomata and the diffusion of dissolved CO_2 (and HCO_3^-) in the mesophyll cells. In the same manner as before, ε_1 can be expressed as the sum of the fractionation at the stomata, ε_s, and fractionation in the aqueous phase, except that in the latter case the equilibrium discrimination between gaseous and dissolved CO_2, ε_{dg}, must be added (Vogel, 1980). Thus

$$\varepsilon_1 = \varepsilon_s \cdot r_s/r_1 + (\varepsilon_{dg} + \varepsilon_i)r_i/r_1.$$

The equilibrium fractionation, ε_{dg}, is $-1.1‰$ (Vogel *et al.*, 1970). In so far as simple gas kinetic theory applies to the diffusion of CO_2 through water, ε_i, the fractionation for diffusion through the cell walls and cell sap, would be expected to be $-3.2‰$ (Vogel, 1980), but direct measurement has produced a value of $-0.7‰$ (O'Leary, 1984) so that this value is preferred.

Equation (3e) thus becomes

$$\varepsilon = \varepsilon_b \cdot r_b/r + \varepsilon_s \cdot r_s/r + (\varepsilon_{dg} + \varepsilon_i)r_i/r + (\varepsilon_{dg} + \varepsilon_c') \cdot r_c/r, \qquad (4a)$$

where $r = r_b + r_s + r_i + r_c$, and $\varepsilon_c' = (\varepsilon_c - \varepsilon_{dg})$ is the fractionation between dissolved CO_2 and the fixed carbon.

Equation (4a) can also be expressed in terms of partial pressures (Farquhar, 1983):

$$\varepsilon = \varepsilon_b \cdot (p_a - p_b)/p_a + \varepsilon_s(p_b - p_i)/p_a + (\varepsilon_{dg} + \varepsilon_i)(p_i - p_c)/p_a + \varepsilon_c \cdot p_c/p_a.$$
$$(4b)$$

Since the diffusional fractionations are similar, viz, $\varepsilon_b = -2.9‰$, $\varepsilon_s = -4.4‰$, and $(\varepsilon_{dg} + \varepsilon_i) = -1.8‰$, the simplified expression,

$$\varepsilon = \varepsilon_d \cdot r_d/r + \varepsilon_c \cdot r_c/r, \qquad (3)$$

can be used with $\varepsilon_d \approx (-3.1 \pm 1.3)‰$ for most applications.

IV. Daylight Respiration of C₃ Plants

The CO_2 produced by both photorespiration and dark respiration during photosynthesis is introduced into the cytoplasm of the mesophyll cells and reabsorbed by the acceptor. The effect which this recycling will have on the overall fractionation can be obtained by following the approach taken previously (Vogel, 1980).[1] For the sake of simplicity we consider the two forms of respiration together and take the CO_2 compensation point as a measure of their combined magnitude. At the compensation point, c_{cp}, the net CO_2 flux from outside the leaf is zero and the assimilation rate is just balanced by the rate of respiratory CO_2 being introduced into the system with flux density, i. Thus

$$i = c_{cp}/r_c.$$

[1] In Vogel (1980) it was erroneously assumed that $c_k'/c_k = R_p$. In fact $c_k'/c_k = R_p \cdot \alpha_r/\alpha_c$.

According to Farquhar *et al.* (1980) this is a good estimate of the respiratory flux, also under conditions of normal CO_2 concentration. In the latter case, the reaction rate (flux density) is

$$j + i = c_c/r_c.$$

So that, subtracting i,

$$j = (c_c - c_{cp})/r_c. \tag{5a}$$

The flux density is also

$$j = (c_a - c_c)/r_d.$$

Eliminating c_c gives

$$j = (c_a - c_{cp})/(r_d + r_c) = (c_a - c_{cp})/r, \tag{5b}$$

where r is the total resistance.

If this relation holds for the ^{12}C molecules, a similar one, with primes, can represent the ^{13}C molecules and, dividing, we obtain

$$R_p = \frac{j'}{j} = \frac{r}{r'} \cdot \frac{R_a - \mu R_r/\alpha_c}{1 - \mu},$$

since the isotope ratio of the respired CO_2, $R_r = i'/i$, and $c'_{cp}/c_{cp} = i' r'_c/i\, r_c = R_r \cdot 1/\alpha_c$, and where $\mu = c_{cp}/c_a$. Also $r/r' = \alpha$, the fractionation factor without respiration, and $R_r/R_p = \alpha_r$, the fractionation factor during decarboxylation.

The total fractionation factor, $\alpha' = R_p/R_a$, thus becomes,

$$\alpha' = \alpha/(1 - \mu + \mu \cdot \alpha\, \alpha_r/\alpha_c),$$

or, in terms of fractionation,

$$\begin{aligned} \varepsilon' &= \varepsilon - \mu(\varepsilon - \varepsilon_c + \varepsilon_r) \\ &= \varepsilon(1 - \mu) + \mu(\varepsilon_c - \varepsilon_r). \end{aligned} \tag{6a}$$

Replacing ε in Eq. (6) with Eq. (3b) we obtain a revised expression for the fractionation of plants with daytime respiration, viz,

$$\varepsilon' = \{\varepsilon_d + \mu(\varepsilon_c - \varepsilon_d)\} \cdot r_d/r + \varepsilon_c \cdot r_c/r - \mu\varepsilon_r. \tag{6b}$$

This is equivalent to replacing ε_d in Eqs. (3a)–(3c) with

$$\varepsilon'_d = \varepsilon_d + \mu(\varepsilon_c - \varepsilon_d).$$

Thus Eq. (3c) can be written:

$$\varepsilon' = \varepsilon'_d + (\varepsilon_c - \varepsilon'_d) \cdot r_c/r - \mu\varepsilon_r. \tag{6c}$$

It does not apply to Eqs. (3d) and (3f) since Eq. (1) has been replaced by Eq. (5), see below.

The value of ε_r is not known, but it is expected, for various reasons, to be small. One of these is that decarboxylation in general is irreversible and in the steady state no fractionation can take place. The various attempts to measure ε_r have produced contradicting results which are discussed by O'Leary (1981).

Equation (6) implies that, even if ε_r is zero, the overall fractionation will be affected by the respiration. In situations where the carboxylation is rate limiting, viz, where $r_c \gg r_d$ the effect will be negligible. In the other extreme where the stomata are nearly closed and $r_d \gg r_c$, ε' will be larger than $\varepsilon_d = -4.4‰$.

Taking the compensation point typically as 50 ppm, $\mu = c_{cp}/c_a = 50/300 = \frac{1}{6}$, the actual fractionation, ε', will be $-8.2‰$ or about 4‰ more negative than anticipated by the simple model, Eq. (3). Thus the most positive δ-value for a C_3 plant with a compensation point of 50 ppm would be $(\delta_{atm} + \varepsilon'_{min}) = (-7.5 -8.2)‰ = -15.7‰$ and not $(-7.5 -4.4)‰ = -11.3\%$. The fractionation, ε', of such a C_3 plant could, therefore, vary between -8.2 and $-27‰$ ($\delta = -15.7$ to $-34.5‰$). Halfway between these two extremes, where $r_d = r_c$, the δ-value becomes $-25.1‰$ or about 2‰ more negative than that without daytime respiration.

Finally, if the two forms of respiration need to be separated, then ε_r must be replaced by $(\varepsilon_{fr} + \rho\varepsilon_{dr})$ in Eq. (6), where ρ is the ratio of the rates of dark respiration in the light to photorespiration, ε_{fr} the discrimination associated with photorespiration, and ε_{dr} that associated with dark respiration.

The equation describing the effect of respiration in terms of concentration (or partial pressures) as presented by Farquhar *et al.* (1982) can be derived from Eq. (6) by eliminating r_c and r from Eq. (3c) with the aid of Eq. (5a) and Eq. (5b). We then obtain

$$\varepsilon' = \varepsilon_d + (\varepsilon_c - \varepsilon_d)c_c/c_a - \mu\varepsilon_r,$$

or using their terminology and replacing c_c/c_a with p_i/p_a,

$$\Delta = a + (b - a)p_i/p_a - f\Gamma/p_a,$$

whereby the last term includes both the photorespiration and the dark respiration in the light for simplicity. We see that the fractionation has a linear relationship with c_c/c_a, as also with r_c/r, but in the former case c_c cannot sink below c_{cp}, while r_c/r can approach zero.

V. Discrimination in C_4 Plants

The isotope discrimination in C_4 plants is somewhat more difficult to explain. The initial carboxylation reaction in this pathway has been found to be one between bicarbonate and the acceptor, phosphoenolpyruvate (PEP) in the presence of PEP carboxylase. This being the case, the fractionation between dissolved CO_2 and HCO_3^- needs to be taken into account. In the cytoplasm the presence of carbonic anhydrase will greatly enhance the establishment of chemical and isotope equilibrium between the CO_2 and HCO_3^-, so that the equilibrium fractionation factor for the reaction

$$CO_2(dis) + H_2O = H^+ + HCO_3^-$$

needs to be added to Eq. (3). For the sake of convenience this discrimination, ε_{bd}, can be combined with ε_{dg}, $\varepsilon_{dg} + \varepsilon_{bd} = \varepsilon_{bg}$ so that Eq. (3c) becomes:

$$\varepsilon = \varepsilon_d + (\varepsilon_{bg} + \varepsilon_4 - \varepsilon_d)\, r_c/r. \tag{7}$$

The equilibrium discrimination between gaseous CO_2 and bicarbonate is $\varepsilon_{bg} = +7.9\%_0$ at 25°C, with a temperature dependence of $-1.1\%_0$ per 10°C (Emrich *et al.*, 1970; Mook *et al.*, 1974). The fractionation that accompanies the kinetic reaction of bicarbonate with PEP carboxylase has also been measured several times, the average value being $\varepsilon_4 = -2.2\%_0$ (O'Leary, 1981), so that the carbon initially bound in the malic acid is expected to be 5.7‰ enriched in ^{13}C with respect to the CO_2 in the leaf. The equation given above would also suggest the overall fractionation in C_4 plants to range from $-3.1\%_0(\varepsilon_d)$ to $+5.7\%_0(\varepsilon_c)$, depending on the relative magnitude of the diffusional resistance r_d and the resistance to carboxylation. This is contrary to experience: The δ-values of C_4 plants growing in nature range from about $-10\%_0$ to $-16\%_0$, which indicates fractionations of $(-10 + 7.5) = -2.5\%_0$ to $-8.5\%_0$. Nowhere are plants found to be enriched in ^{13}C with respect to the atmosphere.

To resolve the discrepancy it has been proposed that leakage of CO_2 from the bundle sheath back to the mesophyll cells creates additional fractionation (O'Leary, 1981; Farquhar, 1983). The malic acid (or aspartic acid) that is initially formed in the process transports the assimilated CO_2 to the bundle sheath cells where decarboxylation makes it available for refixation by ribulose biphosphate in the C_3 mode. This may be seen as a means of concentrating the CO_2 at the specialized chloroplasts for rapid absorption.

The effect of the leakage can be derived by following an approach similar to that used in Section IV. Let the flux of returning CO_2/HCO_3^- from the bundle sheath be

$$i = (c_k - c_c)/r_k,$$

where c_k and c_c are the concentrations in the Kranz cells and the mesophyll cells, respectively, and r_k is the diffusional resistance experienced in this transport.

In the mesophyll cells this flux joins the main influx j to react with PEP carboxylase. Thus

$$j + i = c_c/r_c.$$

Subtracting i from this expression, j becomes

$$j = (c_c r_k + c_c r_c - c_k r_c)/r_c r_k.$$

Following Farquhar (1983) we express the leakage, φ, as the ratio of the leakage rate to that of the reaction rate:

$$\varphi = \frac{i}{j + i} = \frac{r_c}{r_k} \cdot \frac{c_k - c_c}{c_c}.$$

As before, the isotope ratio of the eventually fixed carbon, R_p, can be expressed as the ratio of flux densities for ^{13}C and ^{12}C molecules, viz,

$$R_p = \frac{j'}{j} = \alpha_4 \, \alpha_k \, R_i \cdot \frac{r'_k + r'_c - r'_c \cdot c'_k/c'_c}{r_k + r_c - r_c \cdot c_k/c_c},$$

where $\alpha_4 = r_c/r'_c$ is the fractionation factor for PEP carboxylation, $\alpha_k = r_k/r'_k$ is that for back diffusion, and $R_i = c'_c/c_c$. If we recall furthermore that $R_k = c'_k/c_k$, and that the fractionation factor for C_3 carboxylation, $\alpha_3 = R_p/R_k$, we obtain, after rearrangement and using the expression for φ,

$$\frac{R_p}{R_i} = \frac{\alpha_4 + \alpha_k \varphi \cdot c_c/(c_k - c_c)}{1 - \varphi + (\alpha_k/\alpha_3)\varphi \cdot c_k/(c_k - c_c)}.$$

Since the total fractionation factor for the carboxylation is $\alpha'_4 = R_p/R_i$, and the α's can be replaced by $(1 + \varepsilon)$, and $1/\alpha \approx (1 - \varepsilon)$, we obtain

$$\varepsilon'_4 = [\varepsilon_4 + \varphi(\varepsilon_3 \cdot c_k/(c_k - c_c) - \varepsilon_k)]/[1 + \varphi \cdot c_c/(c_k - c_c)].$$

Following Farquhar (1983) the expression can be simplified to

$$\varepsilon'_4 \approx \varepsilon_4 + \varphi(\varepsilon_3 - \varepsilon_k), \tag{8}$$

since, under normal conditions, $c_k \gg c_c$. The value for the carboxylation fractionation in Eq. (7), ε_4, may now be replaced by ε'_4. Thus for C_4 plants

$$\varepsilon = \varepsilon_d + (\varepsilon_{bg} + \varepsilon_4 + \varphi(\varepsilon_3 - \varepsilon_k) - \varepsilon_d)(c_c/c_a), \tag{9}$$

or in the terminology of Farquhar (1983),

$$\Delta = a + (b_4 + \phi b_3 - a)p_i/p_a,$$

whereby $\varepsilon_k = s$ is ignored.

The implications of this revised expression for C_4 photosynthesis may now be considered. Using the fractionation values already given, viz, $\varepsilon_d \approx -3.1\%$, $\varepsilon_{bg} = +7.9\%$, $\varepsilon_4 = -2.2\%$, $\varepsilon_3 = -27\%$, and assuming $\varepsilon_k = \varepsilon_i = -0.7\%$, we see that with $\varphi = 0.34$ the right-hand term becomes zero and the relative isotope composition of bound carbon will remain constant at $\delta \approx (-7.6 - 3.1) \approx -11\%$, irrespective of the value of c_c (or p_i). Furthermore, to explain the low δ-values observed in nature of $\delta = -15.5\%$ (Fig. 1), φ must be greater than 0.6. Such a value for φ is considerably higher than is to be expected on other grounds (see Farquhar, 1983, p. 214).

A possible explanation for this discrepancy could be that equilibrium between CO_2 and HCO_3^- is not established in the process. The kinetic fractionation during the hydration of CO_2 is -6.9% (Marlier and O'Leary, 1984), so that nonequilibrium would result in a fractionation value, ε_{bg}, intermediate between $+7.9‰$ and $-6.9‰$. The overall effect would thus make the fixed carbon more negative.

VI. Experimental Verification

As was mentioned before, the basic equation, Eq. (3), suggests that the isotope fractionation during photosynthesis should vary between that due to diffusion, when the diffusional process alone is rate limiting, and that caused by the carboxylation reaction, when this reaction dominates the assimilation rate. The changeover between these two extremes is only gradual: the ratio of the two rate-controlling resistances, r_d/r_c, must extend over four orders of magnitude, from $100/1$ to $1/100$, in order to cover the full range (Fig. 3).

An obvious way of verifying the predicted variability of the isotope fractionation, at least qualitatively, is by manipulating the carbon dioxide content of the environment. Since the carboxylation resistance will be greatly increased at high CO_2 concentrations (Eq. (2)), such conditions will shift the rate-controlling step to the carboxylation reaction.

A. Materials and Methods

Growth experiments similar to those described previously (Vogel, 1980) were conducted, albeit in somewhat greater detail. The plants were grown in nutrient solutions which did not contain carbonates, in sealed Plexiglas containers $30 \times 30 \times 50$ cm large. The growth chambers were in the open under a fiberglass shelter and received about 10 to 12 h sunshine per day. The average temperature in the chambers during most of the day ranged

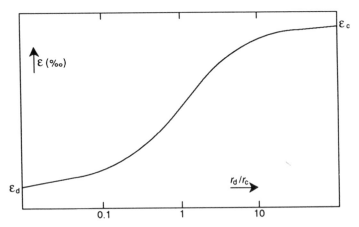

Figure 3.   The variation of the overall fractionation, ε, caused during photosynthesis in relation to the ratio of the carboxylation resistance, r_c, to the diffusional resistance, r_d, according to Eq. (3a).

between 25°C and 30°C. Air mixed with various amounts of CO_2 was bubbled through the nutrient solutions at a rate well in excess of the assimilation rate of the plants. The composition of the gas was continuously monitored by diverting a fraction of the stream to a URAS mk II infrared gas analyzer. The flow rates were adjusted so that there was no visible difference in concentration between in- and outflow. Samples of the gas were collected from time to time for isotope analysis. The experiments lasted from 17 days for the high CO_2 concentrations to 43 days at the lowest value. About halfway through, a plant was removed in order to determine the time taken for the dry weight to double.

Three different species were cultivated: algae which were soon established in the nutrient solution, rye (*Secale cereale* v. Balbo), a C_3 plant, and maize (*Zea mays* v. Mayford), a C_4 species. The latter two were grown from seed germinated in the dark and, in the case of the maize, most of the seed was removed before transfer to the chambers. Growth was allowed to continue until the dry weight of the seedlings had increased at least 10 times to minimize the effect of inherited carbon. The doubling time of the cereals varied between 4 and 12 days.

B. Results

The results for the algae are shown in Fig. 4. In the case of submerged water plants no gaseous phase is involved and all diffusion takes place in aqueous solution. The fractionation for the algae ranges from −3‰ at 0.033 vol % CO_2 in the air to −27‰ at 2.5% CO_2. This variation has been observed by various workers in the past, e.g., Degens *et al.* (1968), Pardue *et al.* (1976), Vogel (1980), Mizutani and Wada (1982). Of interest here, however, is that the shape of the idealized curve shown in Fig. 4 resembles that

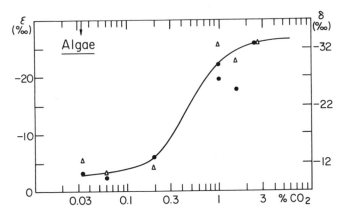

Figure 4. Experimental results for the isotope fractionation, ε, of *Algae* spp. grown in different concentrations of CO_2 expressed as vol% in the air supply. As the experiments were carried out in Pretoria, at an elevation of 1380 m, the concentration values need to be multiplied by 8.66 to convert them to mbar. The different symbols indicate two separate sets of experiments run a year apart.

in Fig. 3, suggesting that the basic equation does indeed describe the fractionation in general terms.

At the lower CO_2 concentrations the fractionation appears to have reached a minimum. The value of $-3‰$ includes the equilibrium fractionation, ε_{dg}, of $-1‰$ between the gaseous and dissolved phase, so that the actual discrimination in the liquid phase is about $-2‰$, slightly larger than that measured by O'Leary (1984) for diffusion of CO_2 through water ($-0.7‰$). The maximum discrimination of $-27‰$ confirms previous observations (Vogel, 1980) and appears to be a good estimate of the fractionation of the RuBP reaction *in vivo*.

The results obtained for the cereals are presented in Fig. 5. The fractionation observed in the rye plants varied between $-17.5‰$, at a CO_2 concentration of 0.022‰, to $-27‰$ at a concentration of 2.5%. The shape of the curve in the figure suggests again that $-27‰$ represents the limiting value of ε, i.e., $\varepsilon_c = -27‰$. The discrimination values produced in the rye plants cover most of the range observed for C_3 plants in nature ($\delta = -22$ to $-34‰$), thus confirming the expectation that the variability within the C_3 category of plants is due to changes in environmental factors rather than to intrinsic differences between species.

The range of the fractionation obtained for the maize was about 14‰ (Fig. 5). At the lower concentrations we again seem to have reached a limiting value of about $-2‰$. This is smaller than the $-4.4‰$ predicted for stomatal diffusion and may indicate that boundary layer effects contributed substantially to the diffusional resistance under the conditions of these experiments.

At high concentrations the approximations in Eq. (8) do not apply. If we use the nonabbreviated equation and assume that the malate maintains a surplus CO_2 content in the bundle-sheath cells of say 3 mbar (Berry and Farquhar, 1978) and that φ is between 0.1 and 0.2, the fractionation would

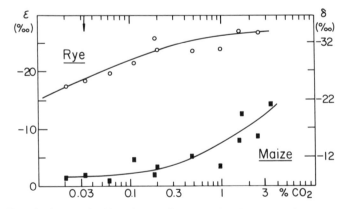

Figure 5. The isotope fractionation, ε, for rye and maize grown from seed in various concentrations of CO_2. The CO_2 content of the air is expressed in vol % and can be converted to mbar by multiplying by 8.66 (see Fig 4).

be about halfway between ε_4 and ε_3, viz., about $-16‰$. The measured values thus do not contradict the model. The large discrimination could also be explained if equilibrium conditions do not exist between CO_2 and HCO_3^-, as mentioned in Section V. Another possibility would be that, at high CO_2 concentrations (high internal partial pressures), the leakage from the bundle sheaths is reversed and part of the CO_2 is directly fixed by the C_3 pathway (Farquhar, 1983).

The discrimination produced in the maize at CO_2 concentrations above 1% is well beyond the maximum observed for C_4 plants ($\delta = -16‰$, Fig. 1), which suggests that some other factor may limit the fractionation under natural conditions to about $-8‰$.

VII. Discussion and Conclusions

A. Aquatics

The basic equation for the isotope fractionation during photosynthesis, Eq. (3), appears to explain adequately the discrimination values observed in aquatics. This includes the temperature dependence of the isotopic composition of marine plankton (Sackett *et al.*, 1965): At low temperatures the carboxylation reaction will be inhibited to a greater extent than will the diffusion process, so that it will become more rate limiting. The explanation of the large ^{13}C depletion of up to $-24.5‰$ in submerged plants growing in hard water springs (Vogel, 1980) is also obvious.

B. C₃ Plants

The δ-values of terrestrial C_3 plants growing under natural conditions range from $-22‰$ to $-34‰$. The model for this category of plants predicts that the ratio of the carboxylation resistance, $r_c(= 1/k)$, to that of diffusion varies between somewhat less than $1/1$ to as much as $30/1$. Expressed in terms of partial pressure of CO_2, this means that p_c/p_a is variable between about 0.5 and 1 in the natural environment. In some laboratory experiments, by, e.g., Hubick *et al.* (1988), the approximate expression, Eq. (3f), in which the partial pressure of CO_2 at the reaction site, p_c, is replaced by the partial pressure in the intercellular spaces, p_i, appears to give a good description of the fractionation. In other cases, such as the three C_3 plants investigated by Evans *et al.* (1986), the observed discriminations at low values of p_i/p_a deviate considerably from those of Eq. (3f), suggesting that $p_c/p_a = \frac{1}{2} p_i/p_a$ at $p_i/p_a = 0.5$. This would imply that the resistance to liquid diffusion becomes significant under these conditions.

C. C₄ Plants

The model for C_4 photosynthesis is less satisfactory. It seems clear that diffusional resistance is responsible for most of the discrimination in C_4 plants, but a quantitative explanation of fractionation values of $-8‰$, corresponding to $\delta = -16‰$, cannot be given. At high CO_2 concentrations, fractionation of $-14‰$ can be achieved in maize (Section VI, Fig. 5), but here the conditions inside the leaf become unclear and the simple equations derived in Section V do not apply. Experiments in which p_i/p_a was changed by other means (light intensity, $p(O_2)$, etc.) failed to produce any change in the isotope ratio of maize and *Amaranthus edulis* (Evans *et al.*, 1986). The same group has recently shown that, under certain conditions, the fractionation can become more positive as p_i/p_a increases (Henderson *et al.*, 1992).

We have also been unable to produce any significant change in the isotope composition of maize grown at high and low light intensity or with high and low nitrate and/or ammonium concentrations in the nutrient solution. Farquhar (1983) presents data of Boag that show an average *in*crease of 2‰ in three C_4 grasses grown in a sodium-deficient culture, so that this method of decreasing the C_4 carboxylation rate also failed to produce ^{13}C depletion in the plant.

D. CAM Plants

CAM plants assimilating CO_2 in the dark are not affected by leakage and their isotope content should be easier to interpret than that of C_4 plants. O'Leary and Osmond (1980) have shown that the CO_2 fixed overnight is the same or slightly enriched with respect to the atmosphere in two succulents. No attempt was made, however, to vary the relative resistances to diffusion and carboxylation, so that the validity of Eq. (9) cannot be assessed. The fact that δ-values less negative than $-10‰$ are not found in nature must thus, for lack of further evidence, be interpreted as the presence of sufficient daytime assimilation along the C_3 pathway.

In conclusion it may be stated that the carbon isotope composition of plants is determined mainly by the relative magnitude of the resistances to diffusion and carboxylation. In the case of submerged water plants the isotope discrimination can vary between that expected for diffusion of CO_2 alone and that produced by the carboxylation reaction. In terrestrial C_3 plants the rate of carboxylation is always the most important limiting step and the variability of the isotopic composition is more restricted. The assimilation rate of C_4 plants, on the other hand, appears to be controlled mainly by the diffusional resistance, and the observed isotope fractionation is close to that predicted for diffusion. The current model used to explain the isotope ratios of C_4 plants, however, still needs further verification. Controlled experiments with CAM plants would probably be useful in this respect.

Acknowledgments

The extensive support of Annemarie Fuls in running the experiments reported here is gratefully acknowledged. A. S. Talma is thanked for supervising the isotope ratio analyses.

References

Bender, M. M. 1971. Variations in the $^{13}C/^{12}C$ ratios of plants in relation to the pathway of photosynthetic carbon dioxide fixation. *Phytochemistry* **10**: 1239–1244.

Berry, J. A., and G. D. Farquhar. 1978. The CO_2 Concentrating Function of C_4 Photosynthesis: a Biochemical Model, pp. 119–131. *In* D. Hall, J. Coombs, and T. Goodwin, (eds.), Proceedings 4th International Congress Photosynthesis. Biochemical Society, London.

Craig, H. 1954. Carbon-13 in plants and the relationship between carbon-13 and carbon-14 variations in nature. *J. Geol.* **62**: 115–149.

Degens, E. T., R. R. L. Guillard, W. M. Sackett, and J. A. Hellebust. 1968. Metabolic fractionation of carbon isotopes in marine plankton. I. Temperature and respiration experiments. *Deep Sea Res.* **15**: 1–9.

Emrich, K., D. H. Ehhalt, and J. C. Vogel. 1970. Carbon isotope fractionation during the precipitation of calcium carbonate. *Earth Planet. Sci. Lett.* **8**: 363–371.

Evans, J. R., T. D. Sharkey, J. A. Berry, and G. D. Farquhar. 1986. Carbon isotope discrimination measured concurrently with gas exchange to investigate CO_2 diffusion in leaves of higher plants. *Aust. J. Plant Physiol.* **13**: 281–292.

Farquhar, G. D. 1980. Carbon isotope discrimination by plants: Effects of carbon dioxide concentrations and temperature via the ratio of intercellular and atmospheric CO_2 concentration, pp. 105–111. *In* G. I. Pearman, (ed.) Carbon Dioxide and Climate: Australian Research. Australian Academy of Science, Canberra.

Farquhar, G. D., M. H. O'Leary, and J. A. Berry. 1982. On the relationship between carbon isotope discrimination and the intercellular carbon dioxide concentration in leaves. *Aust. J. Plant Physiol.* **9**: 121–137.

Farquhar, G. D. 1983. On the nature of carbon isotope discrimination in C_4 species. *Aust. J. Plant Physiol.* **10**: 205–226.

Farquhar, G. D., and R. A. Richards. 1984. Isotopic composition of plant carbon correlates with water-use efficiency of wheat genotypes. *Aust. J. Plant Physiol.* **11**: 539–552.

Farquhar, G. D., S. von Caemmerer, and J. A. Berry. 1980. A biochemical model of photosynthetic CO_2 assimilation in leaves of C_3 species. *Planta (Berl.)* **149**: 88–90.

Henderson, S. A., S. von Caemmerer, and G. D. Farquhar. 1992. Short-term measurements of carbon isotope discrimination in several C_4 species. *Aust. J. Plant Physiol.* **19**: 263–285.

Hubick, K. T., R. Shorter, and G. D. Farquhar. 1988. Heritability and genotype x environment interactions of carbon isotope discrimination and transpiration efficiency in peanut (*Arachis hypogaea* L.). *Aust. J. Plant Physiol.* **15**: 799–813.

Kays, W. M. 1966. Convective Heat and Mass Transfer. McGraw–Hill, New York.

Lommen, P. W., C. R. Schwintzer, C. S. Yocum, and D. M. Gates. 1971. A model describing photosynthesis in terms of gas diffusion and enzyme kinetics. *Planta (Berl.)* **98**: 195–220.

Marlier, J. F., and M. H. O'Leary. 1984. Carbon kinetic isotope effects on the hydration of carbon dioxide and the dehydration of bicarbonate ion. *J. Am. Chem. Soc.* **106**: 5054–5057.

Mizutani, H., and E. Wada. 1982. Effect of high atmospheric CO_2 concentration on $\delta^{13}C$ of algae. *Origins Life* **12**: 377–390.

Mook, W. G., J. C. Bommerson, and W. H. Staverman. 1974. Carbon isotope fractionation between dissolved bicarbonate and gaseous carbon dioxide. *Earth Planet. Sci. Lett.* **22**: 169–176.

O'Leary, M. H. 1981. Carbon isotope fractionation in plants. *Phytochemistry* **20**: 553–567.

O'Leary, M. H. 1984. Measurement of the isotope fractionation associated with diffusion of carbon dioxide in aqueous solution. *J. Phys. Chem.* **88**: 823–825.

O'Leary, M. H., and C. B. Osmond. 1980. Diffusional contribution to carbon isotope fractionation during dark CO_2 fixation in CAM plants. *Plant Physiol.* **66**: 931–934.

Osmond, C. B., M. M. Bender, and R. H. Burris. 1976. Pathways of CO_2 fixation in the CAM plant *Kalanchoë daigremontiana*. III. Correlation with $\delta^{13}C$ value during growth and water stress. *Aust. J. Plant Physiol.* **3**: 787–799.

Pardue, J. W., R. S. Scalan, C. Van Baalen, and P. L. Parker. 1976. Maximum carbon isotope fractionation in photosynthesis by blue-green algae and a green algae. *Geochim. Cosmochim. Acta* **40**: 309–312.

Roeske, C. A., and M. H. O'Leary. 1984. Carbon isotope effects on the enzyme-catalyzed carboxylation of ribulose biphosphate. *Biochemistry* **23**: 6275–6284.

Sackett W. M., W. R. Eckelman, M. L. Bender, and A. W. H. Bé. 1965. Temperature dependence on carbon isotope composition in marine plankton and sediments. *Science* **148**: 235–237.

Smith, B. N., and S. Epstein. 1971. Two categories of $^{13}C/^{12}C$ ratios for higher plants. *Plant Physiol.* **47**: 380–384.

Troughton, J. H. 1972. Carbon isotope fractionation by plants, pp. E40–E57. *In* T. A. Rafter and T. Grant-Taylor, (eds.), Proceedings of the Eighth International Radiocarbon Dating Conference, Lower Hutt, New Zealand, organized by the Royal Society of New Zealand.

Urey, H. C. 1947. The thermodynamic properties of isotopic substances. *J. Chem. Soc.* 562–581.

Vogel, J. C. 1980. Fractionation of the carbon isotopes during photosynthesis, pp. 111–135. *In* Sitzungsberichte der Heidelberger Akademie der Wissenschaften, mathematisch-naturwissenschaftliche Klasse Jahrgang 1980. 3. Abhandlung. Springer-Verlag, Berlin/Heidelberg/New York

Vogel, J. C., P. M. Grootes, and W. G. Mook. 1970. Isotopic fractionation between gaseous and dissolved carbon dioxide. *Z. Phys.* **230**: 225–238.

5

Carbon and Oxygen Isotope Effects in the Exchange of Carbon Dioxide between Terrestrial Plants and the Atmosphere

Graham D. Farquhar and Jon Lloyd

I. Carbon Isotope Effects

Plants differ from the atmosphere in their average relative abundances of carbon isotopes. This variation arises because the kinetic parameters of chemical reactions can be affected by the atomic masses of the compounds involved.

A. The Physical Chemistry of Isotope Effects

The carbon isotope ratio of a compound is the molar abundance ratio $^{13}C : {}^{12}C$. The isotopic composition of a carbon compound is normally expressed as the proportional deviation ($\delta^{13}C$) from some standard,

$$\delta = \frac{R_A}{R_S} - 1, \tag{1}$$

where R_A is the $^{13}C/^{12}C$ ratio in the compound being examined and R_S is the isotope ratio of the standard. Traditionally, R_S for $^{13}C/^{12}C$ determinations has been that of a fossil belemnite from the Pee Dee formation in South Carolina (denoted PDB). The R_S of PDB has traditionally been taken as 0.01124 (Craig, 1957), but more recent analysis suggests that a value of 0.01118 is more appropriate (Bakke *et al.*, 1991).

For plant material δ ranges from -0.0089 to -0.0301, while the δ of atmospheric CO_2, δ_a, had a value of -0.0080 in 1988. Such values are normally presented as "parts per mil" ($= \times 10^{-3}$ or ‰); ‰ is not a unit and δ

is dimensionless. Therefore, using conventional notation, the range of δ_p is from $-8.9‰$ to $-30.1‰$, while that of δ_a is $-8.0‰$.

The partitioning of isotopes between two substances with different isotope ratios is referred to as isotope fractionation. Fractionations are caused by

1. kinetic effects, which depend on the differences in reaction rates and diffusion coefficients of isotopic molecules, and
2. thermodynamic effects in isotope exchange reactions.

Following the convention of Farquhar *et al.* (1989a), carbon isotope effects are defined as the ratio of carbon isotope ratio in reactant (R_r) and product (R_p), i.e.,

$$\alpha = R_r/R_p. \tag{2}$$

1. Kinetic Effects   A kinetic isotope effect occurs when the rate of chemical reaction is sensitive to the atomic mass at a particular position on one of the reacting species. For two competing isotope reactions

$$C^{12} - A \xrightarrow{k_{12}} C^{12} - B \quad \text{and} \quad C^{13} - A \xrightarrow{k_{13}} C^{13} - B,$$

Equation (2) simply means

$$\alpha_{\text{kinetic}} = \frac{k^{12}}{k^{13}}. \tag{3}$$

Included in the kinetic group are diffusion effects. Generally, light isotopes are more mobile than their heavier counterparts. From a consideration of unidirectional diffusion, with source analogous to reactant, α is the ratio of the diffusion coefficients. For two gases, the binary diffusivity (of one gas in another) is inversely proportional to the square root of their reduced masses (Mason and Marrero, 1970), where the reduced mass is defined as $m_1 m_2/(m_1 + m_2)$. (Note that the diffusion coefficient for CO_2 in air is the same as that for air in CO_2.) Thus, the isotope effect during diffusion of CO_2 in air is

$$\alpha_{\text{diff}} = \frac{D_{(C^{12}O_2)}}{D_{(C^{13}O_2)}} = \sqrt{(m_{(C^{13}O_2)}/m_{(C^{12}O_2)})(m_{(C^{12}O_2)} + m_{\text{air}}/m_{(C^{13}O_2)} + m_{\text{air}})}, \tag{4}$$

where D denotes the effective diffusion coefficient and m is the molecular mass. Taking m_{air} as 29, α_{diff} for diffusion through dry air is therefore 1.0044 (Vogel, 1980; Farquhar *et al.*, 1982). Isotope effects on the diffusion of gases should be independent of temperature (Mason and Marrero, 1970).

Diffusion through water depends more on the volume of the solute molecule than on its mass, and so isotopic effects in solution are generally less than those in the gas phase.

2. **Thermodynamic Effects**   In a sample system, at chemical equilibrium,

$$A \underset{k_2}{\overset{k_1}{\rightleftharpoons}} B,$$

where A and B represent different chemical moieties or different phases. We take the left hand side, by convention, to be the "reactant." Following then the convention of Eq. (2), the equilibrium isotope effect, or fractionation factor (α_{eqbm}) is defined as the ratio

$$\alpha_{eqbm} = \frac{R_A}{R_B}. \tag{5}$$

For example, the fractionation factor for the hydration of CO_2

$$CO_2 + H_2O \rightleftharpoons H_2CO_3$$

is

$$\frac{C^{13}/C^{12} \ (CO_2)}{C^{13}/C^{12} \ (H_2CO_3)} = 1/1.011 \text{ at } 25°C$$

(Mook *et al.*, 1974). Thus the isotope effect for the hydration of CO_2 is, by our convention, less than unity.

The balance in any reversible reaction usually depends on temperature, so that equilibrium isotope effects also depend on temperature. In practice enzymatic reactions often contain reversible steps so that fractionations during catalysis can be temperature dependent (see Part I of Appendix in Farquhar *et al.*, 1989a).

3. **Definitions of Isotope Discrimination and Plant Isotopic Composition**
Isotope effects, α, typically have values near unity. It is numerically convenient to deal with Δ of α from unity viz,

$$\Delta = \alpha - 1, \tag{6}$$

where Δ represents the isotopic discrimination. Farquhar and Richards (1984) proposed that plant processes should be analyzed using the convention of Eq. (2), so that, at the whole-plant level

$$\Delta = \alpha - 1 = \frac{R_a}{R_p} - 1, \tag{7}$$

where R_a and R_p are the isotopic abundances of carbon in CO_2 in the air and carbon plant, dry matter, respectively.

Unlike δ_p, Δ is independent of the isotopic composition of the standard and it is also independent of R_a. From Eq. (7) we can obtain an expression for Δ in terms of δ_a and δ_p viz,

$$\Delta = \frac{\delta_a - \delta_p}{1 + \delta_p}. \tag{8}$$

Apart from its advantage over δ_p in that it directly expresses the consequences of biological processes (independent of source air isotopic composition), this definition of Δ also has the advantage that discriminations occurring in series are additive when weighted by the associated changes in CO_2 partial pressure (see for example Eq. (15)). However, for examination of isotopic fractionations in parallel, $(1 - R_p/R_a)$ is more convenient.

B. Processes Affecting Carbon Isotope Exchange between Plants and the Atmosphere

1. Isotopic Composition of the Atmosphere   Measuring stations designed to evaluate long-term trends in concentration and isotopic composition of trace gases have been selected to ensure that such trends are unaffected by diurnal fluctuations in animal (including human) or plant activity. Measurements at such stations show that the $\delta^{13}C$ of atmospheric CO_2 is currently close to $-8‰$ (Goodman and Francey, 1988). This value is slowly becoming more negative as the atmosphere becomes depleted in $^{13}CO_2$ relative to $^{12}CO_2$. The first extensive measurements of δ_a by Keeling (1958, 1961) when compared with more recent data from similar locations (Keeling *et al.*, 1979) indicates that δ_a shifted from $-6.69‰$ in 1958 to $-7.24‰$ in 1978, an average decrease of $0.028‰$ per year. This depletion has arisen as a consequence of the anthropogenic burning of fossil fuels and, to a lesser extent, deforestation. CO_2 from both these sources has an isotopic composition typical of C_3 plant matter (ca $-26‰$). The ice core record shows that δ_a was only about $-6.4‰$ prior to the industrial revolution (Friedli *et al.*, 1986). There is a latitudinal gradient in δ_a, with δ_a at 60°N typically being $0.2‰$ more negative than that at 60°S (Keeling *et al.*, 1989), presumably a consequence of most anthropogenic burning of fossil fuels occurring north of 20°N.

Seasonal variations in δ_a are observed, being the greatest at high northern latitudes. At Point Barrow, Alaska (71°N), δ_a varies from $-8.2‰$ in early spring to $-7.2‰$ in autumn. These changes in δ_a correlate well with seasonal changes in the atmospheric partial pressure of CO_2 (C_a) which are themselves associated with the seasonal patterns of photosynthesis and respiration by the terrestrial biosphere. High C_a occur at the time of maximum net release of CO_2 from the biosphere, with respired CO_2 being depleted in ^{13}C (Keeling *et al.*, 1989). Due to less biospheric activity in the southern hemisphere, the seasonal amplitude in both δ_a and C_a is less than that in the northern hemisphere (Goodman and Francey, 1988).

2. Isotopic Composition of Air within Plant Canopies   Even greater variations in δ_a may occur within, or just above, plant canopies. For rural air within the canopy layer near the Pacific coast of North America, Keeling (1958, 1961) found that his data closely fitted the equation for a single source of isotopic composition, δ_1 and a simple mixing process,

$$\delta_a = \delta_1 + \frac{M}{C_a}, \tag{9}$$

where C_a is the CO$_2$ partial pressure and M is a constant. Such an expression arises if

$$C_a \delta_a = C_o \delta_o + C_l \delta_l, \tag{10a}$$

where C_o is the partial pressure of CO$_2$ (pCO$_2$) in a background gas of composition δ_o, to which an additional pCO$_2$, C_l, with composition δ_l is added. That is to say,

$$C_a = C_o + C_l. \tag{10b}$$

Eliminating C_l from Eqs. (10a) and (10b) and taking δ_l to be equivalent to δ_p, Lancaster (1990) expressed Eqs. (10) as

$$\delta_a = \delta_p + \frac{(\delta_t - \delta_p)C_t}{C_a}, \tag{11}$$

where δ_a is the isotopic composition of the air within the "canopy layer" (see below), and C_t and δ_t represent the pCO$_2$ and isotopic composition of the remaining troposphere, being an infinite background reservoir. The canopy layer does not need to be precisely defined in this case, but it must be close enough to vegetation for diurnal cycles in nearby plants and soil to be measured. From Eq. (11) it can be seen that if the isotopic composition of carbon entering and leaving plants is invariant over time, then a plot of δ_a *versus* $1/C_a$ gives a straight line relationship with a slope $(\delta_t - \delta_p)C_t$ (equal to M in Eq. (9)) and an intercept δ_p. Thus provided that δ_t and C_t remain constant, and that there is no discrimination of CO$_2$ during plant or soil respiration (but see Section I,B,6,b), then from concurrent measurements of δ_a and C_a one can infer δ_p.

Equations (9)–(11) do not consider the possibility of any refixation of respired CO$_2$ during daytime CO$_2$ fixation and a model to estimate this recycling of CO$_2$ from the relationship between C_a and δ_a has been developed by Sternberg (1989). His equation, derived for the special situation where C_a and δ_a are not changing with time is

$$\delta_a = \delta_p + \frac{(\delta_t - \delta_p)(1 - \rho)C_t}{C_a} + \rho \Delta_A, \tag{12}$$

where ρ is the ratio of the uptake of respired CO$_2$ via photosynthesis relative to the loss of respired CO$_2$ out of the canopy layer via turbulent mixing and Δ_A is the isotopic discrimination during photosynthetic CO$_2$ assimilation. As ρ approaches 0, Eq. (12) approaches Eq. (11). For tropical rainforest in Panama, Sternberg (1989) obtained an estimate for ρ of 0.07–0.08, while for tropical forest formations in Trinidad, Broadmeadow *et al.*, (1992) calculated ρ to range from 0.08 to 0.26 depending on time of day and canopy structure.

3. Flow of CO$_2$ across the Laminar Boundary Layer

Equation (4), which describes the ratios of the diffusivities of ^{12}CO$_2$ and ^{13}CO$_2$ in free air, does not apply within a laminar boundary layer. From the Pohlhausen analysis of mass transfer from a plate in laminar parallel flows (Kays, 1966), the appropriate ratio is that in free air to the $2/3$ power. Thus, from Eqs. (4) and (6) it can be seen that the theoretical value for boundary layer discrimination (a_b) is equal to $(1.044^{2/3} - 1)$ or 2.9‰.

4. Transport of CO_2 within the Leaf

a. *Diffusion across the Stomata*   As shown in Eq. (4) the discrimination against $^{13}CO_2$ in free air (usually denoted by the symbol a) has a theoretical value of 4.4‰. For accurate calculations, this value should be corrected for the presence of water vapor (in which the discrimination against $^{13}CO_2$ is 6.5‰; Eq. (4)), but this changes the effective value of a by only about 1%. A more significant problem may occur at low stomatal apertures (less than about 0.5 μm) where the nature of the diffusion of gases is markedly different than that at greater apertures or in free air. Equations such as (4), and indeed all equations used in the calculation of gas exchange parameters, are only valid where the mean free path of molecular motions of the gases involved (λ) is much smaller than the stomatal pore diameter. The mean free path represents the average distance traveled by a gas molecule per unit time divided by the average number of collisions per unit time. That is to say, it is the average distance traveled by a gas molecule before its trajectory is interrupted by another. It is a complicated parameter to determine accurately, particularly in a mixture of compounds, but at 25°C and at atmospheric pressure (10^5 Pa) it is of the order of 0.1 μm. For stomatal pores whose diameter is much greater than this value, the rate of diffusion is dependent on the concentration gradient and on the frequency and type of molecular collision (continuum diffusion), and conventional isotope discrimination and gas exchange equations apply. For stomatal apertures smaller than λ, intermolecular collisions become rare compared to the frequency of collisions with the guard cell walls. In this case, binary diffusion coefficients are inappropriate, with the rate of diffusion of any gaseous species being only proportional to its mean thermal speed, v, which is inversely proportional to its molecular mass and independent of any other gaseous species present. The discrimination associated with this diffusion (referred to as molecular or Knudsen diffusion), a_K is therefore

$$a_K = \sqrt{(m_{C_{13}O_2}/m_{C_{12}O_2})} - 1, \tag{13}$$

or 11.3‰. This is markedly different from the 4.4‰ value appropriate for continuum flow.

We can calculate stomatal conductances at which Knudsen flow becomes important using theoretical equations relating stomatal conductance to stomatal dimensions. For a pore of length l and diameter d (Monteith, 1973)

$$g_s = \frac{(\pi n d^2 D) P}{4RT(l + \pi \, d/8)}, \tag{14}$$

where g_s is the stomatal conductance (mol H_2O m^{-2} s^{-1}), n is the number of stomata per unit leaf area, P is the atmospheric pressure (Pa), R is the universal gas constant (8.314 J mol^{-1} K^{-1}), and T is the temperature (K). Knudsen flow will be more significant for species with leaves having a high frequency of small stomata such as occurs for members of the genus *Citrus* (Kriedemann and Barrs, 1981). From measurements of the frequency and dimensions of stomata for *Citrus mitis*, Davies and Kozlowski (1974) reported that $l = 4.7$ μm and $n = 596 \times 10^6$ m^{-2}. Thus, at 10^5 Pa and 25°C we can calculate the conductance at which Knudsen flow will become important for stomatal diffusion into such leaves, corresponding to apertures 0.5 μm or less. From Eq. (14) this is when g_s is less than about 0.05 mol H_2O m^{-2} s^{-1}. While this is not a value that typically occurs under high photon irradiances and high humidities (Lloyd *et al.*, 1987), for *Citrus* trees in the field, values such as this are observed under conditions of high vapor pressure difference (Syvertsen *et al.*, 1988) and under conditions of low photon irradiance (Syvertsen, 1984). Complicated equations are required for a precise determination of the ratio of molecular to continuum flow as stomatal conductances decrease below this level, but it is clear that for some species at least, Eq. (4) may effectively underestimate the discrimination associated with diffusion into the leaf when stomatal conductances are low. This may be particularly true for leaves under dense canopies, where almost all gas exchange occurs under conditions of low photon irradiance.

At first sight, it might also appear that the partial pressure of CO_2 in the stomatal cavity, C_{st},[1] calculated according to conventional gas exchange equations (Caemmerer and Farquhar, 1981) would also be in error when Knudsen diffusion is occurring. This is because to calculate C_{st} the correct relative diffusivities of CO_2 and H_2O must be used. It turns out however that the ratios of the binary diffusion coefficients CO_2/air and H_2O/air are almost identical to $[m(CO_2)/m(H_2O)]^{-1/2}$. The effect of Knudsen diffusion on gas exchange into and out of leaves was examined in detail by Leuning (1983). He found that even in the case of stomatal apertures being as low as 0.01 μm, C_{st} would be underestimated by only 3%.

b. Diffusion through Intercellular Air Spaces   Early experiments with amphistomatous leaves of cotton and *Xanthium* indicated maximum drawdowns across intercellular air spaces of less than 20 μbar (Farquhar and Raschke, 1978) and it has therefore often been assumed that the partial pressure of CO_2 at the mesophyll cell wall surface, C_w, is close to C_{st} (Farquhar and Caemmerer, 1982). However, three-dimensional modeling studies suggest drawdowns of as much as 100 μbar for hypostomatous leaves under some conditions (Parkhurst, 1986). Parkhurst *et al.* (1988) measured gradients similar to those predicted by the models using amphistomatous leaves that were fed CO_2 on only one surface simulating hypostomatous leaves. The existence of a substantial resistance to CO_2 diffusion in

---

[1]Previously referred to as the "intercellular" partial pressure of CO_2, C_i.

hypostomatous leaves was also indicated by increases in CO_2 assimilation rates of as much as 27% when gas exchange was measured in helox (air with nitrogen replaced by helium) where the diffusivity of CO_2 is 2.3 times greater than that of air. By contrast, the average increase in assimilation for amphistomatous leaves was 7% (Parkhurst and Mott, 1990). Therefore, it seems reasonable to conclude that, particularly for hypostomatous leaves, isotope fractionations associated with diffusion from substomatal cavity to the cell wall of photosynthesizing mesophyll cells could be significant. For *Metrosideros polymorpha*, Vitousek *et al.* (1990) found less negative δ_p with increasing altitude on Mouna Loa (Hawaii), increasing from -30 to $-24‰$. They attributed this result to increases in diffusion resistance as a consequence of increases in leaf thickness. Where low intercellular air space conductances exist, they will be variable spatially, being dependent on the distance of the photosynthesizing cell from the substomatal cavity. Methods of calculating the mean discrimination in such situations are not straightforward (Section I,D,2).

   c. Flow from the Cell Wall Surface to the Sites of Carboxylation   This pathway, sometimes referred to as the liquid phase, involves several resistances in series: diffusion across the water-filled interstices of the cell walls followed by the plasmalemma and chloroplast membrane and finally through the chloroplast stroma to the sites of carboxylation. The total fractionation associated with this pathway is normally taken (Evans *et al.*, 1986) as the sum of the equilibrium fractionation as CO_2 enters solution (1.1‰ at 25°C; Mook *et al*, 1974) and a_l the discrimination occurring during diffusion of dissolved CO_2 in water (0.7‰; O'Leary, 1984). Clearly, given the complexity of the pathway this is somewhat simplistic. It ignores any possible fractionations associated with diffusion across membranes and assumes that the ratio of diffusion coefficients of $^{13}CO_2/^{12}CO_2$ in free water is appropriate in the thin cell wall pores and in the chloroplast stroma (where bicarbonate diffusion actually accounts for most of the transport).

   It is carbonic anhydrase which interconverts CO_2 and bicarbonate. The isotope fractionation for the equilibrium reaction is temperature dependent, being $-9.0‰$ at 25°C (Mook *et al.*, 1974). In this context, however, the effect on the isotopic composition of fixed CO_2 in c_3 plants is unimportant as all CO_2 hydrated by carbonic anhydrase at the chloroplast surface is dehydrated again prior to carboxylation by RuBisCO (but see Section I.5,b).

   From theoretical considerations, Cowan (1986) has shown that, provided there is sufficient carbonic anhydrase in the chloroplast, the stroma itself should provide little barrier to CO_2 diffusion. Isotope fractionations associated with diffusion from mesophyll wall surface to the sites of carboxylation are presumably then associated with resistances in the cell wall or across membranes. Caemmerer and Evans (1991) hypothesized that, for wheat, the wall resistance was a function of the surface area of chloroplast exposed to the intercellular airspace per unit projected leaf area (S_c). Comparing the leaf anatomy of a range of hypostomatous leaf types, Syvertsen,

Lloyd, McConchie, and Farquhar (unpublished) have also found that S_c was correlated with leaf internal resistance and, in addition, found intercellular air space resistances to be important.

5. Fractionations Associated with Carboxylations

a. Ribulose-1,5 Bisphosphate Carboxylase/Oxygenase (RuBisCO) A major kinetic fractionation occurs during the fixation of CO_2 by RuBisCO. Using spinach, Roeske and O'Leary (1984) observed α to be 1.029 ± 0.001 at pH 8.0 and 25°C with respect to dissolved CO_2, which when multiplied by the isotope effect for the dissolution of CO_2 in water (1.0011) makes the discrimination by this enzyme, b_3, 30‰.

b. Phosphoenolpyruvate Carboxylase As for RuBisCO, PEP-c discriminates against ^{13}C, the value for discrimination against $H^{13}CO_3^-$, b_4^*, being 2.0‰ (O'Leary, 1981). However, there are two equilibrium fractionations which need to be taken into account when determining the total discrimination by the enzyme, b_4 (Farquhar, 1983) viz,

$$b_4 = e_s + e_b + b_4^*,$$

where e_s is the equilibrium effect caused by CO_2 going into solution (1.1‰ at 25°C; Mook *et al.*, 1974) and e_b is the equilibrium discrimination effect accompanying the hydration of CO_2 (−9.0‰ at 25°C; Mook *et al.*, 1974). Thus, at 25°C, b_4 is usually taken as ∼ 5.7‰ (O'Leary, 1981).

Fractionations associated with PEP-c must be taken into account for C_3 as well as for C_4 plants (Farquhar and Richards, 1984) as a proportion of the carbon fixed is associated with anaplerotic reactions.

c. Other Carboxylations Plants contain many other carboxylases, most of which function in biosynthetic pathways. However, these carboxylases probably contribute less than 0.05 of the total plant C (Raven and Farquhar, 1990) and are not considered further here.

6. Decarboxylations

a. Glycine Decarboxylation and Photorespiration Being a pyridoxal phosphate-dependent carboxylase (Walker and Oliver, 1986), glycine carboxylase is thought to have a discrimination of around 8‰; a value that compares well with discriminations observed to occur during photorespiration (Rooney, 1988). A direct measurement of α for this enzyme has yet to be undertaken, however.

Farquhar *et al.* (1982) calculated the effect of the photorespiratory release of CO_2 from glycine on the net discrimination observed during CO_2 uptake. This calculation assumed that the isotopic composition of glycine was identical to that of recently assimilated carbon. The effect is given by the term $f\Gamma^*/C_a$ in Eq. (15), where f corresponds to the 8‰ (above) and Γ^* is the CO_2 photocompensation point.

b. Respiratory Release of CO_2 Studies examining the isotopic composition of CO_2 evolved from potato tubers (Jacobson *et al.*, 1970), and from the alga *Chlorella pyrenoidosa* (Abelson and Hoering, 1961), have shown the isotopic composition of respired air to change dramatically with time. Nevertheless, when the alga *Dunaliella* were maintained in darkness for several days, the $^{13}C/^{12}C$ ratio of dry matter became depleted (Degens *et al.*, 1968), indicating that respired air would have been enriched in ^{13}C. This result seems a little surprising given that one-third of the CO_2 evolved during aerobic respiration comes from the pyruvate dehydrogenase step, and, for micro-organisms at least, the CO_2 evolved from this decarboxylation is strongly depleted in ^{13}C relative to the pyruvate carboxyl group (Melzer and Schmidt, 1987).

Farquhar *et al.* (1982) showed that for such respiration in the light that the term eR_d/kC_a is necessary in Eq. (15), where e is the discrimination in the decarboxylation, R_d is the rate of decarboxylation, and k is the carboxylation efficiency.

C. Integration into C_3 Metabolism

1. During Steady-State Photosynthesis   Total discrimination against $^{13}CO_2$ during CO_2 assimilation Δ_A can be calculated by adding all the discriminations associated with diffusion from the atmosphere to the sites of carboxylation and weighting the diffusional discriminations according to the associated drawdown of CO_2, including the discriminations associated with CO_2 fixation itself, and then subtracting the fractionations associated with the release of CO_2 via photorespiration and respiration *viz*

$$\Delta_A = a_b \frac{C_a - C_s}{C_a} + a \frac{C_s - C_{st}}{C_a} + a \frac{C_{st} - C_w}{C_a}$$

$$+ (e_s + a_1)\frac{C_w - C_c}{C_a} + b\frac{C_c}{C_a} - \frac{eR_d/k + f\Gamma^*}{C_a}. \tag{15}$$

Equation (15) indicates that the extent of discrimination against $^{13}CO_2$ during photosynthesis is dependent on a large number of leaf properties. Nevertheless, the fractionation constants are relatively invariant across leaf types, and it therefore seems reasonable to conclude that differences in Δ_A are mainly a consequence of differences in the ratios of CO_2 partial pressures within the leaf.

Equation (15) is a more complete form of the often-used approximate equation for isotope discrimination (Farquhar *et al.*, 1982):

$$\Delta_A = a \frac{C_a - C_{st}}{C_a} + b\frac{C_{st}}{C_a} = a + (b - a)\frac{C_{st}}{C_a}. \tag{16}$$

(C_{st} is a more precise definition for the intercellular partial pressure of CO_2, C_i.) Comparison of the two equations shows that the simpler version ignores fractionations associated with the boundary layer conductance, photorespiration, and respiration, and assumes that there is no drawdown of CO_2 from the stomatal cavity to the sites of carboxylation.

One experimental approach has been to measure Δ_A "on line." In this approach (Evans *et al.*, 1986) the degree of enrichment of δ_a in $^{13}CO_2$ is measured as air passes over a leaf in an open gas exchange system. From simple mass balance considerations, if the partial pressures and isotopic composition of air entering and leaving the chamber are known, then Δ_A can be calculated (Evans *et al.*, 1986). If concurrent measurements of CO_2 assimilation and transpiration are made, then the appropriateness of the simple equation in accounting for the discrimination observed (sometimes referred to as Δ_{obs}) can be determined. When such measurements have been undertaken, substantial differences between Δ_{obs} and the predictions of Eq. (16) have been observed (Evans *et al.*, 1986; Caemmerer and Evans, 1991; Lloyd *et al.*, 1992), with Eq. (16) overestimating discrimination by as much as 8‰.

Fractionations associated with respiration or photorespiration are too small to account for the differences observed, and measurements were made under conditions of a high boundary layer conductance, where $C_a \approx C_s$, so comparisons of Eqs. (15) and (16) show that this difference must be due to C_c being substantially lower than C_{st}. The precise location of the low internal conductance (g_i) responsible for this drawdown remains to be determined, but as mentioned previously, there appears to be both an intercellular air space and a mesophyll cell wall/membrane/chloroplast stroma component (Section I,C,4,c). CO_2 assimilation at saturating photon irradiance and g_i are very well correlated for a range of species and leaf types (Caemmerer and Evans, 1991; Lloyd *et al.*, 1992; Loreto *et al.*, 1992), which means that C_{st}/C_c tends to be a reasonably conservative parameter. At saturating photon irradiances and $C_a = 350$ μbar, the difference between C_{st} and C_c is typically around 60 μbar.

2. Effects of Heterogeneity in Photosynthetic Characteristics   Equation (15) treats the leaf as a homogenous entity. There are, however, variations in light environment, photosynthetic capacity (Terashima and Inoue, 1985), and, if intercellular air space diffusion resistances are high (Section I,C,4,b), variations in C_w (Parkhurst *et al.*, 1988) across a leaf. Clearly then, even if all stomata were to have the same conductance, variations in C_c exist. Variations in stomatal aperture and, hence, conductance are however often observed (e.g., Smith *et al.*, 1989). Thus, considerable heterogeneity in CO_2 partial pressures and, hence, CO_2 assimilation rates can exist within leaves (Terashima *et al.*, 1988). The problem here is that the relationships between the various diffusion conductances and CO_2 assimilation are non-linear (Farquhar and Sharkey, 1982). Farquhar (1989) evaluated the effect of stomatal heterogeneity in stomatal conductance. He showed that estimates of C_{st} from gas exchange measurement are weighted according to stomatal conductance, but that carbon isotope discrimination is weighted according to CO_2 assimilation rate. Thus, if there is variation in g_s, and any sort of barrier to lateral transport within the leaf (e.g., in the case of bundle-sheath extensions), then even if C_c directly underneath each stoma is equivalent to the C_{st} of that stoma, the relationship between Δ_A will deviate from that of Eq. (15). Lloyd *et al.* (1992) examined this effect in more detail and showed that it could cause discrepancies of as much as 2‰ between Eq.

(15) and the actual Δ_A. Variations in photosynthetic capacity and/or g_i also cause similar differences between actual and predicted discriminations.

D. C₄ Metabolism

A theoretical treatment of isotope discrimination during C_4 photosynthesis was presented by Farquhar (1983), who showed that

$$\Delta = a + [b_4 + \phi(b_3 - s) - a]\frac{C_c}{C_a}, \qquad (17)$$

where ϕ is the ratio of the rate of CO_2 leakage from the bundle sheath to the rate of PEP carboxylation (leakiness) and s is the fractionation during this leakage. Measurements of on-line $^{13}CO_2$ discrimination by C_4 plants were made by Henderson *et al.* (1992) who showed that for a number of dicotyledonous and monocotyledonous species of the various C_4 decarboxylation types leakiness is around 0.2, except at low irradiances where it increases. The magnitude of the drop in pCO_2 from the stomatal cavity to the sites of carboxylation within the mesophyll cytoplasm of C_4 plants remains to be determined, but even if this difference were large, only small changes in the calculated leakiness would ensue (Henderson *et al.*, 1992).

There is a significant difference between long-term isotope discrimination (as assessed from carbon isotope composition of dry matter) and short-term "on-line" measures of discrimination (Henderson *et al.*, 1992). The reasons for this remain to be determined, but fractionations postphotosynthesis may be involved. It is presumably this difference that caused Marino *et al.* (1992) to use a value of ϕ considerably greater than 0.2 in their analysis of the relationship between δ_p and δ_a. This relationship allowed them to infer values of δ_a back to 55,000 years B.P.

II. Oxygen Isotope Effects

A. Introduction

Oxygen isotopes are of interest to plant carbon and water relations for at least two reasons. First, it is possible that the $^{18}O/^{16}O$ ratio of organic matter may be useful for determining whether differences between genotypes in $\delta^{13}C$ and C_c/C_a are caused by differences in photosynthetic capacity or in stomatal conductance (Farquhar *et al.*, 1989b).

The $^{18}O/^{16}O$ ratio in atmospheric CO_2 provides additional information about exchange between the atmosphere and water, some of which is via biological activity. Carbonic anhydrase catalyses oxygen isotopic exchange between leaf water and CO_2 but other processes such as exchange with the oceans need to be taken into account (Friedli *et al.*, 1987; Francey and Tans, 1987; Farquhar *et al.*, 1993).

B. Processes Determining $^{18}O/^{16}O$ in Leaf Water

1. Fractionations Associated with Evaporation   The process of evaporation tends to cause enrichment of ^{18}O relative to ^{16}O in leaf water, because the vapor pressure of H_2O^{18} is less than that of H_2O^{16}, and because H_2O^{18} diffuses more slowly from the leaf. The former effect is given by ε^*, the proportional depression of equilibrium vapor pressure by the heavier molecule compared to H_2O^{16}. Bottinga and Craig (1969) showed that

$$\varepsilon^* = [2.644 - 3.206(10^3/T) + 1.534(210^6/T^2)] \times 10^{-3}, \qquad (18)$$

where T is the absolute temperature (K) and so $\varepsilon^* = 9.2‰$ at 25°C and 9.6‰ at 20°C.

The diffusivity of H_2O^{16} in air is 1.028 times that of H_2O^{18} (Merlivat, 1978) so that the kinetic fractionation factor, ε_k, is 28‰ for diffusion in air, as through the stomata. In the leaf boundary layer this term will be 19‰ (assuming the ⅔ power effect discussed earlier), so that for a stomatal resistance, r_s, and a boundary layer resistance, r_b (Farquhar *et al.*, 1989b)

$$\varepsilon_k = \frac{28r_s + 19r_b}{r_s + r_b} \times 10^{-3}. \qquad (19)$$

Following the derivation by Craig and Gordon (1965) for a free water surface, Farquhar *et al.* (1989b) presented an equation which applies to the isotopic composition at the sites of evaporation, Δ_E, expressed in relation to the isotopic composition of the source water, as

$$\Delta_E = \frac{R_E}{R_S} - 1, \qquad (20)$$

where R_E and R_S are the $^{18}O/^{16}O$ ratios at the site of evaporation and for source water, respectively. [N.B. We use the Δ notation although we could retain the δ notation, since the isotope standard is arbitrary, and could be taken as source (or stem) water, to avoid confusion with δ on a SMOW or PDB scale.] It can be thought of as a discrimination as defined earlier (Eq. (5)), the process being evaporation, the output being the isotope ratio of evaporated water, and the "input" being the isotope ratio at the sites of evaporation, R_E. The equation is

$$\Delta_E = \varepsilon_k + \varepsilon^* + (\Delta_V - \varepsilon_k)e_a/e_i, \qquad (21)$$

where e_a and e_i are the vapor pressures in the atmosphere and intercellular air spaces, respectively, and Δ_V is the isotopic composition of water vapor in the air, again relative to that of source water

$$\Delta_V = \frac{R_V}{R_S} - 1. \qquad (22)$$

Equation (21) is a convenient approximation for the mathematically correct form of this particular model

$$\Delta_E = (1 + \varepsilon^*)[1 + \varepsilon_k + (\Delta_V - \varepsilon_k)e_a/e_i] - 1,$$

or, in a form similar to that of Flanagan *et al.* (1991),

$$\frac{R_E}{R_S} = \alpha^* \left[\alpha_k + \left(\frac{R_V}{R_S} - \alpha_k \right) e_a/e_i \right], \tag{23}$$

so that it underestimates Δ_E by about 0.1‰. This imprecision is less than that in measurement uncertainty, and for most applications is outweighed by the convenience of Eq. (21). This is highlighted in the context of the discrepancy between leaf measurements and predictions of the model of about 2‰. Equation (21) is similar to that of Francey and Tans (1987), but their equation is incorrect as it contains a typographical error, with the isotopic composition of atmospheric water vapor (δ_V) inadvertently substituted for the isotopic composition of source water (Δ_S), (P. Tans, personal communication).

Much of the discrepancy between Δ_E and whole leaf-water Δ probably arises because of gradients within the leaf, and these are now examined.

2. Convection and Diffusion   When water at the site of evaporation becomes enriched in ^{18}O because of the above effects, H_2O^{18} will tend to diffuse away from these sites. This effect, in turn, is opposed by convection of source water to the sites. Farquhar (unpublished, and cited by White, 1989) suggested that a useful analog of this effect is of source water, moving in one dimension to a point where evaporation occurs, causing composition to be Δ_E with respect to source water at that point. The solution in the steady state is

$$\Delta_l = \Delta_E e^{-El/CD}, \tag{24}$$

where E (mol m^{-2} s^{-1}) is the rate of water movement, C is the concentration of water (mol m^{-3}), D is the diffusivity of H_2O^{18} in water (m^2 s^{-1}), and l (m) is the distance from the site of evaporation. The average value of Δ_l, over length L, Δ_L, say, is then (Farquhar *et al.*, 1993)

$$\Delta_L = \frac{CD}{EL} \Delta_E (1 - e^{-EL/CD}). \tag{25}$$

The dimensionless term, $EL/(CD)$, is known as the Péclet number (Ikeda, 1983), ℘, (note that E/C is the velocity of the convection stream) and is the ratio of convective to diffusive effects. ℘ is large in the xylem of the stem which means that plant transpiration does not usually enrich the soil (Zimmerman *et al.*, 1967) despite enrichment in the leaves where ℘ is smaller. The pathway of water movement within a leaf is complex, and ℘ is presumably much smaller for the mesophyll cells than for the veins, bundle-sheath extensions, etc. Due to lower transpiration rates ℘ will be smaller in stressed plants than in unstressed ones and may explain the enrichment in stem water observed by Flanagan *et al.* (1991).

Using this notation, Eq. (25) becomes

$$\Delta_L = \frac{\Delta_E(1 - e^{-℘})}{℘}. \tag{26}$$

Obviously, the application of Eq. (26) to whole leaf studies with E taken as the transpiration rate is not straightforward. The effective length, L, will be larger than local anatomical measurements just as the effective length used by Penman and Schofield (1951) to describe the leaf resistance to water vapor diffusion is much larger than the length of the actual path through the stomatal pore.

From Eq. (26), the difference, $\Delta_E - \Delta_L$ is given as a fraction of Δ_E by

$$1 - \frac{\Delta_L}{\Delta_E} = 1 - \frac{(1 - e^{\wp})}{\wp}, \tag{27}$$

which is positively related to E (Farquhar *et al.*, 1993) and is shown in Fig. 1.

Such a trend is apparent in the data of White (1989) and, notably, of Flanagan *et al.* (1991), and may be inferred from Table 1 of Walker *et al.* (1989).

With values of $\wp$ determined for a particular leaf, it should be possible to deduce the effective length, L. White (1989) attempted to do this but incorrectly used Eq. (24) rather than Eq. (25), and thus underestimated L by a factor of about 6. Flanagan (unpublished) used Eq. (25) to calculate L for *Phaseolus vulgarius*, obtaining a value of about 8 mm. As mentioned previously, this length would be appropriately scaled if water were moving from one epidermis to the other across a slab of water. In practice the medium is

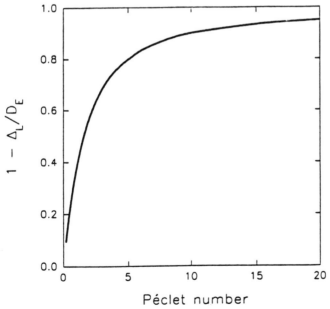

Figure 1. The ratio of the observed isotopic fractionation to that which would occur in the absence of diffusive effects $(\Delta_E - \Delta_L)/\Delta_E$ as a function of the Péclet number, which gives the ratio of convective to diffusional effects (see Eq. (27)).

porous, there is lateral movement of water through veins and bundle-sheath extensions, and the subsequent movement may be largely through cell walls. Local diffusion lengths are therefore much smaller and calculated effective velocities (E/C) much larger than is calculated using the two-dimension analogue. This also means that p will show considerable variation within the leaf, as well as varying between leaves and plants. Flanagan *et al.* (1991) pointed out that the Péclet effect would cause a discrepancy between whole leaf-water Δ and Δ_E if stomatal heterogeneity were of significance. Such a spatial heterogeneity in leaf water has recently been confirmed by Luo and Sternberg (1992).

White (1989) and Leaney *et al.* (1985) discussed the existence of compartmentation within the leaf which would also cause a difference. Leaney *et al.* (1985) presented a compartment model which effectively assigns p values of zero and infinity to two fixed compartments. Equation (26) represents a continuum. Yakir and co-workers (1990a,b) extended the compartmentalization concept theorizing the existence of three isotopically distinct pools: one associated with transpiration, one associated with "metabolism," and another with the vascular conduction of water. There is however still no theory to account for the existence of such discrete isotopic compartments (Yakir, 1992). White (1989) has emphasized that discrepancies could arise if the system were not in a steady state.

C. Exchange of CO_2 with Leaves

When air flows over a leaf which discriminated against ^{13}C, the $^{13}C/^{12}C$ ratio of CO_2 in the air increases. Similarly, CO_2 in the air can become enriched in ^{18}O, the extent of which depends on the $^{18}O/^{16}O$ ratio of water in the chloroplast (Farquhar *et al.*, 1993).

Consider a leaf chamber, with leaf area a (m^2), with a flow of dry air u, and CO_2^{16} concentrations entering and leaving the chamber of c_e and c_o (mol mol^{-1}), respectively. Then

$$\frac{c_e}{c_o} = u + \frac{aA}{c_o}, \tag{28}$$

where A denotes the rate of assimilation of CO_2^{16}. Similarly, for $CO^{16}O^{18}$, with symbols primed

$$\frac{c_o'}{c_e'} = u + \frac{aA'}{c_o'}, \tag{29}$$

so that the fractionation of the air, Δ_a is

$$1 + \Delta_a = \frac{c_e'/c_e}{c_o'/c_o} = \frac{1 + aA'/uc_o'}{1 + aA/uc_o}. \tag{30}$$

To obtain an expression for this effect, we need to develop one for $(c'_o/c_o)/(A'/A)$, which we denote by $1 + \Delta_A$, as it is analogous to the expression for discrimination against CO_2 during photosynthesis. In this case the $^{18}O/^{16}O$ ratio in the "product" is spread among organic compounds, water, etc.

Taking an expression for A as

$$A = \frac{C_a - C_c}{P(r_b + r_s + r_{ias}) + r_w}, \tag{31}$$

where C_c is the CO_2 partial pressure in the chloroplast, P is the atmospheric pressure, and the resistances refer to the boundary layer, stomata, intercellular air spaces (gas phase: m^2 s mol^{-1}), and cell wall (liquid phase: m^2 s Pa mol^{-1}), respectively, it follows that

$$\alpha = 1 + \Delta_A = \frac{1 + \bar{a}}{1 - (C_c/C_a - C_c)\Delta_{ca}}, \tag{32}$$

where $1 + \Delta_{ca} = R_c/R_a$, the subscript c and a referring to CO_2 in the chloroplast and ambient air, respectively, and $\bar{a}$ is the weighted mean value of the discrimination factors occurring during the various diffusion steps, i.e.,

$$\bar{a} = \frac{(C_a - C_s)a_b + (C_s - C_w)a + (C_w - C_c)a_w}{C_a - C_c}, \tag{33}$$

where C_c being the effective CO_2 partial pressure inside the chloroplast, C_s and C_w being those at the leaf surface and wall surface, and a_b, a, and a_w being the discrimination at each step in the catena. We take a to be 8.8‰ (based on the reduced mass of $CO)_2$ and air and recently confirmed by direct measurement (M. Lehman and U. Siegenthaler, personal communication) and a_b to be 5.8‰ (based on the ⅔ power effect discussed earlier). The value of a_w is unknown but presumably small and sums an equilibrium dissolution effect (−0.8‰: Vogel *et al.*, 1970) and fractionation during diffusion through solution. Assuming that the latter is twice that for diffusion of $^{13}CO_2$ in solution [0.7‰ (O'Leary, 1984) to 0.9‰ (Jähne *et al.*, 1987)] we calculate an overall value for a_w of +0.8‰.

One would expect that R_c is *nearly* equal to R_e, the value at isotopic equilibrium with water in the chloroplast of isotope ratio R_w, interconversion being catalyzed by carbonic anhydrase according to $R_e/R_w = \alpha_{BC}$, the fractionation factor of 1.0412 for oxygen isotopic exchange between gaseous CO_2 and liquid water at 25°C (O'Neil *et al.*, 1975). The equilibrium should not, however, be complete as this would require an excess of carbonic anhydrase (Cowan, 1986). We can therefore solve the simplified nonequilibrium case to yield

$$\Delta_A = \frac{\bar{a}(1 + 3\rho) + (C_c/C_a - C_a)(\Delta_{ea} + 3\rho b)}{1 - (C_c/C_a - C_c)\Delta_{ea} + 3\rho(C_a/C_a - C_c)},$$ (34)

where Δ_{ea} is the value of Δ_{ca} which would occur at full equilibrium ($\Delta_{ea} = \alpha_{BC}R_w/R_a - 1$), ρ is the ratio of the rate of carboxylation by RuBisCO to rate of hydration of CO_2 by carbonic anhydrase, and b is the discrimination by RuBisCO against the heavier isotope, which only affects Δ_A when ρ is greater than 0. The value of b is unknown for ^{18}O, but is presumably small as the oxygens of CO_2 do not bind to the active sites of RuBisCO.

For full isotopic equilibrium with chloroplastic water prior to fixation ($\rho = 0$), a good approximation is

$$\Delta_A = \bar{a} + \frac{C_c}{C_a - C_c} \Delta_{ea},$$ (35)

and, at the opposite extreme, with no equilibration prior to fixation ($\rho \to \infty$), Eq. (34) becomes exactly

$$\Delta_A = \bar{a}\frac{C_a - C_c}{C_a} + b\frac{C_c}{C_a}.$$ (36)

Equation (36) is analogous to the equation for fractionation against $^{13}CO_2$ (Farquhar *et al.*, 1982).

The above description ignores dark respiration and photorespiration. To a good approximation these effects are described by the following modification of Eq. (34): where Δ_{mc} is $R_m/R_c - 1$, and R_m is the $^{18}O/^{16}O$ ratio of CO_2 respired from the mitochondria,

$$\Delta_A = \frac{\bar{a}[1 + 3\rho(1 - \Gamma/C_c)] + (C_c/C_a - C_c)[\Delta_{ea} + 3\rho(b + (\Gamma/C_c)\Delta_{mc})]}{1 - (C_c/C_a - C_c)\Delta_{ea} + 3\rho(1 - \Gamma/C_c)(C_a/C_a - C_c)},$$ (37)

and Γ is the CO_2 compensation partial pressure.

Farquhar *et al.* (1993) examined the relationship between Δ_A and C_c/C_a for leaves of several C_3 herbaceous and woody species and found that the values of Δ_A observed are very close to those which would be obtained if the CO_2 within the chloroplast were in full equilibrium with water at the sites of evaporation, suggesting that ρ is not much greater than zero. This is somewhat surprising as ρ should theoretically be greater than zero. For example, from concurrent measurements of RuBisCO and carbonic anhydrase activities in wheat (Evans, 1983), a value of 0.05 can be calculated, in line with theoretical estimates (Cowan, 1986). However, measurements with wheat have also showed Δ_A to be just on, or slightly above, the line for isotopic exchange with chloroplast water at the evaporating surface (Caemmerer and Evans, personal communication).

There are several possible explanations for this discrepancy. If there is heterogeneity in the isotopic composition of chloroplast water and if CO_2 exchange is preferentially with enriched portions of the leaf, then ρ will be underestimated using the on-line trapping technique. δ_E should increase along a file of evaporating cells and this may be the case for water moving from the vascular bundles to the epidermes, although the spatially averaged δ_E will be as calculated in Eq. (21). Electron micrographs of the peach and citrus leaves used by Farquhar *et al.* (1993) showed more chloroplasts per unit cell wall area in the upper palisade layer(s) and if this leads to greater exchange in these cells then ρ will be underestimated. Alternatively, if b for ^{18}O were to be substantial then ρ will appear too low. For example, with a $\rho = 0.05$, and $C_c/C_a = 0.5$, Δ_A would be ~5‰ greater for $b = 30‰$ than for $b = 0‰$. A third possibility is that there is substantial heterogeneity in C_c across the leaf. This would result in the C_c appropriate for $C^{18}O^{16}O$ discrimination analysis being considerably higher than that measured from $^{13}CO_2$ discrimination in a gas exchange system, even after differences between conductance versus assimilation weightings of C_c (Lloyd *et al.*, 1992; Farquhar *et al.*, 1993) are taken into account (Lloyd and Farquhar, unpublished derivation).

The conclusion of Farquhar *et al.* (1993) that the chloroplast water with which CO_2 exchanges is close to that at the leaf evaporating surface differs from that of Yakir *et al.* (1993) who found photorespired CO_2 to be substantially depleted in ^{18}O compared to that in full equilibrium with evaporating surface water. They also found that oxygen evolved during photosynthetic uptake (reflecting chloroplast water) was also less enriched than evaporating surface water, but less so than for photorespired CO_2. The reasons for the differences between the isotopic composition of CO_2 and O_2 in the study by Yakir *et al.* (1993) and also for the differences between Farquhar *et al.* (1993) and Yakir *et al.* (1993) are not clear. However, by extending their model to the global scale, Farquhar *et al.* (1993) showed that chloroplast water at the leaf evaporating surface explains both the overall value for the ^{18}O composition of CO_2 as well as the latitudinal gradient is atmospheric isotopic composition. The isotopic composition of atmospheric oxygen is also consistent with terrestrial plant chloroplast water being the same as that for the leaf evaporating surface (Lloyd, Farquhar, Taylor, and Robertson, unpublished model results).

Unlike $^{13}CO_2$, the $C^{18}O^{16}O$ discrimination during photosynthetic uptake is not reflected in the isotopic composition of organic matter. In the absence of discrimination against ^{18}O during enzymatic fixation by RuBisCO the proportion of oxygen initially fixed with a gaseous isotopic signature is exactly 3ρ of the chloroplast water related one (I. R. Cowan, unpublished result and cf. Eq. (34)). Nevertheless, rapid equilibration occurs between water and the triose phosphates produced immediately after CO_2 fixation

(De Niro and Epstein, 1979). Triose phosphates exported from the chloroplast should therefore rapidly equilibrate with cytosolic water with an equilibrium fractionation of $+27‰$ prior to incorporation into 6 and 12 carbon sugars which themselves exchange their oxygens with water more slowly (Sternberg *et al.*, 1986).

Leaf cellulose is usually less enriched than would be the case for full equilibration with daytime δ_E (Ferhi and Letolle, 1979) indicating that the cytosolic water in which synthesis occurs is less than δ_E due, we suspect, to a proportion of cellulose synthesis from reserves occurring at night when leaf water is less enriched and also as a consequence of a higher mean value of ρ for the cytoplasm than the chloroplast. This makes cytosolic water intermediate between δ_E and δ_S. The latter is consistent with the notion that the bulk of leaf water is in a "symplastic pool" as discussed by Yakir *et al.* (1990a). Subsequent work (Yakir *et al.*, 1990b) showed that the water pool directly involved in cellulose biosynthesis (the "metabolic pool") can be buffered against external effects such as those associated with daily fluctuations in humidity. This is also consistent with our advective/diffusive model of leaf water isotopic composition presented earlier. As the leaf-to-air vapor pressure difference and hence δ_E increase, E also increases and bulk leaf water becomes progressively less enriched with respect to δ_E. Long-term differences in ambient humidity are, however, reflected in the isotopic composition of leaf cellulose (Ferhi and Letolle, 1979).

References

Abelson, P. H., and T. C. Hoering. 1961. Carbon isotope fractionation in formation of amino acids by photosynthetic organisms. *Proc. Natl. Acad. Sci.* **47**(5): 623–632.

Bakke, E. L., D. W. Beaty, and J. M. Hayes. 1991. The Effect of Different Ion Correction Methodologies on $\delta^{18}O$ and $\delta^{13}C$ Results. Paper presented to the Geological Society of America Annual Meeting, San Diego, October 1991

Bottinga, Y., and H. Craig. 1969. Oxygen isotope fractionation between CO_2 and water, and the isotopic composition of marine atmospheric CO_2. *Earth Planet. Sci. Lett.* **8**: 363–342.

Broadmeadow, M. S. J., H. Griffiths, C. Maxwell, and A. M. Borland. 1992. The carbon isotope ratio of plant organic material reflects temporal and spatial variations in CO_2 within tropical forest formations. *Oecologia* **89**: 435–441

Caemmerer, S., and J. R. Evans. 1991. Determination of the average partial pressure of CO_2 in chloroplasts from leaves of several C_3 plants. *Aust. J. Plant Physiol.* **18**: 287–305.

Caemmerer, S., and G. D. Farquhar. 1981. Some relationships between the biochemistry of photosynthesis and the gas exchange of leaves. *Planta* **153**: 376–387.

Cowan, I. R. 1986. Economics of carbon fixation in higher plants, pp. 133–170. *In* T. J. Givnish (ed.), On the Economy of Plant Form and Function. Cambridge Univ. Press, Cambridge.

Craig, H. 1957. Isotopic standards for carbon and oxygen and correction factors for mass-spectrometric analysis of CO_2. *Geochim. Cosmochim. Acta* **12**: 133–149.

Craig, H., and L. I. Gordon. 1965. Deuterium and oxygen-18 variations in the ocean and the marine atmosphere, pp. 9–130. *In* Tongiorgi (ed.), Proceedings of a Conference on Stable Isotopes in Oceanographic Studies and Palaeotemperatures. Spoleto, Italy.

Davies, W. J., and T. T. Kozlowski. 1974. Stomatal responses of five woody angiosperms to light intensity and humidity. *Can. J. Bot.* **52**: 1525–1534.

Degens, E. T., R. R. L. Guillard, W. M. Sackette, and J. A. Hellebust. 1968. Metabolic fractionation of carbon isotopes in marine plankton. I. Temperature and respiration measurements. *Deep Sea Res.* **15:** 1–9.

DeNiro, M. J., and S. Epstein 1979. Relationship between the oxygen isotope ratios of terrestrial plant cellulose, carbon dioxide and water. *Science* **204:** 51–53.

Epstein, S., P. Thompson, and C. J. Yapp. 1977. Oxygen and hydrogen isotopic ratios in plant cellulose. *Science* **218:** 1209–1215.

Evans, J. R. 1983. Photosynthesis and Nitrogen Partitioning in Leaves of *Triticum qestivum* L. and Related species. PhD thesis, Australian National University, Canberra.

Evans, J. R., T. D. Sharkey, J. A. Berry, and G. D. Farquhar. 1986. Carbon isotope discrimination measured concurrently with gas exchange to investigate CO$_2$ diffusion in leaves of higher plants. *Aust. J. Plant Physiol.* **13:** 281–293.

Farquhar, G. D. 1983. On the nature of carbon isotope discrimination in C$_4$ species. *Aust. J. Plant Physiol.* **72:** 245–250.

Farquhar, G. D. 1989. Models of integrated photosynthesis of cells and leaves. *Phil. Trans. R. Soc. London Ser. B.* **323:** 357–367.

Farquhar, G. D., and S. von Caemmerer. 1982. Modelling of photosynthetic response to environmental conditions, pp. 549–587. *In* Lange, O. L., Nobel, P. S., Osmond, C. B., and Ziegler, H. (eds.), Encyclopedia of Plant Physiology, New Series Vol. 12B. Springer-Verlag, Neidelberg.

Farquhar, G. D., and K. Rashke. 1978. On the resistance to transpiration of the sites of evaporation within the leaf. *Plant Physiol.* **61:** 1000–1005.

Farquhar, G. D., and R. A. Richards. 1984. Isotopic composition of plant carbon correlates with water-use efficiency of wheat genotypes. *Aust. J. Plant Physiol.* **11:** 539–552.

Farquhar, G. D., and T. D. Sharkey. 1982. Stomatal conductance and photosynthesis. *Annu. Rev. Plant Physiol.* **33:** 317–345.

Farquhar, G. D., M. H. O'Leary, and J. A. Berry. 1982. On the relationship between carbon isotope discrimination and the intercellular carbon dioxide concentration in leaves. *Aust. J. Plant Physiol.* **9:** 121–137.

Farquhar, G. D., J. R. Ehleringer, and K. T. Hubick. 1989a. Carbon isotope discrimination and photosynthesis. *Annu. Rev. Plant Physiol. Plant Mol. Biol.* **40:** 503–537.

Farquhar, G. D., K. T. Hubick, A. G. Condon, and R. A. Richards. 1989b. Carbon isotope discrimination and water use efficiency. In Rundel, P. W., Ehleringer, J. R., and Nagy, K. A. (eds). Stable Isotopes in Ecological Research. Springer-Verlag, New York, pp. 21–46

Farquhar, G. D., J. Lloyd, J. Taylor, L. B. Flanagan, J. P. Syversten, K. T. Hubick, S. C. Wong, and J. R. Ehleringer. 1993. Vegetation effects on the isotopic composition of oxygen in atmospheric CO$_2$. *Nature* **363:** 439–443.

Ferhi, A., and R. Letolle. 1979. Relation entre le milieu climatique at les teneurs en oxygène-18 de la cellulose des plantes terrestres. *Physiol. Vég.* **17:** 107–117.

Flanagan, L. B., and J. R. Ehleringer. 1991. Stable isotope composition of stem and leaf water: Applications to the study of plant water use. *Funct. Ecol.* **5**.

Flanagan, L. B., J. P. Comstock, and J. R. Ehleringer. 1991. Comparison of modelled and observed environmental influences on the stable oxygen and hydrogen isotope composition of leaf water in *Phaseolus vulgaris* L. *Plant Physiol.* **96:** 588–596.

Francey, R. J., and P. P. Tans. 1987. Latitudinal variation in oxygen-18 in atmospheric CO$_2$. *Nature* **327:** 495–497.

Freidli, H., H. Lotscher, H. Oeschger, U. Siegenthaler, and B. Stauffer. 1986. Ice core record of the ^{13}C/^{12}C ratio of atmospheric CO$_2$ in the past two centuries. *Nature* **324:** 237–238.

Friedli, H., U. Siegenthaler, D. Rauber, and H. Oeschger. 1987. Measurements of concentration, ^{13}C/^{12}C and ^{18}O/^{16}O ratios of trophospheric carbon dioxide over Switzerland. *Tellus* **39B:** 80–88.

Goodman, H. S., and R. F. Francey. 1988. ^{13}C/^{12}C and ^{18}O/^{16}O in baseline CO$_2$, pp. 54–58. *In* Forgan, B. W., and Fraser, P. J. (eds.), Baseline Atmospheric Program (Australia). CSIRO, Canberra.

Guy, R. D., M. F. Fogel, J. A. Berry, and T. C. Hoering. 1987. Isotope fractionation during oxygen production and consumption by plants, pp. 597–600. *In* J. Biggins (ed.), Progress in Photosynthetic Research III. Nijhoff, Dordrecht.

Henderson, S. A., S. von Caemmerer, and G. D. Farquhar. 1992. Short term measurements of carbon isotope discrimination in several C_4 species. *Aust. J. Plant Physiol.* **19:** 263–285.

Ikeda, T. 1983. Maximum Principle in Finite Element Models for Convection-Diffusion Phenomena. North Holland, Amsterdam.

Jacobson, B. S., B. N. Smith, S. Epstein, and G. G. Laties. 1970. The prevalence of carbon-13 in respiratory carbon dioxide as an indicator of the type of endogenous substrate. *J. Gen. Physiol.* **55:** 1–17.

Jähne, B, G. Heinz, and N. Dietrich. 1987. Measurements of the diffusion coefficients of sparingly soluble gases in water. *J. Geophys. Res.* **92:** 10767–10776.

Kays, W. M. 1966. Convective Heat and Mass Transfer. McGraw–Hill, New York.

Keeling, C. D. 1958. The concentration and isotopic abundances of atmospheric carbon dioxide in rural areas. *Geochim. Cosmochim. Acta* **13:** 322–334.

Keeling, C. D. 1961. The concentration and isotopic abundances of carbon dioxide in rural and marine air. *Geochim. Cosmochim. Acta* **24:** 277–298.

Keeling, C. D., W. G. Mook, and P. P. Tans. 1979. Recent trends in the $^{13}C/^{12}C$ ratio of atmospheric carbon dioxide. *Nature* **277:** 121–123.

Keeling, C. D., R. B. Bacastow, A. F. Carter, S. C. Piper, T. P. Whorf, M. Heimann, W. G. Mook, and H. Roeloffzen. 1989. A three dimensional model of atmospheric CO_2 transport based on observed winds. 1. Analysis of observational data, pp. 165–236. *In* D. H. Peterson, (ed.), Aspects of Climate Variability in the Pacific and the Western Americas, American Geophysical Union, Monograph 55, Washington.

Kriedemann, P. E., and H. D. Barrs. 1981. Citrus Orchards, Vol. IV, pp. 325–417. *In* Kozlowski, T. (ed.), Water Deficits and Plant Growth. Academic Press, New York.

Lancaster, J. 1990. Carbon-13 fractionation in Carbon Dioxide Emitting Diurnally from Soils and Vegetation at Ten Sites on the North American Continent. PhD thesis, University of California, San Diego.

Leaney, F. W., C. B. Osmond, G. B. Allison, and H. Zeigler. 1985. Hydrogen isotope composition of leaf water in C_3 and C_4 plants: Its relationship to the hydrogen isotope composition of dry matter. *Planta* **164:** 215–220.

Leuning, R. 1983. Transport of gases into leaves. *Plant Cell Environ.* **6:** 181–194.

Lloyd, J., P. E. Kriedemann, and J. P. Syvertsen. 1987. Gas exchange, water relations and ion concentrations of leaves on salt stressed 'Valencia' orange, *Citrus sinensis* (L.) Osbeck. *Aust. J. Plant Physiol.* **14:** 605–617.

Lloyd, J., J. P. Syvertsen, P. E. Kriedemann, and G. D. Farquhar. 1992. Low conductances for CO_2 diffusion from stomata to the sites of carboxylation in leaves of woody species. *Plant Cell Environ.* **15:** 873–899.

Loreto, F., P. C. Harley, G. Di Marco and T. D. Sharkey. 1992. Estimation of mesophyll conductance to CO_2 flux by three different methods. *Plant Physiol.* **98:** 1437–1443.

Luo, Y-H., and L. Sternberg. 1992. Spatial D/H heterogeneity of leaf water. *Plant Physiol.* **99:** 348–350

Marino, B. D., M. B. McElroy, R. J. Salawich, and W. G. Spalding. 1992. Glacial-to-interglacial variations in the carbon isotopic composition of atmospheric CO_2. *Nature* **357:** 461–466

Mason, E. A., and T. R. Marrero. 1970. The diffusion of atoms and molecules. *Adv. At. Mol. Phys.* **6:** 155–232.

Melzer, E., and H-L. Schmidt. 1987. Carbon isotope effects on the pyruvate dehydrogenase reaction and their importance for relative carbon-13 depletion in lipids. *J. Biol. Chem.* **262**(17): 8159–8164.

Merlivat, L. 1978. Molecular diffusivities of H2180 in gases. *J. Chem. Phys.* **69:** 2864–2871.

Monteith, J. L. 1973. Principles of environmental physics. Arnold, London.

Mook, W. G., J. C. Bommerson, and W. H. Staverman. 1974. Carbon isotope fractionation between dissolved bicarbonate and gaseous carbon dioxide. *Earth Planet. Sci. Lett.* **22:** 169–176.

O'Leary, M. H. 1981. Carbon isotope fractionation in plants. *Phytochemistry* **20:** 553–567.

O'Leary, M. H. 1984. Measurement of the isotopic fractionation associated with diffusion of carbon dioxide in aqueous solution. *J. Phys. Chem.* **88:** 823–825.

O'Neil, J. R., L. H. Adami, and S. Epstein. 1975. Revised Value for the O^{18} fractionation between CO_2 and H_2O at 25°C. *J. Res. U.S. Geol. Surv.* **3,** 623–624.

Parkhurst, D. F. 1986. Internal leaf structure: A three dimensional perspective, pp. 215–249. *In* T. Givnish (ed.), On the economy of plant form and function, Cambridge Univ. Press, New York.

Parkhurst, D. F., and K. A. Mott. 1990. Intercellular diffusion limits to CO_2 uptake in leaves. *Plant Physiol.* **94:** 1024–1032.

Parkhurst, D. F., S-C. Wong, G. D. Farquhar, and I. R. Cowan. 1988. Gradients of intercellular CO_2 levels across the leaf mesophyll. *Plant Physiol.* **86:** 1032–1037.

Penman, H. L., and R. K. Schofield. 1951. Some physical aspects of assimilation and transpiration. *Symp. Soc. Exp. Biol.* **5:** 115.

Raven, J. A., and G. D. Farquhar. 1990. The influence of N metabolism and organic acid synthesis on the natural abundance of C isotopes in plants. *New Phytol.* **116:** 505–529.

Roeske, C. A., and M. H. O'Leary. 1984. Carbon isotope effects on the enzyme catalysed carboxylation of ribulaose bisphosphate. *Biochemistry* **23:** 6275–6284.

Rooney, M. A. 1988. Short Term Carbon Isotopic Fractionation in Plants. Ph.D. thesis, University of Wisconsin, Madison.

Smith, S., J. D. B. Weyers, and W. G. Berry. 1989. Variation in stomatal characteristics over the lower surface of *Commelina communis* leaves. *Plant Cell Environ.* **12:** 653–659.

Sternberg, L. 1989. A Model to estimate carbon dioxide recycling in forests using $^{13}C/^{12}C$ ratios and concentrations of ambient carbon dioxide. *Agric. For. Meterorol.* **48:** 163–193.

Sternberg, L. L., M. J. DeNiro, and R. A. Savidge. 1986. Oxygen isotope exchange between metabolites and water during biochemical reactions leading to cellulose synthesis. *Plant Physiol.* **82:** 423–427.

Syvertsen, J. P. 1984. Light acclimation in citrus leaves. II. CO_2 assimilation and light, water and nitrogen use efficiency. *J. Am. Horticult. Sci.* **109:** 812–817.

Syvertsen, J., J. Lloyd, and P. E. Kreidemann, 1988. Salinity and drought stress effects on ion concentrations, water relations and photosynthetic characteristics of orchard citrus. *Aust. J. Agricult. Res.* **39:** 619–627.

Terashima, I., and T. Inoue. 1985. Vertical gradients in photosynthetic properties of spinach chloroplasts dependent on intra-leaf light environment. *Plant Cell Physiol.* **26:** 781–785.

Terashima, I., S. C. Wong, C. B. Osmond, and G. D. Farquhar. 1988. Characterisation of non uniform photosynthesis induced by abscisic acid in leaves having different mesophyll anatomies. *Plant Cell Physiol.* **29:** 385–394.

Vitousek, P. M., C. B. Field, and P. A. Matson. 1990. Variation in foliar ^{13}C in Hawaiian *Metrosideros polymorpha:* A case of internal resistance? *Oecologia* **84:** 362–370.

Vogel, J. C. 1980. Fractionation of carbon isotopes during photosynthesis. Springer-Verlag, Berlin.

Vogel, J. C., P. M. Grootes and W. G. Mook. 1970. Isotopic fractionation between gaseous and disolved carbon dioxide. *Z. Phys.* **230:** 225–238.

Walker, C. D., and R. C. M. Lance. 1991. The fractionation of ^2H and ^{18}O in leaf water of barley. *Aust. J. Plant Physiol.* **18:** 411–425.

Walker, J. L., and D. L. Oliver. 1986. *J. Biol. Chem.* **261:** 2214–2221.

Walker, C. D., F. W. Leaney, J. C. Dighton, and G. B. Allison. 1989. The influence of transpiration on the equilibrium of leaf water with atmospheric water vapour. *Plant Cell Environ.* **12:** 221–234.

White, J. W. C. 1989. Stable hydrogen isotope ratios in plants. *In* P. W. Rundel, J. R. Ehleringer, and K. A. Nagy (eds.), Stable Isotopes in Ecological Research. Springer-Verlag, New York.

Yakir, D. 1992. Water Compartmentation in Plant Tissue: Isotopic Evidence, pp. 205–221. *In* G. N. Somero, C. B. Osmond, and L. Bolis (eds.), Water and Plant Life. Springer-Verlag, Berlin/Heidelberg.

Yakir, D., M. J. DeNiro, and J. R. Gat. 1990a. Natural deuterium and oxygen-18 enrichment in leaf water of cotton plants grown under wet and dry conditions: Evidence for water compartmentation and its dynamics. *Plant Cell Environ.* **13:** 49–56.

Yakir, D., M. J. De Niro, and J. E. Eprath. 1990b. Effects of water stress on oxygen, hydrogen and carbon isotope ratios in two species of cotton plants. *Plant Cell Environ.* **13:** 949–955.

Yakir, D., J. A. Berry, L. Giles and C. B. Osmond. 1993. The $\delta^{18}O$ of water in the metabolic compartment of transpiring leaves (Chapter 33, this volume).

Zimmerman, U., D. Ehhalt, and K. O. Munnich. 1967. Soil-water movement and evapotranspiration: Changes in the isotopic composition of the water, p. 567. *In* Isotopes in Hydrology. Wien, IAEA, Vienna.

6

Environmental and Biological Influences on the Stable Oxygen and Hydrogen Isotopic Composition of Leaf Water

Lawrence B. Flanagan

I. Introduction

The stable isotopic composition of plant leaf water is altered during transpiration. Water vapor molecules containing the lighter isotopes of hydrogen and oxygen escape from the leaf more readily than do molecules containing deuterium and ^{18}O, so that during transpiration, leaf water becomes enriched in heavy isotope molecules. A model of isotopic fractionation, which was originally developed by Craig and Gordon (1965) for processes occurring during evaporation from the ocean, has been used to model leaf-water isotopic composition. The Craig–Gordon model predicts that the isotopic composition of leaf water is a function of three factors: (a) the isotopic composition of stem water (source water), (b) the isotopic composition of atmospheric water vapor, and (c) the ratio of air and leaf vapor pressure (e_a/e_i).

One potential application of the leaf-water evaporative-enrichment model in plant physiological and ecological studies is in studies of the leaf–air vapor pressure gradient (v). Since v is proportional to the ratio of air and leaf vapor pressures, measurements of the isotopic composition of leaf water and leaf cellulose could potentially be applied in field studies to estimate v (Farquhar et al., 1988), without the need for elaborate micrometerological instrumentation. Variation in v among species or within the canopy of a single crop species can result because of variation in leaf energy budgets and leaf gas exchange characteristics. In addition, estimates of transpiration efficiency, based on leaf carbon isotopic composition, are complicated by variation in v (Farquhar et al., 1988). It would be useful, therefore, to be able to estimate v from isotopic measurements. Before measurements of leaf-water isotopic composition can be used to study v, it is necessary to understand potential environmental and biological factors influencing variation in leaf-water isotopic composition which are not associated with variation in e_a/e_i or v. In the following sections I review the theory of isotopic fractionation during transpiration and then discuss recent studies that examine environmental and biological causes of variation in leaf-water isotopic composition. I conclude by presenting data which suggest that studies of the isotopic composition of leaf water have a role in studies of the isotopic composition of atmospheric CO_2 and large-scale vegetation–atmosphere CO_2 exchange.

II. Isotopic Fractionation during Transpiration

A. Isotope Effects

There are two isotope effects during transpiration: (a) an equilibrium effect resulting from the phase change from liquid water to water vapor, and (b) a kinetic effect caused by the different diffusivities of the light and heavy isotopes of water vapor in air. The equilibrium fractionation factor (α^*) is defined as

$$\alpha^* = R_l/R_v, \tag{1}$$

where R is the molar ratio of the heavy and light isotope (i.e., D/H or $^{18}O/^{16}O$) and the subscripts l and v refer to liquid water and water vapor, respectively. In the above definition it is assumed that the air is saturated with water vapor and that the liquid and vapor are at the same temperature. These conditions are satisfied for a leaf where water vapor in the intercellular air spaces is in equilibrium with leaf cell water. The regression equations listed by Majoube (1971) (see also Gat, 1980) can be used to calculate values for the parameter α^*.

The kinetic fractionation factor (α_k) is defined as the ratio of the diffusion coefficients for water vapor molecules containing the light and heavy isotopes. For fractionation during transpiration, α_k can be defined as

$$\alpha_k = g/g', \tag{2}$$

where g and g' refer to the stomatal conductance to water vapor molecules containing the light and heavy isotopes, respectively. Merlivat (1978) has measured the relative rates of diffusion in air of water vapor molecules containing the light and heavy isotopes of hydrogen and oxygen. The values of α_k for molecular diffusion are H/D = 1.025 and $^{16}O/^{18}O$ = 1.0285 (Merlivat, 1978). It is appropriate to use these values for diffusion through the stomatal pore which is a molecular process (Sharkey et al., 1982). The value for the kinetic fractionation factor should be modified, however, for turbulence in the boundary layer outside the stomatal pore. Based on Pohlhausen analysis, the kinetic fractionation factor in a boundary layer (α_{kb}) is calculated by taking the value for molecular diffusion and raising it to the two-thirds power (Kays and Crawford, 1980). The values for α_{kb} are H/D = 1.017 and $^{16}O/^{18}O$ = 1.0189 (Flanagan and Ehleringer, 1991a; Flanagan et al., 1991b). Farquhar et al. (1988) also present an equation to account for boundary layer effects on leaf-water isotopic enrichment.

B. Model of Isotopic Fractionation during Transpiration

The equilibrium and kinetic fractionation factors can be incorporated into a model of isotopic enrichment during evaporation as originally described by Craig and Gordon (1965). In addition to the two fractionation factors, the model also includes the influence of isotopic exchange between water vapor in the atmosphere and leaf water. Applied to leaf transpiration the model can be expressed in the form (Flanagan and Ehleringer, 1991a, Flanagan et al., 1991b)

$$R_l = \alpha^* \left[\alpha_k R_x \left(\frac{e_i - e_s}{e_i} \right) + \alpha_{kb} R_x \left(\frac{e_s - e_a}{e_i} \right) + R_a \left(\frac{e_a}{e_i} \right) \right], \qquad (3)$$

where R is the molar ratio of the heavy to light isotopes and the subscripts l, x, and a refer to leaf water, stem xylem water, and atmospheric water vapor, respectively; e is the partial pressure of water vapor and the subscripts i, s, and a refer to leaf intercellular air spaces, leaf surface, and ambient air, respectively. Ignoring the influence of boundary layer effects, Eq. (3) can be modified to:

$$R_l = \alpha^* \left[\alpha_k R_x \left(\frac{e_i - e_a}{e_i} \right) + R_a \left(\frac{e_a}{e_i} \right) \right]. \qquad (4)$$

Similar expressions have been developed for isotopic enrichment during leaf transpiration by Dongmann et al. (1974) and White (1983).

Isotopic compositions are expressed in this paper using delta notation in parts per thousand (‰),

$$\delta = \left[\frac{R_{Sample}}{R_{Standard}} - 1 \right] \times 1000, \qquad (5)$$

where R is the molar ratio of heavy to light isotope. All water sample isotopic compositions are expressed relative to standard mean ocean water (SMOW). The absolute ratios for SMOW used in the calculations were D/H = 0.00015576 and $^{18}O/^{16}O$ = 0.0020052 (Ehleringer and Osmond, 1989).

In order to compare the observed isotopic composition of leaf water with that predicted by Eqs. (3) and (4), a parameter, f, is calculated as

$$\delta_{Leaf} = \delta_{Model} \cdot f + \delta_{Stem} \cdot (1 - f), \qquad (6)$$

where δ_{Leaf} is the measured isotopic composition of leaf water, δ_{Model} is the isotopic composition of leaf water predicted by either Eq. (3) or Eq. (4), and δ_{Stem} is the isotopic composition of stem water. The parameter f in Eq. (6) describes how close (as a proportion) the observed leaf water is to that predicted by the evaporative-enrichment model.

C. Conditions for Using the Model

Derivation of the evaporative-enrichment model (Eqs. (3) and (4)) assumes that leaf water is at isotopic steady state (Flanagan *et al.*, 1991b). At isotopic steady state, the isotopic composition of transpiration water is the same as the source or stem water isotopic composition (White, 1988, Flanagan *et al.*, 1991b). Precise application of the model requires that leaf water be at isotopic steady state.

A second consideration in applying the leaf-water-enrichment model in field situations is that it is necessary to measure the isotopic composition of atmospheric water vapor (AWV). This parameter varies during the course of a season as environmental conditions change (White and Gedzelman, 1984). It is not correct to simply assume that AWV is in equilibrium with local groundwater. This assumption may be valid in some humid, continental locations but it is certainly incorrect in arid environments and coastal locations (Gat, 1980). Since routine measurements of the isotopic composition of AWV are not made for most meteorological stations, it is necessary to collect AWV during leaf-water investigations in the field.

A final consideration in applying the evaporative-enrichment model is that it is necessary to measure the isotopic composition of stem water in plants being compared under field conditions. Cooccurring plant species can have different functional rooting patterns and make use of different proportions of summer precipitation and groundwater (Flanagan *et al.*, 1992a; Dawson, Chapter 30, this volume). Since the isotopic composition of summer precipitation and groundwater differ, stem-water values will vary depending on the relative uptake of summer precipitation. Proper interpretation of leaf-water isotopic composition depends on knowledge of the source, stem-water isotopic composition.

D. Effect of Boundary Layer Conductance on Model Predictions

The influence of boundary layer conditions on leaf-water isotopic enrichment can be calculated by comparing values predicted by Eqs. (3) and (4). Equation (4), which does not include boundary layer effects, predicts more isotopic enrichment than does Eq. (3). The magnitude of the difference between the two equations is dependent on boundary layer conductance, leaf temperature, the leaf–air vapor pressure difference, and the particular element being considered.

For D/H ratios, the largest difference between Eqs. (3) and (4) is approximately equal to the precision of the measurement techniques ($\pm 2\permil$) when the boundary layer conductance is twice as large as the stomatal conductance (Fig. 1). The maximal difference increases to approximately $5\permil$,

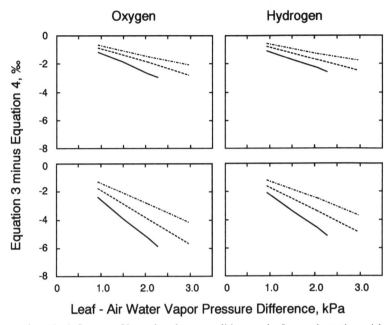

Figure 1. The influence of boundary layer conditions on leaf-water isotopic enrichment. The ordinate shows the difference between leaf-water isotopic compositions predicted by Eqs. (3) and (4). The different lines represent calculations done for different leaf temperatures: solid, 20°C; dashed, 25°C; dot–dash, 30°C. For the two panels in the column labeled oxygen, the top panel shows calculations done with boundary layer conductance two times the value of stomatal conductance and the bottom panel shows calculations done with boundary layer conductance 0.5 times the value for stomatal conductance. A similar format is followed for the two panels in the column labeled hydrogen. From Flanagan *et al.* (1991b).

however, when the boundary layer conductance is reduced to half the stomatal conductance. Such a low boundary layer conductance would be uncommon under field conditions. The influence of boundary layer conductance on deuterium enrichment in leaf water appears negligible, therefore.

Boundary layer effects are quantitatively more important when considering $^{18}O/^{16}O$ ratios. Even when the boundary layer conductance is twice the stomatal conductance, there may be a 3‰ difference between the values predicted by Eqs. (3) and (4), when the leaf–air vapor pressure difference is large (Fig. 1). The maximal difference increases to approximately 6‰ when the boundary layer conductance is reduced to half the stomatal conductance. A slight overestimation of at least 3‰ should be expected, therefore, when using Eq. (4) to predict ^{18}O enrichment in leaf water under field conditions.

Boundary layer conditions have different effects on hydrogen and oxygen isotopic compositions because of differences in the relative magnitudes of the kinetic and equilibrium fractionation factors for the two elements. For hydrogen the equilibrium fractionation factor is quantitatively more important than the kinetic fractionation factor (H/D, $\alpha_k = 1.025$, $\alpha^* = 1.079$ at 25°C). The opposite is true for oxygen, where the kinetic fractionation factor is quantitatively more important ($^{16}O/^{18}O$, $\alpha_k = 1.0285$, $\alpha^* = 1.0094$ at 25°C). Since the influence of boundary layer effects is governed by the magnitude of the kinetic fractionation factor, the oxygen isotopic composition of leaf water should be more sensitive to boundary layer conditions.

III. Environmental and Biological Effects on Leaf-Water Isotopic Composition

A. Comparisons between Observed and Modeled Isotopic Composition

Comparisons have been made between the isotopic enrichment predicted by Eqs. (3) and (4) and that measured in leaf water under a range of conditions: (a) while a leaf was held under constant environmental conditions in gas exchange chamber, and (b) while leaves were exposed to natural diurnal variation in environmental conditions in a glasshouse or in the field.

At steady state under controlled environmental conditions, the observed leaf-water isotopic composition in *Phaseolus vulgaris* was enriched above that of stem water, with the extent of the enrichment dependent on the leaf–air vapor pressure difference and the isotopic composition of atmospheric water vapor. The larger the ν that a leaf is exposed to, the higher the heavy isotopic composition of leaf water. At a constant ν, the observed leaf-water isotopic composition was relatively depleted in heavy isotopes

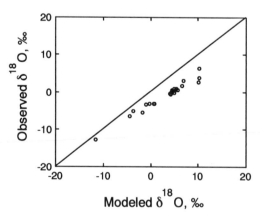

Figure 2.   Comparison of the modeled and observed oxygen isotope composition of water in *Phaseolus vulgarus* leaves. Environmental conditions were as follows: leaf temperature, 30°C; light intensity, 1.3 mmol m^{-2} s^{-1} (400–700 nm); boundary layer conductance to water loss, 2 mol m^{-2} s^{-1}. The stem $\delta^{18}O$ values varied from −14.2‰ to −16.1‰. Modeled values were calculated using Eq. (3). The $\delta^{18}O$ values of atmospheric water vapor varied from −27.6‰ to −13.4‰.

when exposed to AWV that had a low heavy isotope content, and leaf water was relatively enriched in heavy isotopes when exposed to AWV that had a large heavy isotope content. The observed variation in leaf-water isotopic composition agreed well with that predicted by Eq. (3) (Fig. 2). The measured δ values for leaf water, however, were always less than that predicted by the evaporative-enrichment model.

When exposed to natural, diurnal environmental variation in a glasshouse, the oxygen isotopic composition of leaf water in *Cornus stolonifera* was enriched above that of stem water and followed a pattern of diurnal change qualitatively similar to that predicted by Eq. (4) (Fig. 3). The observed hydrogen isotopic composition of leaf water was also enriched above

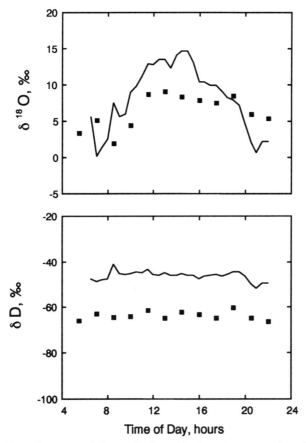

Figure 3. Diurnal pattern of change in the stable isotopic composition of leaf water in *Cornus stolonifera* under glasshouse conditions. The solid line represents the modeled leaf-water isotopic composition calculated with Eq. (4). The squares represent the observed isotopic composition of water extracted from *C. stolonifera* leaves. Stem-water isotopic composition (mean ± SD, $n = 3$): $\delta^{18}O = -15.2‰ \pm 0.2$, $\delta D = -120‰ \pm 1$. Isotopic composition of atmospheric water vapor; $\delta^{18}O = -22.6‰$, $\delta D = -144‰$. From Flanagan and Ehleringer (1991b).

that of stem water but remained relative constant throughout the day in agreement with model calculations. At midday, the model predicted a higher degree of oxygen and hydrogen isotopic enrichment than was actually observed in leaf water (Fig. 3).

The different patterns of diurnal change for hydrogen and oxygen isotopes resulted because of differences in the relative magnitude of the kinetic and equilibrium fractionation factors for the different elements (see Section II, D). During the day, as leaf temperature increases and relative humidity decreases, ν increases and, therefore, the evaporative-enrichment model tends to predict a higher heavy isotope content of leaf water. However, the equilibrium fractionation factor is temperature dependent, i.e., the higher the temperature the lower the value of the equilibrium fractionation factor (Majoube, 1971). For hydrogen isotopes, for which the equilibrium fractionation factor is dominant, the tendency for increased evaporative isotopic enrichment caused by an increased ν was offset by the decrease in the equilibrium fractionation factor as leaf temperature increases during the day. In contrast, the kinetic fractionation factor is dominant for oxygen isotopes so that the decline in the equilibrium fractionation factor with the increase in leaf temperature was not large enough to offset the increase in the ν driving isotopic enrichment via kinetic fractionation.

Under field conditions the isotopic composition of leaf water in the xylem-tapping mistletoe, *Phoradendron juniperinum,* also follows qualitatively similar patterns to that predicted by Eq. (4) (Fig. 4). As observed in the laboratory studies discussed above and several other glasshouse and field studies (Zundel et al., 1978; White, 1983, 1988; Allison et al., 1985; Leaney et al., 1985; Bariac et al., 1989; Walker et al., 1989; Walker and Brunel, 1990; Yakir et al., 1990; Walker and Lance, 1991a), the evaporative-enrichment model Eq. (4) overestimates the degree of isotopic enrichment observed in *Phoradendron* foliage (Fig. 4).

There are at least four possible factors contributing to the difference between the observed and modeled leaf-water isotopic composition. First, the modeled leaf water isotopic compositions were calculated assuming that isotopic steady state had been reached. Only in the controlled environment, gas exchange chamber studies discussed above could the assumption of isotopic steady state be verified. Even when isotopic steady state was verified, a small difference occurred between the modeled and observed isotopic compositions (Fig. 2). In the field, environmental conditions may change too rapidly to allow for isotopic steady state to be attained throughout the entire day. The transpiration rate of a leaf and the leaf-water volume will influence the turnover time of leaf water and, therefore, will affect the time required to reach isotopic steady state. Second, water extracted from leaves by vacuum distillation removes water from leaf vein tissue in addition to water from leaf mesophyll cells. Vein water should be unfractionated and have the same isotopic composition as root and stem water. This would cause the observed isotopic enrichment of leaf water to fall below that predicted by the evaporative enrichment model. However,

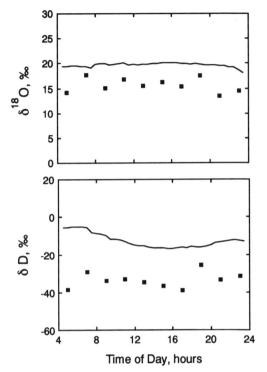

Figure 4. Diurnal pattern of change in the stable isotopic composition of foliage water in a xylem-tapping mistletoe, *Phoradendron juniperinum*, under natural conditions at Coral Pink State Park, Utah. The solid line represents the modeled leaf-water isotopic composition calculated with Eq. (4). The squares represent the observed isotopic composition of water extracted from *P. juniperinum* foliage. Stem-water isotopic composition: $\delta^{18}O = -13.2‰$, $\delta D = -100‰$. Isotopic composition of atmospheric water vapor: $\delta^{18}O = -23.7‰$, $\delta D = -165‰$. Based on data in Flanagan *et al.* (1993a).

vein tissue represents approximately 5% or less of the area of leaf tissues and should only have a minor influence on the isotopic composition of leaf water, therefore. Third, in the glasshouse and field studies, Eq. (4) was used to predict the isotopic composition of leaf water. Equation (4) does not include boundary layer effects which may result in a slight overestimation of leaf-water isotopic enrichment, particularly for oxygen isotopes (see Section II, D). Leaf size and shape, and wind speed are characteristics that will determine boundary-layer conductance. Fourth, the difference between the observed and modeled leaf-water isotopic compositions may result from a shifting balance between the bulk flow of unfractionated water into the leaf and the back diffusion of heavy isotope molecules from sites of evaporative enrichment within the leaf (Farquhar and Lloyd, Chapter 5, this volume; Flanagan *et al.*, 1991b). The transpiration rate of a leaf and the path length for water movement through a leaf are factors which will determine the magnitude of this fourth factor contributing to the difference between the observed and modeled leaf-water isotopic compositions.

Some preliminary support for the "shifting balance" hypothesis has been obtained in controlled environment studies with *P. vulgaris*. The extent of the difference between the modeled and observed isotopic composition of leaf water in *Phaseolus* was a strong linear function of the leaf transpiration rate (Flanagan *et al.*, 1991b). A model of leaf-water isotopic composition incorporating the effect of transpiration rate and water path length is described by Farquhar and Lloyd (Chapter 5, this volume). Based on this model, the effective path length for water movement from the xylem to the evaporative sites in *Phaseolus* leaves is predicted to be approximately 6.3 mm (Fig. 5). Walker *et al.* (1989) and Walker and Brunel (1990) also have observed that the difference between the observed leaf-water isotopic composition and that predicted by the evaporative-enrichment model increased with transpiration rate.

If there is a gradient in the isotopic composition of water from unfractionated water in the leaf veins to maximally enriched water in cells next to the evaporative sites within leaves, different fractions of water pushed out of a leaf using a pressure chamber should have a range of isotopic values. The first water extracted from a leaf should have an isotopic composition similar to stem water. The next series of samples collected should have a progressively more enriched isotopic composition. Such a pattern in the

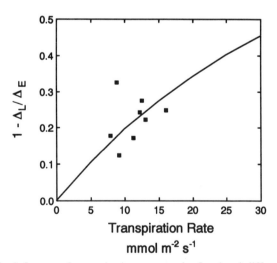

Figure 5. The influence of transpiration rate on the fractional difference between the modeled (Δ_E) and observed (Δ_L) leaf-water discrimination in *Phaseolus vulgaris*. See Farquhar and Lloyd (Chapter 5) for a definition of leaf-water discrimination values and a model of the effects of transpiration rate and path length for water movement between leaf xylem tissue and the sites of evaporative enrichment within leaves. The data points are from Flanagan *et al.* (1991b). The solid line is calculated based on an effective length (L) of 6.3 mm, using a diffusivity of $H_2^{18}O$ in water (D) of 2.485×10^{-9} m^2 s^{-1}.

isotopic values of water extracted using a pressure chamber has been observed by Yakir *et al.* (1989). Yakir and co-workers have argued, based in part on their measurements of water expressed using a pressure chamber, that distinct compartments of water with different isotopic compositions are isolated in different tissues within a leaf (Yakir *et al.*, 1989, 1990). I suggest that their data can be explained by the shifting balance hypothesis. If leaf water is not at isotopic steady state, variation in the isotopic composition of water expressed by a pressure chamber would be maximal.

B. Water Stress Effects

Water stress could potentially affect leaf-water isotopic composition via a number of mechanisms. For example, a reduction in stomatal conductance induced by water stress could result in a higher leaf temperature and larger v which would cause increased leaf-water isotopic enrichment. Alternatively, a reduction in transpiration rate in water-stressed plants could potentially influence the time required to reach isotopic steady state or the balance between the influx of unfractionated xylem water into the leaf and the back diffusion of heavy isotope molecules from the evaporative sites within leaves (Section III, A). The two possible consequences of a reduced transpiration rate would affect leaf-water isotopic composition in contrasting ways, however.

The effect of mild water stress on leaf-water isotopic composition was studied in *C. stolonifera* plants that were grown in a glasshouse. Reduced water applications over a 3-day period resulted in a reduction in midday water potential and an increase in leaf temperature (Table I). Both the hydrogen and oxygen isotopic composition of stem water were enriched in the stressed plants, likely a result of evaporation from the pots receiving reduced water applications.

There was no effect of mild water stress on the hydrogen isotopic composition of leaf water in *Cornus* (Table I). The δD value of leaf water in the stressed plants was 5‰ higher than that of the control plants. The difference in leaf-water values between treatments, however, can be explained by the enriched stem-water δD values in the stressed plants (Table I). Yakir *et al.* (1990) also observed that the hydrogen isotopic composition of leaf water was enriched approximately 5‰ in *Gossypium* plants receiving reduced water applications under field conditions. The isotopic composition of stem water was unfortunately not measured in the study with *Gossypium* plants (Yakir *et al.*, 1990). Without stem-water measurements, it is not possible to determine whether the drought effect was due to enriched stem water values or whether there was an actual effect on leaf-water isotopic composition.

For the oxygen isotopic composition of leaf water in *Cornus*, there was no significant difference between control and stressed plants, despite the fact that the isotopic composition of stem water in the stressed plants was enriched above that observed in the control plants (Table I). This pattern not only contrasts with that observed for the hydrogen data but is different

Table I   Effect of 3 Days of Mild Water Stress on Physiological Characteristics and Leaf- and Stem-Water Isotopic Composition in *Cornus stolonifera* at Midday under Glasshouse Conditions[a]

	Control	Stress	Significance
Stem-water potential (MPa)	-1.5 ± 0.3	-2.5 ± 0.3	*
Leaf temperature (°C)	31.7 ± 3.1	35.7 ± 2.5	*
Stem-water isotopic composition			
δD (‰)	-123 ± 1	-118 ± 2	*
δ^{18}O (‰)	-15.8 ± 0.3	-13.5 ± 0.7	*
Leaf-water isotopic composition			
Observed δD (‰)	-65 ± 2	-60 ± 3	*
Modeled δD (‰)	-45 ± 2	-41 ± 3	*
D/H f value	0.75 ± 0.03	0.76 ± 0.07	NS
Observed δ^{18}O (‰)	7.7 ± 2.3	6.6 ± 2.6	NS
Modeled δ^{18}O (‰)	18.9 ± 0.6	22.2 ± 0.4	*
^{18}O/^{16}O f value	0.67 ± 0.06	0.56 ± 0.08	*

Source: Flanagan and Ehleringer (1991b).

*$P < 0.05$.

[a]Average environmental conditions were air temperature, 27.7°C; relative humidity, 21.5%; photon flux density, 1270 μmol m^{-2} s^{-1}; atmospheric water vapor, δD = -149‰, δ^{18}O = -21.9‰. Modeled leaf-water isotopic compositions were calculated with Eq. (4). Values are the mean $\pm$ standard deviation. $n = 8$ for all values except stem-water potential and stem-water isotopic composition where $n = 4$. Significance tests were based on the result of a Kruskal–Wallis analysis of variance.

from patterns observed in other studies with *Phaseolus* and *Gossypium* plants in which the oxygen isotopic composition of leaf water was enriched in the water-stressed plants (Farris and Strain, 1978; Yakir *et al.*, 1990). If isotopic steady state had not been reached during midday, the leaf-water isotopic composition should have decreased below that predicted by the model, and the discrepancy between the modeled and observed data should have been larger in the stressed plants than in the control plants. A similar pattern should have also occurred for the hydrogen isotope data, unless the time course for isotopic steady state is independent for each isotopic element. At present there is no satisfactory explanation for the significantly lower f value for the ^{18}O/^{16}O data in the water-stressed *Cornus* plants (Table I). More mechanistic studies of the influence of water stress on leaf-water isotopic composition are required.

C. Interspecific Variation

A comparison of four cooccurring plant species under field conditions showed significant differences among the species for the hydrogen and oxygen isotopic composition of leaf water at midday (Fig. 6). The dissimilar leaf-water isotopic compositions occurred despite the fact that the stem-water isotopic composition and the leaf temperature and v were approximately the same for all species (Flanagan *et al.*, 1991a). Differences among species in leaf boundary-layer conductance and the amount of vein tissue (and therefore unfractionated xylem water) were shown not to be the pri-

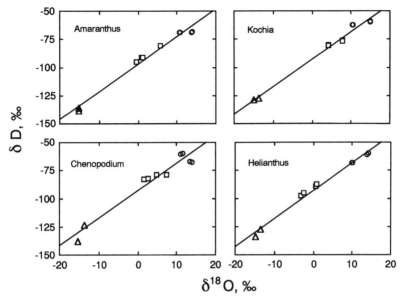

Figure 6.   Relationship between the oxygen and hydrogen isotopic compositions of stem water (triangles), leaf water (squares), and modeled leaf water (circles) at midday under field conditions. Modeled values were calculated with Eq. (4). Data from *Amaranthus retroflexus* plants: $\delta D = 2.46 \ \delta^{18}O - 97.3$, $r^2 = 0.98$, $P < 0.0001$. Data from *Kochia scoparia* plants: $\delta D = 2.44 \ \delta^{18}O - 92.5$, $r^2 = 0.98$, $P < 0.0001$. Data from *Chenopodium album* plants: $\delta D = 2.44 \ \delta^{18}O - 93.1$, $r^2 = 0.95$, $P < 0.0001$. Data from *Helianthus annuus* plants: $\delta D = 2.48 \ \delta^{18}O - 92.7$, $r^2 = 0.99$, $P < 0.0001$. Average environmental conditions were as follows: air temperature, 26°C; leaf temperature, 26°C; relative humidity, 25%; atmospheric water vapor, $\delta^{18}O = -26.6$‰, $\delta D = -202$‰. From Flanagan *et al.* (1991a).

mary causes of the variation in leaf-water isotopic composition. It was suggested that differences in the path length from the xylem tissue to the evaporative sites within leaves may account for the differences observed among species (Flanagan *et al.*, 1991a). It was not possible to completely rule out differences among species in the extent to which leaf water was at isotopic steady state. Three of the four species had equivalent f values calculated for both oxygen and hydrogen isotopic compositions (Flanagan *et al.*, 1991a).

Leaney *et al.* (1985) observed differences in the δD values of leaf water in a comparison of four plant species grown under glasshouse conditions. Two C_4 species (*Amaranthus edulis* and *Panicum maximum*) had significantly higher δD values at midday than did the two C_3 species (*Helianthus annuus* and *Triticum aestivum*). It was suggested that anatomical differences associated with the C_4 photosynthetic pathway may affect water movement through the leaf causing the differences observed in leaf-water isotopic enrichment (Leaney *et al.*, 1985). However, the four plant species studied by Flanagan *et al.* (1991a) also included two C_3 and two C_4 species, and variation among species was not associated with photosynthetic pathway. It is unlikely that the isotopic composition of leaf water is influenced by photosynthetic pathway, but instead is related to leaf characteristics influencing isotopic steady state, boundary-layer conductance, and the path length for water movement through leaves.

D. Intraspecific Variation

Walker and Lance (1991a) compared leaf-water δD and $\delta^{18}O$ values in a range of barley cultivars (*Hordeum vulgare*) grown under field conditions. The cultivars compared exhibited genetic variation for transpiration efficiency, based on measurements of leaf carbon isotopic composition, when grown together under common glasshouse conditions (Walker and Lance, 1991b). There were distinct differences in the hydrogen and oxygen isotopic composition of leaf water among cultivars. The relative rankings were similar for both hydrogen and oxygen isotopes and were consistent over the course of a day. There were no reported differences in the isotopic composition of stem water. Walker and Lance (1991a) suggested that differences in leaf-water isotopic composition reflect differences in canopy temperature or humidity conditions. Since no detailed measurements of canopy environmental conditions were made and porometer measurements of leaf conductance, leaf temperature, and relative humidity in the canopy did not differ significantly among cultivars, it is not possible with the data available at present to determine the mechanisms causing the different leaf-water δ-values.

E. Within-Plant Variation

A few studies have indicated systematic leaf-position or leaf-age variation in the isotopic composition of leaf water. For example, Allison *et al.* (1985) showed that needles collected from the top of 10-year-old *Pinus radiata* trees, grown in a field plantation, had higher δD and $\delta^{18}O$ values than the same age class of needles located in the middle of the tree crown. The canopy position variation in leaf-water isotopic composition was attributed to a gradient in relative humidity within the tree canopy, although no measurements of humidity were presented. In addition, during the Australian summer months (January and February) there was approximately a 5‰ difference between one- and three-year-old needles in the top and middle of the tree canopy for both leaf-water δD and the $\delta^{18}O$ values (Allison *et al.*, 1985). The younger leaf tissue had the more enriched isotopic composition. This pattern was only observed in the summer months, during the spring no differences were apparent among different needle age classes. Walker *et al.* (1989) have also observed that younger leaves of wheat (*T. aestivum*) have higher leaf-water δD values than older tissue. The isotopic composition of foliage water in the xylem-tapping mistletoe, *P. juniperinum*, is much more enriched in deuterium near the tips of the younger branches than in the older foliage branch segments closer to the

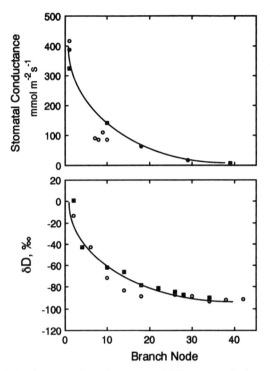

Figure 7. Variation in stomatal conductance and foliage water hydrogen isotopic composition in branch segments of increasing age in the xylem-tapping mistletoe, *Phoradendron juniperinum*. Based on data in Flanagan *et al.* (1993a).

base of the mistletoe, which have an isotopic composition close to that of the stem water of the host tree (Fig. 7). Stomatal conductance, measured at midday, declined abruptly along sequential positions on a mistletoe branch, so that foliage further than approximately 10–15 nodes from the mistletoe tip had very little or no gas exchange (Fig. 7). The sequential decline in stomatal conductance in different positions along a mistletoe branch must have been associated with complete stomatal closure in some sections of the foliage. If no water vapor loss occurs, then there is no opportunity for foliage water to become enriched in heavy isotopes. The reason for stomatal closure on the older mistletoe branch sections is presently unclear, the tissue was green and healthy looking. Similar changes in stomatal opening may occur in the other species discussed above as leaf aging progresses.

An alternative explanation for changes in leaf-water isotopic enrichment associated with leaf position was presented by Walker and Lance (1991a). In barley plants (*H. vulgare*) grown in field plots, the oxygen and hydrogen isotopic composition of water extracted from leaves collected at midday varied with leaf position along the stem. The recently emerged flag leaf had water that was substantially more enriched than the sequence of three leaves lower down the stem, which had more similar leaf-water isotopic compositions. Walker and Lance (1991a) suggested that the recently emerged flag leaf may contain enriched water imported from other leaves in addition to that received from soil water. The flag leaf had isotopic compositions much higher than that predicted by the evaporative-enrichment model, which lends some support for their suggestion. This situation may be special, however, because the flag leaf was likely still undergoing initial expansion and development. The maximal isotopic enrichment observed in mistletoe foliage is similar to that predicted by the evaporative-enrichment model (compare Figs. 4 and 7). In general, I suggest that the progressive isotopic enrichment from older to younger leaves along a plant stem is likely due to complete stomatal closure in portions of older leaves associated with leaf degeneration and/or environmental stress.

Luo and Sternberg (1992) have shown that there can be large variations in the isotopic composition of water in different sections of a leaf. The δD values of water from the main leaf vein showed a progressive increase from the base to the tip of the leaf. The vein-water isotopic composition was always intermediate to δD values for the stem water and mesophyll water. Such a pattern is consistent with the shifting balance hypothesis discussed above (Section III, A; see Farquhar and Lloyd, Chapter 5, this volume). In addition, Luo and Sternberg (1992) observed large spatial variability in δD values within the mesophyll of a large leaf from *Pterocarpus indicus*. They attributed the spatial variation in δD values to patchy stomatal closure across the leaf surface.

IV. Leaf-Water $^{18}O/^{16}O$ Ratio and $C^{18}O^{16}O$ Discrimination during Photosynthesis

In addition to application in studies of ν and transpiration efficiency, the leaf-water evaporative-enrichment model has other potential applications in physiological ecology. Francey and Tans (1987) have documented a latitudinal gradient in the $\delta^{18}O$ values of atmospheric CO_2. They proposed that the gradient results from latitudinal variation in the isotopic composition of environmental waters and an isotopic exchange of oxygen in atmospheric CO_2 with oxygen in leaf (chloroplast) water catalyzed by carbonic anhydrase. This suggestion relies on the assumption that the isotopic composition of leaf water varies in a latitudinal gradient similar to that of environmental waters (a reasonable assumption) and that isotopic exchange of atmospheric CO_2 with leaf water is a major factor affecting the oxygen isotopic composition of atmospheric CO_2. Friedli *et al.* (1987) have independently shown, in theory, that oxygen isotopic exchange with vegetation and soil should be important processes determining the $\delta^{18}O$ value of atmospheric CO_2. The leaf-water evaporative-enrichment model may have applications, therefore, in studies of the isotopic composition of atmospheric CO_2 and large-scale vegetation–atmosphere CO_2 exchange.

In order to test this hypothesis it is necessary to experimentally determine the fractionation processes during vegetation–atmosphere CO_2 exchange. Experiments were conducted to measure the carbon and oxygen isotopic composition of CO_2 before and after it passed over a *Phaseolus* leaf with a known, steady-state leaf-water oxygen isotopic composition. Different plant leaves were maintained at different leaf–air vapor pressure gradients (1.0, 2.0, 3.0 kPa) in a controlled environment gas exchange chamber in order to generate a range of leaf-water $\delta^{18}O$ values. When exposed to an increase in ν there was an expected decline in leaf intercellular/ambient CO_2 ratios and a correlated decline in carbon isotope discrimination as predicted by theory (Farquhar *et al.*, 1989) (Fig. 8). In addition, the leaf-water $\delta^{18}O$ values increased as expected when leaves were exposed to higher ν conditions. In contrast to $^{13}CO_2$ discrimination which declined as ν was increased, the $C^{18}O^{16}O$ discrimination increased with an increase in ν and was positively correlated to the leaf-water $\delta^{18}O$ values (Fig. 8). The observed $C^{18}O^{16}O$ discrimination values were strongly correlated to values predicted by a mechanistic model of isotopic fractionation (Fig. 9), providing support for the role of oxygen isotopic exchange with leaf (chloroplast) water in controlling the $\delta^{18}O$ values of atmospheric CO_2 (see Chapter 5, Farquhar and Lloyd; Farquhar *et al.*, 1993).

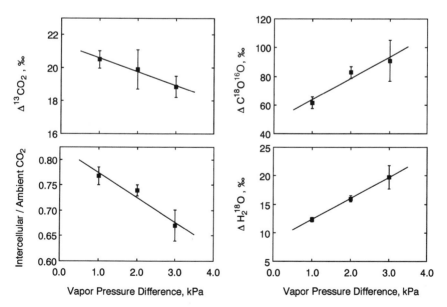

Figure 8. On-line $^{13}CO_2$ and $C^{18}O^{16}O$ discrimination in *Phaseolus vulgaris*. Discrimination values were calculated as described by Farquhar and Lloyd, Chapter 5, this volume. Data points represent means ± 1 SD, $n = 3$. Based on data in Flanagan *et al.* (1993b).

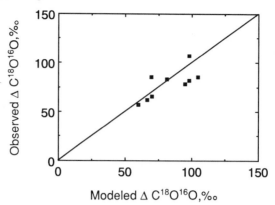

Figure 9. Comparison of modeled and observed $C^{18}O^{16}O$ discrimination in *Phaseolus vulgaris*. Modeled values of $C^{18}O^{16}O$ discrimination were calculated as described by Farquhar and Lloyd (Chapter 5, this volume). Complete isotopic equilibrium between oxygen in CO_2 and chloroplastic water was assumed. Based on data in Flanagan *et al.* (1993b).

V. Conclusions

Under controlled, steady-state conditions, the evaporative-enrichment model provides good prediction of bulk leaf-water isotopic enrichment. Differences between the observed and modeled isotopic compositions are likely related to gradients in the isotopic composition of water from leaf vein tissue through to the evaporative sites within leaves. Use of the evaporative-enrichment model under field conditions is more complicated because of the assumption of isotopic steady state. In addition, precise application of the model requires measurements of the isotopic composition plant stem water and atmospheric water vapor. Application of the model to obtain estimates of the leaf–air vapor pressure difference, in comparisons among leaves on one plant or among different species, is difficult because of variation in leaf-water isotopic enrichment likely associated with patchy stomatal closure and differences in the path length for water movement through leaves. Enrichment of leaf and chloroplast water during transpiration has important effects on $C^{18}O^{16}O$ discrimination during photosynthesis. The model of isotopic fractionation during transpiration will likely become an important tool in studies of the oxygen isotopic composition of atmospheric CO_2 and large-scale vegetation–atmosphere interactions.

References

Allison, G. B., J. R. Gat, and F. W. J. Leaney. 1985. The relationship between deuterium and oxygen-18 delta values in leaf water. *Chem. Geol.* **58:** 145–156.

Bariac, T., S. Rambal, C. Jusserand, and A. Berger. 1989. Evaluating water fluxes of field-grown alfalfa from diurnal observations of natural isotope concentrations, energy budget and ecophysiological parameters. *Agric. For. Meteorol.* **48:** 263–283.

Craig, H., and L. I. Gordon. 1965. Deuterium and oxygen-18 variations in the ocean and the marine atmosphere, pp. 9–130. *In* E. Tongiorgi (ed.), Proceedings of a Conference on Stable Isotopes in Oceanographic Studies and Paleotemperatures, Spoleto, Italy. Lischi & Figli, Pisa.

Dongmann, G., H. W. Nurnberg, H. Forstel, and K. Wagener. 1974. On the enrichment of $H_2{}^{18}O$ in the leaves of transpiring plants. *Radiat. Environ. Biophys.* **11:** 41–52.

Ehleringer, J. R., and C. B. Osmond. 1989. Stable isotopes, pp. 281–300. *In* R. W. Pearcy, J. R. Ehleringer, H. A. Mooney, and P. W. Rundel (eds.), Plant Physiological Ecology. Field Methods and Instrumentation. Chapman & Hall, London.

Farquhar, G. D., K. T. Hubick, A. G. Condon, and R. A. Richards. 1988. Carbon isotope fractionation and plant water-use efficiency, pp. 21–40. *In* P. W. Rundel, J. R. Ehleringer, and K. A. Nagy (eds.), Stable Isotopes in Ecological Research. Springer-Verlag, Berlin.

Farquhar, G. D., J. R. Ehleringer, and K. T. Hubick. 1989. Carbon isotope discrimination and photosynthesis. *Annu. Rev. Plant Physiol. Plant Mol. Biol.* **40:** 503–537.

Farquhar, G. D., J. Lloyd, J. A. Taylor, L. B. Flanagan, J. P. Syvertsen, K. T. Hubick, S. C. Wong, and J. R. Ehleringer. 1993. Vegetation effects on the isotope composition of oxygen in atmospheric CO_2. *Nature* **363:** 439–443.

Farris, F., and B. R. Strain. 1978. The effects of water stress on leaf $H_2{}^{18}O$ enrichment. *Radiat. Environ. Biophys.* **15:** 167–202.

Flanagan, L. B., and J. R. Ehleringer 1991a. Stable isotope composition of stem and leaf water: Applications to the study of plant water-use. *Funct. Ecol.* **5:** 270–277.

Flanagan, L. B., and J. R. Ehleringer. 1991b. Effects of mild water stress and diurnal changes in temperature and humidity on the stable oxygen and hydrogen isotopic composition of leaf water in *Cornus stolonifera* L. *Plant Physiol.* **97:** 298–305.

Flanagan, L. B., J. F. Bain, and J. R. Ehleringer. 1991a. Stable oxygen and hydrogen isotope composition of leaf water in C_3 and C_4 plant species under field conditions. *Oecologia* **88:** 394–400.

Flanagan, L. B., J. P. Comstock, and J. R. Ehleringer. 1991b. Comparison of modeled and observed environmental influences on the stable oxygen and hydrogen isotope composition of leaf water in *Phaseolus vulgaris* L. *Plant Physiol.* **96:** 588–596.

Flanagan, L. B., J. R. Ehleringer, and J. D. Marshall. 1992. Differential uptake of summer precipitation among co-occurring trees and shrubs in a Pinyon-Juniper woodland. *Plant Cell Environ.* **15:** 831–836.

Flanagan, L. B., J. D. Marshall, and J. R. Ehleringer. 1993a. Photosynthetic gas exchange and leaf water stable isotopic enrichment: comparison of a xylem-tapping mistletoe and its host, *Plant, Cell Environ.* **16:** in press.

Flanagan, L. B., S. L. Phillips, J. R. Ehleringer, J. Lloyd, and G. D. Farquhar. 1993b. Effects of leaf water isotopic composition on discrimination against $C^{18}O^{16}O$ during photosynthesis. *Aust. J. Plant Physiol.*, in press.

Francey, R. J., and P. P. Tans. 1987. Latitudinal variation in oxygen-18 of atmospheric CO_2. *Nature* **327:** 495–497.

Friedli, H., U. Siegenthaler, D. Rauber, and H. Oeschger. 1987. Measurements of concentration, $^{13}C/^{12}C$ and $^{18}O/^{16}O$ ratios of tropospheric carbon dioxide over Switzerland. *Tellus* **39B:** 80–88.

Gat, J. R. 1980. The isotopes of hydrogen and oxygen in precipitation, Vol. 1, pp. 24–48. *In* P. Fritz and J-Ch. Fontes (eds.), Handbook of Environmental Isotope Geochemistry. Elsevier, New York.

Kays, W. M., and M. E. Crawford. 1980. Convective heat and mass transfer, 2nd ed. McGraw–Hill, New York.

Leaney, F. W., C. B. Osmond, G. B. Allison, and H. Ziegler. 1985. Hydrogen-isotope composition of leaf water in C_3 and C_4 plants: Its relationship to the hydrogen-isotope composition of dry matter. *Planta* **164:** 215–220.

Luo, Y-H., and L. Sternberg. 1992. Spatial D/H heterogeneity of leaf water. *Plant Physiol.* **99:** 348–350.

Majoube, M. 1971. Fractionnement en oxygene-18 et en deuterium entre l'eau et sa vapeur. *J. Chim. Phys.* **58:** 1423–1436.

Merlivat, L. 1978. Molecular diffusivities of $H_2{}^{18}O$ in gases. *J. Chem. Phys.* **69:** 2864–2871.

Sharkey, T. D., K. Imai, G. D. Farquhar, and I. R. Cowan. 1982. A direct confirmation of the standard method of estimating intercellular CO_2. *Plant Physiol.* **69:** 657–659.

Walker, C. D., and J. P. Brunel. 1990. Examining evapotranspiration in a semi-arid region using stable isotopes of hydrogen and oxygen. *J. Hydrol.* **118:** 55–75.

Walker, C. D., and R. C. M. Lance. 1991a. The fractionation of ^2H and ^{18}O in leaf water of barley. *Aust. J. Plant Physiol.* **18:** 411–425.

Walker, C. D., and R. C. M. Lance. 1991b. Silicon accumulation and ^{13}C composition as indices of water-use efficiency in barley cultivars. *Aust. J. Plant Physiol.* **18:** 427–434.

Walker, C. D., F. W. Leaney, J. C. Dighton, and G. B. Allison. 1989. The influence of transpiration on the equilibrium of leaf water with atmospheric water vapor. *Plant Cell Environ.* **12:** 221–234.

White, J. W. C. 1983. The Climatic Significance of D/H Ratios in White Pine in the Northeastern United States. Dissertation. Columbia University, New York.

White, J. W. C. 1988. Stable hydrogen isotope ratios in plants: A review of current theory and some potential applications, pp. 142–162. *In* P. W. Rundel, J. R. Ehleringer, and K. A. Nagy (eds.), Stable Isotopes in Ecological Research. Springer-Verlag, Berlin.

White, J. W. C., and S. D. Gedzelman. 1984. The isotopic composition of atmospheric water vapor and the concurrent meteorological conditions. *J. Geophys. Res.* **89:** 4937–4939.

Yakir, D., M. J. DeNiro, and P. W. Rundel. 1989. Isotopic inhomogeneity of leaf water: Evidence and implications for the use of isotopic signals transduced by plants. *Geochim. Cosmoschim. Acta* **53:** 2769–2773.

Yakir, D., M. J. DeNiro, and J. R. Gat. 1990. Natural deuterium and oxygen-18 enrichment in leaf water of cotton plants grown under wet and dry conditions: Evidence for water compartmentation and its dynamics. *Plant Cell Environ.* **13:** 49–56.

Zundel, G., W. Miekeley, B. M. Grisi, and H. Forstel. 1978. The $H_2{}^{18}O$ enrichment in the leaf water of tropic trees: Comparison of species from the tropical rain forest and the semi-arid region of Brazil. *Radia. Environ. Biophys.* **15:** 203–212.

II

Ecological Aspects of Carbon Isotope Variation

The contributions in this section focus on the application of carbon and hydrogen isotopes to resolve patterns in ecological systems. Lauteri *et al.* establish the linkages between plant gas exchange and carbon isotopic composition of organic material with different turnover rates, showing the temporal scaling potential of carbon isotopes. In the next two chapters, Griffiths and Broadmeadow, and Jackson *et al.* describe canopy-dependent changes in carbon isotopic composition of leaf material that are dependent on soil-derived carbon dioxide and an approach that allows one to calculate the fraction of plant carbon that is recycled within forest canopies. Livingston and Spittlehouse, and Marshall and Zhang follow with chapters describing variation in carbon isotope composition of forest species and how these variations relate to productivity. These are followed by contributions from Ehleringer, and Rundel and Sharifi illustrating similar patterns in carbon isotope composition and physiological behavior for desert shrub species. Maguas *et al.* describe variation in the carbon isotope composition of lichens and then go on to indicate that these phylogenetic variations in isotope composition are related to modifications in the photosynthetic biochemistry of different lichen groups. In the final two chapters of this section, Patterson and Rundel, and Elsik *et al.* demonstrate the utility of stable isotopes in screening plants for differential sensitivity to atmospheric pollutant stresses.

7

Carbon Isotope Discrimination in Leaf Soluble Sugars and in Whole-Plant Dry Matter in *Helianthus annuus* L. Grown under Different Water Conditions

Marco Lauteri, Enrico Brugnoli, and Luciano Spaccino

I. Introduction

Plant life is characterized by continuous efforts to cope with environmental fluctuations. Water deficit is a common and widely spread experience for most plant communities. Plants are able to adjust the rate of water loss by transpiration through regulation of stomatal aperture, which also affects the rate of CO_2 assimilation and, consequently, the rate of growth. Plant transpiration efficiency (W), the ratio of dry matter produced to water used, is a crucial feature in determining productivity and probability of survival (Cowan, 1986). Improved water-use efficiency has been seen as a desirable trait which may confer drought adaptation and it has been proposed as a selection criterion to improve crop productivity in water-limited environments. However, such applications have been limited because of the intrinsic difficulty in measuring W and by the lack of fast screening methods. A significant breakthrough in this field was represented by the finding that carbon isotope discrimination (Δ) against ^{13}C is negatively related to plant W in C_3 plants.

The fractionation of carbon isotopes during photosynthesis involves biochemical and physical processes and is strongly influenced by the environment (for reviews, see Vogel, 1980; O'Leary, 1981; Farquhar *et al.*, 1989). Farquhar *et al.* (1982) developed a theoretical interpretation of such processes that predicts a linear relationship between Δ and the ratio of intercellular to atmospheric partial pressure of CO_2 (p_i/p_a). Photosynthetic water-use efficiency, the ratio of carbon fixed to water transpired (also called transpiration efficiency) is also dependent, in part, on p_i/p_a, and therefore, indirectly on Δ. Experiments have shown linear negative relationships between Δ measured in whole-plant dry matter and plant water-use efficiency in several plant species (for a recent review, see Farquhar *et al.*, 1989). Consequently, it has been proposed that the analysis of Δ may be used for assessing water-use efficiency in ecophysiological studies and in breeding programs for drought adaptation.

Carbon isotope discrimination in whole-plant dry matter (e.g., structural carbon or certain stored carbohydrates) represents a long-term integration of p_i/p_a, and consequently, of photosynthetic W, over the entire plant life cycle. At the other extreme, the "on-line" analysis of $\delta^{13}C$ of the air CO_2 before and after it passes over a leaf in a gas exchange cuvette gives an instantaneous estimation of the physiological parameters involved at the leaf level (Evans *et al.*, 1986; von Caemmerer and Evans, 1991). Recently, Brugnoli *et al.* (1988) demonstrated that carbon isotope discrimination in the newly fixed carbon (i.e., leaf soluble sugars and starch) correlates with a short-term integration of p_i/p_a, weighted on CO_2 assimilation, over the period of about 1 day, which is a time-scale intermediate between that of Δ in whole-plant dry matter and that measured on-line. A similar short-term method had been previously applied to malate accumulated overnight in crassulacean acid metabolism (CAM) plants (O'Leary and Osmond, 1980).

The analysis of Δ in leaf soluble sugars and starch gives the opportunity to investigate the adaptive response of C_3 plants to environmental fluctuations and to resolve seasonal variations in photosynthetic water-use efficiency, not previously addressed in isotope studies. This can be of particular interest when plants are exposed to drought stress, with fluctuating water availability during ontogeny.

The aim of the present work was to explore the possibility of assessing W on different time scales, using the analysis of carbon isotope discrimination in whole-plant dry matter and in leaf soluble sugars. Water-use efficiency and growth characteristics were measured in four sunflower genotypes exposed to either well-watered or drought conditions. The effects of drought on several physiological parameters including gas exchange characteristics, photosystem II photochemical efficiency, photosynthetic capacity, and short-term and long-term carbon isotope fractionation were studied in order to detect adaptive responses and to determine stomatal and nonstomatal limitations of photosynthesis.

II. Materials and Methods

Four sunflower (*Helianthus annuus* L.) genotypes (Flamme, Gloriasol, Drysol and Romsun HS52) were sown in 50-liter pots containing a mixture of garden soil and sand (⅓, v/v). Plants were grown inside a greenhouse in

Porano (47° 21′ N Lat) under natural daily light period from the 6th of June to the 10th of September. The maximum PFD was approx 1700 μmol m^{-2} s^{-1} (30–42 mol photon m^{-2} day^{-1}). Air temperature was 20°C at night and varied between 28 and 35°C at day. Air relative humidity ranged between 40 and 70%.

Plants were subjected to two different watering regimes. In the well-watered regime the soil was kept as close as possible to field capacity by adding the measured amounts of water transpired by plants on a daily basis. In the water-stress treatment plants were subjected to drought by drastically reducing irrigation, starting from Day 18 after emergence. Drought-stressed plants were irrigated once weekly with variable amounts of nutrient solution corresponding to about one-eighth of the water transpired. Because of the large pot volume, it took several days to significantly deplete the soil water content. In order to determine the amount of water used by plants, pots were periodically weighed and the watering volumes were recorded.

Pots were adapted to minimize evaporative losses from the soil using a layer of gravel and a plastic film. Residual evaporative losses from the soil were measured on several identical pots kept inside the greenhouse without plants. Plant water use was obtained by subtracting evaporative losses from the soil from the total water use. A modified Hewitt nutrient solution (Hewitt and Smith, 1975) with ammonium nitrate in place of potassium nitrate was added weekly to both treatments. The experiment was conducted according to a randomized block design.

Plants were harvested at "seed physiological maturity" and separated into leaves, stems, seeds, and roots, and dry weight of individual fractions and leaf area were measured. Plant water-use efficiency was determined as the ratio of total plant dry weight to water transpired.

Spot gas exchange measurements were taken on several days during ontogeny using a portable open system (Model LCA2, Analytical Development Company, Hoddesdon, UK) modified as previously described (Brugnoli and Lauteri, 1991), in order to avoid the effect of peak broadening due to H_2O and CO_2 cross-sensitivity. Youngest fully expanded and exposed leaves were always used for gas exchange measurements. Gas exchange parameters were calculated according to von Caemmerer and Farquhar (1981). Leaf discs were collected at noon and immediately used for measurements of relative water content (RWC).

Rates of photosynthetic O_2 evolution and photon yield of O_2 evolution were measured using a leaf-disc oxygen electrode (Hansatech, Kings Lynn, Norfolk, UK) as described in detail by Björkman and Demmig (1987). The CO_2 concentration inside the leaf-disc chamber was kept at 10% by flowing a mixture of 10% CO_2 in air, before closing the chamber and recording the rate of O_2 evolution.

The same leaves on which gas exchange had been previously measured were collected at sunset and used for carbon isotope analysis of soluble sugars and insoluble fractions. Leaf soluble sugars were extracted and purified according to Brugnoli et al. (1988). The leaf insoluble fraction remaining after sugar extraction consisted mainly of cellulose and starch. Carbon isotope composition was also measured on whole-plant dry matter from different plant parts (leaves, stems, and roots) collected at the end of the life cycle.

Carbon isotope analysis of soluble and insoluble leaf fractions and of whole-plant dry matter was as previously described (Brugnoli and Lauteri, 1991). Oven-dried or freeze-dried samples were combusted in a Dumas-combustion elemental analyzer (Model Na-1500, Carlo Erba, Milan, Italy). CO_2 was purified in cryogenic traps and then analyzed by a dual-inlet isotope ratio mass spectrometer (VG Isotech, Middlewich, UK). Control of combustion during analysis was achieved using a working standard of sucrose which has a $\delta^{13}C$ of $-25.09 \times 10^{-3} \pm 0.06 \times 10^{-3}$. Carbon isotope ratio of sample CO_2 was compared with that of a reference CO_2 calibrated against PDB. Internal precision of individual measurements of carbon isotope composition was about 0.01×10^{-3}. Carbon isotope composition of greenhouse air CO_2 was measured at different times and was typically -8.0×10^{-3}. Carbon isotope discrimination was determined from the known carbon isotope composition of plant material (δ_p) and source air CO_2 (δ_a) as

$$\Delta = \frac{\delta_a - \delta_p}{1 + \delta_p}$$

according to Farquhar et al. (1989).

III. Results and Discussion

A. Growth Characteristics and Allocation Pattern

Growth of sunflower plants was strongly affected by water availability. As shown in Table I, plant biomass accumulation under water-shortage conditions was reduced by 49% with respect to well-watered plants. Plant biomass accumulation also showed a highly significant variation among genotypes when plants were grown in well-watered conditions, with two genotypes (Gloriasol and Drysol) being the most productive (Table I). However, such genetic variation in plant productivity was not evident under drought conditions. Water shortage caused a 51% decline of leaf area development (Table I) and a concurrent increase of the root-to-shoot ratio (Table I). A strong correlation ($r = 0.95$) was found between plant biomass accumulated at the end of the experiment and leaf surface area (Fig. 1). The most productive genotypes showed the greatest leaf area development (Table I). Thus, the different capacity of biomass production among genotypes may be at least partly explained by differences in leaf surface area and differential allocation of carbon to leaf dry weight at the expense of other organs.

Table I Biomass Accumulation, Leaf Area, Root-to-Shoot Ratio, and Plant Water-Use Efficiency Determined at Physiological Maturity in Four Sunflower Genotypes Grown under Well-Watered and Drought Conditions[a]

Genotype	Total plant biomass (g dry wt)		Leaf area (m^2)		Root/shoot ratio		Plant water-use efficiency [g dry wt (Kg H$_2$O)$^{-1}$]	
	Control	Drought	Control	Drought	Control	Drought	Control	Drought
Flamme	85.5 B	44.6 C	0.233 B	0.122 C	0.075	0.089	2.04	2.18
Gloriasol	96.7 A	46.6 C	0.275 A	0.122 C	0.079	0.083	2.07	2.19
Drysol	94.4 A	45.6 C	0.263 A	0.122 C	0.076	0.087	2.06	2.16
Rom. HS52	81.2 B	47.2 C	0.225 B	0.117 C	0.068	0.082	2.00	2.19
Mean	89.4 A	46.0 B	0.249 A	0.121 B	0.074 B	0.085 A	2.04 B	2.18 A

[a] Mean values and analysis of variance. Each value is the mean of 11 observations. Different capital letters indicate significant difference at the 0.01 probability level. Different bold letters indicate significant differences ($P < 0.01$) between treatment means. Where no letters are reported no significant differences among genotypes were observed.

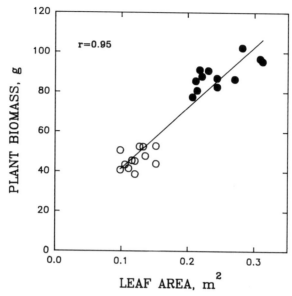

Figure 1. Relationship between plant biomass and leaf area in sunflower plants grown under different water regimes. Measurements were made at the end of the experiment at seed maturity. Symbols: control (●), water stress (○).

B. Stomatal and Nonstomatal Limitations of Photosynthesis and Intrinsic Efficiency of Photochemistry

The drought-induced reduction in growth rate in sunflower was partly explained by a reduction of the rate of net CO_2 uptake (A), mostly caused by partial stomatal closure. As shown in Fig. 2, p_i/p_a and A showed a time-dependent decline in water-stressed plants, with decreasing water availability; concurrently, the leaf RWC varied from 80%, at the beginning of stress treatment, to value as low as 58%, at the end of the experiment. In contrast, in well-watered plants assimilation rate and stomatal conductance re-

Table II Photon Yield of O_2 Evolution on the Basis of Incident Photons (φ_i) and Maximum Rate of O_2 Evolution (P_{max}) at Saturating PFD (2000 μmol m^{-2} s^{-1}) in Sunflower Plants Grown at Different Water Availability[a]

	$\varphi_i \pm$ SE [mol O_2 (mol photons)$^{-1}$] ($n = 3$)	$P_{max} \pm$ SE (μmol m^{-2} s^{-1}) ($n = 5$)
Control	0.099 ± 0.004	44.0 ± 3.1
Drought	0.103 ± 0.003	35.2 ± 2.9

[a]Measurements were made on the genotype Gloriasol on Day 56 and 57 after start of drought stress treatment. Mean values and standard errors (SE).

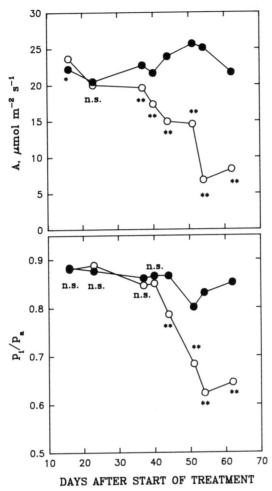

Figure 2. Seasonal variation in CO_2 assimilation rate (A) and ratio of intercellular and atmospheric partial pressures of CO_2 (p_i/p_a) in sunflower plants grown under well-watered and drought conditions. Each data point is the mean of 12 independent measurements across the four genotypes. Within each day, single (*) and double (**) asterisks indicate significant differences among treatments at the 0.05 and at the 0.01 probability levels, respectively; n.s., no significant differences. Symbols: control (●), water stress (○).

mained remarkably constant (Fig. 2) and also RWC remained high (approx 85%).

That the drought-induced reduction of photosynthetic rate was mostly caused by partial stomatal closure was also confirmed by measurements of the rate of O_2 evolution at saturating CO_2. As shown in Table II, water shortage caused no effects on the photon yield of O_2 evolution, measured at strictly rate-limiting PFDs, indicating that the intrinsic efficiency of photochemistry was not affected by drought stress. These results are in close agreement with previous observations on water stress (e.g., Ben *et al.*, 1987; Cornic *et al.*, 1989) and salinity stress (Brugnoli and Lauteri, 1991; Brugnoli and Björkman, 1992). On the other hand, a 20% reduction of the rate of O_2 evolution measured at saturating PFD was found in stressed plants compared to well-watered plants (Table II), indicating a moderate drought-induced nonstomatal effect on photosynthesis. This observation is in agreement with other observations on drought and salinity stress on sunflower and other species (e.g., Ben *et al.*, 1987; Brugnoli and Lauteri, 1991; A. Battistelli, personal communication).

C. The Long-Term Time Scale: Plant Water-Use Efficiency and Δ in Whole-Plant Dry Matter

The drought-induced reduction of stomatal conductance caused a significant decline of p_i/p_a to as low as 0.62 and, consequently, the photosynthetic water-use efficiency (mmol CO_2 fixed/mol H_2O transpired) was enhanced. In agreement with these gas exchange results, plant transpiration efficiency (mg DW/g H_2O used) was increased by 7% in water-stressed compared to well-watered plants (Table I). Although no significant variations in plant W were found among genotypes, W appeared correlated to plant biomass production, particularly in drought-stressed plants ($r = 0.91$, data not shown).

Carbon isotope discrimination measured in whole-plant dry matter was negatively correlated with plant water-use efficiency, as expected for C_3 plants (O'Leary, 1988; Farquhar *et al.* 1982, 1988, 1989). Carbon isotope discrimination in leaves was correlated with Δ of stems ($r = 0.85$) and of root ($r = 0.81$), although roots were always slightly isotopically heavier than stems and leaves (Fig. 3). Figure 4 shows the relationship between Δ in leaves and plant W ($r = 0.85$). This is in agreement with the expected theoretical relationship (see regression equation of Fig. 4) and corroborates the view that Δ is a promising parameter for assessing plant W, and eventually for selecting genotypes adapted to water-limited conditions.

Plant W is determined by a series of physiological parameters such as stomatal conductance, photosynthetic capacity, and respiration losses, integrated over the entire plant life cycle. Other important growth characteristics such as dry matter allocation pattern may be mechanistically related to W (Masle and Farquhar, 1988; Virgona *et al.*, 1990), possibly through the control of stomatal opening mediated by root signals (Davies and Zhang, 1991). Furthermore, W is also directly affected by the environment (i.e., by the vapor pressure difference between leaf and air, which is the driving force for transpiration). As mentioned in the Introduction, these parameters are very rarely at steady state and are indeed subjected to large variations. Such variations are particularly important in stressful environments, where periods with relatively high water availability may be followed by drought with progressively decreasing soil water content. Thus, one would expect large temporal changes in photosynthetic water-use efficiency and growth rate. In water-stressed sunflower plants p_i/p_a, which is negatively

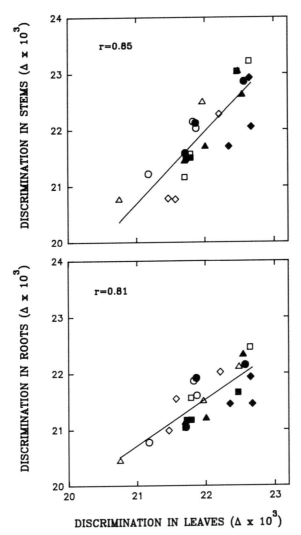

Figure 3. Relationships between carbon isotope discrimination in leaves and in stems (top) and in roots (bottom) in four sunflower genotypes grown under well-watered and drought conditions. Symbols: Drysol ($\bigcirc$, $\bullet$), Flamme ($\square$, $\blacksquare$), Gloriasol ($\triangle$, $\blacktriangle$), Romsun HS52 ($\diamond$, $\blacklozenge$); closed symbols, control; open symbols, water stress.

related to photosynthetic W, declined from 0.88 at the beginning of the treatment to 0.62 at the end of the experiment (Fig. 2). Such short-term variations could not be detected by the analysis of plant W or by the analysis of Δ in whole-plant dry matter, which represents a long-term integration of the fractionation occurring during the plant life, weighted by the rate of CO_2 assimilation. On the other hand, measurements of on-line carbon isotope discrimination in a gas exchange cuvette (Evans $et\ al.$, 1986) are

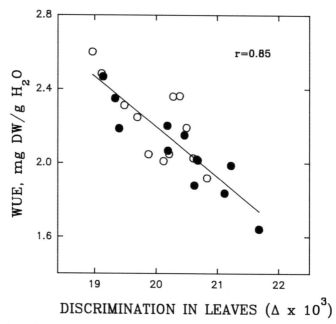

Figure 4. Relationship between plant water-use efficiency (WUE) and carbon isotope discrimination (Δ) in whole-leaf dry matter in sunflower plants grown at different water regimes. Symbols: control (●), water stress (○). Regression equation: $y = 7.65 - 0.27 x$.

difficult to make in the field. The analysis of carbon isotope discrimination in recently fixed carbon such as leaf soluble sugars seems to be more appropriate for this purpose (Brugnoli *et al.*, 1988).

D. The Short-Term Time Scale: Δ in Leaf Sugars and Photosynthetic Water-Use Efficiency

In order to investigate short-term variations of p_i/p_a and of photosynthetic water-use efficiency, carbon isotope composition was measured on soluble sugars extracted from leaves periodically collected during ontogeny. Figure 5 shows RWC, net assimilation rate (A), and Δ in leaf soluble sugars and insoluble fractions as a function of p_i/p_a measured on Day 62 after start of withholding water. The variation of p_i/p_a among treatments was associated with differences in leaf-water status as measured by RWC (Fig. 5, top). In water-stressed plants, leaf RWC was significantly lower than that of control plants. The positive relationship between the rate of CO_2 assimilation and p_i/p_a (Fig. 5, center) is further evidence supporting the view that the drought-induced reduction in the rate of photosynthesis was mainly caused by limitation of CO_2 availability at the carboxylation sites, due to partial stomatal closure. These results are in contrast with those reported by Wise *et al.* (1990) who found a negative relationship between p_i and A in sunflower plants exposed to water deficit. They attributed the increase in p_i with decreasing A to a "down regulation" of photosynthesis involving both

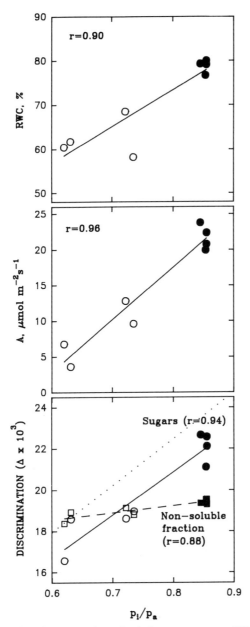

Figure 5. Relationships between p_i/p_a and relative water content (RWC, top), assimilation rate (A, center), and carbon isotope discrimination in leaf soluble sugars and leaf insoluble fraction (bottom) in four sunflower genotypes grown under well-watered and drought conditions. Measurements were made on Day 62 after the start of the stress treatment. Gas exchange measurements were performed on fully exposed leaves (PFD $= 1700 \pm 52$ μmol photons m^{-2} s^{-1}). Each data point is the mean of three independent observations, across the four genotypes. No significant differences were observed among genotypes. In the bottom panel, the solid line represents the regression for Δ in soluble sugars (circles) and the broken line is the regression for Δ in nonsoluble fraction (squares). The dotted line is the theoretical relationship between Δ and p_i/p_a assuming $\Delta = 4.4 + 22.6 \, p_i/p_a$. Symbols: control ($\bullet$, $\blacksquare$), water stress ($\bigcirc$, $\square$).

nonstomatal and stomatal factors. This discrepancy may have several explanations such as different experimental conditions and cultivars. However, possible effects of heterogeneity of stomatal aperture over the leaf surface (Terashima *et al.*, 1988; Terashima, 1992) might be at least partly responsible for the observed difference. Genetic differences in sensitivity and degree of stomatal heterogeneity might also occur (Sharkey *et al.*, 1990). In the present experiment, the fact that p_i declined in water-stressed plants, and the strong relationship between p_i/p_a and Δ in soluble sugars seem to rule out the possibility of significant overestimation of p_i. The positive relationship between Δ in leaf soluble sugars and p_i/p_a ($r = 0.94$; Fig. 5, bottom), found in sunflower plants subjected to different water availability is in agreement with theory (for a review see Farquhar *et al.*, 1989) and with previous results obtained by Brugnoli *et al.* (1988). There may be several explanations for the slight deviation between the observed results (Fig. 5) and the predicted results (Farquhar *et al.*, 1989), including the possibility of substantial resistance to CO_2 transfer from substomatal cavities to the chloroplast and possible effects of respiration and photorespiration (Berry, 1988; Farquhar *et al.*, 1988, 1989). Possible reallocation of previously stored carbohydrates might also affect the isotope composition of leaf soluble fraction. This effect, if present, was not relevant in the present experiment. However, it might be important in early developmental stages of plants, such as in young seedlings, before establishment of full autotrophy (Deléens-Provent and Schwebel-Dugué, 1987) or in other plant tissues where heterotrophic sources of carbon may be relevant (Yakir *et al.*, 1991).

E. Comparisons of Δ in Leaf Soluble Sugars and Insoluble Fraction

Carbon isotope discrimination in leaf insoluble fractions was also related to p_i/p_a ($r = 0.88$; Fig. 5, bottom). However, Δ in leaf insoluble fractions was much less sensitive to change in p_i/p_a than Δ in soluble sugars. This insensitivity can be explained by the fact that the drought-induced decline of p_i/p_a was not prolonged enough to affect the isotopic signature of leaf cellulose, which is the main constituent of the insoluble fraction. On the other hand, even small short-term changes in p_i/p_a may affect significantly the isotope composition of leaf sugars, which have a much faster turnover. These results are in close agreement with previous findings on different species. Björkman and Brugnoli (unpublished results) obtained a very strong relationship between $\delta^{13}C$ of leaf soluble sugars and p_i/p_a, but found that $\delta^{13}C$ of the leaf insoluble fraction was largely insensitive to changes in p_i/p_a measured on a time scale of 1 day in field-grown *Gossypium hirsutum* subjected to water stress. Similarly, Brugnoli and Lauteri (1991) observed higher carbon isotope discrimination in leaves than in seeds in both *G. hirsutum* and *Phaseolus vulgaris* transferred from nonsaline to saline conditions. These differences may be explained by a dramatic decline in p_i/p_a after start of saline treatments, affecting significantly the Δ of seeds but not that of leaves. Recently, Brugnoli and Björkman (1992), working on *G.*

hirsutum continuously grown under different salinity levels and at constant environmental conditions, found very similar relationships between short-term measurements of p_i/p_a and $\delta^{13}C$ of both leaf soluble sugars and the leaf insoluble fraction. This similarity in the isotope composition of different carbon pools (i.e., soluble sugars and cellulose) was explained by a remarkably constant p_i/p_a over time even at the highest salinity concentration.

Carbon isotope discrimination in leaf soluble sugars of water-stressed sunflower plants showed large variations during development of water stress. Figure 6 shows the relationship between the pooled data of Δ in soluble sugars and p_i/p_a ($r = 0.84$) measured on Days 37, 44, 51, and 62 after the beginning of the stress treatment. Both p_i/p_a and Δ in soluble sugars of water-stressed plants showed a dramatic decline during ontogeny, with decreasing water availability, ranging from a mean Δ value of 21.5×10^{-3} on Day 37 to a value as low as 18.2×10^{-3} on Day 62. In contrast, in well-watered plants Δ in soluble sugars showed much smaller

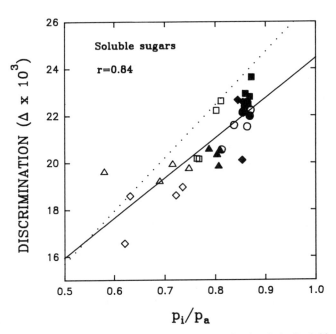

Figure 6. Relationship between carbon isotope discrimination in leaf soluble sugars and p_i/p_a in sunflower plants grown under different water regimes. Carbon isotope discrimination and spot gas exchange measurements were taken on 4 days (37, 44, 51, and 62 days from the beginning of the treatment) at different stages of drought. Each data point is the mean of three individual measurements. No significant differences were observed among genotypes. The dotted line represents the theoretical relationship assuming $\Delta = 4.4 + 22.6\, p_i/p_a$. The solid line is the regression equation for the observed data. Closed symbols, control; open symbols, water stress. Symbols: Day 37 (○, ●), Day 44 (□, ■), Day 51 (△, ▲), Day 62 (◇, ◆).

variation (Fig. 6) likely caused by slight uncontrolled fluctuation of soil water content. These results clearly demonstrate that the analysis of Δ in leaf soluble sugars gives an estimate of p_i/p_a and, therefore, of photosynthetic water-use efficiency, on a time scale of about 1 day.

IV. Summary

Carbon isotope discrimination in whole-plant dry matter appears to be a reliable indicator of plant water-use efficiency in pot-grown *H. annuus* plants. A negative relationship was obtained between plant water-use efficiency and carbon isotope discrimination in structural carbon, both in well-watered and drought conditions, according to theoretical considerations and previous experimental evidence. The drought-induced decline of photosynthesis rate was mainly caused by partial stomatal closure. Consequently, the photosynthetic water-use efficiency in stress plants was enhanced. Carbon isotope discrimination of leaf soluble sugars was always positively correlated with the ratio of intercellular to atmospheric partial pressures of CO_2, on a time scale of about 1 day. The analysis of carbon isotope discrimination in leaf soluble sugars also indicated that the intercellular CO_2 partial pressure declined in water-stressed plants, with the development of drought conditions, causing a consequent increase in photosynthetic water-use efficiency.

Carbon isotope discrimination in leaf soluble sugars is providing new insight into the seasonal variation in water-use efficiency, in response to variation in water availability and fluctuations of other environmental parametes. This appears to be of primary importance for studying short-term water stress which can cause a strong effect on the isotopic signature of leaf soluble sugars and only small influence, if any, on that of whole-plant dry matter. The analysis of Δ in leaf soluble sugars may be used as a convenient substitute for field gas exchange measurements in applications such as ecophysiological studies and breeding programs for selecting genotypes with improved water-use efficiency. This method presents several advantages compared to gas exchange techniques: it is relatively easy to collect samples in the field compared to the time-consuming technical work needed for gas exchange measurements. Furthermore, the estimate of p_i/p_a from Δ is an assimilation-weighted value, while that obtained from gas exchange is a conductance-weighted value (Farquhar, 1989; Farquhar *et al.*, 1989). The assimilation-weighted value of p_i/p_a obtained from carbon isotope discrimination is closer to the capacity-weighted value which is needed for modeling photosynthesis. Therefore, the possible occurrence of spatial heterogeneity of stomatal aperture will have a greater effect on p_i/p_a calculated from gas exchange than on that obtained from carbon isotope discrimination.

Acknowledgments

The authors thank M. C. Guido, A. Battistelli, and M. Guiducci for useful discussions. Research supported by the National Research Council of Italy, special project RAISA, Subproject No. 2, Paper No. 346.

References

Ben, G. Y., C. B. Osmond, and T. D. Sharkey. 1987. Comparison of photosynthetic responses of *Xanthium strumarium* and *Helianthus annuus* to chronic and acute water stress in sun and shade. *Plant Physiol.* **84:** 476–482.

Berry, J. A. 1988. Studies of mechanisms affecting the fractionation of carbon isotopes in photosynthesis, pp. 82–94. *In* P. W. Rundel, J. R. Ehleringer, and K. A. Nagy (eds.), Stable Isotopes in Ecological Research. Springer-Verlag, New York.

Björkman, O., and B. Demmig. 1987. Photon yield of O_2 evolution and chlorophyll fluorescence characteristics at 77K among vascular plants of diverse origins. *Planta* **170:** 489–504.

Brugnoli, E., and O. Björkman. 1992. Growth of cotton under continuous salinity stress: Influence on allocation pattern, stomatal and non-stomatal components of photosynthesis and dissipation of excess light energy. *Planta* **187:** 335–347.

Brugnoli, E., and M. Lauteri. 1991. Effects of salinity on stomatal conductance, photosynthetic capacity, and carbon isotope discrimination of salt-tolerant (*Gossypium hirsutum* L.) and salt-sensitive (*phaseolus vulgaris* L.) C_3 non-halophytes. *Plant Physiol.* **95:** 628–635.

Brugnoli, E., K. T. Hubick, S. von Caemmerer, S. C. Wong, and G. D. Farquhar. 1988. Correlation between the carbon isotope discrimination in leaf starch and sugars of C_3 plants and the ratio of intercellular and atmospheric partial pressure of carbon dioxide. *Plant Physiol.* **88:** 1418–1424.

Cornic, G., J.-L. Le Gouallec, J. M. Briantais, and M. Hodges. 1989. Effect of dehydration and high light on photosynthesis of two C_3 plants (*Phaseolus vulgaris* L. and *Elatostema repens* (Lour.) Hallf.). *Planta* **177:** 84–90.

Cowan, J. R. 1986. Economics of carbon fixation in higher plants, pp. 133–170. *In* T. J. Givnish (ed.), On the Economy of Plant Form and Function. Cambridge Univ. Press, Cambridge, UK.

Davies, W. J., and J. Zhang. 1991. Root signals and the regulation of growth and development of plants in drying soil. *Annu. Rev. Plant Physiol. Plant Mol. Biol.* **42:** 55–76.

Deléens-Provent, E., and N. Schwebel-Dugué. 1987. Demonstration of autotrophic state and establishment of full typical C_4 pathway in maize seedlings by photosynthate carbon isotope composition. *Plant Physiol. Biochem.* **25:** 567–572.

Evans, J. R., T. D. Sharkey, J. A. Berry, and G. D. Farquhar. 1986. Carbon isotope discrimination measured concurrently with gas exchange to investigate CO_2 diffusion in leaves of higher plants. *Aust. J. Plant Physiol.* **13:** 281–292.

Farquhar, G. D. 1989. Models of integrated photosynthesis of cells and leaves. *Phil. Trans. R. Soc. Lond. B* **323:** 357–367.

Farquhar, G. D., M. H. O'Leary, and J. A. Berry. 1982. On the relationship between carbon isotope discrimination and the intercellular carbon dioxide concentration in leaves. *Aust. J. Plant Physiol.* **9:** 121–137.

Farquhar, G. D., K. T. Hubick, A. G. Condon, and R. A. Richards. 1988. Carbon isotope fractionation and plant water-use efficiency, pp. 21–40. *In* P. W. Rundel, J. R. Ehleringer, and K. A. Nagy (eds.), Stable Isotopes in Ecological Research. Springer-Verlag, New York.

Farquhar, G. D., J. R. Ehleringer, and K. T. Hubick. 1989. Carbon isotope discrimination and photosynthesis. *Annu. Rev. Plant Physiol. Plant Mol. Biol.* **40:** 503–537.

Hewitt, E. J., and T. A. Smith. 1975. Plant Mineral Nutrition. English Univ. Press, London.

Masle, J., and G. D. Farquhar. 1988. Effects of soil strength on the relation of water-use efficiency and growth to carbon isotope discrimination in wheat seedlings. *Plant Physiol.* **86:** 32–38.

O'Leary, M. H. 1981. Carbon isotope fractionation in plants. *Phytochemistry* **20:** 553–567.

O'Leary, M. H. 1988. Carbon isotopes in photosynthesis. *Bioscience* **38:** 325–336.

O'Leary, M. H., and C. B. Osmond. 1980. Diffusional contribution to carbon isotope fractionation during dark CO_2 fixation in CAM plants. *Plant Physiol.* **66:** 931–934.

Sharkey, T. D., F. Loreto, and T. Vassey. 1990. Effects of stress on photosynthesis, Vol. IV, pp. 549–556. *In* M. Baltscheffsky (ed.), Current Research in Photosynthesis. Kluwer Academic Publishers, The Netherlands.

Terashima, I. 1992. Anatomy of non-uniform leaf photosynthesis. *Photosynth. Res.* **31:** 195–212.

Terashima, I., S. C. Wong, C. B. Osmond, and G. D. Farquhar. 1988. Characterization of non-uniform photosynthesis induced by abscisic acid in leaves having different mesophyll anatomies. *Plant Cell Physiol.* **29:** 385–395.

Virgona, J. M., K. T. Hubick, H. M. Rawson, G. D. Farquhar, and R. W. Downes. 1990. Genotypic variation in transpiration efficiency, carbon-isotope discrimination and carbon allocation during early growth in sunflower. *Aust. J. Plant Physiol.* **17:** 207–214.

Vogel, J. C. 1980. Fractionation of Carbon Isotopes during Photosynthesis. Springer-Verlag, Berlin.

von Caemmerer, S., and J. R. Evans. 1991. Determination of the average partial pressure of CO_2 in chloroplasts from leaves of several C_3 plants. *Aust. J. Plant Physiol.* **18:** 287–305.

von Caemmerer, S., and G. D. Farquhar. 1981. Some relationships between the biochemistry of photosynthesis and the gas-exchange of leaves. *Planta* **153:** 376–387.

Wise, R. R., J. R. Frederick, D. M. Alm, D. M. Kramer, J. D. Hesketh, A. R. Crofts, and D. R. Ort. 1990. Investigation of the limitations to photosynthesis induced by leaf water deficit in field grown sunflower (*Helianthus annuus* L.). *Plant Cell Environ.* **13:** 923–931.

Yakir, D., C. B. Osmond, and L. Giles. 1991. Autotrophy in maize husk leaves: Evaluation using natural abundance of stable isotopes. *Plant Physiol.* **97:** 1196–1198.

8

Carbon Isotope Discrimination and the Coupling of CO$_2$ Fluxes within Forest Canopies

Mark S. J. Broadmeadow and Howard Griffiths

I. Introduction

The potential for carbon isotope discrimination measurements to integrate processes within tree and forest canopies is relevant to studies which range from plant physiology, through ecology, to metereology and climatic reconstruction. For instance, the interpretation of the carbon isotope ratio versus PDB ($\delta^{13}C$) record of tree rings has been used to estimate past global carbon fluxes (e.g., Francey and Farquhar, 1982; Stuiver and Brazuinas, 1987; Francey and Hubick, 1988; Leavitt and Long, 1991; cf. Stuiver, 1978; Peng *et al.*, 1983; Tans *et al.*, 1990). However, we need to clarify the regulation of initial fractionation processes by environmental factors and source CO$_2$ (Francey *et al.*, 1985; Schleser and Jayasekera, 1985; Medina *et al.*, 1986, 1991; Sternberg, 1989; Sternberg *et al.*, 1989a). There are problems associated with these approaches: the climatologist seeks to use the integrated carbon isotope composition as an annual record of environmental conditions, but should also account for the inherent inter- and intraspecific variation within natural vegetation. Meanwhile the plant physiologist tries too often to scale up measurements from individual days under natural conditions to predict seasonal variations which reflect the long-term coupling between canopy CO$_2$ and H$_2$O fluxes and the atmosphere.

The theoretical background to the relationships between carbon isotope discrimination, the ratio of internal:external partial pressures of CO$_2$ (p_i/p_a), and plant water-use efficiency have been given by Farquhar *et al.* in this volume (Chapter 5, see also Farquhar *et al.*, 1989), together with the potential applications for the evaluation of crop genotypes. However, predictions derived from more homogeneous (and genetically uniform) crop canopies are difficult to translate to the heterogeneous environment found within the canopy of an individual tree or forest canopy. In particular, the carbon isotope composition of plant material (δ_p) may be regulated in part by light limitation of photosynthesis (via p_i/p_a) and in part by the refixation of respiratory CO_2 already depleted in $^{13}CO_2$ with respect to ambient air. A number of studies have shown that carbon isotope discrimination (Δ) in organic material is greater (more negative $\delta^{13}C$ values) low in the canopy (Vogel, 1978; Medina and Minchin, 1980; Medina et al., 1986, 1991; van der Merwe and Medina, 1989; Francey et al., 1985; Schleser and Jayasekera, 1985; Schleser, 1990; Sternberg et al., 1989a; Broadmeadow et al., 1992; Garten and Taylor, 1992).

The proportional input of respiratory CO_2 from soil and decaying leaf litter is likely to vary between temperate and tropical as well as broadleaf and deciduous canopies, although any stratification of CO_2 partial pressure (p_a) and $\delta^{13}C$ (δ_a) are dependent on climatic conditions. The daily and seasonal variations in CO_2 partial pressure within and above forest canopies have been studied with a view to understanding the contribution of tropical forest canopies to global carbon budgets (e.g., Lemon et al., 1970; Allen et al., 1972; Aoki et al., 1975; Wofsy et al., 1988). However, measurements of the $\delta^{13}C$ of source CO_2 (δ_a) have until now been limited by the technical difficulties of collecting sufficient CO_2 in remote locations for isotopic analysis. Following the pioneering work of Keeling (1958, 1961), the comparison between source CO_2 partial pressure and $\delta^{13}C$ within forest canopies has been made with a view to estimating the relative contributions of respiratory CO_2 and variations in p_i/p_a to plant $\delta^{13}C$ (Vogel, 1978; Francey et al., 1985; van der Merwe and Medina, 1989; Sternberg et al., 1989a; Broadmeadow et al., 1992).

Sternberg (1989) used such measurements to model the steady-state gas exchange fluxes for a seasonal rainforest in Panama and showed that 7 to 8% of the respired CO_2 output was reassimilated. In this chapter, we compare published data for gradients of $\delta^{13}C$ and source CO_2 within forest canopies and calculate the proportion of respiratory recycling. We then compare gradients for tropical and temperate canopies and finally use on-line, instantaneous fractionation to demonstrate the potential relationship between discrimination and measured or calculated variations in p_i/p_a for *Picea abies*.

II. Carbon Isotope Composition of Organic Material and Source CO$_2$ within Forest Canopies

The changes in $\delta^{13}C$ found within various tropical and temperate forests have been compiled in Table I, showing a consistent trend toward more negative δ_p values low in the canopy. The magnitude of this depletion was similar for tropical and temperate forests ($-3.6\%o$), although a similar value was found for plant material from canopies of differing exposure ($-3.8\%o$; Ehleringer et al., 1986) and within an isolated lime (*Tilia*) tree

Table I   Changes in Carbon Isotope Composition (δ_p) within Tropical and Temperate Forest Formations

Forest canopy	Canopy height (m)	δ_p (‰)	δ_p gradient (‰ m⁻¹)	Absolute δ_p difference (‰)	Reference
Tropical Formations					
Montane seasonal rainforest,	25	−29.6	0.142	−3.4	Medina et al.,
Puerto Rico	1	−33.0			1991
Mahogany plantation,	25	−30.1	0.154	−3.7	
Puerto Rico	1	−33.8			
Monsoon evergreen	24	−28.6	0.109	−2.4	Ehleringer
broadleaf forest, China	2	−31.0			et al., 1986
(gradients within one	17	−30.1	0.320	−4.8	
canopy)	2	−34.9			
(differing canopy expo-	Open	−27.1		−2.2	
sure)	Mid	−29.3		−1.6	
	Closed	−30.9			
Temperate formations					
Beech (*Fagus*), Germany	20	−28.4	0.157	−3.0	Vogel, 1978
	1	−31.4			
Beech (*Fagus*), Germany	26	−26.8	0.217	−5.0	Schleser and
	3	−31.8			Jayasekera,
					1985
Beech (*Fagus*), Germany	24	−25.2	0.193	−3.1	
	8	−28.3			
Lime (*Tilia*) isolated tree,	15	−25.2	0.164	−2.3	
Germany	1	−27.5		−2.4	
Beech (*Fagus*) Germany	24	−28.6	0.109		Schleser, 1990
	2	−31.0			
	17	−30.1	0.320	−4.8	
	2	−34.9			
Mixed Deciduous Canopy	21	−27.8	0.243	−1.7	Garten and
(*Acer/Quercus*), USA	14	−29.5			Taylor, 1992

(−2.3‰; Schleser and Jayasekera, 1985). The overall δ_p for the tropical forests was generally more negative, with a minimum average value of −33.2‰ low in the canopy as compared to −31.5‰ for temperate canopies (Table I).

The CO₂ partial pressure within the bulk atmosphere (p_a') is steadily increasing as a result of anthropogenic activity, and the carbon isotope composition of atmospheric CO₂ (δ_a') is becoming progressively depleted in ¹³CO₂ (Keeling et al., 1979). Such a relationship also holds within a forest canopy, where local enrichments in p_a as a result of respiratory CO₂ release were shown to be inversely related to δ_a by Keeling (1958, 1961). Using a plot of δ_a against $1/p_a$, we have compiled data from Keeling (1958, 1961) as Fig. 1 and show that for a large number of CO₂ samples collected from a range of habitats, the following relationship holds: $\delta_a = -23.19 + 5035.5/p_a$ ($r^2 = 0.937$). We highlight these data because these papers showed

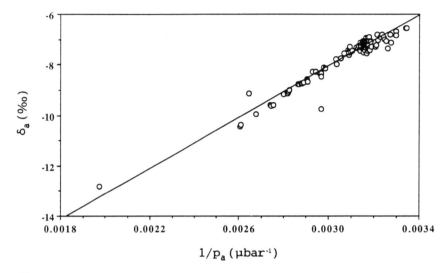

Figure 1.   The relationship between source CO_2 carbon isotope composition (δ_a, ‰) and concentration ($1/p_a$, μbar^{-1}) for air samples collected from a variety of habitats, using data adapted from Table 1 of Keeling (1958) and from Tables 1, 2, 3, and 4 of Keeling (1961).

clearly that there may be diurnal variations in p_a and δ_a within forest canopies: the implications that steady-state conditions of these parameters rarely apply throughout the day have still to be addressed in terms of the effects on δ_p. However, a number of studies have demonstrated the potential stratification in p_a which occurs within tropical forest canopies (Lemon *et al.*, 1970; Allen *et al.*, 1972; Aoki *et al.*, 1975; Wofsy *et al.*, 1988), with p_a at ground level of the Amazonian rainforest often approaching 500 μbar at dawn and dusk (Wofsy *et al.*, 1988).

There have been three sets of studies which have compared δ_p with δ_a and p_a within rainforest canopies for Amazonia (Medina and Minchin, 1980; Medina *et al.*, 1986; van der Merwe and Medina, 1989) and Panama (Sternberg *et al.*, 1989a; Sternberg, 1989), as well as for a Huon pine rainforest in Tasmania (Francey *et al.*, 1985). The changes in these variables have been collated in Table II, where it is apparent that the absolute differences in δ_p associated with neotropical broadleaf rainforests were much greater than that found in the coniferous forest (-4.9 as compared to -1.1‰). From the relationship shown in Fig. 1, the variations in source CO_2 isotope composition of the Amazonian forest (δ_a ranging from -13 to -14.5‰) suggest that p_a is elevated above ambient levels throughout the canopy. However, this reflects the timing of CO_2 collections (dawn and dusk; van der Merwe and Medina, 1989), at the times of maximum respiratory enrichment (Wofsy *et al.*, 1988).

Although Sternberg (1989) pointed out that it is wrong to consider forest canopies to be closed systems, an idea of the proportional contribution of respiratory CO_2 can be obtained from the $\delta^{13}C$ of forest air (δ_a) in relation

Table II Changes in Carbon Isotope Composition of Plant Material (δ_p), Source CO_2 (δ_a), and CO_2 Partial Pressure (p_a) and the Proportion of Respiratory Flux Reassimilated within Forest Canopies

Forest canopy	Canopy height (m)	δ_p (‰)	δ_p diff. (‰)	δ_a (‰)	Resp. CO_2 (%)	p_a (μbar)	Resp. CO_2 $\delta^{13}C$ (litter or canopy) (δ_a, ‰)	Proportion of resp. flux reassim. (%)	Reference
Amazonian rainforest, San Carlos de Rio Negro (Laterite)	30	−28.7		−13.0	22				Medina and Minchin, 1980; Medina et al., 1986; van der Merwe and Medina, 1989
	20	−34.3		−13.7	25				
	5–10	−35.2	−6.5	−13.8	25				
	1	−31.0		−14.5	28	541	−31.0	26	
	litter	−30.5		−15.5	33				
(Podsol)	20	−33.4							
	2–10	−35.4	−4.9						
	1			−14.4	27	508	−29.4	15	
	litter	−29.4							
Panamanian lowland seasonal rainforest	25	−29.6		−8.9	5	349			Sternberg et al., 1989a; Sternberg, 1989
	1	−33.0	−3.4	−10.6	14	375			
	0.5			−11.4	18	390	−30	8	
Tasmanian rainforest (Huon pine)	14.5	−29.2		−7.53		337			Francey et al., 1985
	7	−29.8		−7.57	0.2	338			
	5			−7.80	1	343			
Shaded canopy	1	−31.0	−1.8	−8.37	4	356	−28.2	23	
	13	−28.4							
	7	−28.9							
Semi-exposed canopy	3.5	−29.3	−0.8						
	1	−29.2							
	17	−27.6							
Exposed canopy	14.5	−27.4							
	3	−27.2	−0.7						
	1	−28.3							

to the isotope composition of bulk air ($\delta_{a'}$). For the Panamanian canopy, it was suggested that up to 18% of the CO_2 in air was respiratory in origin, whereas a much smaller proportion was found in the Huon pine canopy (column 5, Table II; Sternberg *et al.* 1989a, Francey *et al.* 1985). Similar calculations for the Amazonian forest show that at dawn, respiratory CO_2 may contribute up to 30%.

In order to account for the fluxes of ambient and respiratory CO_2 in and out of the forest canopy, Sternberg (1989) developed a model utilizing the data for the Panama seasonal rainforest, where stratification of δ_a and p_a was constant within the canopy throughout the day during both wet and dry seasons. By modeling individual CO_2 fluxes, and incorporating terms accounting for the proportion of respired CO_2 reassimilated and for the discrimination against $^{13}CO_2$ associated with photosynthetic CO_2 uptake, the relationship was derived

$$\delta_a = \frac{p_{a'}(\delta_{a'} - \delta_r)(1 - R)}{p_a} + \delta_r + R\Delta,$$

where $p_{a'}$ and $\delta_{a'}$ are the partial pressure and isotope ratio of ambient atmospheric CO_2 (bulk air above the canopy), Δ is the discrimination expressed during photosynthetic CO_2 uptake, δ_r is the isotope ratio of respired CO_2, and R is the proportion of floor-respired CO_2 that is reassimilated. Simple models have in the past assumed that the y-axis intercept is equivalent to δ_r, but this does not take into account the fact that a proportion of the respired CO_2 is reassimilated (R) and that the isotopic signature of this CO_2 will be tempered by Δ. R may thus be estimated from the y-axis intercept of the relationship. However, this estimate of R requires an integrated value for Δ throughout the canopy. Additionally, R may be estimated from the slope of the relationship, with this value independent of Δ. Since canopy Δ is difficult to measure, this chapter concentrates on R obtained from the slope of the relationship between δ_a and $1/p_a$.

In simplistic terms, from measurements of p_a within the canopy compared to values of $\delta_{a'}$ and $p_{a'}$ for bulk air (e.g., $-8\%_0$ and 330 μbar; Sternberg, 1989), we can calculate what the gradient of the relationship between δ_a and $1/p_a$ would be if there was no recycling of respiratory CO_2. The difference between this and the observed gradient indicates the degree to which floor-respired CO_2 is recycled. We have used this approach to analyze the data of Medina and co-workers and Francey *et al.* (1985) for Table II (column 8). We have to assume that the δ_a of respiratory CO_2 is identical to that of the upper canopy, which will contribute most biomass to the litter layer (Sternberg, 1989).

As an example, we use the data of Francey *et al.* (1985) to demonstrate this calculation. They showed that the plot of δ_a versus $1/p_a$ for the measured CO_2 samples within the canopy was $\delta_a = -23.4 + 5349/p_a$. Bulk air $\delta_{a'}$ and $p_{a'}$ were $-7.53\%_0$ and 337 μbar, respectively, with forest air p_a measured as 356 μbar at a height of 1 m within the canopy. If we assume respiratory CO_2 $\delta^{13}C$ to be equivalent to $-28.2\%_0$ (average δ_p for the canopy), the theoretical δ_a (for $p_a = 356$ μbar) when there is no refixation of respiratory CO_2 may be derived from simple proportionality as $-8.63\permil$. Using the above values for δ_a, p_a, $\delta_{a'}$, and $p_{a'}$, the following equation represents the theoretical relationship between δ_a and $1/p_a$ at zero recycling:

$$\delta_a = 6946(1/p_a) - 28.2.$$

Note that at zero recycling, the y-axis intercept is equal to δ_a. As shown by Sternberg (1989), the proportion of the respiratory flux recycled through photosynthesis (R) is as follows:

$$\text{actual slope} = \text{theoretical slope}(1 - R).$$

This treatment yields a value for R of 23% from the data of Francey *et al.* (1985), and thus a much higher proportion of the respiratory flux is refixed than in the Panamanian rainforest canopy. However, respiratory fluxes are likely to be much lower from coniferous litter. From the data of Medina and co-workers, we can calculate that 26 and 15% of the respiratory flux is reassimilated for the laterite and podsol Amazonian forest formations, respectively (Table II). It is important to note that we do not have the actual carbon isotope composition of respiratory CO_2 measured in any of these studies, and variations in this parameter would markedly affect the calculations outlined above.

These calculations suggest that in a range of forest formations, there is significant refixation of respiratory CO_2. However, before scaling up these proportions to provide an estimate of annual respiratory fluxes, the measured gradients in δ_a and p_a must be constant on a daily and seasonal basis, as reported by Sternberg *et al.* (1989a). In the case of Francey *et al.* (1985) the measured gas exchange rates were maximal in the morning, although CO_2 was collected throughout the day. The studies in the Amazonian forests only collected CO_2 at dawn and dusk although δ_a and p_a may vary dynamically throughout the day (Keeling, 1958; Wofsy *et al.*, 1988). Gradients of source CO_2 may also be dissipated by turbulence in response to climatic conditions or by the changing photosynthetic activity within the canopy. We now go on to consider examples where stratification of CO_2 has been shown to be dependent on canopy formation and where the proportion of respiratory flux refixed may not be constant throughout the day.

III. Carbon Isotope Discrimination within Tropical Forest Formations in Trinidad

A. Variations in δ_p with Height

A comparison of the changes in δ_p with height for two contrasting forest formations in Trinidad is shown in Fig. 2, using data from Broadmeadow *et al.* (1992) collected during the dry season in 1990. The forest formation at Simla corresponded to "deciduous seasonal forest" forming a secondary

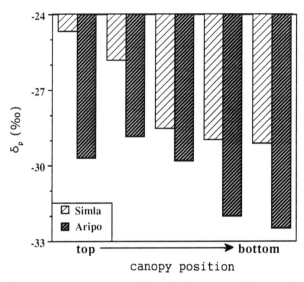

Figure 2.   Mean carbon isotope composition of plant material (δ_p ‰) of between two and six species collected at various heights above ground level (20, 14, 8, 1, and 0.15 m) in two tropical forest canopies in Trinidad. Light hatching, the open canopy at Simla; dark hatching, the closed, moist Aripo habitat. Data replotted from Fig. 1 of Broadmeadow *et al.* (1992).

forest canopy in the Arima Valley, Trinidad. The upper canopy shed leaves during the dry season and the free-draining soil overlay a limestone outcrop. In contrast, the canopy at Aripo, 7.5 km to the east, was in a moist ravine close to a river with species composition that of a nondeciduous "semi-evergreen seasonal forest." Both canopies showed more negative δ_p values low in the canopy, with a difference of −4.6‰ at Simla (−24.6‰ at 20 m to −29.2‰ at 0.15 m) and −2.8‰ at Aripo (−29.7‰ to −32.5‰), corresponding to an overall difference of 4‰ between the two canopies, with discrimination generally greater at Aripo.

B. Diurnal Variations in Source CO_2

Source CO_2 was sampled as described by Broadmeadow *et al.* (1992), with an air intake attached to a rope-pulley system within each canopy enabling rapid and repeated measurement of p_a and collection of CO_2 for subsequent purification (including passing over reduced copper at 600°C to remove N_2O, and the exclusion of water using an acetone coldtrap at −90°C) and isotopic analysis. Sampling was carried out on still days so as to maximize any gradients, and data for the diurnal variations in p_a and δ_a for the Aripo canopy are shown in Figs. 3a and 3b. At dawn, a difference of 34 μbar (377 to 411 μbar) was found between 18 and 0.15 m, which declined to 24 μbar at midday, although stratification was marked throughout the day (Fig. 3a). Analysis of carbon isotope composition at these times showed similar stratification, with δ_a approaching the ambient value of −7.8‰ only

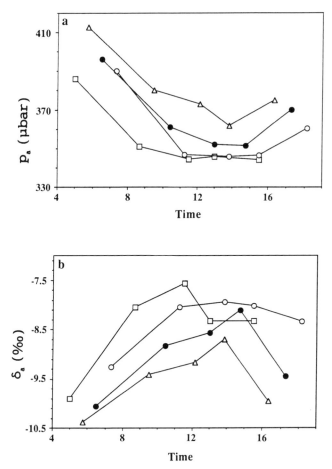

Figure 3. (a) Diurnal course of CO_2 concentration (p_a, μbar) and (b) carbon isotope composition (δ_a, ‰) within the canopy at Aripo collected during March 22, 1990, at 16 m (open triangles), 8 m (open circles), 1 m (closed circles), and 0.15 m (open squares) above ground level. Data replotted from Fig. 2 of Broadmeadow *et al.* (1992).

at midday high in the canopy (Fig. 3b). CO_2 was consistently more depleted in $^{13}CO_2$ at 0.15 m. The dynamic changes in δ_a and p_a indicate that steady-state conditions were not attained within the canopy during the course of the day. A similar situation was found in the more open canopy at Simla, although the gradients in δ_a and p_a were not so marked (Broadmeadow *et al.*, 1992).

C. Proportion of Respiratory Flux Recycled through Photosynthesis

The plot of δ_a versus $1/p_a$ for CO_2 sampled during the course of 1 day for each canopy is shown in Fig. 4. The regression for the data from both sites was $\delta_a = -22.87 + 5073/p_a$ ($r^2 = 0.82$). The distinct nature of the two

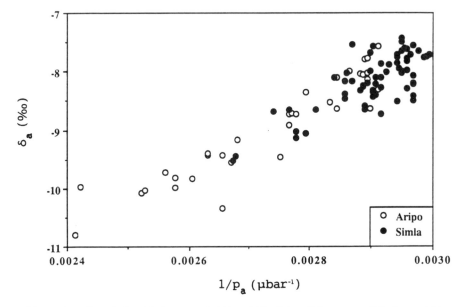

Figure 4.   The relationship between δ_a (‰) and $1/p_a$ (μbar^{-1}) for source CO_2 from within the two canopies during the course of 1 day from 04.00 to 18.00 h in the dry season. Simla (closed circles), CO_2 collected at 20, 14, 8, 1, and 0.15 m above ground level on March 14, 1990; Aripo (open circles), CO_2 collected at 16, 8, 1, and 0.15 m on March 22, 1990. Data replotted from Fig. 3 of Broadmeadow et al. (1992).

canopies was demonstrated by the clusters of data points, and samples from the well-mixed Simla canopy were much closer to ambient values of δ_a and p_a (i.e., $\delta_{a'}$ and $p_{a'}$; Fig. 4).

The proportion of floor-respired CO_2 that is reassimilated may be calculated for the two sites. Assuming bulk air values of -7.8‰ and 334 μbar for $\delta_{a'}$ and $p_{a'}$, and values of -30 and -25.8‰ for δ_r for Aripo and Simla, the slopes of δ_a versus $1/p_a$ corresponding to zero recycling are 7448 and 5892, respectively. From the relationships between δ_a and $1/p_a$ obtained for the two canopies:

$$\delta_a = -24.50 + 5656/p_a \quad \text{(Simla)}$$

$$\delta_a = -24.74 + 6182/p_a \quad \text{(Aripo)}.$$

The slope of the relationship was significantly different for the two sites (comparison of slopes, $P < 0.05$), and R is estimated as 17% for Aripo and 4% for the more open canopy at Simla. However, the model of Sternberg assumes that CO_2 fluxes remain constant throughout the day. This was not the case (Fig. 3), and if the day is divided into three portions (06.00 to 10.00 h, 10.00 to 14.00 h, and 14.00 to 18.00 h), with R recalculated, values of 23, 19, and 8% are obtained.

Analyses of the relationship between carbon isotope composition and partial pressure of source CO_2 therefore provide a useful means of comparing the characteristics of contrasting forest formations. The variations in the proportion of respiratory refixation was probably related to both edaphic factors (higher respiration rates from the moist litter at Aripo) and to canopy structure (more turbulent mixing in the open canopy at Simla). In addition, although p_i/p_a was more likely to be affected by light limitation in the denser canopy at Aripo, plant-water status is likely to have been more limiting at the freely draining Simla site. However, until measurements of δ_a and p_a are made throughout wet and dry seasons, we will not be able to provide a definitive explanation for differences in respiratory fluxes in such canopies.

IV. Carbon Isotope Discrimination within a Temperate Coniferous Canopy

A. Organic Material δ_p and Variations in Source CO_2

Discrimination was lower for temperate and coniferous canopies (Table I), and therefore we have compared gradients within a canopy and for free-standing individuals of Norway spruce (*P. abies*) in a forested area near to Newcastle upon Tyne, United Kingdom. The canopy site was planted in 1961 and had been neither brashed or thinned, leading to a dense canopy structure, with leaf material extending from a height of 13 to 5 m. The free-standing individuals were found in an area of open heath nearby, where the soil was more free draining. Although these trees were younger, photosynthetically active needles extended over a height range similar to that of the closed canopy, from 9 to 1 m above ground level.

Samples of leaf material were collected from evenly spaced points between highest and lowest accessible needles. Both canopy and free-standing individuals showed an absolute difference in δ_p with height of $-1.7‰$, although δ_p at all heights in the canopy was consistently $1‰$ more negative than that for the isolated trees (Fig. 5). In order to investigate these variations, gradients of source CO_2, environmental variables, and leaf components were investigated at each site. No gradient in source CO_2 δ_a was evident adjacent to the isolated trees, and, within the canopy, values of δ_a were constant at $-7.8‰$ for the photosynthetically active portion of the canopy, with no gradient found between 14 and 7 m, although at 1 m, source CO_2 δ_a was $-8.3‰$ (data not shown).

With the development of more sophisticated CO_2 collection techniques (Broadmeadow *et al.*, 1992), a more detailed analysis of CO_2 gradients was made over 2 days in July, 1990, with the combined data shown in Fig. 6. At a height of 0.15 m within the canopy, p_a declined from 419 μbar at dawn to 347 μbar at midday, corresponding to δ_a changing by $2‰$. At the top of the canopy, p_a was 391 μbar at dawn and declined to 330 μbar by late afternoon, with no gradient in δ_a evident in the upper canopy positions. A difference of $0.5‰$ was again found at midday between the upper canopy

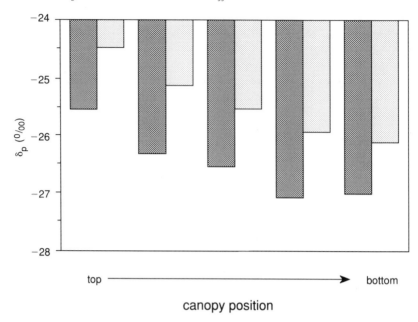

Figure 5.   Mean carbon isotope composition of plant material (δ_p, ‰) of *Picea abies* collected at various heights above ground throughout a dense canopy (hatched bars: 13, 11, 9, 7, and 5 m) and for isolated individuals (open bars: 9, 7, 5, 3, and 1 m).

and 0.15 m. A single regression through these data produced an intercept of −23.6 and slope of 5229 ($r^2 = 0.77$), corresponding to a calculated value for R of 17%. If the data in Fig. 6 are treated in a similar manner to those from the canopy at Aripo, with R calculated for three separate periods through the course of the day, variation in R is again observed. Using the model of Sternberg, R is calculated as 4% (03.00–09.00 h), 20% (09.00–18.00 h), and 30% (18.00–22.00 h) for the early morning, the middle of the day, and the evening periods, respectively (Broadmeadow, 1992). This again indicates the need to measure gas exchange rates throughout the diurnal course when steady-state conditions do not pertain, in order to give an integrated daily value for R. The low proportion of respiratory refixation during the early morning period may correspond to turbulence rapidly dissipating the gradients in δ_a and p_a which had accumulated overnight. Variation was also observed in the y-axis intercept, which is likely to reflect differences in Δ, rather than δ_r.

The small variations in source CO_2 do not explain the overall gradient in δ_p found for the isolated trees, or for the canopy where stratification was minimal in the photosynthetically active regions. Changes in structural components (lipid and lignin) were also insufficient to account for the gradients in δ_p. we must therefore identify those environmental variables which are most likely to affect p_i/p_a, and hence Δ, at both sites. Stomatal

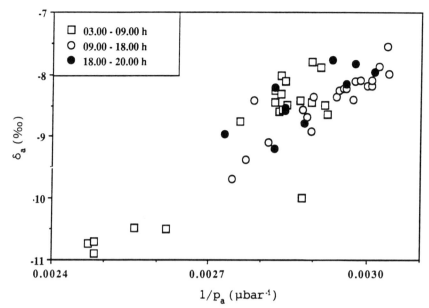

Figure 6. The relationship between δ_a (‰) and $1/p_a$ (μbar^{-1}) for source CO_2 from within the *Picea abies* canopy collected throughout the day on July 13, 1990.

conductance decreased with increasing height for both isolated trees and those within the canopy. However, photosynthetic photon flux density (PPFD) was also reduced low in the canopy, and during the summer declined from 60% transmission in the upper canopy to 1% 4.5 m above the ground and varied depending on aspect for the free-standing trees (Broadmeadow, 1992). As a result of the reduction in PPFD, leaves within the canopy at 8 and 5 m were close to the light compensation point.

Gradients in air temperature (0.58°C m^{-1}) and air vapor pressure (0.005 kPa m^{-1}) were also found within the closed canopy. There were differences in water supply as indicated by measurement of xylem sap pressure with a pressure chamber. These resulted in diurnal changes in water potential from −1.0 (dawn) to −1.7 MPa (midday) at all levels in the free-standing tree. In contrast, xylem sap pressure was stratified within the canopy throughout the day, declining from −0.6 (dawn) to −1.7 MPa (midday) at the top of the canopy while remaining constant around −0.6 to −0.9 MPa at 5 m above ground (Broadmeadow, 1992).

However, these measurements were made on single days under similar still, bright, weather conditions. Photosynthetic CO_2 assimilation may occur throughout the year in a coniferous tree such as *P. abies,* and with turbulent, windy conditions more normally associated with this habitat, we need to make detailed measurements of stomatal conductance throughout the year at each height to determine integrated values of p_i/p_a. The integration provided by carbon isotope composition suggests that PPFD is less limiting for the isolated trees, although differences in water supply (for the tropical canopies, see Section III,C, above) could be significant and should be routinely considered as part of future studies.

B. Instantaneous Fractionation in Response to Variations in Light and CO_2 Regimes

There are inherent difficulties in measuring environmental variables and gas exchange characteristics within forest canopies throughout the year as a means to obtain integrated values of p_i/p_a and explain variations in δ_p. One possible approach is to measure the extent of Δ in the laboratory under controlled environmental conditions. The on-line discrimination technique developed by Evans *et al.* (1986) provides an instantaneous measure of Δ and simultaneous measurement of gas exchange characteristics. However, in view of the observed different patterns of δ_p in juvenile trees (Francey *et al.*, 1985), and with differences in performance in the laboratory as compared to natural conditions, such approaches should be viewed with caution. In order to investigate these variations, and also gain some idea of the relationship between p_i/p_a measured directly and that calculated from Δ (Farquhar *et al.*, 1989), we have carried out a series of measurements of on-line, instantaneous discrimination on young, clonal *P. abies* trees under a range of physiological conditions.

A comparison of light responses was made between plants under "winter" and "summer" conditions, with plants maintained prior to experimentation in an unventilated greenhouse. The instantaneous discrimination measurements were carried out with plants acclimated for several days in a controlled environment chamber (under a 12-h photoperiod, 20/15°C day/night temperature, and PPFD of 400 μmol m^{-2} s^{-1}), using the technique described by Griffiths *et al.* (1990). The data are presented in Fig. 7 as direct measurements of instantaneous Δ in response to increasing PPFD, together with values of Δ calculated from p_i/p_a using the relationship (Farquhar *et al.*, 1989),

$$\Delta = a + (b - a)\frac{p_i}{p_a},$$

where a (fractionation associated with diffusion of CO_2 in air) and b (fractionation expressed during C_3 photosynthesis) correspond to the fractionation against $^{13}CO_2$ associated with diffusion (4.4‰) and assimilation by ribulose bisphosphate carboxylase (RuBisCO) (27‰), including the contribution from phosphoenolpyruvate (PEP)-carboxylase.

Under summer water-stressed conditions (Fig. 7a), there was little change in stomatal conductance as PPFD was increased from 40 to 600 μmol m^{-2} s^{-1}, p_i/p_a declined from 0.9 to 0.7, and the measured instantaneous Δ corresponded well with that predicted from p_i/p_a (Fig. 7a). This contrasted markedly with the performance of the plant acclimated to winter conditions, where stomatal conductance increased from 50 to 200 mmol m^{-2} s^{-1} although p_i/p_a only decreased slightly. As a result, Δ calculated from p_i/p_a declined slightly with PPFD, although measured Δ decreased from

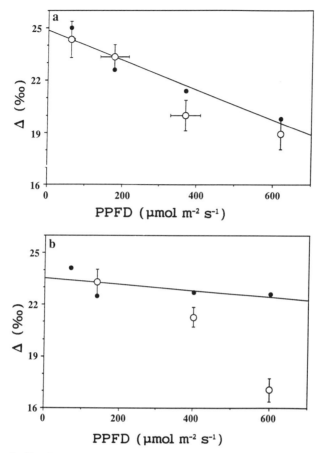

Figure 7. On-line, instantaneous discrimination (Δ, ‰) in shoots of *Picea abies* in response to photosynthetic photon flux density (PPFD, μmol photon m^{-2} s^{-1}) under contrasting environmental conditions. Open circles, measured Δ; closed circles and fitted line, Δ calculated from measured p_i/p_a. (a) Measurements on plants under stressed, summer conditions; (b) winter conditions.

22.3‰ (140 μmol m^{-2} s^{-1}) to 17.1‰ (600 μmol m^{-2} s^{-1}; see Fig. 7b). For both treatments, Δ was inversely related to instantaneous water-use efficiency.

A comparison of CO₂ responses was made for two different periods of acclimation to each CO₂ partial pressure. In the first study, plant material was acclimated for only 30 min. Similar to the summer PPFD response there was little change in stomatal conductance, with the resultant changes in p_i/p_a leading to a slight increase in calculated Δ, although measured Δ increased markedly with CO₂ partial pressure (Fig. 8a). However, following 3 h acclimation to each CO₂ level, p_i/p_a remained constant (around 0.82), and calculated and measured Δ showed a similar slight increase, but with lower measured values of Δ (Fig. 8b).

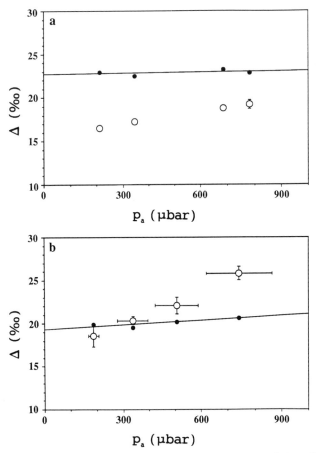

Figure 8. On-line, instantaneous discrimination (Δ, ‰) in shoots of *Picea abies* in response to CO_2 concentration (p_a, μbar). Open circles, measured Δ; closed circles and fitted line, Δ calculated from measured p_i/p_a. Measurements of instantaneous Δ and gas exchange after either (a) long-term (120 min) equilibration to each CO_2 concentration or (b) short-term equilibration.

Measured instantaneous Δ was always consistently related to p_i/p_a, irrespective of experimental conditions. However, the magnitude of the changes in measured Δ was often displaced from those calculated using p_i/p_a. While the causes of these differences remain to be resolved, it may be that respiration of photosynthetic and nonphotosynthetic material is affecting the instantaneous discrimination signal or that mesophyll conductances are varying (von Caemmerer and Evans, 1991). Granted that there may be considerable variations in the responses of plants under differing physiological conditions, these measurements may be used to give some indication of the range of carbon isotope discrimination responses which may occur under natural conditions.

V. Variations in δ_p and δ_a: Implications for the Coupling of CO_2 and H_2O Fluxes within Canopies

In general terms, discrimination is some 3–5‰ greater in the lower reaches of broadleaf canopies, whether tropical or temperate, although Δ is larger in tropical canopies (i.e., δ_p is more negative throughout; Tables I and II). The difference in δ_p with height is less in coniferous canopies, although isolated or exposed trees show smaller Δ whether coniferous or broad-leaved (Tables I and II, Figs 2 and 5; Francey *et al.*, 1985; Schleser and Jayasekera, 1985; Ehleringer *et al.*, 1986). There may also be seasonal variations in Δ (Figs 7 and 8; Lowdon and Dyck, 1974). Young or old coniferous needles show a marked "juvenile effect" (Francey *et al.*, 1985; Broadmeadow, 1992), emphasizing the need for careful selection of leaf age classes in order to determine a mean value of Δ for carbon gain throughout the year.

It appears that the overall differences in δ_p with height can often be related to the regulation of p_i/p_a by gradients in environmental variables (PPFD, leaf-to-air water vapor gradient, temperature) or plant-water status, without recourse to changes in δ_a (Francey *et al.*, 1985; Marek and Pirochtova, 1990; Madhaven *et al.*, 1991). We can identify the contribution that any particular environmental variable makes by use of instantaneous discrimination techniques. For the transition from limiting to saturating PPFD, a change of 8‰ could occur in δ_p which would more than account for the observed differences within canopies (Fig. 7; von Caemmerer and Evans, 1991). This may only in part be related to variations in p_i/p_a, and the magnitude of changes in any CO_2 transfer resistance, or possible interactions with respiration (which may alter the instantaneous discrimination signal), remain to be resolved.

This is not to suggest that respiratory CO_2 does not contribute to variations in δ_p. Stratification in p_a is well documented for forest canopies (Lemon *et al.*, 1970; Allen *et al.*, 1972; Aoki *et al.*, 1975; Wofsy *et al.*, 1988), and is inversely related to δ_a (Keeling 1958, 1961; Francey *et al.*, 1985, van der Merwe and Medina, 1989; Sternberg *et al.*, 1988a, Sternberg 1989). However, the enrichment in $^{13}CO_2$ in ambient air which occurs during photosynthesis (Evans *et al.*, 1986; Griffiths *et al.*, 1990) has not been considered in these approaches. In addition, the absolute proportion of respiratory CO_2 which is refixed or released to the atmosphere remains to be determined and is critical to our understanding of the carbon budgets of forests and variations in δ_a and p_a in the atmosphere on a global basis (Keeling *et al.*, 1979; Stuiver, 1978; Peng *et al.*, 1983; Tans *et al.*, 1990). More studies on the rate of respiratory CO_2 release from soil and litter are required (Schleser and Jayasekera, 1985, estimated this flux to be 5.7 µmol $m^{-2} s^{-1}$ for a beech forest), together with more definitive measurements of the isotope composition of respiratory CO_2. We may then quantify the amount of respiratory CO_2 refixed as a proportion of canopy photosynthesis, to complement the work of Schleser and Jayasekera (1985), who suggested an overall value of 5% for a beech canopy.

The respiratory refixation was estimated to make a 34% relative contribution to the variation in δ_p in comparison to the effect of PPFD limitation by Sternberg *et al.* (1989a), who compared δ_p of exposed and shaded bamboo grown in a glasshouse with understory material. It should be possible to regulate δ_a experimentally, by mixing ambient air with that from "tank" CO_2. This could be used to model the effect of altered δ_a and elevated p_a on δ_p low in the canopy, in response to PPFD limitation. However, there are always problems associated with juvenile effects in seedlings and young trees which may introduce errors when glasshouse results are scaled up to the canopy (Francey *et al.*, 1985; Medina and Minchin, 1980; Medina *et al.*, 1986).

Given the interactions between the effects of environmental limitation and respiratory CO_2 refixation, it has only been possible to extrapolate Δ to estimate canopy water-use efficiency where the contribution from δ_a can be excluded (e.g., Ehleringer *et al.* 1986; Read and Farquhar, 1991; Garten and Taylor, 1992). A consistent relationship between high Δ and more mesic habitats has been found, depending on location within a forest formation and annual rainfall (Fig. 2; Broadmeadow *et al.*, 1992; Garten and Taylor, 1992). In addition, Δ was also correlated with genotypic differences between *Nothofagus* spp. from low and high latitude, reflecting the occurrence of water deficits in each particular region (Read and Farquhar, 1991).

However, there is the intriguing prospect of using several of the isotopes constituting CO_2 and H_2O to integrate canopy processes (Ehleringer and Osmond, 1989; Griffiths, 1991; Farquhar and Lloyd, Chapter 5, this volume). The δD or $\delta^{18}O$ composition of leaf cellulose has been related to the gradient in humidity within a canopy (Sternberg *et al.*, 1989b), although there are problems in analyzing leaf water directly since under natural conditions isotopic equilibrium is not always attained (Flanagan and Ehleringer, 1991; Flanagan *et al.*, 1991; Flanagan, Chapter 6, this volume). However, the potential for analysis of CO_2 or water vapor directly as a means of identifying the variations in stomatal conductance within canopies requires evaluation for crops and natural vegetation.

Although there is much uncertainty surrounding the factors regulating δ_p, there are good prospects for climatologists to use data from isotopic analysis of tree rings in models of paleoenvironmental conditions or in predicting global carbon budgets. These approaches have been used by Peng *et al.* (1983), Stuiver (1978), and Tans *et al.* (1990), based on annual variations in tree ring $\delta^{13}C$. Leavitt and Long (1991) have recently suggested that the seasonal and annual variations in tree ring $\delta^{13}C$ may well correlate with plant-water status, an interesting suggestion in view of the possible relationship between differences in the water supply to the two canopies in Trinidad (Figs. 2 and 4) and for isolated and canopy-forming *P. abies* (Fig. 5). Ultimately, carbon which is deposited annually (whether in tree ring or in sediments) will provide an integrated signal representative of environmental conditions throughout a given time period. Studies of paleoclimatology are using GC-C-IRMS techniques to identify patterns in the historical record of CO_2 partial pressures and specific marker compounds (Jasper and Hayes, 1990). Physiologists should be quantifying the variations in the $\delta^{13}C$ of fractions and compounds within extant plant material as regulated by environmental conditions and provide a basis for paleoclimatic reconstruction and analysis of global carbon budgets.

VI. Summary

In conclusion, it is apparent that measured gradients in δ_a, p_a, and the calculated parameter, R, may be indicative of forest type and thus a useful tool in forest ecology. The model proposed by Sternberg to describe gradients in δ_a and p_a appears to fit the data thus far. However, the question of diurnal fluctuations in δ_a and p_a must be addressed fully in order to provide an accurate estimate for R. Additionally, the collection of respired CO_2 and measurement of δ_a will further validate the model, as alluded to in Sections I and IV, and may provide an estimate of canopy Δ.

The determination of Δ through the use of on-line discrimination techniques enables us to estimate the contribution of gradients in environmental variables (PPFD, p_a, temperature) to the gradient in δ_p and, in conjunction with the measurement of gradients in δ_a, account for the observed gradients in δ_p. With this knowledge of the factors regulating δ_p in forest canopies, we may fully utilize stable isotopes in the study of forest ecophysiology.

Acknowledgments

We are grateful for support in the form of a research studentship (GT4/87/TLS/41) and grant (GR3/7917) from NERC, UK.

References

Allen, L. J., E. Lemon, and L. Muller. 1972. Environment of a Costa Rican Forest. *Ecology* **53:** 102–111.

Aoki, M., Y. Kazutoshi, and K. Hiromichi. 1975. Micrometeorology and assessment of primary production of a tropical rainforest in West Malaysia. *J. Agric. Metereol.* **31:** 115–119.

Broadmeadow, M. S. J. 1992. Carbon Isotope Discrimination in Forest Canopies. Ph.D. thesis. University of Newcastle Upon Tyne, UK.

Broadmeadow, M. S. J., H. Griffiths, C. Maxwell, and A. M. Borland. 1992. The carbon isotope ratio of plant organic material reflects temporal and spatial variations in CO_2 within tropical forest formations in Trinidad. *Oecologia* **89:** 435–441.

Caemmerer, S. von, and J. R. Evans. 1991. Determination of the average partial pressure of CO_2 in leaves of several C₃ plants. *Aust. J. Plant Physiol.* **18:** 287–305.

Ehleringer, J. R., and C. B. Osmond. 1989. Stable Isotopes, pp. 281–299. *In* R. W. Pearcy, J. R. Ehleringer, H. A. Mooney, and P. W. Rundel (eds.), Plant Physiological Ecology, Chapman & Hall, London.

Ehleringer, J. R., Z. F. Lin, C. B. Field, and C. Y. Kuo. 1986. Leaf carbon isotope ratio and mineral composition in subtropical plants along an irradiance cline. *Oecologia* **72:** 109–114.

Evans, J. R., T. D. Sharkey, J. A. Berry, and G. D. Farquhar. 1986. Carbon isotope discrimination measured concurrently with gas exchange to investigate CO_2 diffusion in leaves of higher plants. *Aust. J. Plant Physiol.* **13:** 281–292.

Farquhar, G. D., J. R. Ehleringer, and K. T. Hubick. 1989. Carbon isotope discrimination and photosynthesis. *Annu. Rev. Plant Physiol. Plant Mol. Biol.* **40:** 503–537.

Flanagan, L. B., and J. R. Ehleringer. 1991. Stable isotope composition of stem and leaf water: Applications to the study of plant water use. *Funct. Ecol.* **5:** 270–277.

Flanagan, L. B., J. F. Bain, and J. R. Ehleringer. 1991. Stable oxygen and hydrogen isotope composition of leaf water in C_3 and C_4 plant species under natural conditions. *Oecologia* **88:** 394–400.

Francey, R. J., and G. D. Farquhar. 1982. An explanation of the $^{13}C/^{12}C$ variations in tree rings. *Nature* **297:** 28–31.

Francey, R. J., and K. T. Hubick. 1988. Tree ring carbon isotope ratios re-examined. *Nature* **333:** 712.

Francey, R. J., R. M. Gifford, T. D. Sharkey, and B. Weir. 1985. Physiological influences on carbon isotope discrimination in the huon pine (*Lagarostrobos franklinii*). *Oecologia* **44:** 241–247.

Garten, C. T., Jr., and G. E. Taylor, Jr. 1992. Foliar $\delta^{13}C$ within a temperate deciduous forest: Spatial, temporal, and species sources of variation. *Oecologia* **90:** 1–7.

Griffiths, H. 1991. Applications of stable isotope technology in physiological ecology. *Funct. Ecol.* **5:** 254–269.

Griffiths, H., M. S. J. Broadmeadow, A. M. Borland, and C. S. Hetherington. 1990. Short-term changes in carbon isotope discrimination identify transitions between C_3 and C_4 carboxylation during crassulacean acid metabolism. *Planta* **181:** 604–610.

Jasper, J. P., and J. M. Hayes. 1990. A carbon isotope record of CO_2 levels during the late Quaternary. *Nature* **347:** 462–464.

Keeling, C. D. 1958. The concentration and isotopic abundances of carbon dioxide in rural areas. *Geochim. Cosmochim. Acta* **13:** 322–334.

Keeling, C. D. 1961. The concentration and isotopic abundances of carbon dioxide in rural and marine air. *Geochim. Cosmochim. Acta* **24:** 277–298.

Keeling, C. D., W. G. Mook, and P. P. Tans. 1979. Recent trends in the $^{13}C/^{12}C$ ratio of atmospheric carbon dioxide. *Nature* **277:** 121–123.

Leavitt, S. W., and A. Long. 1991. Seasonal stable-carbon isotope variability in tree rings: Possible paleoenvironmental signals. *Chem. Geol.* (Isotope Geochemistry Section) **87:** 59–70.

Lemon, E., L. H. Allen, and L. Muller. 1970. Carbon dioxide exchange of a tropical rainforest. Part II. *BioScience* **20:** 1054–1059.

Lowden, I. A., and W. Dyck. 1974. Seasonal variation in the isotope ratios of carbon in maple leaves and other plants. *Can. J. Earth Sci.* **11:** 79–88.

Madhaven, S., I. Treichel, and M. H. O'Leary. 1991. Effects of relative humidity on carbon isotope fractionation in plants. *Bot. Acta* **104:** 292–294.

Marek, M., and M. Pirochtova. 1990. Response of the ratio of intercellular CO_2 concentration (c_i/c_a ratio) to basic microclimatological factors in an oak-hornbeam forest. *Photosynthetica* **24:** 122–129.

Medina, E., and P. Minchin. 1980. Stratification of $\delta^{13}C$ values of leaves in Amazonian rainforests. *Oecologia* **45:** 355–378.

Medina E., G. Montes, E. Cuevas, and Z. Roksandic. 1986. Profiles of CO_2 and $\delta^{13}C$ values of the upper Rio Negro Basin, Venezuela. *J. Trop. Ecol.* **2:** 207–217.

Medina, E., L. Sternberg, and E. Cuevas. 1991. Vertical stratification in closed and natural plantation forests in the Luquillo mountains, Puerto Rico. *Oecologia* **87:** 369–372.

Merwe, N. J. van der, and E. Medina. 1989. Photosynthesis and $^{13}C/^{12}C$ ratios in Amazonian rainforests. *Geochim. Cosmochim. Acta* **53:** 1091–1094.

Peng, T.-H., W. S. Broecker, H. D. Freyer, and S. Trumbore. 1983. A deconvolution of the tree ring based $\delta^{13}C$ record. *J. Geophys. Res.* **88:** 3609–3620.

Read, J., and G. D. Farquhar. 1991. Comparative studies in *Nothofagus* (Fagaceae). I. Leaf carbon isotope discrimination. *Funct. Ecol.* **5:** 684–695.

Schleser, G. H. 1990. Investigations of the $\delta^{13}C$ pattern in leaves of *Fagus sylvatica* L. *J. Exp. Bot.* **41:** 565–572.

Schleser, G. H., and R. Jayasekera. 1985. $\delta^{13}C$ variations of leaves in forests as an indication of reassimilated CO_2 from the soil. *Oecologia* **65:** 436–542.

Sternberg, L. da S. L. O'R. 1989. A model to estimate carbon dioxide recycling in forests using $^{13}C/^{12}C$ ratios and concentrations of forest air. *Agric. For. Meteorol.* **48:** 163–173.

Sternberg, L. S. L., S. S. Mulkey, and S. J. Wright. 1989a. Ecological interpretation of leaf carbon isotope ratios: Influence of respired carbon dioxide. *Ecology* **70:** 1317–1324.

Sternberg, L. S. L., S. S. Mulkey, and S. J. Wright. 1989b. Oxygen isotope ratio stratification in a tropical moist forest. *Oecologia* **81:** 51–56.

Stuiver, M. 1978. Atmospheric carbon dioxide and carbon reservoir changes. *Science* **199:** 253–258.

Stuiver, M., and T. F. Brazuinas. 1987. Tree cellulose $^{13}C/^{12}C$ isotope ratios and climate change. *Nature* **328:** 58–60.

Tans, P. P., I. Y. Fung, and T. Takahashi. 1990. Observational constraints on the global atmospheric CO_2 budget. *Science* **247:** 1431–1438.

Vogel, J. C. 1978. Recycling of carbon in a forest environment. *Oecologia Plant.* **13:** 89–94.

Wofsy, S. C., R. C. Harris, and W. A. Kaplan. 1988. Carbon dioxide in the atmosphere over the Amazon Basin. *J. Geophys. Res.* **93:** 1377–1387.

9

Environmental and Physiological Influences on Carbon Isotope Composition of Gap and Understory Plants in a Lowland Tropical Forest

Paula C. Jackson, Frederick C. Meinzer, Guillermo Goldstein, Noel M. Holbrook, Jaime Cavelier, and Fermin Rada

I. Introduction

The stable carbon isotope composition of plant tissue ($\delta^{13}C$) is determined both by the isotopic composition of the CO_2 source and by discrimination against the heavier isotope ^{13}C during photosynthetic CO_2 fixation (see O'Leary, Chapter 3, and Farquhar and Lloyd, Chapter 5, this volume). In C_3 plants this discrimination has two main components: one associated with diffusion of CO_2 through the stomata and one due to discrimination against ^{13}C by the primary carboxylating enzyme, RuBisCO. The balance between carboxylation and stomatal limitation of CO_2 diffusion is reflected in the ratio of intercellular to atmospheric partial pressure of CO_2 (p_i/p_a). When stomatal limitation of CO_2 diffusion is small, p_i/p_a will be large and biochemical discrimination against ^{13}C by RuBisCO predominates. Because the biological effects on carbon isotope composition are usually of more interest than source CO_2 effects, it is often more informative to define tissue carbon isotope composition in terms of a discrimination value (Δ) which takes temporal and spatial variation in the $\delta^{13}C$ of the source CO_2 into account (see Farquhar, Chapter 5, this volume). If variation in the isotopic composition of the CO_2 source is known, then analyses of foliar Δ can be used as a powerful tool to understand integrated responses of leaf gas exchange (p_i/p_a) to variations in light, water availability, and other environmental factors (e.g., Farquhar *et al.*, 1982a; Winter *et al.*, 1982; Guy *et al.*, 1986; Hubick *et al.*, 1988; Zimmerman and Ehleringer, 1990).

In closed tropical forests, vertical stratification in the isotopic composition of the source CO_2 may confound the interpretation of foliar $\delta^{13}C$ values (Medina and Minchin, 1980; Farquhar *et al.*, 1989; Sternberg *et al.*, 1989); as a result of soil respiration and relatively poor ventilation, CO_2 concentrations may build up in the understory (Allen and Lemon, 1976; Medina and Minchin, 1980). Since this respiratory CO_2 is derived from material already depleted in ^{13}C, $\delta^{13}C$ values for air in dense tropical forests (Fig. 1) should lie somewhere between the bulk atmospheric value (-7.8 to -8%) and the value for CO_2 derived from soil respiration (-25 to -28.3%). For example, air $\delta^{13}C$ values should become progressively more negative as the soil surface is approached, particularly in the understory (Fig. 1). By contrast, in gaps, openings in the forest created by one or more canopy tree falls, greater air mixing may result in air $\delta^{13}C$ values closer to those of the bulk atmosphere. In forests with pronounced wet and dry seasons the $\delta^{13}C$ value of the air should be less negative during the dry season when soil respiration is reduced.

Foliar $\delta^{13}C$ values of tropical forest plants will thus be a function both of physiological processes leading to discrimination against ^{13}C and of the magnitude of refixation of respiratory CO_2. Despite considerable spatial and temporal heterogeneity of $\delta^{13}C$ values in air of tropical forests, recent studies suggest that variations in the $\delta^{13}C$ of the air can be predicted from measurements of the ambient CO_2 concentration alone, providing reason for optimism concerning the utility of foliar $\delta^{13}C$ analyses in forest species (Sternberg *et al.*, 1989; Broadmeadow *et al.*, 1992).

Bulk atmosphere: -7.8 to -8.0 ⁰/oo

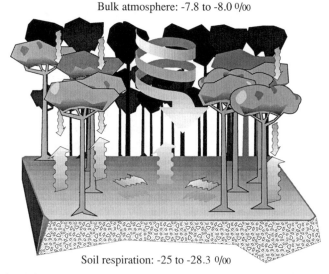

Soil respiration: -25 to -28.3 ⁰/oo

Figure 1.   Schematic representation of CO_2 mixing processes in a tropical forest showing the two principal CO_2 sources: the bulk atmosphere and soil respiration (decomposer respiration and root respiration) and their respective $\delta^{13}C$ values.

In this study foliar $\delta^{13}C$ and photosynthetic gas exchange of high light-requiring species growing in gaps and of shade-tolerant shrub species growing in the understory and in gaps in a lowland tropical forest were examined. Our principal objective was to determine the extent to which the variation in foliar $\delta^{13}C$ observed could be attributed to differences in internal physiological features or to differences in the isotopic composition of the source CO_2. The $\delta^{13}C$ of the air was estimated from the ambient CO_2 concentration using a recently published model (Sternberg *et al.*, 1989). Additional objectives included comparing leaf gas exchange responses of shade-tolerant shrub species growing in the understory and the same species growing in gaps and comparing gas exchange responses of shade-tolerant and high light-requiring species both growing in gaps.

II. Materials and Methods

The study was carried out in a semi-evergreen, moist tropical forest on Barro Colorado Island (BCI), Panama (9° 09'N, 79° 51'W). Mean annual rainfall is approximately 2600 mm and is characterized by a marked seasonality, with a dry season from mid-December until the beginning of May when rainfall is only 160–260 mm (Windsor, 1990). The results presented here were obtained during the dry season of 1991 (February–March).

Piper cordulatum C.DC. (Piperaceae) and *Psychotria limonenesis* Krause (Rubiaceae) were chosen as representative shade-tolerant shrub species. Both are common in the forest understory on BCI. *Palicourea guianensis* Aubl. (Rubiaceae) and *Cecropia obtusifolia* Bertol. (Moraceae) were chosen as the representative high light-requiring species. These species occur only in gaps. Two gap and two understory sites were selected. Both gaps were relatively small, between 8 and 12 m in diameter. The understory sites were located within 100 m of a gap and were slightly larger than the gap sites. All study sites were in secondary growth forest. Three individuals of each species were selected per site (with the exception of *C. obtusifolia* for which only one individual was found) and four of the youngest fully expanded leaves were marked for gas exchange measurements. All gas exchange measurements were made by sealing the gas exchange cuvette over the same premarked area of each leaf. Gas exchange measurements were made between 0900 and 1600 h during 3 to 4 days at each site using a LI-COR LI-6200 portable photosynthesis system. Photosynthetic photon flux density (PPFD) was measured with a quantum sensor during the gas exchange measurements. These data were used to calculate overall average PPFD values for each type of site.

For determination of foliar $\delta^{13}C$ the delineated portions of the leaves used for gas exchange measurements were pooled per plant (with the exception of *C. obtusifolia*), oven-dried at 40°C, and finely ground. Subsamples were combusted and the relative abundance of ^{13}C and ^{12}C in the CO_2 released by the combustion was analyzed by mass spectrometry. Stable carbon isotope composition was expressed as the $^{13}C/^{12}C$ ratio relative to that of the Pee Dee belemnite standard with a precision of 0.2‰. The resulting $\delta^{13}C$ values were used to estimate carbon isotope discrimination (Δ) as

$$\Delta = (\delta^{13}C_a - \delta^{13}C_p)/(1 + \delta^{13}C_p), \tag{1}$$

where a and p represent air and plant, respectively.

Air $\delta^{13}C$ was calculated using the equation developed by Sternberg *et al.* (1989) for data collected at BCI,

$$\delta^{13}C_a = 6703(1/[CO_2]) - 28.3, \tag{2}$$

where $[CO_2]$ is the ambient CO_2 concentration in μliter liter^{-1}. The prevailing ambient CO_2 concentration at each site was estimated by circulating ambient air through the infrared gas analyzer of the portable photosynthesis system used for the leaf gas exchange measurements. This procedure was carried out at frequent intervals between the leaf measurements. Our preliminary data concerning the relationship between $\delta^{13}C_a$ and the ambient CO_2 concentration were consistent with the model presented above and with findings by other authors (see Griffiths *et al.*, Chapter 14, this volume). This suggests that variation in the isotopic composition in the source CO_2 can be adequately predicted from logistically and technically simpler measurements of ambient CO_2 concentration alone. Nevertheless, ambient CO_2 concentration at a height of about 1 m was only 353 μliter liter^{-1} ($\delta^{13}C_a$ of -9.3‰) in the present study compared with about 376 μliter liter^{-1} ($\delta^{13}C_a$ of -10.5‰) in the study by Sternberg *et al.* (1989).

III. Results

Foliar $\delta^{13}C$ values differed among high light-requiring and shade-tolerant species and among individuals of the same species growing in gap and understory sites (Table I). The least negative foliar $\delta^{13}C$ values were observed in the high light-requiring species *C. obtusifolia* and *P. guianensis* (mean = -30.8‰). Foliar $\delta^{13}C$ of the shade-tolerant species *P. cordulatum* and *P. limonensis* growing in a gap site were less negative (mean = -33.3‰) than in individuals of these two species growing in understory sites (mean = -35.2‰).

Measurements of the ambient CO_2 concentration in gap and understory sites indicated that $\delta^{13}C_a$ was 0.9 to 1.6‰ more negative than the value of -8‰ generally attributed to the bulk atmosphere (Table I). However, consistent differences in $\delta^{13}C_a$ in gap and understory sites were not observed. A comparison of foliar $\delta^{13}C$ values with those of the air for shade-tolerant species growing in gap and understory sites showed that only 20 to 35% of the difference in foliar $\delta^{13}C$ could be attributed to differences in the isotopic composition of the air. This was more clearly seen by using $\delta^{13}C_a$ to calculate Δ, a procedure which allows direct comparison of physiological

Table I   Gas Exchange and Carbon Isotope Composition of Leaves and Air for Plants Growing in Forest Gap and Understory Sites[a]

Site	Species	$\delta^{13}C_p$ (‰)	$\delta^{13}C_a$ (‰)	Δ (‰)	A/g (μmol/mol)
Gap 2	C. obtusifolia	-30.5 ± 0.7	-9.6 ± 0.1	21.6 ± 0.9	54.1 ± 9.7
Gap 2	P. guianensis	-31.2 ± 0.5	-9.6 ± 0.1	22.3 ± 0.6	60.4 ± 7.8
	mean	-30.8	-9.6	22.0	57.3
Understory 1 and 2	P. cordulatum	-35.9 ± 0.2	-9.4 ± 0.1	27.5 ± 0.3	8.6 ± 1.8
Gap 1	P. codulatum	-33.4 ± 0.1	-8.9 ± 0.1	25.3 ± 0.02	42.8 ± 5.6
	difference	-2.5	-0.5	2.2	-34.2
Understory 1 and 2	P. limonensis	-34.6 ± 0.2	-9.4 ± 0.1	26.1 ± 0.4	9.5 ± 2.0
Gap 1	P. limonensis	33.2 ± 0.4	-8.9 ± 0.0	25.1 ± 0.4	40.7 ± 5.9
	difference	-1.4	-0.5	1.0	-31.2

[a]Data are means ± 1 SE for four measurements on three plants per species and site (with the exception of C. obtusifolia for which averages are for four measurements on one plant). $\delta^{13}C_a$ values were estimated using ambient CO_2 concentrations and the equation $\delta^{13}C_a = 6703(1/[CO_2]) - 28.3$ from Sternberg et al. (1989). Values of Δ were calculated from $\delta^{13}C_a$ and $\delta^{13}C_p$ as described in the text.

influences on tissue carbon isotope composition. A relatively large range in Δ was observed, from 22‰ in high light-requiring species to 26.8‰ in shade-tolerant individuals growing in the understory. Independent estimates of intrinsic water-use efficiency expressed as the ratio of CO_2 assimilation to stomatal conductance (A/g) determined from leaf gas exchange measurements (see Meinzer et al., Chapter 22, this volume) during the dry season (Table I) were highly correlated with foliar Δ ($r^2 = 0.76$). The patterns of Δ and A/g observed suggested that intrinsic water-use efficiency was highest in the high light-requiring species, intermediate in shade-tolerant species growing in gaps, and lowest in shade-tolerant species growing in the understory.

An additional manifestation of the strong correlation between A/g and Δ was a linear relationship between Δ and p_i/p_a (Fig. 2). There was good agreement between the higher values of Δ and p_i/p_a exhibited by shade-tolerant species growing in the understory and the model proposed by Farquhar et al. (1982b). At lower values of Δ and p_i/p_a, however, the deviation between observed and predicted values became increasingly large. Although the difference in slopes between predicted and observed Δ values was not statistically significant, this increase in deviation may have resulted from an underestimation of $\delta^{13}C_a$ in gaps or from the difference in time scale involved in the types of measurements. Δ values represent integrated physiological responses of leaves, whereas gas exchange measurements represent instantaneous responses to prevailing conditions. Field determinations of $\delta^{13}C_a$ for both sites should be made in order to resolve this question.

Because p_i/p_a and therefore Δ is determined largely by the ratio A/g, A

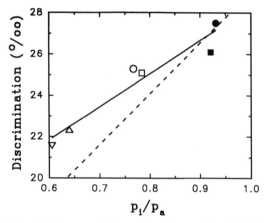

Figure 2. Relationship between foliar carbon isotope discrimination and the ratio of intercellular to ambient partial pressure of CO_2 (p_i/p_a) for high light-requiring and shade-tolerant species growing in gap (open symbols) and understory (closed symbols) sites. (∇) *Cecropia obtusifolia,* ($\triangle$) *Palicourea guianensis,* ($\bigcirc$) *Piper cordulatum,* ($\square$) *Psychotria limonensis.* The dashed line represents the theoretical relationship between discrimination and p_i/p_a ($\Delta = a + (b - a)p_i/p_a$; with a = 4.4‰ and b = 29‰), proposed by Farquhar *et al.* (1982b).

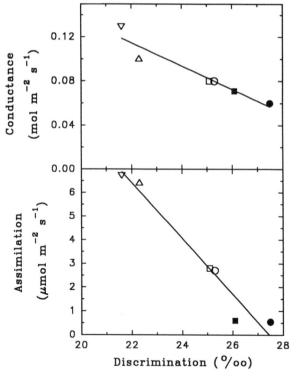

Figure 3. Stomatal conductance and assimilation in relation to foliar carbon isotope discrimination for plants growing in forest gap and understory sites. Symbols are as defined in Fig. 2.

and g were examined individually as functions of Δ in order to determine the nature of the adjustments in leaf gas exchange characteristics responsible for the variation in Δ observed. Both A and g were greatest in the high light-requiring species and declined linearly with increasing Δ (Fig. 3). The 5.9‰ range of Δ was associated with a greater than 10-fold variation in A and only a 2-fold variation in g. This indicates that the lower Δ of plants growing in gaps, particularly in the high light-requiring species, was primarily due to large increases in A. The increase in A associated with higher PPFD in gaps more than compensated for the increase in g, which, in the absence of a change in A, would have raised Δ. Average PPFD during the gas exchange measurements was an order of magnitude higher in the gaps ($288\ \mu\text{mol m}^{-2}\,\text{s}^{-1}$) than in the understory ($28\ \mu\text{mol m}^{-2}\,\text{s}^{-1}$). The relationship between A and Δ and between g and Δ for all species and sites suggested that the physiological basis for the adjustments in gas exchange to differences in the light regime was similar for both shade-tolerant and high light-requiring species.

IV. Discussion

The relatively abrupt decline in wind speed within the canopy of tropical forests (e.g., Roberts *et al.*, 1990) and the release of CO_2 depleted in ^{13}C from the soil may cause the isotopic composition of the source CO_2 near the forest floor to differ substantially from that above the canopy. Within tree fall gaps, however, greater turbulence and hence mixing of the bulk air with the respired CO_2 derived from the forest floor biomass might be expected to result in values of $\delta^{13}C_a$ closer to those of the bulk atmosphere than in the adjacent understory. The 0.5‰ difference in $\delta^{13}C_a$ between gap site 1 and nearby understory sites 1 and 2 were consistent with this idea (Table I). Nevertheless, values of $\delta^{13}C_a$ in gaps may be as negative as those in understory sites (gap site 2, Table I). This is probably a consequence of differences in air turbulence patterns associated with gap size and also of horizontal heterogeneity in the amount of CO_2 released by decomposition processes.

When differences in the isotopic composition of the air among sites were taken into account, it appeared that 65 to 80% of the difference between gap and understory in foliar $\delta^{13}C$ of the shade-tolerance species *P. cordulatum* and *P. limonensis* was the result of internal adjustments in leaf gas exchange characteristics. Contrasting results for forest species growing in the Luquillo Mountains of Puerto Rico were obtained by Medina *et al.* (1991) who reported the expected vertical stratifcation in foliar $\delta^{13}C$ values but no difference in foliar $\delta^{13}C$ between plants growing in gap and understory sites.

The physiological mechanisms responsible for the decrease in foliar Δ in the gap appeared to be similar for both *P. limonensis* and *P. cordulatum*. The lower Δ values measured under higher light availability in the gap were the result of large increases in A (Fig. 3). Although g also increased under higher PPFD in the gap, the large increase in the photosynthetic rate and probably in photosynthetic capacity more than compensated for the higher g. Decreases in Δ and or increases in A with increased light availability have been reported for several tropical forest species (Pearcy, 1987; Mulkey, 1986; Strauss-Debendetti and Bazzaz, 1991). The relatively low intrinsic water-use efficiency observed in understory plants is consistent with the idea that under these conditions of low and unpredictable light availability maximization of carbon gain is a more important constraint than is maintenance of high efficiency of water use through stomatal restriction of gas exchange. Higher A in shade-tolerant plants growing in gaps may reflect adjustments in photosynthetic capacity linked to higher foliar nitrogen content (Vitousek and Denslow, 1986). However, in a study of the relationship between irradiance levels and carbon isotope discrimination in the orchid *Catasetum viridiflavum* the 4‰ decrease in Δ observed at high irradiance was attributed to increased stomatal limitation of photosthesis rather than to changes in photosynthetic capacity (Zimmerman and Ehleringer, 1990). This conclusion was based on measurement of small variation in leaf nitrogen content rather than on measurements of leaf gas exchange.

Recent studies of the relationship between Δ and leaf gas exchange characteristics indicate that there are several mechanisms by which intrinsic water-use efficiency (A/g) can be regulated, each with different implications for plant carbon balance. For example, genotypic variation in A/g in peanut results from variation in photosynthetic capacity at similar levels of g (Wright, Chapter 17, this volume). In contrast, higher intrinsic water-use efficiency (lower Δ) in wheat genotypes resulted from covariance of g and photosynthetic capacity, with the relative reduction in g being greater (Condon *et al.*, 1987). Water-limited coffee plants responded similarly; both A and g were positively correlated with Δ, but the relative change in g was larger (Meinzer *et al.*, 1990). On the other hand, in Hawaiian *Metrosideros polymorpha* populations growing at different levels of soil moisture availability, decreases in Δ were associated with maintenance of nearly constant photosynthetic rates while g declined (Meinzer *et al.*, 1992). In the present study, the increase in A/g (decrease in Δ) in the two shade-tolerant species *P. cordulatum* and *P. limonensis*, associated with higher irradiance in tree fall gaps resulted from an increase in A that compensated for a smaller increase in g. Thus, increases in A/g (as inferred from foliar Δ) associated with increases in light or nutrient availability may reflect an overall increase in rates of carbon accumulation and total gas exchange. This contrasts with increased stomatal limitation of gas exchange and reduced rates of carbon accumulation usually associated with decreases in foliar Δ of water-limited plants.

Shade-tolerant and high light-requiring species differ with respect to a large suite of morphological and physiological features. For example, the high light-requiring species depend on high irradiance and high red to far-red light ratios for germination, exhibit rapid leaf turnover and rapid growth rates, and invest relatively small amounts of carbon in secondary metabolites (e.g., Denslow *et al.*, 1990). Despite these differences, the shade-tolerant and high light-requiring species exhibit the same relationship between Δ and leaf gas exchange characteristics (Fig. 3).

V. Summary

Our results indicate that interpretation of carbon isotope composition of leaf tissue in tropical forest plants may be confounded by spatial and temporal variation in the isotopic composition of the source CO_2. Concurrent site-specific measurements of either ambient CO_2 concentration or $\delta^{13}C_a$ are therefore necessary in these environments.

Our estimated values of $\delta^{13}C_a$ in gap sites were not consistently higher than those of understory sites, possibly due to small gap size and heterogeneity in the amount of CO_2 released by decomposition processes.

The correlation between *A*, *g*, and Δ across light environments and species in the present study suggests that regulation of the relationship between CO_2 acquisition and transpirational losses through biochemical and biophysical responses to light availability represents an important ecological and evolutionary constraint for these species.

Acknowledgments

This project was funded in part by a Mellon Foundation Grant to G. Goldstein and R. Meinzer, a National Science Foundation Fellowship to P. Jackson, and support from the Ecological Research Division of the Office of Health and Environmental Research, U.S. Department of Energy.

References

Allen, L. H., Jr., and R. Lemon. 1976. Carbon dioxide exchange and turbulence *in* a Costa Rican tropical rain forest, pp. 265–308. *In* J. L. Monteith (ed.), Vegetation and the Atmosphere. 2. Case Studies. Academic Press, London.

Broadmeadow, M. S. J., H. Griffiths, C. Maxwell, and A. M. Borland. 1992. The carbon isotope ratio of plant organic material reflects temporal and spatial variations in CO_2 within tropical forest formations in Trinidad. *Oecologia* **89:** 435–441.

Condon, A. G., R. A. Richards, and G. D. Farquhar. 1987. Carbon isotope discrimination is positively correlated with grain yield and dry matter production in field-grown wheat. *Crop Sci.* **27:** 996–1001.

Denslow, J. S., J. C. Schultz, P. M. Vitousek, and B. R. Strain. 1990. Growth responses of tropical shrubs to treefall gap environments. *Ecology* **71:** 156–179.

Farquhar, G. D., M. C. Ball, S. von Caemmerer, and Z. Roksandic. 1982a. Effect of salinity and humidity on δ¹³C value of halophytes–evidence for diffusional isotope fractionation determined by the ratio of intercellular/atmospheric CO_2 under different environmental conditions. *Oecologia* **52:** 121–124.

Farquhar, G. D., M. H. O'Leary, and J. A. Berry. 1982b. On the relationship between carbon isotope discrimination and intercellular carbon dioxide concentration in leaves. *Aust. J. Plant Physiol.* **9:** 121–137.

Farquhar, G. D. J. R. Ehleringer, and K. T. Hubick. 1989. Carbon isotope discrimination and photosynthesis. *Annu. Rev. Plant Physiol. Mol. Biol.* **40:** 503–537.

Guy, R. D., D. M. Reid, and H. R. Krouse. 1986. Factors affecting $^{13}C/^{12}C$ ratios of inland halophytes I. Controlled studies on growth and isotopic composition of *Puccinellia nuttalliana. Can. J. Bot.* **64:** 2693–2699.

Hubick, K. T., R. Shorter, and G. D. Farquhar. 1988. Heritability and genotype x environment interactions of carbon isotope discrimination and transpiration efficiency in peanut (*Arachis hypogaea* L.). *Aust. J. Plant Physiol.* **15:** 799–813.

Medina, E., and P. Minchin. 1980. Stratification of $\delta^{13}C$ values of leaves in Amazonian Rain Forests. *Oecologia* **45:** 377–378.

Medina, E., G. Montes, E. Cuevas, and Z. Roczandic. 1986. Profiles of CO_2 concentration and $\delta^{13}C$ values in tropical rain forests of the upper Rio Negro Basin, Venezuela. *J. Trop. Ecol.* **2:** 207–217.

Medina, E., L. Sternberg, and E. Cuevas. 1991. Vertical stratification of $\delta^{13}C$ values in closed natural and plantation forests in the Luquillo mountains, Puerto Rico. *Oecologia* **87:** 369–372.

Meinzer, F. C., G. Goldstein, and D. A. Grantz. 1990. Carbon isotope discrimination in coffee genotypes grown under limited water supply. *Plant Physiol.* **92:** 130–135.

Meinzer, F. C., P. W. Rundel, G. Goldstein, and M. R. Sharifi. 1992. Carbon isotope composition in relation to leaf gas exchange and environmental conditions in Hawaiian *Metrosideros polymorpha* populations. *Oecologia* **91:** 305–311.

Mulkey, S. S. 1986. Photosynthetic acclimation and water-use efficiency of three species of understory herbaceous bamboo (Gramineae) in Panama. *Oecolgia* **70:** 514–519.

Pearcy, R. W. 1987. Photosynthetic gas exchange responses of Australian tropical forest trees in canopy gap and understory micro-environments. *Func. Ecol.* **1:** 169–178.

Roberts, J., O. M. R. Cabral, and L. F. De Aguiar. 1990. Stomatal and boundary layer conductances in an Amazonian Terra Firme rain forest. *J. Appl. Ecol.* **27:** 336–353.

Sternberg, L. S., S. S. Mulkey, and S. J. Wright. 1989. Ecological interpretation of leaf carbon isotope ratios: Influence of respired carbon dioxide. *Ecology* **70:** 1317–1324.

Strauss-Debenedetti, R., and F. A. Bazzaz. 1991. Plasticity and acclimation to light in tropical Moraceae of different successional positions. *Oecologia* **87:** 377–387.

Vitousek, P. M., and J. Denslow. 1986. Nitrogen and phosphorous availability in treefall gaps of a lowland tropical rainforest. *J. Ecol.* **74:** 1167–1178.

Windsor, D. M. 1990. Climate and moisture variability in a tropical forest: Long-term records from Barro Colorado Island, Panama. Smithsonian Contributions to the Earth Sciences Number 29, Smithsonian Institution Press, Washington, DC.

Winter, K., J. A. M. Holtum, G. E. Edwards, and M. H. O'Leary. 1982. Effect of low relative humidity on $\delta^{13}C$ value in two C_3 grasses and in *Panicum milioides*, a C_3-C_4 intermediate species. *J. Exp. Bot.* **33:** 88–91.

Zimmerman, J. K., and J. R. Ehleringer. 1990. Carbon isotope ratios are correlated with irradiance levels in the Panamanian orchid *Catasetum viridiflavum. Oecologia* **83:** 247–249.

10

Carbon Isotope Fractionation in Tree Rings in Relation to the Growing Season Water Balance

N. J. Livingston and D. L. Spittlehouse

I. Introduction

Plant dry matter accumulation and cummulative transpiration are usually well correlated because both CO_2 uptake and water vapor loss take place via stomata. In trees significant correspondence between seasonal and annual variations in water availability and tree height and annual diameter growth have been reported (Spittlehouse, 1985; Giles *et al.*, 1985; Carter and Klinka, 1990; Robertson *et al.*, 1990). Additionally, correlations between temperature and rainfall and annual stem increment (tree ring width) have been used to reconstruct historic climatic conditions (Fritts, 1976).

Farquhar *et al.* (1982, 1988) have shown that the carbon isotopic composition of plant material, or more specifically $\delta^{13}C$, also depends upon the weather and soil moisture conditions experienced by the plant during growth. A number of researchers have related $\delta^{13}C$ measurements to plant water-use efficiency and productivity (Condon *et al.*, 1987; Hubick and Farquhar, 1989; Virgona *et al.*, 1990).

Measurements of $\delta^{13}C$ in tree rings have been used to study aspects of the global carbon balance (e.g., Peng *et al.*, 1983; Stuiver *et al.*, 1984) and to assess the affects of air pollution on tree growth (Martin and Sutherland, 1990). Most studies have found large temporal variations in tree ring $\delta^{13}C$ with a general downward trend starting in the mid-19th century with the increased burning of fossil fuels and consequent introduction of ^{13}C-depleted CO_2 into the atmosphere (Keeling *et al.*, 1979). However, in some cases, these global trends are masked by local influences or conditions (Leavitt and Long, 1983).

There have been numerous studies that have related annual variations in the isotope discrimination in tree rings to temperature and precipitation during the year (Tans and Mook, 1980; Francey and Farquhar, 1982; Freyer and Belacy, 1983; Leavitt and Long, 1983, 1989). Leavitt and Long (1991) related seasonal variations in $\delta^{13}C$ to solar radiation and water availability. However, this was only done for a few years when detailed environmental and physiological measurements were made. Longer periods need to be studied to provide clearer relationships between carbon isotope ratio and the water balance to remove or factor out the effect of climate from $\delta^{13}C$ reconstruction (Leavitt and Long, 1983).

The objective of our study was to relate tree ring $\delta^{13}C$ values to water availability or deficits predicted from a forest water balance model. It was felt that the combination of $\delta^{13}C$ and stem increment data with simulated site water balance, over a 17-year period, could provide new insights into tree water use and growth.

II. Methods

A. Water Balance Model

Previous studies of tree rings and climate have involved correlations of temperature and precipitation with tree ring width. These have usually been for sites with extreme climates and isolated trees (Fritts, 1976). An effective water balance model, based on physical processes, should allow both the incorporation of site factors, particularly soil water availability, and the integration of environmental variables. This is not an easy task for isolated trees because of the difficulty in defining an individual's root zone and estimating or calculating the energy absorbed by a single tree. In stands, where individual trees are not considered, the task is easier. However, in such cases, there is typically a relatively small interannual variation in ring width. Spittlehouse (1985), Giles et al. (1985), and Robertson et al. (1990) have shown that estimates of transpiration and water deficit from a stand daily water balance model correlate well with growth data from individual trees from within the stand.

It is possible that models that incorporate estimates of photosynthesis would be more effective than water balance models in assessing site productivity. However, such models require considerably more plant and weather data. Even then, they are still not always successful (Price and Black, 1989) and require an even greater level of complexity if attempts are made to partition assimilates.

The water balance model used in this study is described in detail in Spittlehouse and Black (1981) and Spittlehouse (1985, 1989). The model treats the forest canopy as a single layer over a single slab of soil. It is only applicable to established stands where the trees fully exploit the root zone.

The model determines the daily root zone water content as

$$W_i = W_{i-1} + P_i - E_{t,i} - I_i - D_i - R_i, \qquad (1)$$

where W is the water content of the root zone for day i, P is rainfall, E_t is the combined transpiration from the vegetation and evaporation from the soil surface, I is the rainfall interception, D is drainage, and R is runoff, all in (mm). E_t is the lesser of the energy and soil-limited rates. For convenience, all energy terms are expressed in water equivalent units.

The energy available to evaporate water (E_{max}) is given by

$$E_{max} = \alpha(s/[s + \gamma])(R_n - G), \qquad (2)$$

where α is an experimentally determined constant (dimensionless) and is directly related to plant parameters such as stomatal resistance and leaf area index (McNaughton and Spriggs, 1989), R_n is the daily net radiation (mm day^{-1}), G is the soil heat flux (mm day^{-1}) estimated as a fraction of R_n, s and γ are, respectively, the slope of the saturation vapor pressure curve (kPa °C^{-1}) and the psychrometric constant (kPa °C^{-1}) at the daily mean air temperature. Daily net radiation is calculated from measured solar radiation, or sunshine hours, and maximum and minimum air temperature.

The ability of the soil to supply water to the trees (E_s) is

$$E_s = b\theta_e, \qquad (3)$$

where b is an experimentally determined constant (mm day^{-1}) and θ_e is the fraction of extractable water in the root zone (dimensionless).

Rainfall interception is calculated from

$$\begin{aligned} I &= fP^h &&\text{for } P > p \\ I &= P &&\text{for } P \leq p, \end{aligned} \qquad (4)$$

where f, h (dimensionless), and p (mm day^{-1}) are experimentally determined constants. The evaporation of intercepted water (E_i) is calculated from Eq. (1) with a value of a for a wet canopy (α_w).

Total evapotranspiration (E_T) is given by

$$E_T = E_t(1 - I/E_i) + I. \qquad (5)$$

Drainage from the root zone is calculated as a function of water storage in the root zone. Runoff did not occur at the sites discussed in this chapter.

B. Site Description

The study area is located near Courtenay, on the east side of Vancouver Island, British Columbia, Canada (49° 51'N, 125° 14'W). The elevation is 150 m above sea level with an 8% northeast-facing slope. The topography is gently undulating with ridges of 20–30 m relief. The region is in a rain shadow and has warm dry summers. Winter and early spring are invariably wet and the growing season starts in April with the soil root zone water storage fully recharged.

The water balance model was developed and tested on data collected over four growing seasons (Spittlehouse and Black, 1981). The values for α, α_w, b, f, h, and p are 0.8, 2, 10, 0.6, 0.6, and 0.3 mm day^{-1}, respectively. The drainage function is $D = 100 \, (\theta/0.3)^{14.8}$ mm day^{-1}, where θ is the volumetric water content of the root zone. Accumulated, simulated evapotranspiration was within 10% of that measured with water balance and energy balance techniques. The model calibration coefficients are typical for forests on the east coast of Vancouver Island (Spittlehouse, 1989).

Water balance simulations were conducted for the April to October period from 1964 to 1981. Root zone water content at the beginning of April was set to 0.26. This a reasonable assumption because of the rainy winter and early spring in this area. The daily weather data (solar radiation, air temperature, and rainfall) were from nearby climate stations maintained by Environment Canada. The suitability of the data was verified by comparison with onsite data (Spittlehouse, 1985).

The prevailing winds at the site are westerly. Extensive mature forests and the Pacific ocean are upwind of the study area. Downwind, a logging road and the nearest town are about 1 and 30 km away from the study area, respectively. Consequently, there should be minimal contamination of the site by anthropogenic sources of carbon. There might be some contamination from CO_2 from soil and understory respiration, but cyclic enrichment in ^{13}C is usually small in the tree canopy (Francey and Farquhar, 1982).

The forest consists of Douglas-fir (*Pseudotsuga menziesii* (Mirb.) Franco) trees that were planted in 1952 and an understory of salal (*Gaultheria shallon* (Pursh)). In 1964 the trees would have been 2 to 3 m tall. In 1981 they were 12 to 15 m tall with an effectively closed canopy. It is likely that, for the period under study, the trees and understory exploited the whole of the root zone. As the trees grew and shaded the salal they would have used proportionally more of the available soil water, but the total water use at the site would not have changed. Spittlehouse (1989) suggests that the combined understory and tree water use, and thus the rate of soil water depletion, would be similar for much of the life of the stand.

Two sites within the study area were selected: An upper site, where the soil, a gravelly sandy loam (Orthic Humo-Feric Podzol) is 0.75 ± 0.25 m deep over sandstone bedrock and contains about 20% coarse fragments and a lower site that has soil 1 m deep with no coarse fragments. The lower site is approximately 0.5 km downslope from the upper site and has about twice the soil water storage capacity. Tree growth is substantially greater at the lower site.

C. Tree Growth Analysis

Annual basal area increment and samples for analysis of $\delta^{13}C$ were obtained from cores for the years 1963 to 1980. Two cores were taken at 1.3 m, parallel to the slope, from each of nine trees representing a range of bole sizes. Annual tree ring width was used to calculate the annual basal area increment (BAI) for each tree. The long-term growth trend was removed to produce a BAI index for 1964 to 1980. The BAI data were fit to the curve $y = ax^b$, where y is annual increment and x is a year number. The first few rings close to the pith were not included. The BAI index for a year is calculated as the BAI divided by the BAI value predicted from the trend curve for that year. This procedure results in values that have a mean of 1 and a relatively homogeneous variance over time (Fritts, 1976). The deviation of the index from unity indicates whether the trees grew less or more in that year than might have been expected under average conditions. Analysis was done on the averages of the individual cores used for the carbon isotope analysis and on the average for the nine trees. Comparison of the BAI index series for the subsamples with those for the sites gave a ratio of 1.01 ± 0.2, $r^2 = 0.652$, for the lower site, and 1.0 ± 0.02, $r^2 = 0.889$ for the upper site. The results presented below use the BAI of the subsample.

D. Water Balance Analysis

Water balance data are summarized as accumulated transpiration and accumulated transpiration deficit. The former is transpiration accumulated daily to a specific end point. The latter is an accumulation of the daily difference between the energy-limited rate (Eq. (2)) and actual transpiration (the lesser of Eqs. (2) and (3)). Accumulated transpiration and transpiration deficits were compared to the annual BAI indices and $\delta^{13}C$ values. Various start and end points for accumulation and deficits based on reasonable biological criteria were tested. No attempt was made to separate tree from understory water use nor to quantify soil evaporation which was likely less than 10% of the total water use (Kelliher *et al.*, 1986). Despite these simplifications, simulated transpiration can be considered as a good index of actual tree transpiration.

Growth of the Douglas fir trees typically begins in April and ends in the later summer. Spittlehouse (1985) used April 1 as a starting point for accumulating transpiration. The end point for transpiration was either the end of September, or, if earlier, 15 days after soil water potential fell below -0.5 MPa. Accumulations from April 1 to the end of July and end of August were also tested in this study.

E. Isotopic Analysis

Carbon isotope ratio analysis was done on four cores from three trees for the upper site, and three cores from two trees from the lower site. Tree rings from cores were separated into 1-year intervals and milled to ensure sample homogeneity. Resins and oils were removed from the wood with toluene/ethanol in a soxhlet apparatus. Cellulose was then isolated using the sodium chlorite method described by Green (1963).

Two samples per ring were ground in a cyclone mill (<0.2 mm) and analyzed for atom % ^{13}C on a VG-SIRA 12 isotope ratio mass spectrometer (Isotech, Middlewhich, UK). Procedures used for isotopic analysis are described by Boutton (1990) and Swerhone *et al.* (1991).

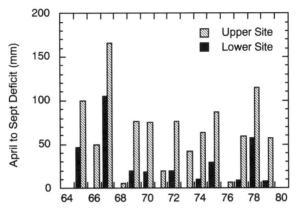

Figure 1. Growing season water deficits for upper and lower sites for 1964 to 1980. There were no deficits in 1964 and 1980.

III. Results

A. The Water Balance

The complete water balance data for the upper site are presented in Spittlehouse (1985). The energy-limited transpiration rates and rainfall were the same for the upper and lower sites. Figure 1 shows the large year-to-year variation in transpiration deficit from April to the end of September for both sites. Transpiration deficits typically occurred during the period from June until September, though in most years heavy rain in late August brought soil water to field capacity. Transpiration deficits at the lower site occurred later and were less intense than those at the higher site because of

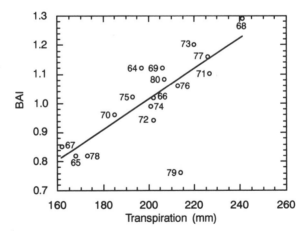

Figure 2. Basal area increment (BAI) index the upper site versus April to July accumulated transpiration (T). The equation of the regression line is BAI $= 0.005T - 0.0034$ $(r^2 = 0.551)$. Numbers by the data points indicate the year.

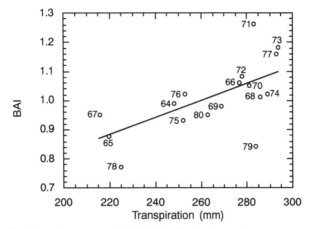

Figure 3. Basal area increment (BAI) index for the lower site versus transpiration (T) accumulated from April to August or a soil-limited situation (15 days drier than -0.5 MPa), whichever comes first. The equation of the regression line is BAI $= 0.003T + 0.219$ ($r^2 = 0.391$). Numbers by the data points indicate the year.

the former's greater water storage capacity. Year-to-year variations in deficits were mainly a function of the year-to-year variations in rainfall.

Accumulated energy-limited transpiration was relatively conservative, varying from 275 to 325 mm. Accumulated actual transpiration for the upper and lower sites is shown in Figs. 2 and 3, respectively. The lower site had a greater accumulated transpiration before transpiration became soil limited.

B. Tree Growth

The relations between BAI index and accumulated transpiration over various periods are summarized in Table I. For the upper site, with 1979 data

Table I Coefficient of Determination (r^2) between Basal Area Increment (BAI) Index and Accumulated Transpiration for Lower and Upper Sites[a]

	Coefficient of determination	
Accumulation period	Upper site	Lower site
April to July	0.551*** (0.832)***	0.001 (0.004)
April to August	0.547*** (0.754)***	0.234* (0.350)*
April to PSI	0.583*** (0.716)***	0.361*** (0.607)***
April to PSI/August	0.577*** (0.729)***	0.391** (0.577)***

[a]Values in parentheses are for correlations without data for 1979. Accumulations are for April 1 to the end of July, to the end of August, to after 2 weeks with soil water potentials less than -0.5 MPa (PSI), or August or PSI, whichever comes first.
*, **, or *** indicate significance at $P \leq 0.05$, 0.01, or 0.001, respectively.

Table II   Coefficient of Determination (r^2) between Basal
Area Increment (BAI) Index and Accumulated Water Deficit
for the Upper and Lower Sites[a]

	Coefficient of determination	
Deficit period	Upper site	Lower site
April to July	0.488*** (0.646)***	0.331* (0.473)**
April to August	0.494** (0.601)***	0.224* (0.300)**
April to September	0.443** (0.565)***	0.206 (0.279)*

[a]Values in brackets are for correlations without data for 1979. Accumulations are
for April 1 to the end of July, the end of August, and the end of September.
    *, **, or *** indicate significance at $P \leq 0.05$, 0.01, or 0.001, respectively.

excluded, transpiration accumulated between April 1 and the end of July
(Fig. 2) accounted for the greatest variability in BAI ($r^2 = 0.832$). For the
lower site, accumulated transpiration to the end of August or to the time, if
earlier, when transpiration was soil limited (Fig. 3) gave the highest coefficient of determination ($r^2 = 0.607$). Starting accumulations on May 1 rather
than April 1 gave poorer correlations. Generally, accumulated deficit
(Table II) explained less of the variability in BAI than accumulated transpiration.

C. Carbon Isotope Ratios

There was no systematic temporal variation in $\delta^{13}C$ for the years 1964 to
1981. While there was significant intertree variability, differences between
individual trees were generally maintained over the measurement period

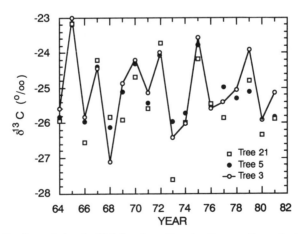

Figure 4.   Yearly variation in $\delta^{13}C$ for three Douglas-fir trees from the lower site from
1964 to 1981. Each point is the average of two cores taken from each tree.

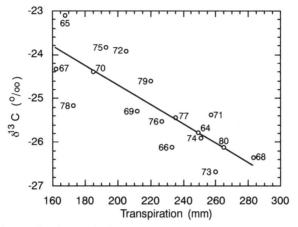

Figure 5. Accumulated transpiration (T) for the lower site between April 1 and the end of August or, if earlier, whenever soil water potential fell below -0.5 MPa for longer than 15 days, versus tree ring $\delta^{13}C$. Values of $\delta^{13}C$ are the averages of two cores from 2 or 3 trees. The equation of the regression line is $\delta^{13}C = -20.23 - 0.22253T$ ($r^2 = 0.677$). Numbers by the data points indicate the year.

(Fig. 4). Since these trees were exposed to the same environment it is likely that the intertree variability reflected genetic differences between trees.

There was generally good correlation between commulative transpiration and $\delta^{13}C$ (Fig. 5 and Table III). Cummulative deficit explained significantly less of the variability in $\delta^{13}C$ (Table IV). As with the correlations between BAI and accumulated transpiration, highest coefficients of determination for accumulated transpiration and $\delta^{13}C$ were for data from the lower site and when transpiration was accumulated between April 1 and the end of August or, if earlier, between April 1 and the time when transpiration had been soil limited for longer than 15 days.

Table III Coefficient of Determination (r^2) between
Tree Ring $\delta^{13}C$ and Accumulated Transpiration for
Lower and Upper Sites[a]

	Coefficient of determination	
Accumulation period	Upper site	Lower site
April to July	0.513	0.386
April to August	0.612	0.527
April to PSI	0.681	0.660
April to PSI/August	0.658	0.677

[a]Accumulations are for April 1 to the end of July, to the end of August, to after 2 weeks with soil water potentials less than -0.5 MPa (PSI), or August or PSI, whichever comes first.
All coefficients were significant as $P \leq 0.001$

Table IV Coefficient of Determination (r^2) between
Tree Ring δ^{13}C and Accumulated Water Deficit
for Lower and Upper Sites[a]

	Coefficient of determination	
Deficit period	Upper site	Lower site
April to June	0.316*	0.130
April to July	0.519***	0.521***
April to August	0.462**	0.493***
April to September	0.451**	0.405**

[a]Accumulations are for April 1 to the end of June, the end of July, the
end of August, and the end of September.
*, **, or *** indicate significance at $P \leq 0.05$, 0.01, or 0.001, respectively.

There was relatively poor correlation between BAI and δ^{13}C. The coefficients of determination for the upper and lower site were 0.41 and 0.48, respectively (Fig. 6).

IV. Discussion

Both δ^{13}C and BAI were better correlated with accumulated transpiration than seasonally accumulated deficits. This is because, for wet years with no deficits, the water balance model does not distinguish years with different

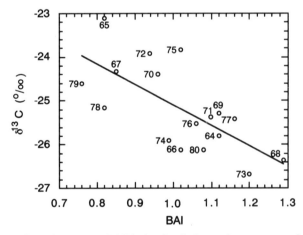

Figure 6. Basal area increment (BAI) index for the lower site versus tree ring δ^{13}C. Values of δ^{13}C are the averages of two cores from two or three trees. The equation of the regression line is δ^{13}C = −20.397 −4.6931BAI (r^2 = 0.478). Numbers by the data points indicate the year.

atmospheric demands. Nonetheless, Spittlehouse (1985) and Giles *et al.* (1985) have shown that the mean deficit over a number of years is a suitable variable for comparing sites.

The most appropriate period over which to accumulate transpiration differed between sites. This was likely related to differences in soil water storage. The April to July accumulation period was appropriate for the upper (and drier site) where tree diameter measurements indicated that over 80% of the stem diameter increment at 1.3 m occurred during this period even in wet years (Spittlehouse, 1985). However longer accumulation periods were more suitable for the lower and wetter site.

Correlations between transpiration and BAI and $\delta^{13}C$ were generally poorer for the lower than upper site. This might be related to the small number of samples taken at the lower site. However, the BAI index of the sampled trees compared well with that for the whole site. An additional factor might be that the water balance approach does not account for the fact that production of photosynthate for stem growth in August may be limited by high air temperature and increased photorespiration. Price and Black (1989, 1991) found a net loss of carbon dioxide from the canopy during such conditions, even though there was still substantial transpiration. This effect would not have been apparent at the higher site which usually had severe deficits and little transpiration in August. Spittlehouse (1985) indicated that in years when transpiration deficits occurred early in the summer there was a lower earlywood BAI index.

Leavitt and Long (1991) analyzed the intraring variation in $\delta^{13}C$. They found that the lowest values occurred during the latest phase of the growth of the ring, and this was probably related to reduction in moisture stress. Such an effect could have influenced our results.

It is probable that photosynthates produced after the soil had wetted in the fall may go into storage in the roots or leaves to aid overwinter maintenance respiration. Fritts (1976) notes that previous years' weather conditions can influence growth in subsequent years. However, no lag effects were found for previous years' transpiration for the upper site (Spittlehouse, 1985).

Data from both sites indicated that 1979 was an anomalous year in that stem increment was about 25% less than would be expected from the water balance data (Figs. 2 and 3). However $\delta^{13}C$ values for that year were well related to the modeled transpiration (Fig. 5). This suggests that the discrepancy between BAI and cummulative transpiration was not related to damage to photosynthetic mechanisms in the foliage. Spittlehouse (1985) suggested that the reduced stem increment could have resulted from below-normal winter temperatures that might have killed a substantial amount of foliage. However, the 1980 data do not indicate such effects, so large-scale reductions in foliage can probably be ruled out. It is possible that a lower amount of photosynthate was allocated to stem growth in 1979, perhaps due to an increase in the amount allocated to root growth. The low winter temperatures may have caused damage to the root system through soil freezing, and there may have been a preferential allocation of photosynthate to regrowth of roots the following summer.

V. Conclusions

Accumulated transpiration was positively correlated with an index of annual basal area increment and changes in carbon isotope discrimination. This is indicative of the strong linkage between plant-water status and photosynthesis. A simple daily water balance model using limited weather data adequately modeled transpiration to explain up to 80% of the annual variation in carbon isotope discrimination and tree growth at two sites.

References

Boutton, T. W. 1990. Stable carbon isotopes of natural materials. 1. Sample preparation and mass spectrometric analysis, pp. 157–171. *In* D. C. Coleman and B. Fry (eds.), Carbon Isotope Techniques in Biological Sciences. Academic Press, San Diego.

Carter, R. E., and K. Klinka. 1990. Relationships between growing-season soil water-deficit, mineralizable soil nitrogen and site index of coastal Douglas-fir. *For. Ecol. Manage.* **30:** 301–311.

Condon, A. G., R. A. Richards, and G. D. Farquhar. 1987. Carbon isotope discrimination is positively correlated with grain yield and dry matter production in field-grown wheat. *Crop Sci.* **27:** 996–1001.

Farquhar, G. D., M. H. O'Leary, and J. A. Berry. 1982. On the relationship between carbon isotope discrimination and the intercellular carbon dioxide concentration of leaves. *Aust. J. Plant Physiol.* **9:** 121–137.

Farquhar, G. D., K. T. Hubick, A. G. Condon, and R. A. Richards. 1988. Carbon isotope fractionation and plant water use efficiency, pp. 21–40. *In* P. W. Rundel, J. R. Ehleringer, and K. A. Nagy (eds.), Stable Isotopes in Ecological Research. Ecological Studies 68. Springer-Verlag, New York.

Francey, R. J., and G. D. Farquhar. 1982. An explanation of the $^{13}C/^{12}C$ variations in tree rings. *Nature* **290:** 232–235.

Freyer, H. D., and N. Belacy. 1983. $^{13}C/^{12}C$ records in northern hemispheric trees during the past 500 years—anthropogenic impact and climatic superpositions. *J. Geophys. Res. C* **88:** 6844–6852.

Fritts, H. D. 1976. Tree Rings and Climate. Academic Press, New York.

Giles, D. G., T. A. Black, and D. L. Spittlehouse. 1985. Determination of growing season water deficits on a forested slope using water balance analysis. *Can. J. For. Res.* **15:** 107–114.

Green, J. W., 1963. Wood Cellulose, pp. 9–21. *In* R. L. Whistler (ed.), Methods of Carbohydrate Chemistry. Academic Press, New York.

Hubick, K. T., and G. D. Farquhar. 1989. Carbon isotope discrimination and the ratio of carbon gained to water lost in barley cultivars. *Plant Cell Environ.* **12:** 795–804.

Keeling, C. D., W. G. Mook, and P. P. Tans. 1979. Recent trends in $^{13}C/^{12}C$ ratio of atmospheric carbon dioxide. *Nature* **277:** 121–123.

Kelliher, F. M., T. A. Black, and D. T. Price. 1986. Estimating the effects of understory removal from a Douglas-fir forest using a two-layer canopy evapotranspiration model. *Water Resource Res.* **22:** 1891–1899.

Leavitt, S. W., and A. Long. 1983. An atmospheric $^{13}C/^{12}C$ reconstruction generated through removal of climate effects from tree-ring $^{13}C/^{12}C$ measurements. *Tellus Ser. B* **35:** 92–102.

Leavitt, S. W., and A. Long. 1988. Intertree variability of ^{13}C in tree rings, pp. 95–104. *In* P. W. Rundel, J. R. Ehleringer, and K. A. Nagy (eds.), Stable Isotopes in Ecological Research. Ecological Studies 68. Springer-Verlag, New York.

Leavitt, S. W., and A. Long. 1989. Drought indicated in carbon-13/carbon-12 ratios of southwestern tree rings. *Water Res. Bull.* **25:**341–347.

Leavitt, S. W., and A. Long. 1991. Seasonal stable-carbon isotope variability in tree rings: Possible paleoenvironmental signals. *Chem. Geol.* **87:** 59–70.

Martin, B., and E. K. Sutherland. 1990. Air pollution in the past recorded in width and stable carbon isotope composition of annual growth rings of Douglas-fir. *Plant Cell Environ.* **13:** 839–844.

McNaughton, K. G., and T. W. Spriggs. 1989. An evaluation of the Priestley and Taylor equation and the complimentary relationship using results from a mixed-layer model of the convective boundary layer, pp. 89–104. *In* T. A. Black, D. L. Spittlehouse, M. D. Novak, and D. T. Price (eds.), Estimation of Areal Evapotranspiration. International Association of Hydrological Science, Wallingford, UK.

Peng, T-H., W. S. Broecker, H. D. Freyer, and S. Trumbore. 1983. A deconvolution of the tree ring based ^{13}C record. *J. Geophys. Res.* **88:** 3609–3620.

Price, D. T., and T. A. Black. 1989. Estimation of forest transpiration and CO_2 uptake using the Penman-Monteith equation and a physiological photosynthesis model, pp. 213–227. *In* T. A. Black, D. L. Spittlehouse, M. D. Novak, and D. T. Price (eds.), Estimation of Areal Evapotranspiration. International Association of Hydrological Science, Wallingford, UK.

Price, D. T., and T. A. Black. 1991. Effects of summertime changes in weather and root-zone soil water storage on canopy CO_2 flux and evapotranspiration of two juvenile Douglas-fir stands. *Agric. Forest Meterol.* **53:** 303–323.

Robertson, E. O., L. A. Jozsa, and D. L. Spittlehouse. 1990. Estimating Douglas-fir wood production from soil and climate data. *Can. J. Forest Res.* **20:** 357–364.

Spittlehouse, D. L. 1985. Determining the year-to-year variation in growing season water use of a Douglas-fir stand, pp. 235–254. *In* B. A. Hutchison and B. B. Hicks (eds.), The Forest Atmosphere Interaction, Reidel, Dordrecht, Holland.

Spittlehouse, D. L. 1989. Estimating evaporation from land surfaces in British Columbia, pp. 245–256. *In* T. A. Black, D. L. Spittlehouse, M. D. Novak, and D. T. Price (eds.), Estimation of Areal Evapotranspiration. International Association of Hydrological Science, Wallingford, UK.

Spittlehouse, D. L., and T. A. Black. 1981. A growing season water balance model applied to two Douglas-fir stands. *Water Resources Res.* **17:** 1651–1656.

Stuiver, M., R. L. Burk, and P. D. Quay. 1984. $^{13}C/^{12}C$ ratios in tree rings and the transfer of biospheric carbon to the atmosphere. *J. Geophys. Res.* **89:** 11731–11748.

Swerhone, G. D. W., K. A. Hobson, C. van Kessel, and T. W. Bouton. 1991. An economical method for the preparation of plant and animal tissue for $\delta^{13}C$ analysis. *Commun. Soil Sci. Plant Anal.* **22:** 177–190.

Tans, P. P., and W. G. Mook. 1980. Past atmospheric CO_2 levels and the $^{13}C/^{12}C$ ratios in tree rings. *Tellus* **32:** 268–283.

Virgona, J. M., K. T. Hubick, H. M. Rawson, G. D. Farquhar, and R. W. Downes. 1990. Genotypic variation in transpiration efficiency, carbon isotope discrimination and carbon allocation during early growth in sunflower. *Aust. J. Plant Physiol.* **17:** 207–214.

11

Carbon and Water Relations in Desert Plants: An Isotopic Perspective

James R. Ehleringer

Deserts have fascinated ecologists for many decades, not only because of the extremes in abiotic conditions and the harshness of the environmental conditions to which plants must be adapted, but also because of the diversity of physiological and morphological patterns evolved by plants that persist in such regions. Water and the discussions of water use have been pervasive in our thinking about desert plants and of how they have adapted to the overall low precipitation amounts and prolonged soil moisture deficits that characterize these environments (Walter and Stadelmann, 1974; MacMahon and Schimpf, 1981). The historical view was that the primary selective force in plant adaptation to deserts was abiotic extremes of high temperature and low precipitation; competitive interactions were thought to be essentially nonexistent (Shreve and Wiggins, 1964). This view has given way to a broader view that competitive interactions (especially for water) may be as important in structuring desert communities (Fonteyn and Mahall, 1978; Fowler, 1986) as the adaptations that allow plants to persist through environmental extremes. In this paper, the role that stable isotopes have played in increasing our understanding of the water and carbon relations of desert plants is examined from both abiotic and biotic interaction perspectives.

I. Desert Climate and Life Form Variation

On a global basis, deserts share one feature in common: limited precipitation. Long-term average precipitation in deserts is less than 250 mm annually. Most years are below average (arithmetic mean), since annual precipitation amounts in deserts are not normally distributed, but instead follow a gamma distribution (McDonald, 1956). The coefficient of variation of annual precipitation is a measure of the unpredictability of that precipitation; it increases exponentially with decreases in mean annual precipitation (Fig. 1). What this means for plant performance in aridlands is that perennial plants must not only be adapted to low precipitation amounts, but also, as average precipitation decreases, they must be able to tolerate droughts of increasing frequency and duration.

Under such water-limited conditions, we might expect selection for tolerance of water deficits, perhaps efficiency in water use, or drought escape by completion of the growth cycle before onset of drought. Vegetation com-

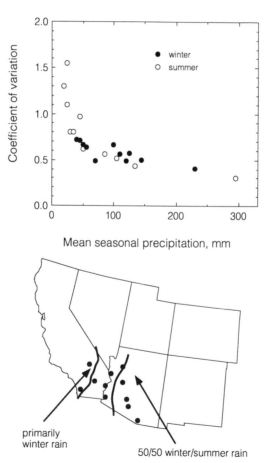

Figure 1. Coefficients of variation of summer precipitation and winter precipitation as a function of long-term mean values (1929–1988) for different aridland locations in the southwestern United States. Based on data from U.S. Weather Bureau records.

position and the morphological or physiological characteristics of these species that allow them to persist in arid climates vary from strongly convergent to highly variable between desert locations, reflecting both abiotic and edaphic constraints (Walter and Stadelmann, 1974; MacMahon and Schimpf, 1981; Smith and Nowak, 1990). While much of the metabolic activity of desert plants is driven by the availability of soil moisture, it is to a large extent the seasonality, duration, and predictability of that precipitation that influences variation in species composition and physiological diversity.

II. Carbon Isotope Composition as a Measure of the Set Point for Gas Exchange Activity

Gas exchange metabolism at the leaf level can be considered as consisting of a number of trade-offs. As stomata open to allow increased CO_2 diffusion rates, outgoing H_2O diffusion rates are also increased. Where a specific plant operates in this continuum between maximizing photosynthetic rate and minimizing transpiration will depend on both environmental conditions and genetic constraints. At either of the extremes, two alternative patterns of water use exist for desert plants (Mulroy and Rundel, 1977; Mooney and Gulmon, 1982). In the first, plants possess a rapid growth rate and high rates of gas exchange during moist periods of the year and complete seasonal growth before the onset of drought. These plants tend to have limited tolerance to water deficits. In contrast, the second category of plants have a prolonged period of growth and lower rates of gas exchange, and maintain activity longer into the drought period. These plants tend to be tolerant of water deficit. Perhaps an oversimplification, but examples of these contrasting patterns would be annuals versus perennials or drought-deciduous versus evergreen perennials. There is no *a priori* reason why these contrasting patterns, albeit perhaps of less magnitude, cannot occur intraspecifically or even within a single breeding population. In this chapter, I first consider interspecific aspects of variation and then consider intraspecific aspects of variation and their natural selection consequences.

Strong trade-offs will exist between photosynthesis (A) and transpiration (E) of the dominant desert landscape species as stomata open and close. Perhaps contrary to popular perception, the preponderance of desert species have C_3 photosynthesis, not C_4 or CAM photosynthesis (Stowe and Teeri, 1978; Teeri *et al.*, 1978; Winter and Troughton, 1978; Winter, 1981). The C_4 photosynthetic pathway is common only in those perennials of saline habitats (especially in the Chenopodiaceae) and in summer-active herbaceous species (grasses, annuals) in areas receiving reliable summer monsoonal rains.

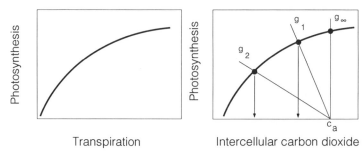

Figure 2. (Left) The expected relationships between photosynthetic rate (A) and transpiration (E) of C_3 photosynthetic leaves as stomata open or close under high light. (Right) The dependence of photosynthesis on intercellular CO_2 (c_i) in a C_3 plant and its relationship to stomatal conductance. The slope of the line emerging from the ambient CO_2 concentration (c_a) and intersecting the $A - c_i$ response curve is the stomatal conductance. The point of intersection of the two curves is the operational c_i value and photosynthetic rate. The extent of photosynthetic stomatal limitation is the reduction in photosynthesis from that occurring with infinite stomatal conductance ($c_i = c_a$) to the observed photosynthetic rate.

Associated with differences in photosynthetic rate of C_3 species will be differences in instantaneous water-use efficiency (A/E), depending on where individual leaves are operating on the photosynthesis versus transpiration curve (Fig. 2). Just where a leaf is operating on this curve also defines $\partial A/\partial E$ (Cowan and Farquhar, 1977); A/E is also closely related to the extent of stomatal limitation of photosynthetic rate (Sharkey, 1985). Thus, while many of the contributions to this volume have focused on carbon isotope composition (especially carbon isotope discrimination, Δ) and its exclusive application to water-use efficiency, the other water-related gas exchange parameters should not be ignored. If natural selection favors some parameter related to long-term intercellular CO_2 concentration (as measured by carbon isotope composition), it will be difficult to distinguish among the above, closely related characters, a point worth keeping in mind when interpreting the carbon isotope composition data that follow.

A more productive approach might be to consider long-term intercellular CO_2 concentration as being an indicator of the set point for gas exchange metabolism, reflecting overall trade-offs between carbon gain and water loss and associated characters that go along with having either a higher or lower rate of gas exchange activity. The Δ value then becomes a convenient measure of long-term intercellular CO_2. Many features other than simply leaf-level physiology may show strong correlations with the Δ value, including aspects of water-conducting capacity, root/shoot surface areas, and mineral nutrition. As seen below, plants with lower Δ values not only tend to have higher water-use efficiencies, but also lower photosynthetic rates, longer life expectancies, and greater survival under long-term stress conditions. On the other hand, plants with higher Δ values tend to possess an opposite set of metabolic and life-history characteristics.

III. Δ as a Reliable Indicator of Intercellular CO₂ Concentration and Water-Use Efficiency

In C_3 desert plants (Fig. 3), carbon isotope discrimination (Δ) is related to intercellular CO_2 concentration (c_i) as predicted by a model developed by Farquhar *et al.* (1982). These on-line gas exchange data suggest that there is no reason to suspect that desert species behave any different from theoretically expected relationships for C_3 species and, therefore, that it is possible from field observations of Δ in leaves to infer long-term c_i/c_a values.

Direct extrapolation of carbon isotope discrimination data to water-use efficiency among plants assumes that leaf temperatures are equal. Rarely is this exactly satisfied; more typically, leaf temperatures are within several degrees of each other, depending on leaf size, thermal load, and transpiration rate. Ehleringer *et al.* (1992) evaluated the consequence of leaf temperature differences between plants and how this might affect interpretation of rankings of water-use efficiency based on Δ values alone. They calculated a near-linear relationship between leaf temperature differential and the difference in isotopic composition necessary to distinguish differences in water-use efficiency. When leaf temperature differentials between plants were ≤2.5°C, a 1‰ difference in Δ was sufficient to correctly and unambiguously rank plants with respect to water-use efficiency, indicating the extent to which changes in intercellular CO_2 can offset possible differences in the evaporative gradient among plants when calculating water-use efficiencies. As leaf Δ values among different species often range 4‰ on a

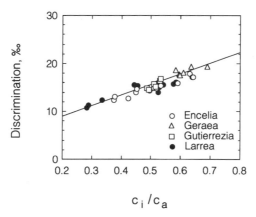

Figure 3. Observed carbon isotope discrimination values from on-line gas exchange measurements as a function of the simultaneously measured ratio of intercellular to ambient CO_2 concentrations for *Encelia farinosa* (drought-deciduous shrub), *Geraea canescens* (annual), *Gutierrezia sarothrae* (winter-deciduous shrub), and *Larrea tridentata* (evergreen shrub). Line through the data represents the C_3 carbon isotope discrimination model, $\Delta = a + (b - a) \, c_i/c_a$, where a is 4.4‰ and b is 27‰. From Ehleringer *et al.* (1992).

given sampling date, Ehleringer *et al.* (1992) concluded that Δ values appeared to be a feasible approach for ranking relative short- or long-term water-use efficiency differences among aridland plants.

IV. Desert Environments Are Characterized by Low Δ Values

Surveys of the carbon isotope composition of aridland plants have indicated that most are C_3 species and are usually characterized by low c_i values (reviewed in Ehleringer, 1989); absolute Δ values appear related to individual species longevity (Ehleringer and Cooper, 1988; Ehleringer and Cook, 1991). In the driest desert environments, such as the Atacama Desert of northern Chile, Δ values are quite low with effective c_i values as low as 125 μl liter^{-1} (Fig. 4). Such low c_i values might once have been thought to occur only in C_4 species, and they indicate that stomata are very nearly closed during the main periods of carbon gain. Such low Δ values have now been found in a large number of species from deserts throughout the world (see Rundel, Chapter 12, this volume; Winter and Troughton, 1978; Winter, 1981; Ehleringer, 1989; Ehleringer *et al.*, 1992).

The Δ values of perennial plants from the Sonoran Desert of North America are generally not as low as those observed in plants from the Atacama Desert, likely reflecting the shorter drought duration that plants experience in the Sonoran Desert. Whereas precipitation in the Atacama Desert comes primarily during the winter and plants experience frequent extended interannual droughts, most parts of the Sonoran Desert are less arid and plants there experience drought periods that typically last less than a single year.

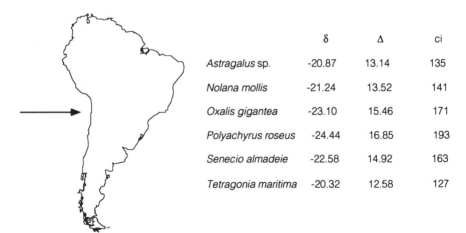

	δ	Δ	c_i
Astragalus sp.	-20.87	13.14	135
Nolana mollis	-21.24	13.52	141
Oxalis gigantea	-23.10	15.46	171
Polyachyrus roseus	-24.44	16.85	193
Senecio almadeie	-22.58	14.92	163
Tetragonia maritima	-20.32	12.58	127

Figure 4. Carbon isotope discrimination values and corresponding calculated intercellular carbon dioxide concentrations for several species common to the Atacama Desert of northern Chile. Based on data from Ehleringer *et al.* (1993).

One pattern that emerges from studies of Sonoran Desert plant species is that carbon isotope discrimination at the species level is inversely related to life expectancy. Ehleringer and Cooper (1988) observed that longer-lived species (>50 years) had the lowest Δ values among species within the community, irrespective of the plant age when measurements were made. Short-lived perennial species (2–5 years) had carbon isotope values that were 3–5‰ lighter. Species with intermediate life expectancies (10–40 years) were intermediate in Δ value. Recently Ehleringer *et al.* (1993b) extended this observation by evaluating Δ values in five different communities across the Colorado Plateau, Mojave, and Sonoran Deserts. These observations confirmed the earlier pattern; while there is overlap among categories, there remains a clear pattern with Δ values of long-lived species being lowest and short-lived highest (Fig. 5). All of these observations represent mean Δ values ($n = 5$–10 individuals per species) for mature individuals of a species. It is worth noting that seedlings and young juveniles often do have higher carbon isotope discrimination values than adults, although relative ranking among species remains consistent (Sandquist *et al.*, in press).

Between slope and wash microhabitats, Ehleringer and Cooper (1988) observed gradients in Δ values for a species, suggesting higher Δ values for individuals occupying the wettest microhabitats (wash). These patterns were initially attributed to acclimation to soil moisture variability. However, recently Schuster *et al.* (unpublished data) have shown that there is significant local genetic differentiation across microhabitats in the desert, specifi-

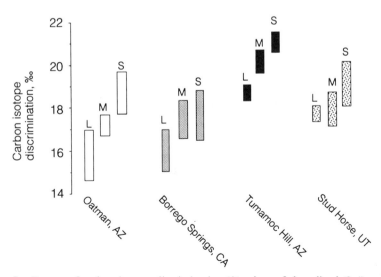

Figure 5.   Ranges of carbon isotope discrimination (Δ) values of short-lived (2–5 years), medium-lived (10–40 years), and long-lived (>50 years) species at four different sites in the deserts of western North America. Data are from Ehleringer and Cooper (1988) and Ehleringer *et al.* (1993b).

cally between a wash and adjacent slope microhabitats. Carbon isotope discrimination is a highly heritable character in desert plants (Schuster *et al.*, 1992a) and the higher Δ values of wash plants likely represent local adaptation, although the significance of these Δ value differences is not yet well established.

The Δ surveys described above represent a snapshot picture in time, but these observations are consistent with longer-term observations. Ehleringer and Cook (1991) analyzed Δ values of species across years at the same site and observed a significant, consistent ranking of differences among species (Fig. 6). The slope of the relationship was not 1 : 1, likely reflecting an acclimation by the species to site water deficits in the different years. Likewise, the Δ values of individuals are not constant through the season, but vary with soil moisture deficit (Smith and Osmond, 1987; DeLucia and Heckathorn, 1989; Ehleringer *et al.*, 1992).

The absolute water-use efficiency through the growing season is very much influenced by both changes in c_i (measured by Δ) and by temperature changes, which influences the leaf-to-air water vapor gradient. Changes in the evaporative gradient in deserts are sufficiently large that plants are not able to maintain a constant water-use efficiency throughout the year (Ehleringer *et al.*, 1992). Instead, absolute water-use efficiency fluctuates substantially in response to seasonal temperature changes between winter, spring, and summer months. However, it is doubtful that there is any substantive advantage to the plant to maintain a constant water-use efficiency. What is likely to be of greater importance to plant fitness is its performance relative to others in the community. Thus, it is the relative ranking of Δ values that provides greater insight into any fitness component that might be related to water-use efficiency. Generally, the relative

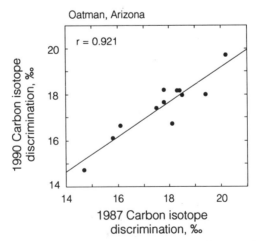

Figure 6. Mean carbon isotope discrimination (Δ) values for several species in 1987 and 1990 at a desert site near Oatman, Arizona. From Ehleringer and Cook (1991).

ranking among plants remain constant over time (Johnson *et al.*, 1990; Ehleringer and Cook, 1991; Ehleringer, 1993).

V. Interpopulation-Level Variation in Carbon Isotope Discrimination

As clearly described in many chapters within this volume, cultivar-level variation in carbon isotopic composition is known to exist within crop species. While such variation may represent the product of agricultural breeding efforts, the variation is no doubt reflective of the level of variation to be

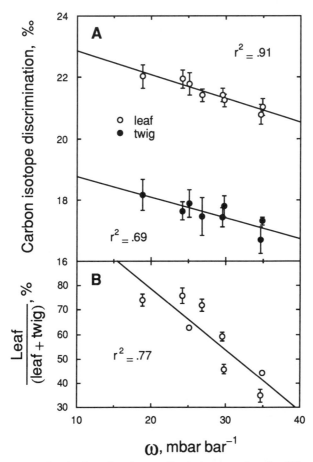

Figure 7. (A) Leaf and twig carbon isotope discrimination values for different ecotypes of *Hymenoclea salsola* field grown under uniform conditions plotted as a function of ω, the average leaf-to-air water vapor gradient weighted for those times of the year when soil moisture is available. (B) Leaf photosynthetic area as a fraction of total photosynthetic area (leaf and twig) plotted against ω. Data are from Comstock and Ehleringer (1992).

expected in native species. Interpopulation variability (i.e., ecotypic) has been examined in a limited number of species and is thought to reflect differences in environmental conditions to which plants are adapted. Comstock and Ehleringer (1992) have shown ecotypic variation in *Hymenoclea salsola,* a common shrub in the Mojave and Sonoran Deserts.

Under common garden conditions, the isotopic variation was greater than 2‰ and was negatively related to ω (Comstock and Ehleringer, 1992), the average leaf-to-air water vapor gradient weighted for periods when soil moisture was available (Fig. 7). *H. salsola* has both photosynthetic twigs and leaves, with twigs always having a greater water-use efficiency (Comstock and Ehleringer, 1988). The fraction of leaf-to-twig photosynthetic areas is also negatively related to ω, resulting in plants from drier habitats (atmospheric drought) having both higher water-use efficiencies at the leaf level as well as a greater allocation to the more water-use efficient twig tissues in these environments. Overall, this results in a combined morphological–physiological progression toward canopies of greater water-use efficiency in climates with drier atmospheric conditions. The implication of the Comstock and Ehleringer (1992) study is that the seasonality of soil moisture inputs is important in affecting absolute Δ values; in habitats where precipitation occurs during the hotter summer months, plants had lower Δ values than plants in sites receiving equivalent amounts of precipitation during cooler winter–spring periods of the year.

VI. Carbon Isotope Discrimination and Competition for Water

If Δ values are used to evaluate water-use efficiency, any significance of possible water-use efficiency differences should depend on knowing whether plants are competing for the same water source. That is, variation in the patterns of water use will likely be dependent on whether adjacent plants within a community are competing for the same limiting resource. It can be argued that efficient use of a resource, such as water, may only be adaptive if plants exert some control over the rates of soil water extraction from the soil volume in which their roots are located. If plants are competing for the same limiting water source, there may be selection against conservative use of this resource, and capture of the resource as fast as possible and perhaps even wasteful consumption of the water since it cannot be internally stored (except in succulent crassulacean acid metabolism (CAM) plants). Unfortunately insufficient data are available to assess possible intraspecific variations in the use of soil moisture. However, a number of data sets exist utilizing isotopic analyses of xylem sap, which allow a quantitative evaluation of the extent to which neighboring plants use the same or different soil moisture sources (see Dawson, Chapter 30, and Thorburn, Chapter 32, in this volume). When δD xylem sap analyses are combined with water-stress estimates, they provide a quantitative measure of both water used by neighboring species and of the water-stress impact on plant performance due to the presence of that neighbor.

In the Colorado Plateau Desert of southern Utah, there appears to be life-form-dependent variation in sources of moisture used during the summer months (Fig. 8). Annual and herbaceous perennial vegatation appear to use the moisture from summer convection storms, whereas woody perennial species either use none or only a limited fraction of the summer-input moisture (Ehleringer *et al.*, 1991). One possible conclusion from these data is that woody and herbaceous perennials (as functional groups) were not in direct competition with each other. However, during the spring growing season and during stress periods of the year (late spring and fall), the hydrogen isotope ratios of xylem sap of woody and herbaceous perennials were similar, suggesting use of the same general water source (Ehleringer *et al.*, 1991). This does not demonstrate that they are competing for the same water on a microscale, it simply says that they are exploiting the same general moisture stratum in the soil. Yet plants in these arid zones are thought to compete for limited soil moisture (Fowler, 1986); it may be that there is competition for soil moisture among species within a life form during the summer and among species across all life forms at

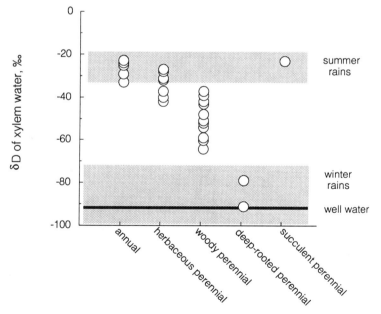

Figure 8. Hydrogen isotope ratios of xylem sap in mid-summer of different species co-occurring at Stud Horse Point, Utah, a Colorado Plateau desert site. Species are functionally grouped according to life form. Also shown are the range of summer and winter rain hydrogen isotope ratio values as well as the value for groundwater at this site. Based on data in Ehleringer *et al.* (1991).

other times of the year. As of yet, however, there is no direct experimental evidence to support these anticipated competitive patterns. High water-use efficiency (such as associated with a reduced rate of water use) might not have high survival value in such competitive environments, unless there were other overriding factors selecting for high water-use efficiency (or something else related to low Δ values). Long-term survival through extended drought periods is one such consideration and is discussed in the following section. During summer months, when herbaceous perennials have access to surface moisture minimally used by woody perennials, it is interesting to note that the Δ values of species within the community were positively associated with the utilization of summer moisture inputs. This is the water-use pattern expected if herbaceous plants were not competing with woody perennials, but with the main loss of surface moisture due to abiotic physical factors.

VII. Selection for Variation in Carbon Isotope Discrimination Values

To better understand the possible significance of Δ, it is necessary to focus on the extent of the variation found at the individual level, where natural selection is operating. Limited population level data are available, but recent studies in desert plants (Schuster *et al.*, 1992b; Ehleringer, 1992) indicate that there can be significant intrapopulational variation in Δ values (Fig. 9). In a study of the extent of population-level variance in Δ values of warm and cold desert ecosystems, Schuster *et al.* (1992b) observed that variance was greater in the shorter-lived species than in the longer-lived species. If long-term survival is related to a plant's Δ value through water-use efficiency, stomatal limitation, $\partial A/\partial E$, or other water-related characters as mentioned earlier, than both lower Δ values and a narrower variance in values would be expected in that long-lived population. This is exactly the pattern observed in the species comparisons, where variances in Δ values were 0.82 versus 0.28, and 0.92 versus 0.47 for the shorter- versus longer-lived species, respectively, and population mean values of the shorter-lived species were 1.1 and 2.2‰ higher than those of the longer-lived species in the cold- and warm-desert habitats, respectively. Moreover, the longer-lived species in each comparison were characterized by a regular spatial distribution, consistent with the notion that plants were no longer competing with each other (Fonteyn and Mahall, 1978). In contrast, the shorter-lived species in each of the comparisons had clumped spatial distributions.

Variation in Δ values at the population level is consistent over time (Fig. 10). Leaf Δ values from *Encelia farinosa* shrubs in the Sonoran Desert changed between successive sampling periods during the spring, but there was a high degree of uniformity in the relative rankings of those Δ values through time. Johnson *et al.* (1990) showed that cloned grasses also maintained the same relative rankings of Δ values when grown under contrast-

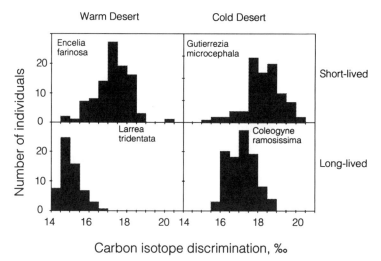

Figure 9.   Frequency distribution of carbon isotope discrimination values of subpopulations of species occurring in the Colorado Plateau (cold desert) and Sonoran Desert (warm desert) that differ in their life expectancy. Data are from Schuster *et al.* (1992b).

ing soil moisture regimes in a rain-out shelter. Furthermore, many studies have reported significant correlations in the rankings of Δ values between years (Johnson *et al.*, 1990; Ehleringer and Cook, 1991). The importance of these observations is that snapshot evaluations of Δ values at a particular

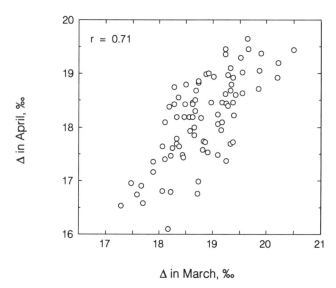

Figure 10.   Carbon isotope discrimination values for individuals of *Encelia farinosa* measured under natural conditions in early March and again on newly produced leaves in late April. Data are from Ehleringer (1993).

point in time may be useful in extrapolating to long-term patterns in Δ values.

The Δ value appears to be a highly heritable character in native species, with heritabilities of up to 0.7 reported in the literature (Geber and Dawson, 1990; Schuster *et al.*, 1992a). Similar values have been reported for agricultural species by others in this volume. A large additive genetic variance in a native species may be indicative of frequent changes in the direction of the selective pressures in the natural environment. Either spatial (e.g., microsite) or temporal (year-to-year) variability in the direction of the selective pressures could result in a broad population-level variance, if either of the two tails of the Δ distribution had a selective advantage over the other under a particular set of environmental conditions.

Longer-lived organisms appear to have lower Δ values. This is consistent in both interspecies comparisons (Ehleringer and Cooper, 1988; Smedley *et al.*, 1991; Ehleringer *et al.*, 1992) and intraspecies comparisons (Ehleringer *et al.*, 1990; Richards and Condon, Chapter 29, this volume; Hall *et al.*, Chapter 23, this volume). Within *E. farinosa* populations, there is a weak but significant correlation with plant size (Ehleringer, 1993). Since reproductive output is also correlated with plant size (Ehleringer and Clark, 1988), it would appear that either low Δ results in larger plants or some critical factor eliminates high Δ value plants from the population over extended time periods. Long-term drought may be such a selective pressure.

Under natural field conditions, *E. farinosa* plants compete for water (Ehleringer, 1984). Plants with neighbors had greater water deficits (as measured by plant-water potential), maintained lower leaf areas and growth rates, and had lower reproductive rates than plants whose neighbors had been removed. Related to this, it may be that high Δ plants in the population are poorer competitors for limited soil moisture. In an experiment to evaluate the competitive and growth relationships of different Δ value genotypes within the population, Ehleringer (1993) removed neighbors from around individuals having differing Δ values. The rankings of Δ values among *E. farinosa* individuals were maintained between pre- and postneighbor-removal periods (Fig. 11). Growth rate in response to neighbor removal was proportional to Δ value (Fig. 12); no differences in growth rates were observed for similar control plants. Such a positive response suggests an adaptive value for high Δ plants in environments of high resource availability, such as years with above-average precipitation or in microhabitats without neighbors. On the other hand, it is not clear whether there are disadvantages to having a high Δ value.

The experiment presented in Fig. 12 was repeated on an adjacent hillside in 1987–1991. However, after the experiment was underway there was a period of unusually long and severe drought throughout the deserts of the western United States (1988–1990). Mortality in *E. farinosa* populations that had been monitored since the early 1980s was 50–80%. Ehleringer (1993) noted that mortality was dependent on the presence of

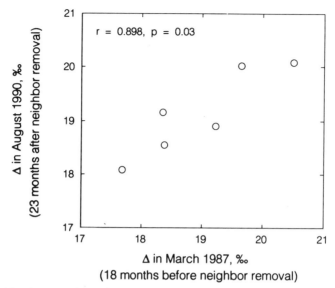

Figure 11. A comparison of carbon isotope discrimination values measured on *Encelia farinosa* individuals in the field 18 months before neighbors had been removed and 23 months following neighbor removal. Data are from Ehleringer (1993).

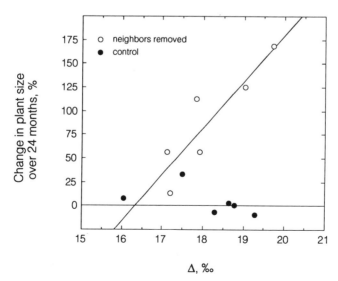

Figure 12. Growth rates of *Encelia farinosa* individuals following neighbor removal and of control plants whose neighbors were not removed as a function of the carbon isotope discrimination value of the plant. Data are from Ehleringer (1993).

neighbors. Those *E. farinosa* individuals whose neighbors had been removed (prior to the onset of the extended drought) survived the drought period, irrespective of their Δ value. However, for control plants whose neighbors had not been removed, low Δ plants had a significantly higher probability of surviving the drought than did high Δ plants. The mechanism whereby the low Δ plants survived the extended drought is unknown, but this pattern is consistent with community-level observations that long-lived species have lower Δ values than short-lived components of the community.

An implication of these results is that where desert habitats are spatially heterogeneous and/or temporally variable, natural selection against both high and low Δ individuals may occur. The deserts of North America are spatially heterogeneous (e.g., wash and slope microhabitats). Schuster *et al.* (unpublished) have shown that *E. farinosa* individuals occupying wash microhabitats are genetically distinct from those on adjacent slope microhabitats. Plants in the wash microhabitat have higher Δ values, but this may reflect either the greater water availability in this microhabitat (possible acclimation response) or the decreased life expectancy of plants occupying areas subject to frequent flash flood events (possible genetic response). The temporal variability of soil moisture availability and the likelihood for extended drought periods may select against high Δ individuals. It is possible that field observations indicating a tendency for low Δ values to be associated with the larger individuals in a population may reflect drought-induced loss of high Δ individuals over time. Long-term observations and experimental manipulations are necessary to evaluate this hypothesis.

VIII. Summary

There has been a recent tendency to exclusively associate Δ values with water-use efficiency in C_3 plants. Yet it is clear that water-use efficiency, stomatal limitations to photosynthesis, $\partial A/\partial E$, and other possible water-relations parameters will be closely linked with Δ, and it will be difficult to differentiate among these as to which is the most important parameter with respect to adaptation. A more productive approach might be to consider Δ as an indicator of the set point for gas exchange metabolism, reflecting overall tradeoffs between carbon gain and water loss.

Plants in aridlands have low Δ values. However, substantial variation in Δ values does exist within plant communities, and it is negatively correlated with the life expectancy of a species. At the population level of a species, high Δ values appear to be associated with high growth rate under noncompetitive conditions, whereas low Δ values appear to be related to the ability to persist through long-term drought conditions. Variation in Δ within a population may be related to selection for either tail of the distribution, driven by both spatial and temporal variability in habitat quality and water availability. In the future, greater emphasis should be placed on understanding the dynamics of populations in desert environments, where episodic events play a major role in structuring populations and landscapes.

References

Comstock, J. P., and J. R. Ehleringer. 1988. Contrasting photosynthetic behavior in leaves and twigs of *Hymenoclea salsola,* a green-twigged, warm desert shrub. *Am. J. Bot.* **75:** 1360–1370.

Comstock, J. P., and J. R. Ehleringer. 1992. Correlating genetic variation in carbon isotopic composition with complex climatic gradients. *Proc. Natl. Acad. Sci.* **89:** 7747–7751.

Cowan, I. R., and G. D. Farquhar. 1977. Stomatal functions in relation to leaf metabolism and environment, pp. 471–505. *In* Integration of Activity in the Higher Plant. Symposium of the Society of Experimental Biology, Vol. 31.

Dawson, T. E., and J. R. Ehleringer. 1991. Streamside trees that do not use stream water. *Nature* **350:** 335–337.

DeLucia, E., and S. A. Heckathorn. 1989. The effect of soil drought on water-use efficiency in a contrasting Great Basin Desert and Sierran montane species. *Plant Cell Environ.* **12:** 935–940.

Ehleringer, J. R. 1984. Intraspecific competitive effects on water relations, growth, and reproduction in *Encelia farinosa. Oecologia* **63:** 153–158.

Ehleringer, J. R. 1989. Carbon isotope ratios and physiological processes in aridland plants, pp. 41–54. *In* P. W. Rundel, J. R. Ehleringer, and K. A. Nagy (eds.), Stable Isotopes in Ecological Research. Ecological Studies Series. Springer-Verlag, New York.

Ehleringer, J. R. 1993. Carbon isotope variation in *Encelia farinosa:* Implications for competition and drought survival. *Oecologia,* in press.

Ehleringer, J. R., and C. Clark. 1988. Evolution and adaptation in *Encelia* (Asteraceae), pp. 221–248. *In* L. Gottlieb and S. Jain (eds.), Plant Evolutionary Biology. Chapman & Hall, London.

Ehleringer, J. R., and C. S. Cook. 1991. Carbon isotope discrimination and xylem D/H ratios in desert plants, pp. 489–497. Stable Isotopes in Plant Nutrition, Soil Fertility, and Environmental Studies. IAEA, Vienna.

Ehleringer, J. R., and T. A. Cooper. 1988. Correlations between carbon isotope ratio and microhabitat in desert plants. *Oecologia* **76:** 562–566.

Ehleringer, J. R., J. W. White, D. A. Johnson, and M. Brick. 1990. Carbon isotope discrimination, photosynthetic gas exchange, and water-use efficiency in common bean and range grasses. *Acta Oecol.* **11:** 611–625.

Ehleringer, J. R., S. L. Phillips, W. F. S. Schuster, and D. R. Sandquist. 1991. Differential utilization of summer rains by desert plants: Implications for competition and climate change. *Oecologia* **88:** 430–434.

Ehleringer, J. R., S. L. Phillips, and J. P. Comstock. 1992. Seasonal variation in the carbon isotopic composition of desert plants. *Funct. Ecol.* **6:** 396–404.

Ehleringer, J. R., P. W. Rundel, B. Palma, and H. A. Mooney. 1993a. Carbon isotope ratios of plants of the Atacama Desert, in review.

Ehleringer, J. R., D. R. Sandquist, and S. L. Phillips. 1993b. Patterns of life form and carbon isotope discrimination across deserts in western North America, in review.

Farquhar, G. D., and R. A. Richards. 1984. Isotopic composition of plant carbon correlates with water-use efficiency of wheat genotypes. *Aust. J. Plant Physiol.* **11:** 539–552.

Farquhar, G. D., M. H. O'Leary, and J. A. Berry. 1982. On the relationship between carbon isotope discrimination and the intercellular carbon dioxide concentration in leaves. *Aust. J. Plant Physiol.* **9:** 121–137.

Farquhar, G. D., J. R. Ehleringer, and K. T. Hubick. 1989. Carbon isotope discrimination and photosynthesis. *Annu. Rev. Plant Physiol. Mol. Biol.* **40:** 503–537.

Fonteyn, P. J., and B. E. Mahall. 1978. Competition among desert perennials. *Nature* **275:** 544–545.

Fowler, N. 1986. The role of competition in plant communities in arid and semi-arid regions. *Annu. Rev. Ecol. Syst.* **17:** 89–110.

Geber, M. A., and T. E. Dawson. 1990. Genetic variation in and covariation between leaf gas exchange, morphology, and development in *Polygonum arenastrum*, an annual plant. *Oecologia* **85:** 53–158.

Johnson, D. A., K. H. Asay, L. L. Tieszen, J. R. Ehleringer, and P. G. Jefferson. 1990. Carbon isotope discrimination—potential in screening cool-season grasses for water-limited environments. *Crop Sci.* **30:** 338–343.

MacMahon, J. A., and D. J. Schimpf. 1981. Water as a factor in the biology of North American desert plants, pp. 114–171. *In* D. D. Evans and J. L. Thames (eds.), Water in Desert Ecosystems. Dowden, Hutchinson and Ross, Inc., Stroudsburg, PA.

McDonald, J. E. 1956. Variability of Precipitation in an Arid Region: A Survey of Characteristics for Arizona. University of Arizona Institute of Atmospheric Physics Tech. Report No. 1.

Mooney, H. A., and S. L. Gulmon. 1982. Constraints on leaf structure and function in reference to herbivory. *Bioscience* **32:** 198–206.

Mulroy, T. W., and P. W. Rundel. 1977. Annual plants: Adaptations to desert environments. *Bioscience* **27:** 109–114.

Sandquist, D. R., W. S. F. Schuster, L. A. Donovan, S. L. Phillips, and J. R. Ehleringer. 1993. Differences in carbon isotope discrimination between seedlings and adults of southwestern desert perennial plants. *Southwest. Nat.* in press.

Schuster, W. S. F., S. L. Phillips, D. R. Sandquist, and J. R. Ehleringer. 1992a. Heritability of carbon isotope discrimination in *Gutierrezia microcephala*. *Am. J. Bot.* **79:** 216–221.

Schuster, W. S. F., D. R. Sandquist, S. L. Phillips, and J. R. Ehleringer. 1992b. Comparisons of carbon isotope discrimination in populations of aridland plant species differing in lifespan. *Oecologia* **91:** 332–337.

Sharkey, T. D. 1985. Photosynthesis in intact leaves of C$_3$ plants: Physics, physiology and rate limitations. *Bot. Rev.* **51:** 53–105.

Shreve, F., and I. L. Wiggins. 1964. Vegetation and Flora of the Sonoran Desert. Stanford Univ. Press, Stanford.

Smedley, M. P., T. E. Dawson, J. P. Comstock, L. A. Donovan, D. E. Sherrill, C. S. Cook, and J. R. Ehleringer. 1991. Seasonal carbon isotope discrimination in a grassland community. *Oecologia* **85:** 314–320.

Smith, S. D., and R. S. Nowak. 1990. Ecophysiology of plants in the intermountain lowlands, pp. 179–241. *In* C. B. Osmond, L. F. Pitelka, and G. M. Hidy (eds.), Plant Biology of the Basin and Range. Springer-Verlag, New York.

Smith, S. D., and C. B. Osmond. 1987. Stem photosynthesis in a desert ephemeral, *Eriogonum inflatum*: Morphology, stomatal conductance and water-use efficiency in field populations. *Oecologia* **72:** 533–541.

Stowe, L. G., and J. A. Teeri. 1978. The geographic distribution of C$_4$ species of the Dicotyledonae in relation to climate. *Am. Nat.* **112:** 609–623.

Teeri, J. A., L. G. Stowe, and M. D. Murawski. 1978. The climatology of two succulent plant families: Cactaceae and Crassulaceae. *Can. J. Bot.* **56:** 1750–1758.

Walter, H., and E. Stadelmann. 1975. A new approach to the water relations of desert plants, Vol. 2, pp. 213–310. *In* G. Brown (ed.), Desert Biology. Academic Press, New York.

Winter, K. 1981. C$_4$ plants of high biomass in arid regions of Asia—occurrence of C$_4$ photosynthesis in Chenapodiaceae and Polygonaceae from the Middle East and USSR. *Oecologia* **48:** 100–106.

Winter, K., and J. H. Troughton. 1978. Photosynthetic pathways in plants of coastal and inland habitats of Israel and the Sinai. *Flora* **167:** 1–34.

12

Carbon Isotope Discrimination and Resource Availability in the Desert Shrub *Larrea tridentata*

Philip W. Rundel and M. Rasoul Sharifi

I. Introduction

Recent advances in understanding factors which determine the carbon isotopic composition of plant tissues have resulted in the use of stable isotope measurements to assess plant physiological processes and their interaction with the environment. Stable carbon isotope determinations provide time-integrated measures of plant physiological activities and plant interaction with the environment (Rundel *et al.*, 1988). Because the technique yields data which integrate activity over seasonal and possibly longer time periods, minor and transient environmental variations are minimized (see Chapters 3 and 5 by Farquhar and O'Leary).

Carbon isotope ratios ($\delta^{13}C$) in plant leaf tissues provide a potentially valuable means of estimating the assimilation weighted mean daytime concentration of intercellular CO_2 (c_i) at the phenological stage when that foliar carbon was fixed. Farquhar *et al.* (1982) established a theoretical relationship between carbon isotope ratio ($\delta^{13}C$) expressed in ‰ and c_i where ambient CO_2 (c_a) remained relatively constant during leaf formation:

$$\delta^{13}C = -12.2 - 24.6 c_i/c_a.$$

How can this relationship be applied in understanding the response of desert plants to environmental stress? A significant question for consideration is the physiological mechanisms by which widespread and ecologically successful species balance photosynthetic uptake and water loss across gradients of resource availability. If photosynthetic capacity (A) and leaf conductance (g) change linearly across such gradients and the slope of A and g pass through the zero intercepts, then intrinsic water-use efficiency (A/g) and c_i/c_a remain constant. Under these conditions, $\delta^{13}C$ will remain constant. Long-term variations in photosynthetic capacity and maximal stomatal conductance are commonly positively correlated (Körner *et al.*, 1979; Hall and Schulze, 1980; Schulze and Hall, 1982; Farquhar and Sharkey, 1982). If a plant species responds to stress by changing either A or g to a greater degree than the other, such that the A-to-g relationship is curvilinear or linear without passing through the origin, intrinsic water-use efficiency is altered and $\delta^{13}C$ should change.

The objective of our studies has been to look at the application of carbon isotope techniques to understanding the mode of physiological adaptation used by *Larrea tridentata*, an ecologically important warm-desert shrub, in responding to broad environmental gradients of water and nutrient availability. This chapter deals with genetic and phenotypic components of variation of $\delta^{13}C$ within field populations of *L. tridentata*. Experimental manipulations of water and nutrient availability are described as an approach to assess the phenotypic plasticity in A and g, as well as $\delta^{13}C$ under conditions of differential resource availability and changing environmental conditions.

II. Physiological Ecology of *Larrea tridentata*

L. tridentata, commonly known as creosote bush, is a widespread evergreen shrub throughout the warm desert regions of the Southwestern United States and Mexico. *L. tridentata* exhibits remarkable ecological dominance throughout the Mojave, Sonoran, and Chihuahuan Desert regions over large gradients in elevation, substrate conditions, and climatic seasonality (Barbour, 1969; Mabry *et al.*, 1977). In the Mojave Desert and western Sonoran Desert of California, *Larrea* grows under winter rainfall regimes, while in the eastern Sonoran and Chihuahuan Deserts it grows with summer precipitation. It ranges from the extreme environment of Death Valley, California, where mean July temperatures reach 47°C and summer vapor pressure deficits approach 10 kPa, to less xeric habitats where more mesic communities replace desert shrubland. Overall, nevertheless, *Larrea* ecosystems share conditions of sporadic discontinuous precipitation, low soil nitrogen, high solar irradiance, and high extremes of summer temperature (Shreve and Wiggins, 1964; Noy-Meir, 1973; Bailey, 1981; Ehleringer, 1985).

Larrea, as an evergreen shrub, is unusual in its desert habitats where the great majority of woody species are drought or winter deciduous. Numerous studies have examined the physiological adaptations of *L. tridentata* to water stress (Syvertsen *et al.*, 1975; Syvertsen and Cunningham, 1977; Odening *et al.*, 1974; Oechel, 1972; Mooney *et al.*, 1977; Meinzer *et al.*,

1986; Lajtha and Whitford, 1989). Oechel *et al.* (1972) found that *Larrea* growing in the Sonoran Desert remained metabolically active throughout the year, utilizing photosynthesis for the continuous production of new leaves and shoots. Odening *et al.* (1974) showed that *Larrea* had a high protoplasmic tolerance to water stress and was able to maintain high net photosynthesis rates at low water potentials. Seasonal acclimation in temperature responses of photosynthetic processes in *Larrea* was documented by Mooney *et al.* (1977). Meinzer *et al.* (1986) found that *Larrea* was able to maintain a constant turgor pressure over a broad range of leaf-water potentials. Seasonal and annual differences in soil water supply have been found to cause significant changes in the phenological events and growth activity of *L. tridentata* (Sharifi *et al.*, 1988, 1990). Water-use efficiency and nitrogen-use efficiency were found to be inversely correlated in *L. tridentata* by Lajtha and Whitford (1989).

In previous investigations we found that *Larrea* underwent periods of active leaf flush and growth (up to twice annually), depending on environmental conditions (Sharifi *et al.*, 1988). Under favorable conditions, active growth was continuous. Thus, *Larrea* is able to exploit a wide range of xerophytic habitats, through a combination of drought tolerance and leaf persistence through drought periods and the ability to adapt growth activity to ambient environmental conditions.

III. Field Studies and Laboratory Experiments

Our field studies have been carried out at a number of sites across the Southwestern United States within the Mojave, Sonoran, and Chihuahuan Desert regions. Data from four of these sites are reviewed in this chapter:

1. Mojave Desert: Nevada Test Site gradient from 1470 to 640 m elevation, near Las Vegas; natural and roadside populations surveyed in 1987.
2. Mojave Desert: Honda Site near California City at 1000 m elevation; natural and irrigated population analyzed in 1990.
3. Sonoran Desert: Living Desert Reserve, near Palm Desert, California, at 70 m elevation; natural populations studied in 1984, 1985, 1988 and 1990; irrigated populations in 1984, 1985 and 1990.
4. Chihuahuan Desert: Jornada Experimental Range, New Mexico; natural populations at about 1300 m surveyed in 1986.

Greenhouse and growth chamber experiments were carried out at the University of California, Los Angeles, from 1989 to 1992. Greenhouse plants were used for experimental studies with water and nitrogen availability. Effects of high temperature and vapor pressure deficit conditions were investigated using controlled growth chambers. Details of these experiments are given by Sharifi and Rundel (1993).

IV. Components of Variation in Gas Exchange and $\delta^{13}C$

A. Field Populations

Carbon isotope values of leaves from mature *L. tridentata* shrubs within a population at Living Desert Reserve showed significant patterns of change between four sampling dates over a 9-month period. Mean $\delta^{13}C$ for this leaf population changed from $-25.2‰$ in December 1984 to $-23.6‰$ in September 1985 (Fig. 1). From May to September when leaves were fully mature, there was no significant change in $\delta^{13}C$.

Short-term changes in the availability of water to shrubs have an effect on the $\delta^{13}C$ of *Larrea* leaves in field populations. We compared $\delta^{13}C$ for leaf populations of *L. tridentata* at Living Desert Reserve in 1984 and 1985, years of normal precipitation, and found mean values of $-24.5‰$. With the onset of extended drought conditions in California in 1987, these desert areas experienced reductions in soil water availability. Mean $\delta^{13}C$ for leaf populations, produced in 1988 and 1990, were -22.4 and $-22.0‰$, respectively, as the drought continued (Fig. 2). Root/shoot balance changes slowly with altered water availability in *Larrea*, and thus the mature shrubs respond to drought by a physiological adjustment in their water-use efficiency. *Larrea* leaves in 1990 with a mean $\delta^{13}C$ of $-22.0‰$ would be operating with an intrinsic water-use efficiency (A/g) 20% higher than that in the predrought leaves.

Long-term change in water availability to *Larrea*, however, may not produce the expected change in $\delta^{13}C$ of leaf tissues. Although *Larrea* shrubs along roadsides in the Mojave Desert experience improved water relations from collected runoff and grow to 6–25 times the size of adjacent populations away from the road (Johnson *et al.*, 1975), mean values of leaf $\delta^{13}C$ in

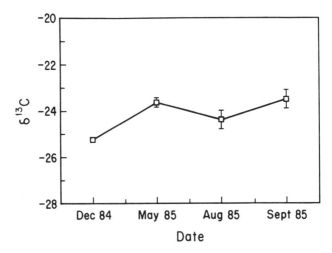

Figure 1.   Mean and standard deviation $\delta^{13}C$ for a population of leaves of *Larrea tridentata* from December 1984 to September 1985 at Living Desert Reserve. Data for 10 shrubs.

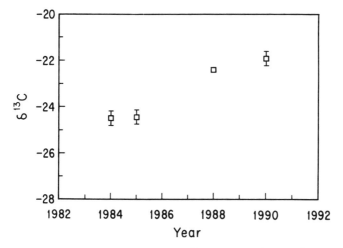

Figure 2.   Mean and standard deviation of $\delta^{13}C$ in summer leaves of *Larrea tridentata* at Living Desert Reserve for 2 years with normal precipitation (1984 and 1985) and 2 subsequent years (1988, 1990) of extreme drought. Data for 10 shrubs.

roadside and nonroadside populations have not been found to be significantly different. With chronic conditions of variable water availability throughout the life of these plants, we hypothesize that root/shoot ratios retain a balance appropriate to maintaining a similar water-use efficiency in both populations.

Short-term irrigation experiments in both the Mojave and Sonoran Deserts have enabled us to look at phenotypic response in the $\delta^{13}C$ and water-use efficiency (W) of *L. tridentata*. Leaves of irrigated field populations of *Larrea* at Living Desert Reserve in 1984 and 1985 exhibited a change in $\delta^{13}C$ values by about 2‰ compared to control populations (Fig. 3). This phenotypic change was comparable to that which we found in control populations between more typical years (1984, 1985, 1988). Under the extreme drought conditions of 1990, this treatment effect increased to 5‰, which corresponds to a decrease in intrinsic water use efficiency of about 34% in the irrigated plants. Drought-stressed control shrubs in the Mojave Desert averaged 21.3‰ in 1990 compared to −24.5‰ in irrigated populations (Fig. 3). We calculate from the isotope values that these Mojave Desert controls had intrinsic water-use efficiencies which were 26% greater than those in adjacent irrigated shrubs.

Despite the range of $\delta^{13}C$ ratios encountered in irrigated and control populations of *L. tridentata* at Living Desert Reserve in 1990, a highly significant linear relationship between the assimilation and conductance rates of mature leaves was present ($r = 0.97$). Maximum assimilation rates measured were about 18 μmol m^{-2} s^{-1} in irrigated shrubs and 5 μmol m^{-2} s^{-1} in control shrubs. The ratios of A/g and c_i/c_a were also significantly related to $\delta^{13}C$ as theory would predict (Fig. 4).

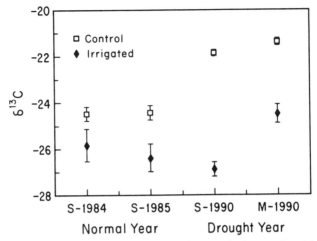

Figure 3. Mean and standard deviation of $\delta^{13}C$ in paired control and irrigated shrub populations of *Larrea tridentata*. Data are shown for 2 normal years (1984 and 1985) and 1 drought year (1990) at Living Desert Reserve in the Sonoran Desert (S) and for 1990 at the Honda Site near California City in the Mojave Desert (M).

B. Controlled Growth Experiments

Greenhouse and growth chamber experiments allow a means of assessing aspects of the phenotypic and genetic plasticity of W in *L. tridentata*. We have used three types of experiments in which we have varied water availability, nutrient availability, and vapor pressure deficit (vpd). In the green-

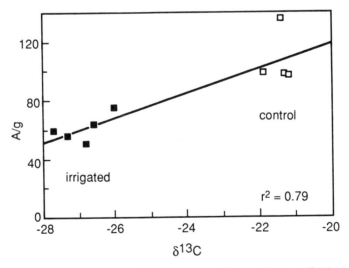

Figure 4. Relationship of the intrinsic water-use efficiency (A/g) to $\delta^{13}C$ for control and irrigated populations of *Larrea tridentata* at Living Desert Reserve in 1990.

house irrigation and nitrogen fertilization experiment, 2-year-old *Larrea* seedlings were subjected to three soil water treatments (daily, weekly, and biweekly irrigation) and five nitrogen fertilization treatments (equivalent to 50, 100, 200, and 500 kg ha^{-1} h^{-1} and control).

Photosynthetic capacity of *Larrea* leaves grown under greenhouse conditions was controlled by water availability. Mean assimilation rates were about 10 μmol m^{-2} s^{-1} in our greenhouse plants watered daily, but dropped to about 7 μmol m^{-2} s^{-1} in plants watered weekly, and 6 μmol m^{-2} s^{-1} in plants watered biweekly. These assimilation levels were intermediate between those we measured in irrigated and control field populations at Living Desert Reserve in 1990.

The δ^{13}C values of leaf tissues in our greenhouse irrigation experiments ranged from a mean of -26‰ in plants watered biweekly to -31‰ in plants watered daily. This range of 5‰ was comparable in magnitude to that observed at Living Desert in the summer of 1990, but δ^{13}C was proportionally more negative. Greenhouse plants watered biweekly had values of δ^{13}C comparable to control plants at Living Desert in 1990, while *Larrea* watered daily had more negative values of δ^{13}C than those ever encountered in the field. These lower values of δ^{13}C corresponded to lower intrinsic water-use efficiencies, as indicated by the linear relationship between A/g and δ^{13}C in greenhouse plants (Fig. 5).

The photosynthetic capacity of well-watered *L. tridentata* can be controlled by nitrogen availability. The addition of ammonium nitrate equivalent to a range of 0–500 kg N ha^{-1} year^{-1} to pots increased maximum assimilation rate from 4 to more than 13 μmol m^{-2} s^{-1} (Fig. 6). These fertilization regimes were reflected in the increased nitrogen content of

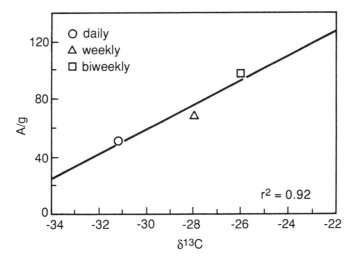

Figure 5.   Relationship of intrinsic water-use efficiency (A/g) to δ^{13}C for *Larrea tridentata* in greenhouse irrigation experiments.

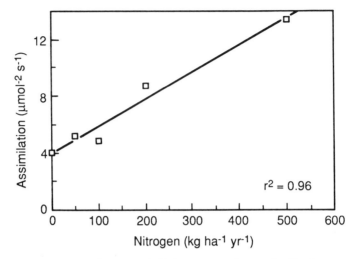

Figure 6. Relationship of mean assimilation rate to nitrogen fertilization treatment for seedlings of *Larrea tridentata* under greenhouse conditions.

leaf tissues (data not shown). Unlike the situation with water availability, the substantial differences in assimilation rate due to nitrogen availability did not alter W or $\delta^{13}C$ significantly. All treatment plants had carbon isotope ratios of -30 to $-31‰$, comparable to the well-watered greenhouse plants in the previous experiments.

When gas exchange data from field shrubs of *L. tridentata* and greenhouse-grown plants from experimental studies were combined, assimila-

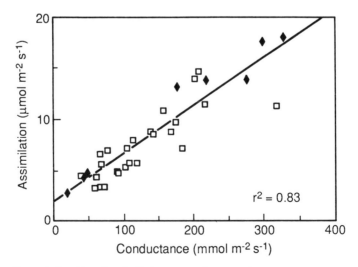

Figure 7. Relationship of assimilation to conductance for *Larrea tridentata* under controlled growth and field conditions. Field populations are shown by the solid diamonds, while open squares indicate greenhouse plants from experimental studies.

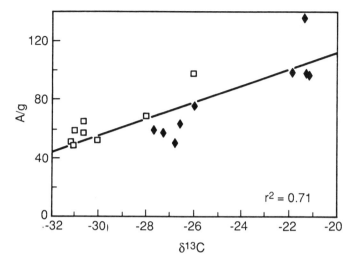

Figure 8. Relationship of intrinsic water-use efficiency (A/g) to $\delta^{13}C$ in *Larrea tridentata* under both field and controlled growth conditions. Field populations are indicated by the solid diamonds.

tion rates showed a highly significant correlation with leaf conductance (Fig. 7). Thus, despite sharply contrasting conditions of plant age and growing conditions, the organizational "rules" for plant gas exchange did not change. The intrinsic water-use efficiency (A/g) was linearly related to $\delta^{13}C$ for all of these data (Fig. 8). Nevertheless, the significant relationship between assimilation rate and $\delta^{13}C$ seen in field populations was not present under greenhouse conditions (Fig. 9). This absence of correlation

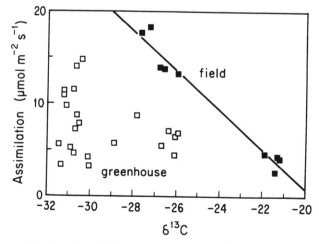

Figure 9. Relationship of assimilation rate to $\delta^{13}C$ in field and greenhouse populations of *Larrea tridentata*.

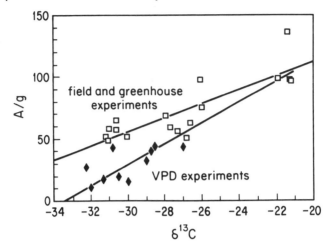

Figure 10. Comparison of the relationships of intrinsic water-use efficiency (A/g) to $\delta^{13}C$ in field and greenhouse experiments and in growth chamber vpd experiments with *Larrea tridentata*. See text for discussion.

resulted from leaf nitrogen contents and vpd environments which varied more substantially under controlled growth conditions than in the field.

These fundamental physiological relationships between intrinsic water-use efficiency (A/g) and c_i/c_a ratio were altered, however, when environmental vpd was modified. We grew seedlings of *L. tridentata* for 8–12 weeks under a series of controlled vpd conditions with constant day and night temperatures of 45 and 30°C (Sharifi and Rundel, 1993). Daytime vpd conditions were set at 2.9, 4.8, and 7.8 kPa in individual experiments.

When the relationship between A/g and $\delta^{13}C$ in the vpd experiments was compared with data from field plants and other experimental plants, plants in the vpd experiments had lower intrinsic water-use efficiency (A/g) at a given value of $\delta^{13}C$ and this difference increased with lower A/g ratio (Fig. 10). The slope of c_i/c_a for the vpd experiments was greater than that for the field and greenhouse plants studies, reflecting a lower W when vpd decreased. Small vpd produced tissues with c_i/c_a ratios of around 0.90 compared to ratios of only 0.65–0.75 in well-watered greenhouse experiments. It is not surprising that changing vpd environment sharply impacted $\delta^{13}C$ through its effect on leaf conductance.

V. Summary

Patterns of distribution of $\delta^{13}C$ in leaf tissues of *L. tridentata* demonstrate a strong genetic component to the control of gas exchange parameters and thus W in this species. Our analyses of multiple populations of *L. tridentata* from widely ranging sites in the Sonoran, Mojave, and Chihuahuan Deserts have found a relatively small degree of variation in $\delta^{13}C$ and thus W despite substantial differences in water availability and seasonality among these sites (Fig. 11). Values of $\delta^{13}C$ in leaf tissues range from -21.6 to $25.8‰$ in our samples of individual shrubs from these regions in years with normal precipitation. For the Sonoran and Mojave Deserts where our sample sizes are larger, intrapopulational variation in $\delta^{13}C$ was often as great as interpopulational differences. Roadside shrubs, which grow many times larger because of improved water relations, were not significantly different in $\delta^{13}C$ from adjacent populations of *Larrea* away from the road (Fig. 11). We have observed no significant change in $\delta^{13}C$ among *Larrea* shrubs, across substantial environmental gradients of water availability from wash woodland to rocky bajada. There is also an evident phenotypic component of physiological response to short-term changes in water availability brought on by drought or augmented water resources. Under natural conditions of multiyear drought we have observed a phenotypic plasticity in the physiological nature of gas exchange characteristics of *Larrea* sufficient to increase mean $\delta^{13}C$ value $2-3‰$. Similarly, field plants have responded to irrigation over 2 years by decreasing $\delta^{13}C$ value $2-3‰$ in field plants and water-use efficiency declined. Despite the strong linear relationship between assimilation and conductance over substantial changes in environmental condition the relative homeostasis in gas exchange characteristics and W in *L. tridentata* under controlled-growth experiments changed greatly from that in the

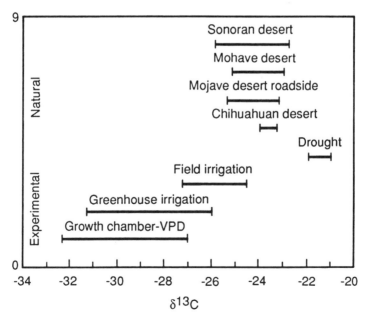

Figure 11. Distribution ranges of $\delta^{13}C$ in *Larrea tridentata* under field conditions and controlled growth conditions.

field. The $\delta^{13}C$ ratio of drought-stressed greenhouse plants was comparable to that of irrigated plants in the field, while well-watered greenhouse plants had much more negative values of $\delta^{13}C$ (Fig. 11). Thus the phenotypic plasticity of W and $\delta^{13}C$ under greenhouse and growth chamber conditions extended far beyond that encountered in the field.

These data suggest the importance of understanding of phenotypic components of water-use efficiency in plant species. Despite a relatively conservative response to gradients of water availability under field conditions, *L. tridentata* has a broad potential for phenotypic flexibility in changing W.

Acknowledgments

This research was supported by NSF Grant BSR 89 07831 and by the Ecological Research Division of the Office of Health and Environmental Research, U.S. Department of Energy.

References

Bailey, H. P. 1981. Climatic features of deserts, pp. 13–42. *In* D. D. Evans and J. L. Thames (eds.), Water in Desert Ecosystems. Hutchinson and Ross, Stroudsburg, PA.

Barbour, M. G. 1969. Age and space distribution of the desert shrub *Larrea divaricata*. *Ecology* **50:** 679–685.

Ehleringer, J. 1985. Annuals and perennials of warm deserts, pp. 162–180. *In* B. F. Chabot and H. A. Mooney (eds.), Physiological Ecology of North American Plant Communities. Chapman & Hall, New York.

Farquhar, G. D., and T. D. Sharkey. 1982. Stomatal conductance and photosynthesis. *Annu. Rev. Plant Physiol.* **33:** 317–345.

Farquhar, G. D., M. H. O'Leary, and J. A. Berry. 1982. On the relationship between carbon isotope discrimination and the intercellular carbon dioxide concentration in leaves. *Aust. J. Plant Physiol.* **9:** 121–127.

Farquhar, G. D., J. R. Ehleringer, and K. T. Hubick. 1989. Carbon isotope discrimination and photosynthesis. *Annu. Rev. Plant Physiol. Plant Mol. Biol.* **40:** 503–537.

Hall, A. E., and E.-D. Schulze. 1980. Stomatal responses to environment and a possible inter relationship between stomatal effects on transpiration and CO_2 assimilation. *Plant Cell Environ.* **3:** 467–474.

Johnson, H. B., F. C. Vasek, and T. Yonkers. 1975. Productivity, diversity and stability in Mojave Desert roadside vegetation. *Bull. Torrey Bot. Club* **102:** 106–115.

Körner, C. H., J. A. Scheel, and H. Bauer. 1979. Maximum leaf diffusive conductance in vascular plants. *Photosynthetica* **13:** 445–482.

Lajtha, K., and W. G. Whitford, 1989. The effect of water and nitrogen amendments on photosynthesis, leaf demography, and resource-use efficiency in *Larrea tridentata*, a desert evergreen shrub. *Oecologica* **80:** 341–348.

Mabry, T. J., J. H. Hunziker, and D. R. DiFeo (eds.). 1977. Creosote Bush: Biology and Chemistry of *Larrea* in New World Deserts. Dowden, Hutchinson and Ross, Stroudsburg, PA.

Meinzer, F. C., P. W. Rundel, M. R. Sharifi, and E. T. Nilsen. 1986. Turgor and osmotic relations of the desert shrub *Larrea tridentata*. *Plant Cell Environ.* **9:** 467–475.

Meinzer, F. C., M. R. Sharifi, E. T. Nilsen, and P. W. Rundel. 1988. Effects of manipulation of water and nitrogen regime on the water relations of the desert shrub *Larrea tridentata*. *Oecologia* **77:** 480–486.

Meinzer, F. C., C. S. Wisdom, A. Gonzalez-Coloma, P. W. Rundel, and L. M. Shultz. 1990. Effects of leaf resin on stomatal behavior and gas exchange of *Larrea tridentata* (DC) Cov. *Funct. Ecol.* **4:** 579–584.

Mooney, H. A., O. Björkman, and G. J. Collatz, 1977. Photosynthetic acclimation to temperatue in the desert shrub *Larrea divaricata.* I. Carbon dioxide exchange characteristics of intact leaves. *Plant Physiol.* **61:** 406–410.

Noy-Meir, I. 1973. Desert ecosystems: Environment and producers. *Annu. Rev. Ecol. Syst.* **4:** 25–52.

Odening, W. R., B. R. Strain, and W. C. Oechel. 1974. The effect of decreasing water potential on net CO_2 exchange of intact desert shrubs. *Ecology* **55:** 1086–1095.

Oechel, W. C., B. R. Strain, and W. R. Odening. 1972. Tissue water potential, photosynthesis, ^{14}C-labeled photosynthate utilization, and growth in the desert shrub *Larrea divaricata Cov. Ecol. Monogr.* **42:** 127–141.

Rundel, P. W., J. R. Ehleringer, and K. A. Nagy. 1988. Stable Isotopes in Ecological Research. Springer-Verlag, New York.

Schulze, E-D., and A. E. Hall. 1982. Stomatal responses, water loss and CO_2 assimilation rates of plants in contrasting environments, pp. 181–230. *In* O. L. Lange, P. S. Nobel, C. B. Osmond, and H. Ziegler (eds.), Plant Physiological Ecology. Encyclopedia of Plant Physiology, Vol. 12B. Springer-Verlag, Heidelberg, Germany.

Sharifi, M. R., and P. W. Rundel. 1993. The effect of atmospheric saturation deficit on carbon isotope discrimination in the desert shrub *Larrea tridentata* (creosote bush). *J. Exp. Bot.* **44:** 481–487.

Sharifi, M. R., F. C. Meinzer, E. T. Nilsen, P. W. Rundel, R. A. Virginia, W. M. Jarrell, D. J. Herman, and P. C. Clark. 1988. Effect of resource manipulation on the quantitative phenology of *Larrea tridentata* (creosote bush) in the Sonoran Desert of California. *Am. J. Bot.* **75:** 1163–1174.

Sharifi, M. R., F. C. Meinzer, P. W. Rundel, and E. T. Nilsen. 1990. Effect of manipulating soil water and nitrogen regimes on clipping production and water relations of creosote bush, pp. 245–249. *In* Symposium on Cheatgrass Inversion, Shrub Die-Off and Other Aspects of Shrub Biology and Management. USDA Forest Service General Technical Report INT-276.

Shreve, F., and I. L. Wiggins. 1964. Vegetation and Flora of the Sonoran Desert. 2 vols. Stanford Univ. Press, Stanford, CA.

Syvertsen, J. P., and G. L. Cunningham. 1977. Rate of leaf production and senescence and effect of leaf age on net gas exchange of creosote bush. *Photosynthetica* **11:** 161–166.

Syvertsen, J. P., G. L. Cunningham, and T. V. Feather. 1975. Anomalous diurnal patterns of stem water potentials in *Larrea tridentata. Ecology* **56:** 1423–1428.

13

Altitudinal Variation in Carbon Isotope Discrimination by Conifers

John D. Marshall and Jianwei Zhang

I. Introduction

Many conifer species are broadly distributed both geographically and altitudinally in western North America. Native genotypes vary considerably across these broad ranges (Rehfeldt, 1984), presumably because genetic differentiation to local conditions has occurred. The presumption of "adaptedness" (cf., Rehfeldt, 1989, 1992) is circumstantially supported by the numerous forestry studies documenting reduced vigor when genotypes are grown far from their source (Wright, 1976) and more directly by correlations between common garden performance and environment (e.g., altitude, latitude, aspect) at the seed source (Monserud and Rehfeldt, 1990).

Instantaneous water-use efficiency (A/E), defined as the molar ratio of photosynthesis relative to transpiration, can be analyzed in terms of local adaptation. Generally speaking, high A/E would be most beneficial on driest sites and least beneficial where water requirements of vegetation are exceeded by precipitation. In fact, high A/E might be detrimental on the wettest sites because it frequently comes at the expense of rapid growth (Condon et al., 1987; Austin et al., 1990; Johnson et al., 1990; Ehdaie et al., 1991), nitrogen-use efficiency (Field et al., 1983; Field and Mooney, 1986), or competitiveness (Cohen, 1970; DeLucia and Schlesinger, 1991). Given these tradeoffs, one would predict high A/E only under conditions of low water availability.

Other variables that might influence A/E change with altitude as well. For instance, the decrease in barometric pressure with altitude (approximately 12% km^{-1}; Friend et al., 1989) results in proportional differences in CO_2 partial pressure that may be important. In addition, the partial pressure of water vapor is expected to decline, temperature declines, and satu-

ration water vapor pressure declines. Finally, the decline in barometric pressure would increase diffusion coefficients, resulting in higher stomatal conductance for a given stomatal aperture (Gale, 1972; Smith and Dono-hue, 1991). Consequently, several interacting environmental characteristics change with altitude.

In this chapter we summarize carbon isotope discrimination (Δ) from common garden and *in situ* studies of two broadly distributed conifer species: Douglas-fir (*Pseudotsuga menziesii* (Mirb.) Franco) and western larch (*Larix occidentalis* Nutt.). As explained in earlier chapters (Farquhar and Lloyd, Chapter 5) and (O'Leary, Chapter 3), Δ can be used to screen plants growing in the same aerial environment for differences in A/E. Our recent results are discussed with reference to earlier work, mostly on herbaceous plants, in an attempt to explain the observed patterns where experimental data are not yet available. Genetic data were collected from common garden studies and compared to data from *in situ* sampling. By combining the two kinds of data, we conclude that A/E is a strongly plastic trait, varying consistently in response to variables correlated with altitude.

II. Carbon Isotope Discrimination and Water-Use Efficiency in Conifers

Use of Δ to compare A/E among plants requires several assumptions: first, that leaf temperatures, and therefore leaf-to-air vapor pressure differences, are similar among the plants being compared; second, that $\delta^{13}C$ of source CO_2 is identical among plants being compared; and third, that biosynthetic fractionation is similar among the plants. These assumptions are particularly well met by the "needle" leaves of conifers because they are narrow and their boundary layer conductances are therefore high. It is unlikely that the leaf temperatures would deviate significantly from air temperature, except under unusual circumstances (e.g., krummholz; Hadley and Smith, 1987). Therefore, one can justify the assumption that leaf-to-air vapor pressure difference is equal to vapor pressure deficit of the atmosphere among all conifer leaves at a site. On a larger scale, the roughness of conifer canopies, particularly near the top, leads to good mixing between the canopy atmosphere and the ambient atmosphere (Jarvis and McNaughton, 1986), which allows one to assume that the CO_2 used in photosynthesis is isotopically identical to that of CO_2 in the bulk atmosphere. However, when comparing leaves from different sites, particularly if the sites differ in leaf-to-air vapor pressure difference, it is necessary to keep in mind that Δ is correlated with A/g; the relation with A/E holds up only as long as the leaf-to-air vapor pressure difference remains unchanged. As we discuss the common garden studies below we treat Δ as a measure of A/E. We discuss it as a measure of A/g in the *in situ* studies along altitudinal gradients.

Calculated values of the carbon isotope discrimination of conifer leaves from the literature are presented in Table I. Data from tree rings have not

Table I Carbon Isotope Discrimination of Conifer Leaves[a]

Species	Δ mean (SE)	Location	Reference
Juniperus osteosperma	12.4(0.2)[b]	Nevada	DeLucia et al., 1988
Pinus monophylla	12.5(0.4)[b]		DeLucia et al., 1988
Pinus ponderosa	14.1(ND)[b]		DeLucia et al., 1988
Pinus jeffreyi	14.9(0.3)[b]		DeLucia et al., 1988
Pinus edulis	12.9(0.2)[b]	Arizona/New Mexico	Leavitt and Long, 1986
Juniperus monosperma	14.7(0.3)[b]	Arizona	Leavitt and Long, 1982
Juniperus deppeana	14.6(0.1)[b]		Leavitt and Long, 1982
Juniperus osteosperma	15.2(0.1)[b]		Leavitt and Long, 1982
Juniperus osteosperma	15.8(0.3)	Utah	Schulze and Ehleringer, 1984
Juniperus osteosperma	16.9(0.9)	Utah	Marshall and Ehleringer, 1990
Pinus edulis	16.6(ND)	New Mexico	Lajtha and Barnes, 1991
Juniperus monosperma	16.9(ND)		Lajtha and Barnes 1991
Abies lasiocarpa	15.5(ND)	Alberta	Gower and Richards, 1990
Pinus albicaulis	16.6(ND)	Washington	Gower and Richards, 1990
Tsuga mertensiana	17.2(ND)	Washington	Gower and Richards, 1990
Larix lyallii	17.2(ND)	Alberta/Washington	Gower and Richards, 1990
Pinus contorta	17.7(ND)	Washington	Gower and Richards, 1990
Pinus ponderosa	17.9(ND)	Oregon	Gower and Richards, 1990
Larix occidentalis	19.4(ND)	Washington/Oregon	Gower and Richards, 1990
Pinus massoniana (sun)	18.9(0.6)	China	Ehleringer et al., 1986
Pinus taeda and P. echinata	19.5(ND)[b]	Tennessee	Garten and Taylor, 1992
Lagarostrobos franklinii	19.5(ND)[b]	Tasmania	Francey et al., 1985
Lagarostrobos franklinii	20.4(ND)		Francey et al., 1985
Picea abies (sun)	20.3(0.1)	Bavaria	Gebauer and Schulze, 1991
Larix	20.5(ND)	Bavaria	Vogel, 1978

[a]Where $\delta^{13}C$ was published, Δ has been calculated using the equation $\Delta_{plant} = (\delta^{13}C_{air} - \delta^{13}C_{plant})/(1 + \delta^{13}C_{plant})$, assuming $\delta^{13}C_{air} = -8\permil$ and $\delta^{13}C_{plant}$ is expressed relative to PDB.

[b]Determined on cellulose, which is usually heavier than whole tissue (i.e., discrimination is lower) by 1–2‰ (Leavitt and Long, 1982).

Table II Altitudinal Variation in Δ *in Situ*, from Published Observations

Vegetation type	Intercept	Slope	r_2	Significance	Reference
Herbs	19.1	−0.03	0.15	<0.001	Körner et al., 1991
Nardus stricta					Friend et al., 1989
1986	18.5	−1.46	0.27	NS	
1987	18.4	0.92	0.67	0.05	
Vaccinium myrtillus					Friend et al., 1989
1986	20.4	−1.25	0.16	NS	
1987	20.9	0.56	0.91	0.01	
Forbs	20.8	−0.77	0.37	<0.0001	Körner et al., 1988
Shrubs	20.4	−0.41	0.18	0.009	Körner et al., 1988
Trees	21.3	−0.87	0.34	0.002	Körner et al., 1988
Overall	20.8	−0.69	0.35	<0.0001	Körner et al., 1988
Metrosideros polymorpha	20.7	−1.31	0.45	0.012	Vitousek et al., 1988
Metrosideros polymorpha					Vitousek et al., 1990
Young lava, wet	21.4	−1.4	0.80	<0.001	
Old lava, wet	22.9	−1.9	0.73	<0.001	
Young lava, dry	20.1	−0.16	0.01	NS	
Old lava, dry	18.8	0.54	0.17	NS	
Pseudotsuga menziesii	20.9	−1.87	0.74	0.0002	Marshall and Zhang, unpublished
Larix occidentalis	22.9	−1.46	0.56	0.010	Marshall and Zhang, unpublished

been included because there are substantial and variable differences between leaves and tree rings (Leavitt and Long, 1982, 1986; Francey, 1986). The isotopic data in Table I are heavily weighted toward species growing in the semiarid woodlands of western North America, and, therefore, one must use care in drawing generalizations regarding the conifers from these data. However, given that all the data lie below the median for C_3 plants as a whole, which lies at about 20.6‰ (O'Leary, 1988), it seems likely that conifers as a rule maintain higher A/g than other C_3 plants.

III. Altitudinal Variation in Carbon Isotope Discrimination

Past surveys of carbon isotope discrimination of plants along altitudinal gradients have found decreased discrimination (indicative of increased A/g) at higher altitudes. This increase, especially when combined with the aforementioned decrease in the leaf-to-air water vapor gradient, should result in a pronounced increase in A/E with altitude. Decreased Δ values were first detected in a global survey across a 5600-m gradient that included forbs, shrubs, and trees (Körner *et al.*, 1988). Subsequent work found the same pattern in a single tree species (*Metrosideros polymorpha*) distributed across a 1600-m gradient in Hawaii (Vitousek *et al.*, 1988; 1990). However, a later study, in which sampling was designed so that sites varied in atmospheric pressure while temperature remained constant, found weaker relationships with altitude (Körner *et al.*, 1991). Still another study described a reversal in the direction of the relationship between isotope discrimination and altitude in succeeding years (Friend *et al.*, 1989). Regression coefficients describing these relationships are summarized in Table II.

We collected and analyzed foliage from two conifers, Douglas-fir (*P. menziesii* (Mirb.) Franco) and western larch (*L. occidentalis* Nutt.). Leaf samples were collected across a 3000-m altitudinal gradient at 15 sites along the western flank of the northern Rockies, from southern Idaho into eastern Washington and British Columbia. There was a pronounced decrease in Δ with altitude (Figs. 1 and 2). Slopes of the regressions were steep and coefficients of determination were very high. It appears that altitude is indeed correlated with carbon isotope discrimination in conifers. But it is highly improbable that altitude per se causes this effect; more likely some correlate of altitude is the direct cause. In the following sections we summarize available evidence on variables that might fill this role.

IV. Genetic Variation in Carbon Isotope Discrimination in Conifers

Strong altitudinal gradients in Δ raise the question of whether the gradients can be attributed to genetic differences among populations from the

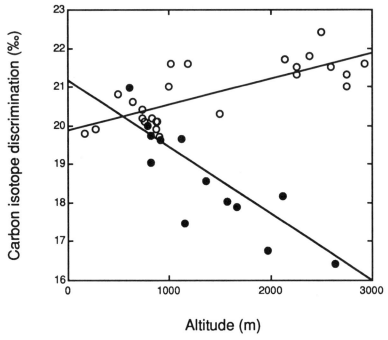

Figure 1. Variation in leaf Δ among Douglas-fir genotypes in a common garden (●) experiment (at age 15 years) and among leaf samples collected *in situ* (○). The common garden was located at approximately 600 m altitude and included genotypes from most of the natural distribution of Douglas-fir. The *in situ* data were collected along the west flank of the northern Rocky Mountains, from southern Idaho into southern British Columbia. Common garden Δ (‰) = 19.9 + 0.00066* (altitude (m)), r^2 = 0.58. *In situ* Δ (‰) = 21.1 − 0.00188* (altitude (m)), r^2 = 0.74.

different altitudes. This possibility seems particularly likely given classic ecotype studies (Clausen *et al.* 1941). However, genetic variation cannot be distinguished from environmental variation *in situ*. Planting of seed from different sources into uniform blocks within a common garden experiment is necessary to minimize environmental variation and make it possible to assign any remaining variation either to genotype or to the interaction between genotype and environment.

Three recent studies have evaluated genetic contributions to altitudinal variation in Δ. Morecroft and Woodward (1990) grew *Nardus stricta* collected at different altitudes in controlled environments varying in temperature, frost frequency, and barometric pressure. They were unable to identify genotypic differences among populations. Read and Farquhar (1991) grew accessions of 22 species of *Nothofagus* in controlled environments and measured Δ. They observed pronounced differences among species; tropical species showed less discrimination than those from higher latitudes. Altitude of origin and Δ values were correlated in the controlled environment; species from higher altitudes discriminated less than those from

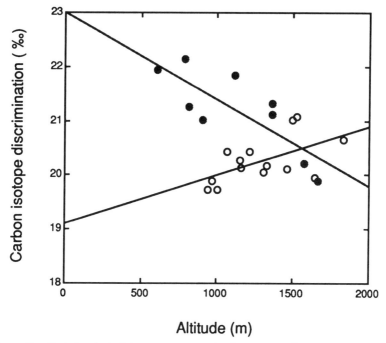

Figure 2. Variation in leaf Δ among western larch genotypes in a common garden (○) experiment (using 2-year-old seedlings) and among samples collected *in situ* (●). The common garden was located at approximately 800 m altitude and included genotypes from throughout the northern Rocky Mountains. The *in situ* data were collected along the west flank of the northern Rocky Mountains, from southern Idaho into southern British Columbia. Common garden Δ (‰) = 19.1 + 0.00089* (altitude (m)), $r^2 = 0.32$. *In situ* Δ (‰) = 23.0 − 0.00161* (altitude (m)), $r^2 = 0.64$.

lower altitudes (Fig. 3). This relationship is similar to the *in situ* data collected in other studies.

We have recently measured Δ in leaves of 15-year-old Douglas-fir trees grown in a common garden from seed collected across most of the natural distribution (Zhang *et al.*, 1992). Seed sources from the western Cascades and the Coast Range of Oregon and Washington were excluded because they would not survive the winters at the planting site in the Trinity Valley, in southeastern British Columbia, at 600 m altitude.

Isotopic discrimination was closely related to altitude of origin (Figs. 1 and 2); however, the relationship was opposite that of the *Nothofagus* data described above (Fig. 3). Moreover, it was opposite the *in situ* data for Douglas fir (Fig. 1). The isotope data were supported by instantaneous gas exchange data (Zhang *et al.*, 1993).

We have obtained similar results from experiments conducted on 2-year-old seedlings of western larch grown in an open-walled rain shelter at the University of Idaho Forest Research Nursery in Moscow, Idaho, at 800 m altitude (Zhang and Marshall, 1993). The seedlings were grown from seed

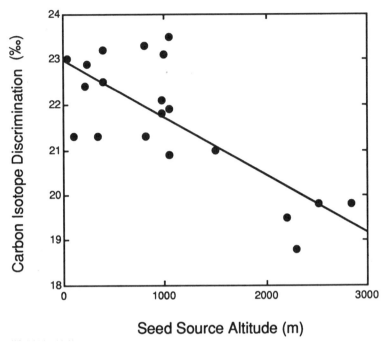

Figure 3.   Variation in leaf Δ among *Nothofagus* spp. collected at a range of altitudes and grown in a common environment. Data from a species occurring on ultrabasic soil were deleted, as were unreplicated samples (Read and Farquhar, 1991). Δ (‰) = 23.0 − 0.00127* (altitude (m)), $r^2 = 0.58$.

collected across the northern Rockies and were watered frequently. The altitudinal pattern was similar to that in Douglas-fir; discrimination tended to be higher among genotypes collected at high altitude (Fig. 2). This trend was statistically significant even though the seed was collected across an altitude range of less than 1000 m. Transpiration efficiency (g dry mass/kg water transpired) was also significantly related to altitude of origin ($r = -0.54$, $P < 0.05$). High-altitude genotypes had lower A/E by both measures—exactly the opposite of what we observed *in situ*, but exactly the same trend observed with Douglas-fir. We concluded that the genetic effect opposes the altitude effect, leading to partial homeostasis of populations *in situ*.

V. Plasticity of Δ across Altitudinal Gradients

If population differences explained the *in situ* altitude gradient, the two regression lines in Figs. 1 and 2 would be indistinguishable. In contrast, if the patterns were determined exclusively by environment, the common garden data would not have a significant slope. Apparently, both genotype and environment influence Δ. Large differences were observed between Δ values obtained *in situ* and values obtained from populations found at similar altitude and grown in the common garden, particularly in the genotypes from the highest altitudes (Figs. 1 and 2). The magnitude of the difference is illustrated by the following example. Carbon isotope discrimination in trees from greater than 2000 m averaged about 21.5‰ in the common garden. At this altitude the *in situ* data were below 16‰. This difference represents a halving of A/g caused by bringing the trees down from high altitude to the common garden.

Both genotype and environment contribute to the variation observed *in situ*. Our sampling methods do not allow us to analyze the interaction between genotype and altitude, which would be necessary for a rigorous analysis of plasticity. Previous experiments have found primarily joint effects of genotype and environment on Δ; however, in these experiments the environment varied due to differences in water availability (Condon *et al.*, 1987; Hubick *et al.*, 1988; Virgona *et al.*, 1990) rather than altitude. Further work examining common gardens planted at several altitudes will be necessary to analyze the interactions.

VI. Potential Mechanisms for the Plastic Response

Plants retain considerable potential for plasticity in variables that would result in changes in Δ. We discuss them following the path of a carbon dioxide molecule from the ambient atmosphere to the chloroplast. First, CO_2 must diffuse across the boundary layer, the layer of relatively still air at the surface of the leaf. However, as mentioned earlier, the boundary layer conductance is high around needle leaves.

After crossing the boundary layer, CO_2 must diffuse through stomata. Here one would expect considerable potential for variability. However, the lower CO_2 partial pressure at high altitude (Friend *et al.*, 1989) is approximately offset by increased diffusion coefficients, resulting in little difference in the rate at which CO_2 would diffuse through stomata (Gale, 1972; Smith and Donahue, 1991). Stomatal conductance might also increase with altitude due to increased stomatal density (number of stomata/m^2 leaf surface) under decreasing CO_2 partial pressures (Woodward and Bazzaz, 1988). However, the plastic response of stomatal density is opposite that needed to explain the *in situ* values.

Once inside the leaf, the CO_2 molecule diffuses down a concentration gradient to the choroplast. Although intercellular CO_2 concentration, as calculated from gas exchange data, is the value at the base of the stomata, inside the leaf, Δ is determined by the CO_2 concentration at the chloroplast. The steepness of the gradient from stomata to chloroplasts almost certainly varies among leaves (von Caemmerer and Evans 1991). It may be negligible for typically thin-leaved crop plants and herbs, but may be significant in thick leaves, e.g. sclerophylls (Vitousek *et al.*, 1990). The gradient should be steepest in sclerophyllous, hypostomatous leaves with high photosynthetic rates, e.g., *Metrosideros* (Vitousek *et al.*, 1988, 1990). To cause the altitudinal gradient in individual species reported here, leaf thickness would have to increase with altitude, which is in fact what is observed (Körner *et al.*, 1986, 1989; Körner and Diemer, 1987; Vitousek *et al.* 1988, 1990).

Finally, leaves of many species adjust morphologically and physiologically in a manner that results in greater assimilation of CO_2 at a given intercellular CO_2 concentration. Körner and Diemer (1987) described this response in paired comparisons of herbs from the same genus growing at altitudes of 600 and 2600 m. As an example, data from *Ranunculus* spp. are presented (Fig. 4). The increase in carboxylation capacity with altitude was sufficient to maintain similar photosynthetic rates at similar conductances despite lower CO_2 partial pressures. Such an increase in carboxylation capacity, expressed per unit leaf area, could result either from increased nitrogen concentration per unit leaf weight or from a thickening of the leaf at constant nitrogen concentration.

Whatever the physiological mechanisms underlying the shift in Δ, the environmental stimulus inducing the response remains to be identified. Körner *et al.* (1991) have suggested variation in oxygen partial pressures as a cause of the altitudinal Δ increase. Low temperature may also cause the phenotypic shift in Δ because lateral leaf expansion is generally more sensitive to low temperatures than is photosynthesis, leading to thicker leaves with higher carboxylation capacities at high altitudes (Körner *et al.*, 1989). If so, the increase in A/E with altitude would result from a developmental constraint rather than an adaptive character (Körner and Menendez-Riedl, 1989).

The implications of these data to forestry remain unclear given the many gaps in our knowledge. However, the notion of adaptedness of a genotype,

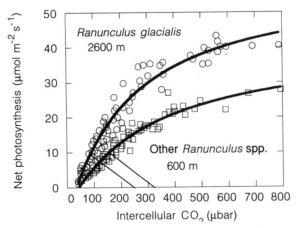

Figure 4.   Net photosynthetic rates in relation to intercellular CO_2 concentrations within leaves of *Ranunculus* spp. grown at 600 and 2600 m altitude (redrawn from Körner and Diemer, 1987). Straight lines represent leaves of equal conductance at the two altitudes.

insofar as it implies specialization in response to life at a particular altitude, must account for the pronounced variation in A/E. If increased A/E results in similarly increased biomass production from a finite water supply, selection for A/E might enhance forest productivity in the dry environments in which forestry is often practiced in western North America.

VII. Summary

Surveys of *in situ* variation in Δ have revealed pronounced decreases in Δ with altitude in two conifer species of fundamentally different leaf structure and physiology. Common garden studies using genotypes from across the natural distribution of these species detected variation in Δ among genotypes from different altitudes; however, the variation in the common garden studies opposed the *in situ* differences. Therefore, we conclude that genotypic differences exist, but they offset, rather than explain, *in situ* altitudinal gradients. Further work is necessary to determine whether observed phenotypic plasticity can be explained by low-temperature-induced increases in carboxylation capacity among conifers.

Acknowledgments

The authors acknowledge helpful reviews of earlier versions of the manuscript by Paul Doescher and Ron Robberecht. The manuscript was written with the help of funding from a McIntire-Stennis grant to the University of Idaho and a University of Idaho Seed Grant.

References

Austin, R. B., P. Q. Craufurd, M. A. Hall, F. Acevedo, B. da Silveira Pinheiro, and E. C. K. Ngugi. 1990. Carbon isotope discrimination as a means of evaluating drought resistance in barley, rice and cowpeas. *Bull. Soc. Bot. Fr.* 137, *Actual. Bot.* **1:** 21–30.

Caemmerer, S. von, and J. R. Evans. 1991. Determination of the average partial pressure of CO_2 in chloroplasts from leaves of several C_3 plants. *Aust. J. Plant Physiol.* **18:** 287–305.

Clausen, J., D. D. Keck, and W. M. Hiesey. 1941. Experimental Studies on the Nature of Species. IV. Genetic Structure of Ecological Races. Carnegie Institute of Washington Publication, no. 242. Washington, DC.

Cohen, D. 1970. The expected efficiency of water utilization in plants under different competition and selection regimes. *Israel J. Bot.* **19:** 50–54.

Condon, A. G., R. A. Richards, and G. D. Farquhar. 1987. Carbon isotope discrimination is positively correlated with grain yield and dry matter production in field-grown wheat. *Crop Sci.* **27:** 996–1001.

DeLucia, E. H., and W. H. Schlesinger. 1991. Resource-use efficiency and drought tolerance in adjacent Great Basin and Sierran plants. *Ecology* **72:** 51–58.

DeLucia, E. H., W. H. Schlesinger, and W. D. Billings. 1988. Water relations and the maintenance of Sierran conifers on hydrothermally altered rock. *Ecology* **69:** 303–311.

Ehdaie, B., A. E. Hall, G. D. Farquhar, H. T. Nguyen, and J. G. Waines. 1991. Water-use efficiency and carbon isotope discrimination in wheat. *Crop Sci.* **31:** 1282–1288.

Ehleringer, J. R. 1988. Changes in leaf characteristics of species along elevational gradients in the Wasatch Front, Utah. *Am. J. Bot.* **75:** 680–689.

Ehleringer, J. R., C. B. Field, Z. F. Lin, and C. Y. Kuo. 1986. Leaf carbon isotope and mineral composition in subtropical plants along an irradiance cline. *Oecologia* **70:** 520–526.

Field, C., and H. A. Mooney. 1986. The photosynthesis-nitrogen relationship in wild plants, pp. 25–55. *In* T. J. Givnish (ed.), On the Economy of Plant Form and Function. Cambridge Univ. Press, Cambridge.

Field, C., J. Merino, and H. A. Mooney, 1983. Leaf age and seasonal effects on light, water, and nitrogen use efficiency in a California shrub. *Oecologia* **56:** 348–355.

Francey, R. J. 1986. Carbon isotope measurements in baseline air, forest canopy air and plants. *In* J. R. Trabalka and D. E. Reichle (eds.), The changing carbon cycle: A global analysis. Springer-Verlag, New York.

Francey, R. J., R. M. Gifford, T. D. Sharkey, and B. Weir. 1985. Physiological influences on carbon isotope discrimination in huon pine (*Lagarostrobos franklinii*). *Oecologia* **66:** 211–218.

Friend, A. D., F. I. Woodward, and V. R. Switsur. 1989. Field measurements of photosynthesis, stomatal conductance, leaf nitrogen and $\delta^{13}C$ along altitudinal gradients in Scotland. *Funct. Ecol.* **3:** 117–122.

Gale, J. 1972. Availability of carbon dioxide for photosynthesis at high altitudes: theoretical considerations. *Ecology* **53:** 494–497.

Garten, C. T., Jr., and G. E. Taylor, Jr. 1992. Foliar $\delta^{13}C$ within a temperate deciduous forest: spatial, temporal, and species sources of variation. *Oecologia* **90:** 1–7.

Gebauer, G., and E.-D. Schulze. 1991. Carbon and nitrogen isotope ratios in different compartments of a healthy and a declining *Picea abies* forest in the Fichtelgebirge, NE Bavaria. *Oecologia* **87:** 198–207.

Gower, S. T., and J. H. Richards. 1990. Larches: Deciduous conifers in an evergreen world. *Bioscience* **40:** 818–826.

Hadley, J. L., and W. K. Smith. 1987. Influence of krummholz mat microclimate on needle physiology and survival. *Oecologia* **73:** 82–90.

Hubick, K. T., R. Shorter, and G. D. Farquhar. 1988. Heritability and genotype × environment interactions of carbon isotope discrimination and transpiration efficiency in peanut. *Arachis hypogea* L. *Aust. J. Plant Physiol.* **15:** 799–813.

Jarvis, P. G., and K. G. McNaughton. 1986. Stomatal control of transpiration: Scaling up from leaf to region. *Adv. Ecol. Res.* **15:** 1–49.

Johnson, D. A., K. H. Asay, L. L. Tieszen, J. R. Ehleringer, and P. G. Jefferson. 1990. Carbon isotope discrimination: Potential in screening cool-season grasses for water-limited environments. *Crop Sci.* **30:** 338–343.

Körner, Ch., and M. Diemer. 1987. *In situ* photosynthetic responses to light, temperature and carbon dioxide in herbaceous plants from low and high altitude. *Funct. Ecol.* **1:** 179–194.

Körner, Ch., and S. P. Menendez-Riedl. 1989. The significance of developmental aspects in plant growth analysis, pp. 141–157. *In* H. Lambers *et al.* (eds.), Causes and Consequences of Variation in Growth Rates and Productivity of Higher Plants. SPB Academic Publishing bv, The Hague, The Netherlands.

Körner, Ch., P. Bannister, and A. F. Mark. 1986. Altitudinal variation in stomatal conductance, nitrogen content and leaf anatomy in different life forms in New Zealand. *Oecologia* **69:** 577–588.

Körner, Ch., G. D. Farquhar, and Z. Roksandic. 1988. A global survey of carbon isotope discrimination in plants from high altitude. *Oecologia* **74:** 623–632.

Körner, Ch., M. Neumayer, S. P. Menendez-Reidl, and A. Smeets-Scheel. 1989. Functional morphology of mountain plants. *Flora* **182:** 353–383.

Körner, Ch., G. D. Farquhar, and S. C. Wong. 1991. Carbon isotope discrimination by plants follows latitudinal and altitudinal trends. *Oecologia* **88:** 30–40.

Lajtha, K., and F. J. Barnes. 1991. Carbon gain and water use in pinyon pine–juniper woodlands in northern New Mexico: Field versus phytotron chamber measurements. *Tree Physiol.* **9:** 59–67.

Leavitt, S. W., and A. Long. 1982. Evidence for $^{13}C/^{12}C$ fractionation between tree leaves and wood. *Nature* **298**: 742–744.

Leavitt, S. W., and A. Long. 1986. Stable-carbon isotope variability in tree foliage and wood. *Ecology* **67**: 1002–1010.

Marshall, J. D., and J. R. Ehleringer. 1990. Are xylem-tapping mistletoes partially heterotrophic? *Oecologia* **84**: 224–228.

Monserud, R. A., and G. E. Rehfeldt. 1990. Genetic and environmental components of variation in site index in inland Douglas-fir. *For. Sci.* **36**: 1–9.

Morecroft, M. D., and F. I. Woodward. 1990. Experimental investigations on the environmental determination of $\delta^{13}C$ at different altitudes. *J. Exp. Bot.* **41**: 1303–1308.

O'Leary, M. H. 1988. Carbon isotopes in photosynthesis. *Bioscience* **38**: 328–336.

Read, J., and G. Farquhar. 1991. Comparative studies in *Nothofagus* (Fagaceae). I. Leaf carbon isotope discrimination. *Funct. Ecol.* **5**: 684–695.

Rehfeldt, G. E. 1984. Microevolution of conifers in the northern Rocky Mountains: A view from common gardens, pp. 32–46. *In* Proceedings, Eighth North American Forest Biology Workshop, Logan, Utah.

Rehfeldt, G. E. 1989. Ecological Adaptations in Douglas-fir *Pseudotsuga menziesii* var. *glauca*: A synthesis. *For. Ecol. Manage.* **28**: 203–215.

Rehfeldt, G. E. 1992. Breeding strategies for *Larix occidentalis*: Adaptation to the biotic and abiotic environment in relation to improving growth. *Can. J. Forest Res.* **22**: 5–13.

Schulze, E.-D., and J. R. Ehleringer. 1984. The effect of nitrogen supply on growth and water-use efficiency of xylem-tapping mistletoes. *Planta* **162**: 268–275.

Smith, W. K., and R. A. Donahue. 1991. Simulated influence of altitude on photosynthetic CO_2 uptake potential in plants. *Plant Cell Environ.* **14**: 133–136.

Virgona, J. M., K. T. Hubick, H. M. Rawson, G. D. Farquhar, and R. W. Downes. 1990. Genotypic variation in transpiration efficiency, carbon-isotope discrimination and carbon allocation during early growth in sunflower. *Aust. J. Plant Physiol.* **17**: 207–214.

Vitousek, P. M., P. A. Matson, and D. R. Turner. 1988. Elevational and age gradients in hawaiian montane rainforest: Foliar and soil nutrients. *Oecologia* **77**: 565–570.

Vitousek, P. M., C. B. Field, and P. A. Matson. 1990. Variation in foliar $\delta^{13}C$ in Hawaiian *Metrosideros polymorpha*: A case of internal resistance? *Oecologia* **84**: 362–370.

Vogel, J. C. 1978. Recycling of carbon in a forest environment. *Oecologia Plant.* **13**: 89–94.

Woodward, F. I., and F. A. Bazzaz. 1988. The responses of stomatal density to CO_2 partial pressure. *J. Exp. Bot.* **39**: 1771–1781.

Wright, J. W. 1976. Introduction to Forest Genetics. Academic Press, New York.

Zhang, J. and J. D. Marshall. In review. Water-use efficiency differences among populations of western larch, a deciduous conifer.

Zhang, J., J. Marshall, and B. Jacquish. 1992. Genetic differentiation in carbon isotope discrimination and gas exchange in *Pseudotsuga menziesii*: A common-garden experiment. *Oecologia* **93**: 80–87.

14

Characterization of Photobiont Associations in Lichens Using Carbon Isotope Discrimination Techniques

C. Máguas, H. Griffiths, J. Ehleringer, and J. Serôdio

I. Introduction

Each lichen comprises a symbiotic association between photobiont (chlorophyte algae and/or cyanobacteria) and mycobiont partners. Uptake of CO_2 by lichens may be relatively transient, a function of thallus water content, and regulated by variations in atmospheric humidity, dewfall, and precipitation. Without the possibility of regulating water content by stomata, poikilohydric lichens must steer between Scylla and Charybdis: sufficient water to reactivate photobiont photosynthesis and CO_2 uptake limited by diffusion in a water-saturated thallus.

While groups of lichens may be readily distinguished in terms of lifeform, as well as the primary photobiont, some lichens containing a green algal phycobiont also have a subsidiary association with a cyanobacteria, which are limited to specialized structures (cephalodia). Major phycobiont genera include *Trebouxia* and *Myrmecia*, supplying the mycobiont with carbohydrates in the form of ribitol (Honegger, 1991). *Nostoc* is the most common cyanobiont genus, which furnishes the mycobiont with glucans and fixed nitrogen (Honegger, 1991). Work on the photosynthetic characteristics and water relations of lichens has shown that those primarily containing green algal phycobionts can be distinguished from those solely with cyanobacteria (cyanobiont) (Lange and Ziegler, 1986; Lange *et al.*, 1988).

Associations with phycobiont alone (or containing cephalodia) can reactivate photosynthesis from water vapor in air, while the cyanobiont, surrounded by a gelatinous sheath, requires rewetting with liquid water (Lange *et al.*, 1988; Bilger *et al.*, 1989). These groups have previously also

201

been distinguished by means of carbon isotope ratio. Lichens classified to date as containing green algae have been shown to have carbon isotope composition (δ^{13}C) values which range from -18 to $-34‰$ (Lange and Ziegler, 1986), and for the species listed in Lange (1988), the mean δ^{13}C was -29.6 ± 3.6 ($n = 16$). The δ^{13}C of cyanobiont lichens ranged from -14 to $-28‰$, with a mean value of $-23.5 \pm 1.7‰$ ($n = 20$; Lange, 1988). The lower carbon isotope discrimination (Δ) in the latter group has been related to a high diffusion resistance through the thallus depending upon water content (Lange and Ziegler, 1986; Lange, 1988). However, CO_2 compensation points suggest "C_4-like" characteristics, which have also been related to the action of a CO_2-concentrating mechanism (CCM) in the photobiont in lichens (Raven et al., 1990; Raven 1991; Griffiths et al., 1993). Such a CCM was originally reported for microalgae and cyanobacteria (Badger et al., 1978) and has been suggested to occur in endosymbiotic associations of fungus and cyanobacterium (Geosiphon pyriforme; Kluge et al., 1991).

In this chapter we reevaluate the variations in Δ for lichens and show that the following three groups may be distinguished on the basis of photobiont association: phycobiont, phycobiont plus cephalodia, and cyanobiont. In addition, on-line, instantaneous carbon isotope discrimination, thallus CO_2 compensation points, and maximum photosynthetic rates are used to characterize the various lichen groups. Finally we consider the interactions between diffusion resistance imposed by thallus water content, refixation of respiratory CO_2 derived from the mycobiont, and the possible activity of any CO_2-concentrating mechanism in the photobiont.

II. Carbon Isotope Discrimination and Gas Exchange Methodology

A. Lichen Material

The lichen species were collected in Portugal at Serra da Arrábida (mediterranean ecosystem), Serra de Aires e Candeeiros (oak forest), and in Jura, Scotland, for Ramalina fastigiata samples. At Serra da Arrábida the lichen material was collected on north and south slopes at different altitudes: north slope from 110 to 430 m and south slope from 150 to 420 m.

B. Carbon Isotope Discrimination in Organic Material

δ^{13}C and Δ were determined using standard mass spectrometric techniques at the University of Newcastle upon Tyne, United Kingdom, and the University of Utah. Calculations of Δ in organic material assumed a δ^{13}C of source air of $-8‰$ versus PDB standard, using the relationship of Farquhar et al. (1989).

C. On-Line, Instantaneous Carbon Isotope Discrimination

Photosynthesis in stored lichen material was reactivated using standard techniques by spraying with deionized water, with thalli acclimated for 3 days and gently cleaned of soil or any other debris prior to experimentation, under a photosynthetic photon flux density (PPFD) of 30 μmol m^{-2} s^{-1} at a temperature of 15°C. For measurement of CO_2 uptake and respiration, pieces of lichen thalli, weighing approximately 0.5 g dry wt, were placed in a water-jacketed Plexiglas chamber at 15°C and PPFD of 120 μmol m^{-2} s^{-1}. For each experiment thallus water content was determined from the mean fresh weight before and after the CO_2 gas exchange measurements, and the oven-dried weight at the end of the experiment and was expressed as a percentage of thallus dry weight. CO_2 gas exchange was determined using an ADC LCA2 IRGA in differential mode, and the air was supplied using two mass flow controllers (Brooks Instruments, Cheshire, UK), from a cylinder of CO_2 (Distillers Ltd., UK) and from CO_2-free compressed air. Photosynthetic maximum rates (V_{max}, nmol CO_2 g^{-1} dry wt s^{-1}) in the light were determined 15–20 min after chamber closure, over a period of about 90 min, with measurements of dark respiration used to calculate gross thallus photosynthetic rate. The CO_2 was collected for isotopic analysis downstream of the cuvette in a liquid N_2 cold trap in a cryodistillation line, as described by Griffiths *et al.* (1990). Collections were made over 10–15 min, with CO_2 repurified to remove H_2O and N_2O by passage through two −90°C cold traps and over a reduced copper furnace at 600°C. Samples of reference CO_2 were collected at regular intervals and were purified and analyzed in the same way. The precision of these measurements as determined on 10 replicate CO_2 samples was 22.0 ± 0.2‰.

During on-line discrimination, the CO_2 leaving the cuvette is enriched in $^{13}CO_2$, and the discrimination expressed by the thallus is derived using the expression of Evans *et al.* (1986), whereby

$$\Delta = \frac{\xi(\delta_o - \delta_e)}{\delta_o - \xi(\delta_o - \delta_e)}.$$

The concentration difference of CO_2 external (c_e) and leaving the chamber (c_o) was maintained at 2–3 Pa with the value of ξ calculated from

$$\xi = \frac{c_e}{c_e - c_o}.$$

Measurements of the isotopic composition of the CO_2 entering and leaving the cuvette (δ_o, δ_e, respectively) were made as described above. It is important to note that the source CO_2 used in the majority of this study had a $\delta^{13}C$ value of −22‰, as opposed to bulk-air CO_2 which is currently −8‰. It should be noted that the use of a commercial source of CO_2 ($\delta^{13}C \approx$ −22‰) for gas exchange measurements could in theory increase instantaneous Δ by up to 4‰ (I.R. Cowan, personal communication), although no differences were noted in on-line Δ when source CO_2 was varied from −22 to −42‰ in a later part of this study.

D. Thallus CO_2 Compensation Point

Sections of lichen thallus were placed in a glass water-jacketed chamber (volume, 100 cm^3; temperature, 15°C; and PPFD, 120 μmol m^{-2} s^{-1}) as part of a closed system gas exchange system. CO_2 uptake was determined using an ADC 225 mk III IRGA, with an internal purge system used to reduce CO_2 partial pressure initially to speed the approach of the steady-state compensation point and prevent large changes in thallus water content.

III. Characterization of Three Lichen Groups as Determined by Photobiont Association

A. Carbon Isotope Discrimination in Organic Material

The three groups of lichens characterized in this study (Table I) form part of a broad survey of Δ, from which selected or related species are subsequently used for more detailed studies of gas exchange and on-line discrimination characteristics. The associations with green algae as phycobiont (G), taken from a total of 8 genera, all contained chlorophyte algae in the genus *Trebouxia,* one of the most commonly found phycobiont genera. For those lichens with green algal phycobiont which also associate with cyanobacteria (G + cephal), the cephalodia may form an integral structure within the medulla or be associated with the thallus surface (James and Henssen, 1975). The cyanobacteria contribute little fixed carbon to the association but are differentiated to fix atmospheric nitrogen. There are some 21 genera containing cephalodia, including *Peltigera* (James and Henssen, 1975), although *Lobaria* was the only genus represented at the main study site in Serra da Arrábida (Table I). Many of these genera also contain lichen associations where cyanobacteria are the only photobiont (cyanobiont, CB), with *Nostoc* found in the lichens sampled as part of this survey (Table I).

Table I   List of Lichen Species, Photobiont Type, and Their Algal Genus, upon Which the Carbon Isotope Discrimination Analyses Were Performed[a]

Lichen genera	Photobiont	Algae genera
Anaptichia	Phycobiont	*Trebouxia*
Evernia	Phycobiont	*Trebouxia*
Hypogymnia	Phycobiont	*Trebouxia*
Parmelia	Phycobiont	*Trebouxia*
Physcia	Phycobiont	*Trebouxia*
Ramalina	Phycobiont	*Trebouxia*
Usnea	Phycobiont	*Trebouxia*
Xanthoria	Phycobiont	*Trebouxia*
Lobaria	Phycobiont + cephal	*Myrmecia* + *Nostoc*
Leptogium	Cyanobiont	*Nostoc*
Lobaria	Cyanobiont	*Nostoc*
Nephroma	Cyanobiont	*Nostoc*
Peltigera	Cyanobiont	*Nostoc*

[a] Phycobiont + cephal = phycobiont + cephalodia

A number of species from each of these groups were collected from the Macchia vegetation and oak trees at the Serra de Arrábida, Portugal. Carbon isotope discrimination was lowest in the phycobiont associations, with a mean value of 13.4 ± 1.5‰ for 18 species (Fig. 1), at the bottom of the range normally associated with C_3 higher plants. There were also statistically significant differences within this group for two well-represented genera, *Parmelia* and *Ramalina* (Fig. 1, species number 4–7 and 8–14, respectively). The three species of *Lobaria* (G + cephal) showed the highest discrimination, with a mean value of 23.2‰, while lichens with cyanobacteria as cyanobiont formed an intermediate group in terms of Δ, with a mean value of 15.2‰ for the 3 species sampled (Fig. 1). The distinction between Δ in these two groups (G + cephal and CB) corresponds with the $\delta^{13}C$ data in Lange and Ziegler (1986) and Lange (1988).

B. Gas Exchange and Instantaneous Discrimination in Lichens

Carbon isotope discrimination in organic material was reanalyzed for the specific samples used to determine gas exchange and instantaneous dis-

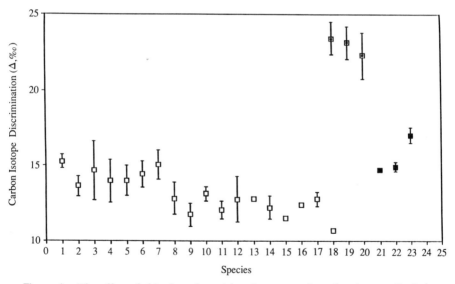

Figure 1. The effect of altitude and north/south exposure in carbon isotope discrimination in a range of lichen species was studied in Serra da Arrábida. Using a two-way layout ANOVA model (analysis of variance), it was found that 90% of the total variability present on the data could be explained by species alone. The most important difference was found between *Lobaria* spp. (23.2 ± 2.2‰) and all the other genera. Significant differences were also found between *Ramalina* spp. (12.5 ± 1.0‰) and *Parmelia* spp. (14.2 ± 1.1‰) but never within each genus. Species list: 1, *Anaptychia ciliata*; 2, *Evernia prunastri*; 3, *Hypogymnia physodes*; 4, *Parmelia caperata*; 5, *Parmelia hypotropa*; 6, *Parmelia reticulata*; 7, *Parmelia sulcata*; 8, *Ramalina calicaris*; 9, *Ramalina canariensis*; 10, *Ramalina everneoides*; 11, *Ramalina farinacea*; 12, *Ramalina fastigiata*; 13, *Ramalina lusitanica*; 14, *Ramalina pusilla*; 15, *Physcia adscendens*; 16, *Physcia leptalaea*; 17, *Usnea* sp.; 18, *Xanthoria parietina*; 19, *Lobaria amplissima*; 20, *Lobaria laetivirens*; 21, *Lobaria pulmonaria*; 22, *Leptogium furfuraceum*; 23, *Nephroma laevigatum*; 24, *Peltigera canina*.

crimination characteristics. This data, presented as part of Table II, independently confirmed the three groups identified in Fig. 1. However, there were additional differences between the groups in terms of on-line Δ values, maximum photosynthetic rates (V_{max}), and thallus CO_2 compensation point when measured at optimal thallus water content (Table II).

The low values of Δ in organic material of phycobiont lichens (15.0‰) contrasted with the higher values of on-line, instantaneous Δ found for the same group (22.2‰; Table II). In terms of higher plant photosynthesis, while instantaneous Δ is suggestive of the C_3 pathway, the thallus compensation point (1.98 Pa) is more likely to be associated with a CCM. Lichens containing green algae plus cephalodia were consistently "C_3-like" with high values of instantaneous Δ and organic material Δ (26.7 and 24.7‰, respectively) as well as CO_2 compensation point (Table II). Maximum photosynthetic rates were on average slightly higher for the phycobiont plus cephalodia group (4.73 as opposed to 3.09 nmol CO_2 g^{-1} dry wt s^{-1}) (Table II). In contrast, organic material Δ was low in cyanobiont associations, with instantaneous values lower than those in organic material Δ (12.0 and 16.3‰, respectively), and the thallus compensation point was the lowest of the three groups (Table II). The responses of lichens containing cyanobacteria consistently suggested the activity of a CCM, while the average maximum rate of net CO_2 uptake (13.89 nmol CO_2 g^{-1} dry wt s^{-1}) was higher than that in the phycobiont groupings.

C. Contribution of Thallus Respiration to Instantaneous Discrimination

It is important to note that measurements of photosynthetic characteristics of lichens reflect a combination of photobiont and mycobiont processes. Rates of dark respiration may be equivalent to 50‰ of net CO_2 uptake, and if the thallus desiccates, a high proportion of the recently fixed carbon may be respired quite rapidly (Farrar, 1975; Lange and Ziegler, 1986; Lange, 1988). Since most of the respiration is derived from the mycobiont, it is likely that similar rates occur in the light. Accordingly, estimates of thallus CO_2 compensation point (see Table II) are likely to overestimate the value expressed by the photobiont, because of respiratory CO_2 release by the fungal partner. In the same way, the on-line instantaneous isotope signature of CO_2 leaving the gas exchange cuvette will be altered and reflects the proportion of gross photosynthetic processes. In effect the phycobiont is supplied with CO_2 from two different sources: one representing ambient CO_2 ($\delta^{13}C$ of $-8‰$), the other from respiratory CO_2 (similar to whole thallus $\delta^{13}C$).

Respiration was shown to comprise 50 and 31‰ of net CO_2 uptake for *Lobaria pulmonaria* and *Lobaria scrobiculata*, respectively (Lange and Ziegler, 1986), although this proportion was reduced when measured under the experimental conditions used in this study to 23 and 15‰ of gross photosynthetic rate (data not shown). The proportion of respiratory CO_2 leaking out of the thallus, and the carboxylation strength of the photobiont, will be

Table II Carbon Isotope Discrimination in Organic Material (Δ), Instantaneous Carbon Isotope Discrimination (Δ_i), Maximum Photosynthetic Rate (V_{max}), and Compensation Point (Comp. pt.) for Different Lichen Associations with Phycobiont (G), Phycobiont + Cephalodia (G + cephal), and Cyanobiont (CB)

Group	Species	Δ (‰)	Δ_i (‰)	V_{max} (nmol CO_2 g^{-1} dry wt s^{-1})	Comp. pt. (Pa)
G	Evernia prunastri	16.7	22.4	3.72	2.50
	Parmelia reticulata	16.3	20.0	3.60	1.93
	Ramalina fastigiata	12.1	22.9	1.96	1.52
	Mean	15.0 ± 2.5	22.2 ± 0.5	3.09 ± 1.14	1.98 ± 0.49
G + Cephal	Lobaria amplissima	23.5	28.2	3.03	7.53
	Lobaria pulmonaria	24.0	27.1	6.00	5.35
	Sticta aurata	26.5	33.8	5.18	5.67
	Mean	24.7 ± 1.6	26.7 ± 3.6	4.73 ± 1.52	6.18 ± 1.12
CB	Lobaria scrobiculata	16.7	11.3	8.84	1.51
	Peltigera canina	15.9	12.6	18.88	1.35
	Mean	16.3	12.0	13.89	1.43

related to the magnitude of the diffusion resistance operating within the thallus. Interpretation of on-line discrimination in lichens should be viewed with caution until the contribution from respiratory processes can be quantified.

IV. Evaluation of Lichen Physiology Using Carbon Isotope Discrimination and Gas Exchange Techniques: Discussion

It is apparent that the two groups of lichens previously identified in terms of carbon isotope composition should now be revised. Highest carbon isotope discrimination is shown by those associations consisting of phycobiont + cephalodia, while those comprising phycobiont or cyanobiont alone show lower discrimination. The morphological and physiological correlates of these separate categories remain to be determined with particular reference to the different rehydration mechanisms found in phycobiont and cyanobiont lichens (Lange and Ziegler, 1986; Lange, 1988; Lange et al., 1986, 1989). More detailed analyses of ultrastructural modifications to the mycobiont during cycles of (de)hydration, and also in response to aging across a thallus, are required. Phycobiont cells tend to be smaller with higher photosynthetic activity in actively growing regions, and the mycobiont may be denser in old regions of the thallus (Hill, 1985). This may be related to a 3–4‰ gradient in $\delta^{13}C$ with age across large thalli of L. pulmonaria (C. Máguas and E. Brugnoli, unpublished data).

There are difficulties in characterizing the in vivo photosynthetic competence of the photobiont when lichenized. It has been assumed that thallus respiration rates were constant under light and dark conditions and that photosynthetic responses and CO_2 compensation points were true indicators of photobiont respiratory activity. Another possible approach would be to evaluate the photosynthetic characteristics of photobiont cells when cultured in vitro, although carboxylation conductance and source–sink relationships may change photobiont responses in the absence of the mycobiont partner (Richardson, 1973). Finally, the large proportion of respiratory CO_2, which may be fixed in the light, could alter the on-line discrimination signal and the relationship between carboxylation strength and diffusion resistance within the thallus now needs to be investigated.

Although the traditional view is that diffusion resistance predominantly regulates Δ in lichens (Lange and Ziegler, 1986; Lange, 1988; Lange et al., 1988) we may also interpret the data presented above in terms of the activity of a CO_2-concentrating mechanism in the photobiont (Badger et al., 1978; Raven et al., 1990; Raven, 1991; Griffiths et al., 1993; Palmqvist et al., 1993). In theoretical terms, it is not possible to distinguish between a low Δ associated with either diffusion resistance or any CO_2-concentrating mechanism activity (Sharkey and Berry, 1985; Raven, 1991). Despite this difficulty, changes in instantaneous Δ were consistent with the induction of such a mechanism in the chlorophyte Chlamydomonas reinhardtii (Sharkey and Berry, 1985). In addition, low organic material Δ was associated with the induction/repression of a CO_2-concentrating mechanism in *Chlorella emersonii* (Beardall *et al.*, 1982).

Measurements of on-line discrimination and CO_2 compensation points are consistent with the operation of a CO_2-concentrating mechanism in cyanobiont lichens. Free-living cyanobacteria have an obligate requirement for a CO_2-concentrating mechanism in view of the kinetic properties of cyanobacterial RuBisCO (Espie *et al.*, 1991; Raven, 1991). Such a mechanism could account for the C_4-like Δ characteristics and thallus compensation points in the lichenized cyanobacteria (Table II, Fig. 1).

Cyanobacteria isolated from lichens did have the capacity to induce a CO_2-concentrating mechanism when cultured *in vitro* (Griffiths *et al.*, 1993). The Δ of organic material from cultures of *Nostoc* was lower than that of *Coccomyxa sp.* (the dominant phycobiont in *Peltigera aphthosa*), suggesting that the Chlorophyte mechanism was less efficient or more easily repressed (Griffiths *et al.*, 1993). However, the only measurement of the *in vitro* fractionation factor for cyanobacterial RuBisCO is lower than that for Chlorophytes (Guy *et al.*, 1987), which would also result in lower Δ in cyanobiont associations.

Having observed the higher values of V_{max} for cyanobiont associations (when expressed per unit thallus dry weight), this evidence intuitively would not seem to be consistent with higher diffusion resistances. A more detailed analysis of field and laboratory studies showed that cyanobiont lichens have generally higher maximum rates of CO_2 uptake at higher thallus water contents that found for phycobiont lichens (Griffiths *et al.*, 1993).

While variations in the expression of a CO_2-concentrating mechanism could account for the different categories of CO_2 compensation points found for the three lichen groups (Table II), the extent of photorespiration *in vivo* remains to be determined. Further work is also required to investigate the basis for the consistent C_3-like characteristics of phycobiont plus cephalodia lichens, as determined from organic material Δ, instantaneous Δ, and CO_2 compensation point (Fig. 1, Table II) as compared to the lower organic Δ of those solely containing a phycobiont. It is tempting to speculate that the improved nitrogen supply from cephalodia could alter photosynthetic characteristics in lichenized phycobionts, as found for *C. emersonii* (Beardall *et al.*, 1982).

In order to place these observations in the context of the ecological niches occupied by each group, we contend that cyanobiont associations are *more* often restricted to moist, shaded habitats (James and Henssen, 1975; Millbank, 1985; Honnegar, 1991). While this observation may be criticized in that cyanobiont lichens are found in exposed, desert areas (but usually where runoff of water occurs following rainfall) we suggest that these are the exceptions. Thus the potential for a higher thallus water content and need for liquid water to reactivate photosynthesis provides a more stable microclimate for the operation of the cyanobacterial CO_2-concentrating mechanism. While the strategy for CO_2 uptake following rehydration from atmospheric water vapor is excluded from cyanobiont lichens, the higher carboxylation conductances, low photorespiration, and proportionally lower dark respiration rates result in higher rates of CO_2 assimilation over longer periods, once rewetted. The CO_2 uptake characteristics of phycobiont lichens, which are found in more exposed habitats where photosynthesis is more transiently reactivated, are more C_3-like with low carboxylation conductances. While the variation in gas exchange characteristics between phycobiont ± cephalodia and cyanobiont lichens will provide intriguing lines of research in the future, carbon isotope discrimination techniques have proved to be a powerful means of distinguishing the three groups of lichens.

V. Summary

Analysis of the $\delta^{13}C$ values of organic material in lichens shows that there are three categories, with carbon isotope discrimination related to photobiont association. Assuming a source air of $\delta^{13}C$ of $-8‰$, discrimination (Δ) was lowest in lichens containing only a single photobiont, such as the groups with green algal phycobiont (13.4‰) or cyanobiont (15.2‰), and highest in associations containing phycobiont together with cyanobacteria in cephalodia (23.2‰). Additional measurements of on-line discrimination, photosynthetic rates, and CO_2 compensation points are presented for each group, and consideration is given to the problems associated with the likely occurrence of respiration rates derived from the mycobiont in the light.

Lichens containing phycobiont plus cephalodia were consistently C_3-like, with high on-line discrimination and CO_2 compensation points. These lichens had maximum photosynthetic rates slightly higher than those of lichens with a single phycobiont (4.73 and 3.09 nmol CO_2 g^{-1} dry wt s^{-1}), although in the latter group on-line discrimination and CO_2 compensation points were lower. Maximum photosynthetic rates were higher in cyanobiont lichens (13.89 nmol CO_2 g^{-1} dry wt s^{-1}), which suggests that diffusion resistances are not higher in cyanobiont lichens. When compared with low CO_2 compensation points and low on-line discrimination, these high carboxylation conductances are more likely to be associated with the activity of a CO_2-concentrating mechanism in lichens solely containing cyanobacteria.

Acknowledgments

Cristina Máguas has been supported by JNICT—Science Programme studentship (BD/153/90-RN). We thank Mark Broadmeadow for his assistance throughout all the experimental work conducted in Newcastle, as well as valuable discussions. Professor Fernando Catarino has provided enthusiastic support for all of these investigations.

References

Badger, M. R., A. Kaplan, and J. A. Berry. 1978. A mechanism for concentrating CO_2 in *Chlamydomonas reinhardtii* and *Anabaena variabilis* and its role in photosynthetic CO_2 fixation. *Yearb. Carnegie Inst. Washington* **77**: 251–266.

Beardall, J., H. Griffiths, and J. A. Raven. 1982. Carbon isotope discrimination and the CO_2 accumulating mechanism in *Chlorella emersonii. J. Exp. Bot.* **33**: 729–737.

Bilger, W., S. Rimke, U. Schreiber, and O. L. Lange. 1989. Inhibition of energy-transfer to photosystem II in lichens by dehydration: different properties of reversibility with green and blue-green phycobionts. *J. Plant Physiol.* **134**: 261–268.

Espie, G. S., A. G. Miller, R. A. Kandasamy, and D. T. Canvin. 1991. Active HCO_3-transport in cyanobacteria. *Can. J. Bot.* **69**: 936–944.

Evans, J. R., T. D. Sharkey, J. D. Berry, and G. D. Farquhar. 1986. Carbon isotope discrimination measured concurrently with gas exchange to investigate CO_2 diffusion in leaves of higher plants. *Aust. J. Plant Physiol.* **13**: 281–292.

Farquhar, G. D., J. R. Ehleringer, and K. T. Hubick. 1989. Carbon isotope discrimination and photosynthesis. *Annu. Rev. Plant Physiol. Plant Mol. Biol.* **40**: 503–537.

Farrar, J. F. 1975. The lichen as an ecosystem: Observation and experiment, pp. 385–407. *In* D. H. Brown, D. L. Hawksworth, and R. H. Bailey (eds.), Lichenology: Progress and problems—Proceedings of an International Symposium Held at University of Bristol. Academic Press, London.

Griffiths, H., M. S. J. Broadmeadow, A. M. Borland, and C. S. Hetherington. 1990. Short-term changes in carbon-isotope discrimination identify transitions between C3 and C4 carboxylation during Crassulacean acid metabolism. *Planta* **181**: 604–610.

Griffiths, H., J. A. Raven, and C. Máguas. 1993. A comparison of carbon accumulation strategies in lichens containing phycobiont and cyanobiont, submitted for publication.

Guy, R. D., M. F. Fogel, J. A. Berry, and T. C. Hoering. 1987. Isotope fractionation during oxygen production and consumption by plants, pp. 597–600. *In* T. Biggins (ed.), Progress in Photosynthesis Research III. Nijhoff, Dordrecht.

Hill, D. J. 1985. Changes in photobiont dimensions and numbers during co-development of lichen symbionts, pp. 303–319. *In* D. H. Brown (ed.), Lichen Physiology and Cell Biology. Plenum Press, London.

Honnegar, R. 1991. Functional aspects of the lichen symbiosis. *Annu. Rev. Plant Physiol. Plant Mol. Biol.* **42**: 553–578.

James, P. W., and A. Henssen. 1975. The morphological and taxonomic significance of cephalodia, pp. 27–77. *In* Lichenology: Progress and Problems—Proceedings of an International Symposium Held at the University of Bristol. Academic Press, London.

Kluge, M., D. Mollenhauer, and R. Mollenhauer. 1991. Photosynthetic carbon assimilation in *Geosiphon pyriforme* (Kutzing) F.v. Wettstein, an endosymbiotic association of fungus and cyanobacterium. *Planta* **185**: 311–315.

Lange, O. L. 1988. Ecophysiology of photosynthesis: Performance of poikilohydric lichens and homoiohydric mediterranean sclerophylls. *J. Ecol.* **76**: 915–937.

Lange, O. L., and H. Ziegler. 1986. Different limiting processes of photosynthesis in lichens, pp. 147–161. *In* R. Marcelle, H. Clitjers, and M. Van Poucke (eds.), Biological Control of Photosynthesis. Nijhoff, Dordrecht.

Lange, O. L., E. Kilian, and H. Ziegler. 1986. Water vapor uptake and photosynthesis in lichens: Performance differences in species with green and blue-green algae as phycobionts. *Oecologia* **71**: 104–110.

Lange, O. L., T. G. A. Green, and H. Ziegler. 1988. Water status related photosynthesis and carbon isotope discrimination in species of the lichen genus *Pseudocyphellaria* with green and blue-green photobionts and in photosymbiodemes. *Oecologia* **75**: 494–501.

Lange, O. L., W. Bilger, S. Rimke, and U. Schreiber. 1989. Chlorophyll fluorescence of lichens containing green and blue-green algae during hydration by water vapor uptake and by addition of liquid water. *Bot. Acta* **102**: 306–313.

Millbank, J. W., 1985. Lichens and plant nutrition. *Proc. R. Soc. Edinburgh Sect. B* **85:** 253–261.

Palmqvist, K., C. Máguas, M. R. Badger, and H. Griffiths. 1993. Assimilation, accumulation and isotope discrimination of inorganic carbon in lichens: Evidence for the operation of a CO_2 concentrating mechanism. *Cryptogenic Botany,* in press.

Raven, J. A. 1991. Implications of inorganic carbon utilization: Ecology, evolution and geochemistry. *Can. J. Bot.* **69:** 908–924.

Raven, J. A., and G. D. Farquhar. 1990. The influence of N metabolism and organic acid synthesis on the natural abundance of C isotopes in plants. *New Phytol.* **116:** 505–529.

Raven, J. A., A. M. Johnston, L. L. Handley, and S. G. McInroy. 1990. Transport and assimilation of inorganic carbon by *Lichina pygmaea* under emersed and submersed conditions. *New Phytol.* **114:** 407–417.

Richardson, D. H. S. 1973. Photosynthesis and carbohydrate movement, pp. 249–285. *In* V. Ahmadjian and M. E. Hale (eds.), The Lichens. Academic Press, London.

Sharkey, T. D., and J. A. Berry. 1985. Carbon isotope fractionation of algae as influenced by an inducible CO_2 concentrating mechanism, pp. 389–402. *In* William J. Lucas and Joseph A. Berry (eds.), Inorganic Carbon Uptake by Aquatic Organisms. American Society of Plant Physiologists, Rockville, MD.

15

Carbon Isotope Discrimination and Gas Exchange in Ozone-Sensitive and -Resistant Populations of Jeffrey Pine

Mark T. Patterson and Philip W. Rundel

I. Introduction

Gaseous air pollutants, such as ozone (O_3), adversely impact the growth and physiology of many plant species (Darrall, 1989). Anthropogenic ozone in the southern Sierra Nevada of California reaches phytotoxic levels during the summer and fall months (Miller *et al.*, 1972; Warner *et al.*, 1982; Pedersen and Cahill, 1989) and Jeffrey pine (*Pinus jeffreyi* Grev. & Balf.), a common species growing in the mixed conifer belt of the Sierra, is among the first to exhibit visible injury symptoms. The expression of ozone-induced visible injury within a population of Jeffrey pine is highly variable, ranging from individuals that show hypersensitive responses to others that show no visible injury. The mechanism underlying this variability is still unclear, but it is presumably related to genotypic or microenvironmental differences between trees. The control of pollutant entry into leaf tissues by stomata has been suggested to be important in determining the degree of ozone-induced physiological injury and therefore sensitivity to gaseous pollutants (Turner *et al.*, 1972; McLaughlin and Taylor, 1980; Temple, 1986; Maier-Maercker and Koch, 1991). The variation in ozone injury within a population of Jeffrey pine may reflect different intrinsic capacities for gas exchange between individual trees. Understanding the physiological factors influencing pollutant entry into plant tissues can lead to a more predictive understanding of ozone resistance and sensitivity in Jeffrey pine.

A major controlling factor influencing gaseous flux into and out of leaves is stomatal conductance. Stomata regulate the diffusion of not only CO_2

and H_2O but also O_3 and other gaseous pollutants. The internal dose of ozone, rather than the ambient dose to which the plant is exposed, has been implicated as the most important factor determining ozone-induced injury in many plant species (Reich, 1987). Interspecific comparisons of external ozone dose responses can be ambiguous because species can differ in patterns of gas exchange, antioxidant production, acclimation to ozone, and repair of ozone-damaged tissues (Heath, 1980, 1987). Within a species, however, one would expect individuals with a greater integrated stomatal conductance to be more predisposed to ozone injury because they would experience a greater internal dose. This may explain some of the variability that has been observed within certain populations of ozone-stressed plant species (Coyne and Bingham, 1981; Boyer *et al.,* 1986, Duriscoe, 1987; Berrange *et al.,* 1991).

Ozone uptake by leaf tissues is mainly determined by ambient O_3 concentration and stomatal conductance (Laisk *et al.,* 1989). Many plant species have shown reductions in stomatal conductance in response to acute ozone fumigations (Rich and Turner, 1972; Keller and Halser, 1984; Martin *et al.,* 1988; Roper and Williams, 1989; Matyssek *et al.,* 1991). However, with low levels of ozone exposure, stomatal conductance may not change significantly and may even increase (Zimmermann *et al.,* 1988; Sesak and Richarson, 1989; Maier-Maercker and Koch, 1991). Little is known about the mechanism leading to variation in sensitivity within a species, but individuals with higher values of stomatal conductance will have higher pollutant uptake rates (Reich, 1987; Laisk *et al.,* 1989). Under field conditions where plants are exposed to long-term low levels of ozone, the severity of ozone-induced injury should be positively correlated with stomatal conductance. This is especially important in long-lived tissues, such as Jeffrey pine needles, in which chronic exposure to ozone can result in large accumulated doses.

Foliar stable carbon isotope ratio ($\delta^{13}C$) can be used to estimate mean intercellular CO_2 concentrations and, potentially, to evaluate internal ozone dose. Discrimination against the naturally occurring stable isotope $^{13}CO_2$ occurs during both the diffusion and carboxylation components of gas exchange. The overall effects are integrated over the period in which carbon is assimilated into leaf tissues resulting in a relationship between $\delta^{13}C$ and mean intercellular CO_2 concentration (Farquhar *et al.,* 1982). Stable carbon isotope discrimination (Δ), determined from foliar $\delta^{13}C$ relative to the $\delta^{13}C$ of the atmosphere, is related to intercellular CO_2 concentrations (c_i) such that higher c_i lead to greater Δ. Stomatal conductance, along with photosynthetic activity, regulates c_i by controlling the rate at which CO_2 diffuses along a concentration gradient into leaf tissues where the CO_2 is ultimately fixed in the photosynthetic carbon reduction cycle.

Discrimination is affected by relative changes in stomatal conductance and photosynthesis. When comparing individuals with similar photosynthetic capacities, Δ can be positively correlated to leaf conductance (Ehleringer, 1990). Different intercellular CO_2 concentrations, and therefore Δ values, between ozone-resistant and -sensitive trees may result from differences between the trees in stomatal limitation to photosynthesis. Conversely, if stomatal conductance is similar, ozone-induced differences in photosynthetic capacity can lead to higher values of c_i and Δ. Ambient ozone may affect the processes involved in both stomatal control and photosynthesis which emphasizes the need for gas exchange data to interpret the physiological mechanisms involved in the variation in Δ in a field setting. Populations of Jeffrey pine exhibiting a wide range of ozone injury provide a natural experiment with which to test the hypothesis that ozone sensitivity is related to greater inherent stomatal conductance in a field situation.

This paper addresses the differences in O_3 susceptibility between individuals in a population of Jeffrey pine growing in an ozone-stressed environment by utilizing a combination of stable isotope and gas exchange measurements. It is hypothesized that higher Δ values will reflect higher stomatal conductances in ozone-sensitive individuals whereas lower Δ will occur in resistant individuals primarily from limitations on gaseous diffusion of O_3 through stomata. Gas exchange measurements will aid in determining if sensitive trees differ from resistant trees in stomatal behavior and in carbon assimilation capacity.

II. Materials and Methods

A. Study Site and Plant Material

The study site was located on the western slope of the Sierra Nevada in Sequoia National Park at an elevation of 6700 feet. Jeffrey pine grows in the mixed conifer forest on soils composed mainly of decomposed granite. Most of the precipitation occurs in the winter and spring months as snow and averages 100 cm annually. During the period of study only one significant rainfall occurred, that in mid-September. Summer ozone levels frequently peak above 0.1 ppm and 24-h means range between 0.06 and 0.07 ppm-h. Cumulative ozone doses in this area during the period between May and September are high enough to produce damage in sensitive conifer species such as Jeffrey pine (Miller *et al.*, 1972; Williams, 1980).

Trees of approximately the same age (20–50 years), canopy position (open), and size (3–5 m) were used in the study. Individual study trees were placed into resistant and sensitive categories based on their relative expression of recognized ozone-induced visible injury symptoms. Trees defined as sensitive exhibited chlorotic mottle on young needles and retained only 1 or 2 years of needles. Resistant trees showed little or no chlorosis and retained 5 or 6 years of needles. Similar symptomatology attributed to ozone injury has been described elsewhere in both controlled fumigation studies and in field sites experiencing chronic ozone exposure (Miller *et al.*, 1972; Temple, 1988). Ozone-induced injury varied in this site such that approximately 10% of the individuals exhibited severe symptoms, 10% showed little or no symptoms, and the rest showed intermediate symptomatology. Resistant and sensitive trees grew, in many cases, side by side and there were no clear physical or topographical differences between the sites of the trees differing in sensitivity.

B. Gas Exchange Measurements

Gas exchange measurements were taken using the Li-Cor Li-6200 portable gas exchange system configured in a closed mode. One or two fascicles of needles on three south-facing branches on each tree were sealed in a ¼ liter cuvette and were allowed to draw down ambient CO_2 concentration approximately 5 ppm. Photosynthetic rates were determined from the rate of CO_2 depletion and expressed on a total surface area basis. Stomatal conductances were calculated from steady-state transpiration rates, leaf temperatures, and ambient vapor pressure deficits between leaf and air in the cuvette. Intercellular CO_2 concentrations (c_i) were calculated using equations described by von Caemmerer and Farquhar (1981). Gas exchange measurements typically took 30–40 s and leaf and cuvette conditions never varied more than 1–2°C or 3% RH from ambient environmental conditions.

Jeffrey pine needles occur in distinct whorls on the branches that flush annually beginning in late June and ending by the end of August. Needle age was determined by position on the branch relative to the current season's foliage. Up to 30 plants in both resistant and sensitive categories were sampled. Measurements were taken between 10:00 and 12:00 PM in full sunlight at 2- to 4-week intervals throughout the study (April–October). After gas exchange measurements were taken, fresh weight of each needle was determined and total surface area was calculated geometrically. Needles were oven-dried at 60°C for 48 h for dry weight determination and then ground to a fine powder for subsequent isotope analysis.

C. CO₂ Response Curves and Stomatal Limitation

CO_2 response curves (A–c_i) for three trees in each category were performed by sealing three to five fascicles of needles in a 1-liter cuvette and enriching it with CO_2. The needles were allowed to slowly draw down CO_2 concentrations (Davis *et al.*, 1987) so that photosynthetic rates and c_i could be determined at approximately 50-ppm intervals. CO_2 responses were determined on 1- and 2-year-old needles both at the beginning of the season and after 2 months of ambient ozone exposure. Regressions of the initial linear portion of the response from three different trees were made to compare carboxylation efficiencies. Biochemical and stomatal limitations to photosynthesis were then estimated from the CO_2 response curves (Sasek and Richardson, 1989; Assman, 1988).

D. Stable Carbon Isotope Discrimination (Δ)

Needles collected from trees at the end of the growing season were used in the isotope analysis. Ground needle tissue from three branches was pooled into age categories for each tree. Six trees in both sensitive and resistant categories were sampled. Samples were analyzed for stable carbon isotope composition (δ) using mass spectrometry. Discrimination values (Δ) were calculated as

$$\Delta = \frac{\delta_a = \delta_p}{1 + \delta_p},$$

where δ_a is the isotopic composition of ambient air taken to be $-8.0\%o$ and δ_p is the isotopic composition of the sample. Theoretical Δ values were calculated from the equation that describes the relationship between the ratio of intercellular to ambient CO_2 concentrations (c_i/c_a) and Δ (see O'Leary (Chapter 3) and Farquhar and Loyd (Chapter 5), this volume):

$$\Delta = 4.4 + 22.6(c_i/c_a).$$

E. Statistical Analyses

Two-tailed Student's t tests were used to determine statistical significance between mean photosynthetic, conductance and discrimination values of sensitive and resistant trees. The slopes of the $A–c_i$ response curves were compared using F statistics (Zar, 1980). All significant values reported are at the 0.95 confidence level.

III. Results

Photosynthetic rates were highest at the beginning of the growth season in both sensitive and resistant categories. By the end of June, photosynthesis steadily dropped to a low in September but recovered after the first rainfall of the season in late September (Fig. 1). There was no significant difference

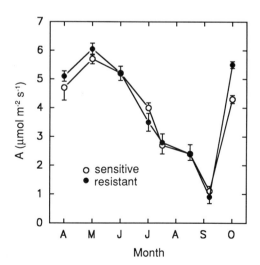

Figure 1.   Mean photosynthetic rates ($\pm$SE, $n = 25$–30) for sensitive and resistant Jeffrey pine. Data were collected on attached needles in full sunlight between 10:00 AM and 12 noon.

between the sensitive and resistant trees in mean photosynthetic rates until the end of the season. After the first rain mean photosynthetic rates of resistant trees recovered to almost the season high whereas sensitive trees exhibited photosynthetic rates that were significantly lower than both resistant trees and the seasonal high in their category.

Stomatal conductance values closely paralleled photosynthesis in both groups of trees (Fig. 2). However, unlike photosynthesis, instantaneous stomatal conductance was significantly higher in sensitive trees at the beginning of the season. This difference was not apparent in mid season when both sensitive and resistant trees had similar stomatal conductances, nor was there a clear separation after the precipitation event in late September although sensitive trees showed a lower mean conductance than resistant trees. Intercellular CO_2 concentrations of sensitive needles were approximately 20 ppm higher than those of resistant trees until August when stomatal conductance was less than half of early season values and c_i were the lowest of the season (Fig. 3). A peak in c_i in September reflected very low photosynthetic values or net respiration of needles at this point in the season.

Carbon isotope discrimination values were consistently higher in sensitive trees; however, the means of resistant and sensitive trees were not statistically different at any age (Fig. 4). Resistant trees showed a trend toward greater Δ with needle age and the increase was significant (i.e., slope significantly different from zero, $P < 0.05$). Δ was correlated with c_i/c_a in 1-year-old needles (Fig. 5) as predicted by theory and, although the regression slope was shallower than the theoretical slope, it was not signifi-

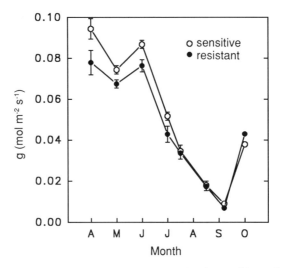

Figure 2.   Mean stomatal conductance ($\pm$SE, $n = 25$–30) for sensitive and resistant Jeffrey pine. Data were collected on attached needles in full sunlight between 10:00 AM and 12 noon.

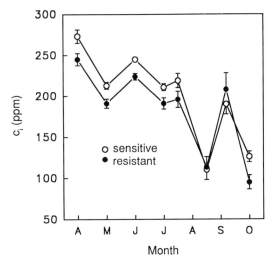

Figure 3.   Mean intercellular CO_2 concentrations (c_i) ($\pm$SE, $n = 25$–30) for sensitive and resistant Jeffrey pine. Data were collected on attached needles in full sunlight between 10:00 AM and 12 noon.

cantly different. Δ values for sensitive trees were, on average, higher than values for resistant trees but fell on the same regression line.

When needles of the same age were compared, photosynthetic responses to c_i were always greater in resistant trees (Fig. 6). The initial slope of 1-year-old needles was 20% greater in resistant trees indicating a carboxyl-

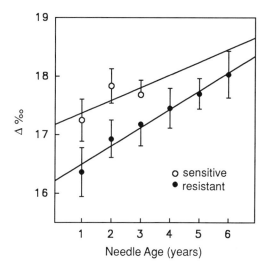

Figure 4.   Mean stable carbon isotope discrimination values (Δ) ($\pm$SE, $n = 6$) in needles of different ages from resistant and sensitive Jeffrey pine. Lines are regressions drawn through all data in each sensitivity category.

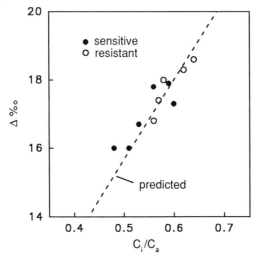

Figure 5. Relationship between Δ and mean c_i/c_a for 1-year-old resistant and sensitive Jeffrey pine needles. Broken line is theoretical relationship described by $\Delta = 4.4 + 22.6(c_i/c_a)$.

ation efficiency greater than that in needles of the same age from sensitive individuals. An even greater difference in carboxylation efficiencies (initial slopes) was found between 2-year-old needles from resistant and sensitive trees. After 6 weeks of exposure to ambient ozone, carboxylation efficiencies decreased in 1-year-old needles of both resistant and sensitive individuals (15 and 35%, respectively). Stomatal limitations decreased photosynthesis 39% in resistant trees in June compared to a 32% limitation calculated in sensitive trees (Table I). By mid-July stomatal limitations increased to 53% in resistant trees but only to 44% in sensitive trees.

Table I Stomatal and Nonstomatal Limitation to Photosynthesis in Ozone-Sensitive and Resistant Jeffrey Pine at Early and Mid-Season Dates

	Needle age	Stomatal (%)[a]		Nonstomatal (%)[b]	
		Sensitive	Resistant	Sensitive	Resistant
1 June					
	1 year	32	39	16	0
	2 years	44	37	86	26
15 July					
	1 year	43	52	35	13
	2 years	36	55	84	25

[a] Relative to c_i of 350 ppm.
[b] Relative to 1 year resistant c_i of 350 ppm.

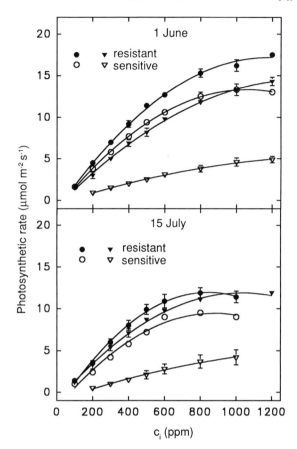

Figure 6. A–c_i response curves for 1 (circles)- and 2 (triangles)-year-old needles from resistant and sensitive trees. Each data point represents a mean from three trees ±SE.

IV. Discussion

The greater discrimination values in sensitive trees support the hypothesis that trees with larger integrated conductances are more prone to ozone injury. Although there were no differences in net photosynthetic rates between sensitive and resistant categories, sensitive trees consistently maintained higher c_i and stomatal conductances throughout much of the growing season. Greater Δ in sensitive trees suggest that they operate at a lower intrinsic water-use efficiency (A/g) and agrees with the instantaneous values measured throughout much of the growing season.

The photosynthetic capacity of sensitive Jeffrey pine needles is lower than that of resistant needles in all age categories. This is especially evident in 2-year-old needles from sensitive trees which operate at only around

20% of the resistant category's capacity. These results are consistent with other work which has shown reduction in photosynthesis in ozone-stressed plants (Reich, 1987) and is also indicative of the especially sensitive nature of the photosynthetic apparatus. Photosynthesis is probably among the first physiological processes to be damaged by ozone and physiological injury occurs in Jeffrey pine before the onset of visible injury symptoms which usually develop, at this site, on sensitive needles in their second year of growth. As the season progressed reductions in photosynthetic capacity occurred in both sensitive and resistant trees. This reduction was more pronounced in sensitive trees where the photosynthetic capacity of 1-year-old needles dropped below that of 2-year old resistant needles by mid season (Fig. 6). The photosynthetic capacity of resistant needles was less in their second year of growth and may reflect an age-related physiological decline. The larger decline of photosynthetic capacity in sensitive needles suggests that injury to the photosynthetic apparatus was occurring due to something other than normal environmental stresses or phenological processes. When Δ is graphed against c_i/c_a (Fig. 5), sensitive and resistant trees fall out on the same regression line but are separated into different areas along the line (resistant in the lower half and sensitive in the upper half). This suggests that, in part, these two categories of trees differ in their stomatal limitations to CO_2 and ozone diffusion.

Sharkey (1985) suggests that plants may exhibit less stomatal limitation to photosynthesis in response to stress. Indeed stomatal limitation in sensitive plants was smaller compared to that in resistant individuals in three of four cases (Table I). Stomata did not close as much in sensitive trees which, consequently, may have partially compensated for their reduced carboxylation efficiency. By operating at a higher c_i, sensitive trees continued to maintain net photosynthetic rates similar to those of resistant trees in 1-year-old needles, but lost more water and processed more ozone as a consequence of their higher stomatal conductances.

The greater instantaneous A/g in resistant trees would have resulted in a higher water-use efficiency (W). This is also evident in the smaller Δ values in resistant trees indicating a greater long-term W (Farquhar *et al.*, 1989). The premature abscission of needles in sensitive trees may be related to a loss of water regulation by damaged guard cells. Greater water loss from sensitive trees because of higher stomatal conductances might place the whole plant under water stress inducing needle abscission. Heath *et al.* (1982) concluded that ozone-induced injury to cell membranes can result in altered water relations of a cell ultimately simulating a drought-like stress. This sort of injury could potentially affect both the function of mesophyll cells responsible for carbon fixation and the guard cells which regulate water loss and CO_2 uptake. An ozone-induced disruption in water balance due to the loss in stomatal control has also been implicated in other conifers (Maier-Maercker and Koch, 1991).

The positive correlation of Δ with age in resistant trees suggests that W is lower in older needles. Jeffrey pine needles do change physiologically with age as photosynthetic rates drop approximately 10% every year (Patterson and Rundel, 1990). However, the accumulation of metabolites or secondary tissues that have different isotopic signatures from the bulk leaf tissue may influence leaf isotope ratios (DeNiro and Epstein, 1977; Benner *et al.*, 1987). The addition of secondary phloem in conifer needles is known for many Sierran species including Jeffrey pine (Ewers, 1982) and may account for the age-related differences in Δ. The accumulation of isotopically lighter compounds, such as cellulose, and the reduction of heavier compounds, such as starch, coupled with an increasing c_i would result in larger Δ values in older needles.

Stable carbon isotope ratios can provide a valuable tool in assessing plant responses to environmental stresses such as air pollution. Ozone has been shown to induce stomatal closure in many plant species (e.g., Keller and Halser, 1984; Matyssek *et al.*, 1991), which is consistent with the increased $\delta^{13}C$ values (lower Δ) found in several species of both trees and herbs in response to ozone (Greitner and Winner, 1988; Martin *et al.*, 1988; Saurer *et al.*, 1991). Ozone-induced reductions in stomatal conductance should play an important role in limiting ozone entry into plant tissues thus reducing injury and operating as a mechanism of ozone resistance. Within a species, greater resistance should be conferred upon those individuals that operate at lower stomatal conductances. Indeed, this study indicates that within a population of ozone-stressed Jeffrey pine, trees appearing to be the most sensitive to ozone operate at stomatal conductances higher than those in the most resistant trees. These results were supported by both gas exchange measurements and discrimination values suggesting that stable carbon isotope discrimination techniques may provide a powerful tool in predicting relative sensitivities of some plant species to gaseous air pollutants.

V. Summary

Stable carbon isotope discrimination (Δ) and gas exchange measurements were used to test the hypothesis that, within a population of Jeffrey pine, ozone-sensitive individuals operate at higher stomatal conductances than ozone-resistant individuals. Although net photosynthetic rates in 1-year-old needles did not differ significantly, instantaneous stomatal conductances were consistently higher in sensitive trees. This was consistent with the larger Δ values measured in sensitive needles; however, the differences in Δ were not statistically significant between resistant and sensitive categories which probably reflects the small differences in Δ ($\approx 1.0‰$) and the small sample size ($n = 6$).

Photosynthetic capacity was lower in sensitive trees although they operated at a higher c_i which was largely due to smaller stomatal limitations on photosynthesis. Although photosynthetic capacity in sensitive trees declined by midseason, stomatal limitation varied such that c_i was always 15–

20 ppm greater than that in resistant trees. This suggests that stomatal control is not impacted by ozone at first but that sensitive trees operate at intrinsically higher conductances which may predispose them to ozone injury.

References

Assmann, S. M. 1988. Stomatal and non-stomatal limitations to carbon assimilation: An evaluation of the path dependent method. *Plant Cell Environ.* **11:** 577–582.

Benner, R., M. L. Fogel, E. K. Sprague, and R. E. Hodson. 1987. Depletion of ^{13}C in lignin and its implications for carbon isotope studies. *Nature* **329:** 708–710.

Berrange, P., D. F. Karnosk, and J. P. Bennett. 1991. Natural selection for ozone tolerance in *Populus tremuloides:* An evaluation of nationwide trends. *Can. J. For. Res.* **21:** 1091–1097.

Boyer, J. N., D. B. Houston, and K. F. Jensen. 1986. Impacts of chronic SO_2, O_3, and $SO_2 + O_3$ exposures on photosynthesis of *Pinus strobus* clones. *Eur. J. For. Pathol.* **16:** 293–299.

von Caemmerer, S., and G. D. Farquhar. 1981. Some relationships between the biochemistry of photosynthesis and the gas exchange of leaves. *Planta* **153:** 376–387.

Coyne, P., and G. Bingham. 1981. Comparative ozone dose response of gas exchange in a ponderosa pine stand exposed to long-term fumigations. *J. Air Pollut. Control. Assoc.* **31:** 38–41.

Darrall, N. M. 1989. The effects of air pollutants on physiological processes in plants. *Plant Cell Environ.* **12:** 1–30.

Davis, J. E., T. J. Arkebauer, J. M. Norman, and J. R. Brandle. 1987. Rapid field measurement of the assimilation rate *versus* intercellular CO_2 concentration relationship in green ash (*Fraxinus pennsylvanica* Marsh.): The influence of light intensity. *Tree Physiol.* **3:** 387–392.

DeNiro, M. J., and S. Epstein. 1977. Mechanism of carbon isotope fractionation associated with lipid synthesis. *Science* **197:** 261–263.

Duriscoe, D. M. 1987. Evaluation of Ozone Injury to Selected Tree species in Sequoia and Kings Canyon National Parks. Final report No.122. Holcomb Research Institute, Butler University, Indianapolis, IN.

Ehleringer, J. R. 1990. Correlations between carbon isotope discrimination and leaf conductance to water vapor in common beans. *Plant Physiol.* **93:** 1422–1425.

Ewers, F. W. 1982. Secondary growth in needle leaves of *Pinus longaeva* (Bristlecone pine) and other conifers: Quantitative data. *Am. J. Bot.* **69:** 1552–1559.

Farquhar, G. D., M. H. O'Leary, and J. A. Berry. 1982. On the relationship between carbon isotope discrimination and the intercellular carbon dioxide concentration in leaves. *Aust. J. Plant Physiol.* **9:** 121–137.

Farquhar, G. D., K. T. Hubick, A. G. Condon, and R. A. Richards. 1989. Carbon isotope fractionation and plant water-use efficiency. *In* P. W. Rundel, J. R. Ehleringer and K. A. Nagy (eds.), *Stable Isotopes in Ecological Research.* Springer-Verlag, New York.

Greitner, C. S., and W. E. Winner. 1988. Increase in $\delta^{13}C$ values of radish and soybean plants caused by ozone. *New Phytol.* **108:** 489–494.

Heath, R. L. 1980. Initial events in injury to plants by air pollutants. *Annu. Rev. Plant Physiol.* **31:** 395–431.

Heath, R. L. 1987. Oxidant air pollutants and plant injury. *In* D. W. Newman and K. G. Wilson (eds.), Models in Plant Physiology and Biochemistry. CRC Press, Boca Raton, FL.

Heath, R. L., P. E. Frederick, and P. E. Chimiklis 1982. Ozone inhibition of photosynthesis in *Chlorella sorokiniana. Plant Physiol.* **69:** 229–233.

Keller, T., and R. Halser. 1984. The influence of a fall fumigation with ozone on the stomatal behaviour of spruce and fir. *Oecologia* **64:** 284–286.

Laisk A., O. Kull, and H. Moldau. 1989. Ozone concentration in leaf intercellular air spaces is close to zero. *Plant Physiol.* **90:** 1163–1167.

Maier-Maerker, U., and W. Koch. 1991. Experiments on the control capacity of stomata of *Picea abies* (L.) Karst. after fumigation with ozone and in environmentally damaged material. *Plant Cell Environ.* **14:** 175–184.

Martin, B., A. Bytnerowicz, and Y. R. Thorstenson. 1988. Effects of air pollutants on the composition of stable isotopes, δ^{13}, of leaves and wood, and on leaf injury. *Plant Physiol.* **88:** 218–223.

Matyssek, R., M. S. Günthardt-Goerg, T. Keller, and C. Scheidegger. 1991. Impairment of gas exchange and structure in birch leaves (*Betula pendula*) caused by low ozone concentrations. *Trees* **5:** 5–13.

McLaughlin, S. B., and G. E. Taylor. 1980. Relative humidity: Important modifier of pollutant uptake by plants. *Science* **211:** 167–169.

Miller, P. R., M. H. McCutchan, and H. P. Milligan. 1972. Oxidant air pollution in the central valley, Sierra Nevada foothills, and Mineral King Valley of California. *Atmos. Environ.* **6:** 623–633.

Patterson, M. T., and P. W. Rundel. 1990. Ozone Impacts on the Photosynthetic Capacity of Jeffrey Pine in Sequoia National Park. Final Report, Product NPS/AQD 90-005.

Pedersen, B. S., and T. A. Cahill. 1989. Ozone at a remote high altitude site in Sequoia national park, California. *In* R. K. Olsen and A. S. Leflon (eds.), *Effects of Air Pollution on Western Forests.* AWMA, Pittsburgh.

Reich, P. B. 1987. Quantifying plant response to ozone: A unifying theory. *Tree Physiol.* **3:** 63–91.

Rich, S., and N. C. Turner. 1972. Importance of moisture on stomatal behavior of plants subjected to ozone. *J. Air Pollut. Control. Assoc.* **22:** 718–721.

Roper, T. R., and L. E. Williams. 1989. Effects of ambient and acute partial pressures of ozone on leaf net CO_2 assimilation of field-grown *Vitis vinefera* C. *Plant Physiol.* **91:** 1501–1506.

Sasek, T. W., and C. J. Richardson. 1989. Effects of chronic doses of ozone on loblolly pine: Photosynthetic characteristics in the third growing season. *For. Sci.* **35:** 745–755.

Saurer M., J. Fuhrer, and U. Siegenthaler. 1991. Influence of ozone on the stable carbon isotope composition, $\delta^{13}C$, of leaves and grain of spring wheat (*Triticum aestivum* L.). *Plant Physiol.* **97:** 313–316.

Sharkey, T. D. 1985. Photosynthesis in intact leaves of C_3 plants: Physics, physiology and rate limitations. *Bot. Rev.* **51:** 53–105.

Temple, P. J. 1986. Stomatal conductance and transpirational responses of field-grown cotton to ozone. *Plant Cell Environ.* **9:** 315–321.

Temple, P. J. 1988. Injury and growth of Jeffrey pine and Giant Sequoia in response to ozone and acidic mist. *Environ. Exp. Bot.* **28(4):** 323–333.

Turner, N. C., S. Rich, and H. Tomlinson. 1972. Stomatal conductance, fleck injury and growth of tobacco cultivars varying in ozone tolerance. *Phytopathology* **62:** 63.

Warner, T. E., D. W. Wallner, and D. R. Volger. 1982. Ozone injury to ponderosa and Jeffrey pines in Sequoia National Parks. *In* C. van Riper, L. Whittig, and M. Murphy (eds.), Proceedings First Biennial Conference of Research in California's National Parks. University of California, Davis.

Williams, W. 1980. Air pollution disease in the Californian forests: A baseline for smog disease on ponderosa and Jeffrey pines in sequoia and Los Padres National Forests, California. *Environ. Sci. Technol.* **14:** 179–182.

Zar, J. H. 1980. *Biostatistical Analysis.* Prentice–Hall, Englewood Cliffs, NJ.

Zimmerman, R., R. Oren, E. D. Schulze, and K. S. Werk. 1988. Performance of two *Picea abies* (L.) Karst. stands at different stages of decline. II. Photosynthesis and leaf conductance. *Oecologia* **76:** 513–518.

16

Carbon Isotope Composition and Gas Exchange of Loblolly and Shortleaf Pine as Affected by Ozone and Water Stress

Christine G. Elsik, Richard B. Flagler, and Thomas W. Boutton

I. Introduction

Ozone (O_3), a photochemical oxidant, has been recognized as the most phytotoxic of the widespread air pollutants (Reich, 1987) and has been implicated as a factor in the recent forest growth decline in the southeastern United States (Sheffield and Cost, 1987). Injury to plant tissue results from oxidation of biological compounds by O_3 and the free radicals it forms, impacting several biochemical and physiological processes, and leading to alterations in growth and biomass allocation (Guderian et al., 1985). Air pollution is among the environmental factors affecting leaf internal carbon dioxide concentration (c_i) through effects on rates of net photosynthesis (A) and stomatal conductance (g), justifying stable carbon isotope analysis as a tool in the study of plant response to air pollution. Stable carbon isotope analysis has been used previously in the study of pollutant effects on trees and crop plants (Freyer, 1979; Greitner and Winner, 1988; Martin et al., 1988; Becker et al., 1989; Boutton and Flagler, 1990; Martin and Sutherland, 1990; Saurer et al., 1991; Taylor, 1991).

Growth declines resulting from elevated O_3 levels have been correlated with decreased photosynthesis in crop and tree species (Reich and Amundson, 1985). Whether the O_3-induced decrease in A results from stomatal or nonstomatal limitations remains controversial, but information on the effect of O_3 on c_i can clarify the mechanism (Runeckles and Chevone, 1992). Ozone may decrease A through increased mesophyll resistance, resulting in increased c_i (Reich, 1987). Alternatively, O_3 may affect guard cells directly, decreasing g, causing decreased c_i, and resulting in diminished A (Moldau

227

et al., 1990). There is no direct evidence supporting an immediate effect of O_3 on guard cell function in tree species (Chappelka and Chevone, 1992); however, stable carbon isotope analysis has indicated stomatal limitation to *A* in tree and crop species (Greitner and Winner, 1988; Martin *et al.*, 1988; Boutton and Flagler, 1990; Saurer *et al.*, 1991; Taylor, 1991).

Since O_3 enters the leaf through the stomata, environmental variables which affect *g*, such as water stress, alter plant response to O_3. Stomatal regulation can be a protective mechanism against both drought and air pollution; stomatal closure minimizes water loss (Teskey and Hinckley, 1986) and reduces O_3 injury through decreased O_3 uptake (Harkov and Brennan, 1980; Olszyk and Tibbitts, 1981). Alternatively, O_3 exposure may alter plant response to water stress by modifying *g* (Reich and Lassoie, 1984). The effects of O_3 on *g* in tree species are inconsistent. Ozone may increase stomatal sensitivity to vapor pressure deficit (Chappelka *et al.*, 1988), preventing possible drought injury, or may reduce stomatal responsiveness and increase transpiration (Keller and Hasler, 1984; Reich and Lassoie, 1984), increasing the likelihood of desiccation during periods of drought.

Ozone and water deficit can alter transpiration efficiency (*W*), the ratio of biomass produced to total water transpired. As a result of the different diffusive conductances for carbon dioxide and water vapor, reduced *g* decreases transpiration to a greater extent than carbon dioxide uptake (Nobel, 1991). Moderate water deficit generally increases *W* by inducing partial stomatal closure. Ozone also alters *W* through effects on *g* and the biochemical reactions of photosynthesis. Ozone-induced stomatal closure, without a reduction in *A*, would result in increased *W*, while decreased stomatal sensitivity to water deficit or decreased *A* would cause a reduction in *W*. Since *W* is related to integrated c_i, carbon isotope analysis may be used to assess *W* (Farquhar *et al.*, 1989b).

The purpose of this study was to investigate the effects of O_3 and water deficit and their interaction on $\delta^{13}C$ and c_i of loblolly and shortleaf pines. Seedlings were exposed to different levels of O_3 and soil moisture in open-top chambers during their first growing season. Stable carbon isotope composition was determined in order to assess integrated gas exchange characteristics. In addition, conventional gas exchange methods were used to measure *A*, *g*, and c_i.

II. Experimental Methods

A. Study Area

The research site was located in the USDA Forest Service Stephen F. Austin Experimental Forest (31° 30' N latitude, 94° 46' w longitude), roughly 12 km southwest of Nacogdoches, Texas. The mean annual maximum and minimum temperatures are 24.2°C and 11.2°C, respectively, and the mean annual precipitation is 115.6 cm. The area immediately surrounding the site consists of mature loblolly and shortleaf pine forest.

B. Plant Material

Seeds from one half-sib shortleaf pine (*Pinus echinata* Mill.) family (S2PE-3) and one half-sib loblolly (*P. taeda* L.) pine family (GR1-8) were stratified for 90 days and then sown, in January 1990, in 7-liter pots containing a fritted clay medium that had been leached with reverse osmosis water. Until treatments began, the seedlings were maintained in a greenhouse and were fertilized weekly beginning 12 weeks after the sowing date with 15 : 30 : 15 (N : P : K) and a micronutrient mix supplemented with chelated iron.

C. Ozone Exposure Chambers

The seedlings were exposed to O_3 in 10 cylindrical open-top field chambers 3 m in diameter and 2.5 m in height (Heagle *et al.*, 1973). Chambers were equipped with a fixed cap to exclude ambient rainfall. Air was forced through a plenum surrounding the lower portion of each chamber at approximately 60 m^3 min^{-1}, during the hours 0600 to 2400 CST daily.

Ozone was generated from O_2 by a corona discharge type generator and was metered to chambers through needle valves. Air from inside each chamber was sampled through Teflon tubing. The O_3 concentration was monitored continually with a uv-photometric O_3-specific analyzer on a time-shared basis in each chamber. The O_3 monitors were calibrated with a uv-photometry transfer standard.

D. Experimental Design and Treatment Regimes

The experimental design was a split–split plot conducted within a completely randomized design. The whole plots were five levels of O_3; the subplots were two water regimes; the sub-subplots were two species. Each treatment combination was replicated twice, requiring 10 chambers. Each treatment combination included 18 seedlings per replication, for a total of 720 seedlings.

Seedlings were placed in the chambers and O_3 treatments were initiated on 25 June 1990, approximately 24 weeks after sowing, when substantial secondary needle tissue had developed in both species. The five O_3 treatments ranged from a subambient level to 2.5 times ambient O_3 concentration. Charcoal filters were used to remove O_3 for the subambient treatment (CF); the ambient treatment consisted of nonfiltered air (NF). The three O_3 addition treatments were 1.7, 2.0, and 2.5 times the ambient O_3 concentration (1.7×, 2.0×, and 2.5×, respectively), and fluctuated as a proportion of ambient O_3, during the hours 0800 to 2000 CST.

All seedlings were watered daily to field capacity with reverse osmosis water to maintain the ψ_{soil} at -0.08 MPa until 10 August 1990. After this date, the two water regimes, well-watered (WW) and water-stressed (WS), were imposed. The WW and WS treatments received water whenever the soil volumetric water content was less than 34% (ψ_{soil} = -0.08 MPa) and 28% (ψ_{soil} = -0.3 MPa), respectively. The intent was to the allow WW seedlings to experience virtually no water stress, while the WS seedlings undergo mild water deficit. Soil moisture was characterized with a Trase System soil moisture device (Soil Moisture Equipment Corp., Santa Barbara, CA) based on time domain reflectometry. A moisture retention curve for the medium was produced, using the pressure plate technique, so that volumetric water content could be related to soil water potential. Trase measurements were taken daily for two randomly selected seedlings per plot to determine whether that plot needed to be watered, at which time reverse osmosis water was added to field capacity. All treatments were watered with fertilizer solution once every 2 weeks on a day that water addition was necessary.

E. Carbon Isotope Composition

Carbon isotope composition was determined for foliage and stems of five seedlings per plot. Carbon isotope composition, expressed as $\delta^{13}C$, was determined using the technique of Boutton (1991). Five milligrams of dried tissue, ground to pass a 40-mesh screen, was combusted to CO_2 in sealed quartz tubes at 850°C. The CO_2 was purified cryogenically and analyzed on a VG-903 dual-inlet, triple collector, gas isotope ratio mass spectrometer (VG Isogas, Middlewich, UK). Precision was approximately 0.1‰ for all $\delta^{13}C$ measurements.

F. Gas Exchange Characteristics

Gas exchange characteristics (A, g, and c_i) were measured biweekly with a portable photosynthesis system (LI-6200, LI-COR, Inc., Lincoln, NE) equipped with a 0.25-liter leaf chamber. Measurements were made on detached fascicles from the oldest flush of each seedling. Previous studies have shown that gas exchange characteristics of shortleaf pine are not significantly affected by fascicle detachment for up to 90 s (Lock and Flagler, 1992), and that photosynthetic rates of loblolly pine are not affected by detachment of fascicles from branches which had been removed from the tree for up to 30 min (Ginn et al., 1991). During each sample period, measurements were made on one fascicle per seedling and two seedlings per treatment combination, on all treatment combinations. The replications were measured on 2 consecutive days, between 1000 and 1400 h CST. During measurements, the leaf chamber was kept in a light box, so that photosynthetically active radiation reaching the chamber would be kept constant at approximately 1300 μmol m^{-2} s^{-1} using two 300-Watt cool beam lamps (General Electric, Cleveland, OH). The light box was equipped with an air blower for cooling and supplying fresh ambient air to the leaf chamber.

G. Data Analysis and Statistics

Response variables were analyzed by analysis of variance (ANOVA) for differences due to O_3, water regime, and interactions for species separately. Ozone effects were broken down to linear and curvilinear orthogonal contrasts. Regression analysis was performed on the moisture treatments separately, using the seasonal sum of hourly O_3 averages between the hours 0800 and 2000 CST (12 h sum zero) as the regressor.

III. Results

A. Ozone Exposures and Meteorology

During the exposure period (25 June–31 October) the mean ambient 12 h d^{-1} O_3 concentration for the hours 0800 to 2000 CST was 0.047 ppm. The highest ambient 1-h peak O_3 concentration was 0.109 ppm. The federal secondary ambient air quality standard for O_3 of 0.120 ppm was exceeded by the three O_3 addition treatments. Ozone exposure statistics for the five treatments are given in Table I. Seasonal and diurnal mean O_3 concentration trends have been illustrated in a previous paper (Elsik *et al.*, 1992). The average temperature and relative humidity during the treatment period (25 June–31 October) was 19.2°C and 81.7%, respectively. The average daily maximum and average daily minimum temperatures were 27.1 and 13.1°C, respectively. The maximum and minimum 1-h temperatures were 33.1, and −0.6°C, respectively. The average daily maximum and average daily minimum relative humidities were 97.2 and 52.9%, respectively. The maximum and minimum 1-h relative humidities were 99.5 and 28.8%, respectively.

Table I Ozone Exposure Statistics for Ambient Air and Loblolly and Shortleaf Pine Exposed to Five Levels of O_3 in Open-Top Chambers in East Texas[a]

Ozone level[b]	12 h day^{-1} O_3 concn (ppm)		1 h day^{-1} peak O_3 concn (ppm)		Sum zero[c] (ppm-h)
	Seasonal mean	Highest	Seasonal mean	Highest	
CF	0.005	0.018	0.013	0.066	9.6
NF	0.037	0.076	0.058	0.108	58.1
1.7×	0.078	0.161	0.123	0.252	118.3
2.0×	0.099	0.218	0.157	0.348	148.7
2.5×	0.114	0.238	0.178	0.357	170.5
AA	0.047	0.079	0.069	0.109	44.1

[a]Each value is the mean from two chambers except for AA values, which are the mean from one ambient air monitor, for the daily period 0800 to 2000 h CST from 25 June to 31 October 1990.

[b]Ozone levels CF, NF, 1.7×, 2.0×, 2.5×, and AA are charcoal filtered, nonfiltered, 1.7 times ambient, 2.0 times ambient, 2.5 times ambient, and ambient air, respectively.

[c]Seasonal sum of hourly O_3 averages between the hours 0800 and 2000 CST from 25 June to 31 October 1990.

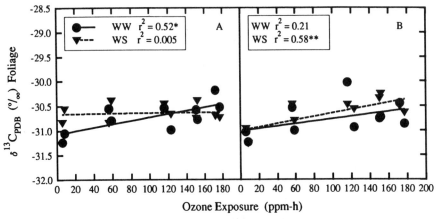

Figure 1. Foliar $\delta^{13}C$ of loblolly (A) and shortleaf (B) pine versus O_3 exposure (O_3 sum zero) for well-watered (WW) and water-stressed (WS) seedlings. Significance at the 0.05 and 0.01 levels are signified by * and **, respectively ($n = 10$).

B. Carbon Isotope Composition

Foliar $\delta^{13}C$ increased with O_3 exposure in both species ($P < 0.1$) (Table II). Increased $\delta^{13}C$ indicates decreased long-term internal CO_2 concentration, and, thus, increased W with elevated O_3. Stem $\delta^{13}C$ increased significantly ($P < 0.05$) with O_3 exposure in loblolly pine only (Table II). A tendency for $\delta^{13}C$ to increase with O_3 exposure was observed in shortleaf pine stems, but this trend was not significant. A consistent increase in $\delta^{13}C$ attributable to moisture deficit was observed only in stem tissue of shortleaf pine ($P < 0.1$) (Table II). Linear regressions for separate moisture regimes are shown in Figs. 1 and 2 for foliage and stem, respectively. Regression analysis indi-

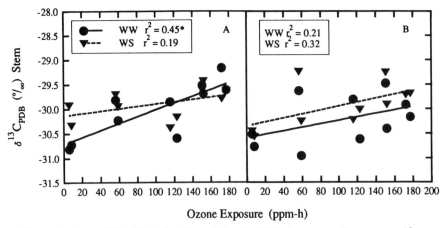

Figure 2. Stem $\delta^{13}C$ of loblolly (A) and shortleaf (B) pine versus O_3 exposure (O_3 sum zero) for well-watered (WW) and water-stressed (WS) seedlings. Signficance at the 0.05 level is signified by * ($n = 10$).

Table II Carbon Isotope Composition of Foliage and Stem Tissue of Loblolly and Shortleaf Pine Seedlings Exposed to O_3 and Water Deficit[a]

		$\delta^{13}C$(‰)			
		Loblolly pine		Shortleaf pine	
Ozone level[b]	Water regime[c]	Foliage	Stem	Foliage	Stem
CF	WW	−31.65 (0.09)	−30.77 (0.21)	−31.14 (0.11)	−30.65 (0.12)
NF	WW	−30.68 (0.12)	−30.03 (0.21)	−30.77 (0.23)	−30.30 (0.63)
1.7×	WW	−30.77 (0.22)	−30.23 (0.75)	−30.48 (0.45)	−30.22 (0.41)
2×	WW	−30.68 (0.10)	−29.60 (0.09)	−30.75 (0.01)	−29.95 (0.47)
2.5×	WW	−30.36 (0.17)	−29.38 (0.22)	−30.66 (0.21)	−30.05 (0.13)
CF	WS	−30.70 (0.14)	−30.14 (0.25)	−31.12 (0.15)	−30.52 (0.06)
NF	WS	−30.63 (0.22)	−29.83 (0.12)	−30.65 (0.16)	−29.75 (0.54)
1.7×	WS	−30.57 (0.10)	−30.28 (0.11)	−30.54 (0.04)	−30.13 (0.10)
2.0×	WS	−30.58 (0.18)	−29.52 (0.09)	−30.32 (0.05)	−29.59 (0.33)
2.5×	WS	−30.73 (0.02)	−29.71 (0.09)	−30.59 (0.06)	−29.71 (0.03)
		ANOVA[d]			
O_3		0.24	0.04	0.19	0.53
Linear		0.06	0.01	0.07	0.17
Quadratic		0.45	0.81	0.10	0.55
Residual		0.57	0.08	0.95	0.75
Water regime		0.43	0.25	0.24	0.05
Interaction		0.26	0.42	0.54	0.68

[a]Values are means and standard errors (in parentheses) of 10 samples.

[b]Ozone levels CF, NF, 1.7×, 2.0×, and 2.5× are charcoal filtered, nonfiltered, 1.7 times ambient, 2.0 times ambient, and 2.5 times ambient, respectively.

[c]Water regimes WW and WS are well-watered and water-stressed, respectively.

[d]Tabular values are probability levels associated with ANOVA. Degrees of freedom associated with the sources of variation are O_3, 4 df; O_3 contrasts, 1 df; water regime, 1 df; interaction, 4 df.

Table III Net Photosynthesis (A), Stomatal Conductance (g), and Internal CO_2 Concentration (c_i) of Loblolly Pine Seedlings Exposed to O_3 and Water Deficit[a]

Ozone level[b]	Water regime[c]	13 October 1990			27 October 1990		
		A (μmol m^{-2} s^{-1})	g (mol m^{-2} s^{-1})	c_i (ppm)	A (μmol m^{-2} s^{-1})	g (mol m^{-2} s^{-1})	c_i (ppm)
CF	WW	4.33 (0.17)	0.129 (0.003)	271 (4)	4.34 (0.23)	0.116 (0.002)	265 (2)
NF	WW	4.08 (0.12)	0.112 (0.021)	268 (16)	3.76 (0.16)	0.090 (0.009)	278 (18)
1.7×	WW	4.02 (0.20)	0.111 (0.008)	267 (2)	3.42 (0.16)	0.090 (0.015)	271 (21)
2×	WW	3.10 (0.93)	0.078 (0.027)	285 (18)	2.24 (0.35)	0.062 (0.004)	288 (15)
2.5×	WW	2.53 (0.33)	0.060 (0.002)	272 (11)	2.50 (0.87)	0.088 (0.029)	301 (6)
CF	WS	4.51 (0.13)	0.092 (0.000)	245 (0)	4.25 (0.76)	0.095 (0.020)	257 (5)
NF	WS	5.70 (0.97)	0.109 (0.022)	242 (6)	4.77 (0.54)	0.104 (0.025)	257 (11)
1.7×	WS	4.47 (0.12)	0.089 (0.004)	243 (7)	3.87 (0.64)	0.075 (0.017)	267 (29)
2×	WS	3.69 (0.70)	0.073 (0.014)	256 (11)	2.55 (0.20)	0.047 (0.014)	255 (35)
2.5×	WS	3.25 (0.20)	0.077 (0.015)	267 (9)	2.65 (0.74)	0.067 (0.008)	271 (10)
				ANOVA[d]			
O_3		0.11	0.22	0.65	0.10	0.27	0.87
Linear		0.02	0.04	0.25	0.02	0.09	0.36
Quadratic		0.25	0.59	0.64	0.91	0.42	0.80
Residual		0.62	0.78	0.72	0.43	0.46	0.95
Water regime		0.02	0.08	0.00	0.14	0.17	0.02
Interaction		0.39	0.06	0.24	0.59	0.59	0.47

[a]Values are means and standard errors (in parentheses) of four measurements taken during the last two sample periods.
[b]Ozone levels CF, NF, 1.7×, 2.0×, and 2.5× are charcoal filtered, nonfiltered, 1.7 times ambient, 2.0 times ambient, and 2.5 times ambient, respectively.
[c]Water regimes WW and WS are well-watered and water-stressed, respectively.
[d]Tabular values are probability levels associated with ANOVA. Degrees of freedom associated with the sources of variation are O_3, 4 df; O_3 contrasts, 1 df; water regime, 1 df; interaction, 4 df.

Table IV Net Photosynthesis (A), Stomatal Conductance (g), and Internal CO_2 Concentration (c_i) of Shortleaf Pine Seedlings Exposed to O_3 and Water Deficit[a]

Ozone level[b]	Water regime[c]	13 October 1990			27 October 1990		
		A (μmol m^{-2} s^{-1})	g (mol m^{-2} s^{-1})	c_i (ppm)	A (μmol m^{-2} s^{-1})	g (mol m^{-2} s^{-1})	c_i (ppm)
CF	WW	5.51 (0.14)	0.169 (0.008)	275 (0)	4.54 (0.05)	0.128 (0.016)	273 (12)
NF	WW	5.42 (0.43)	0.154 (0.006)	274 (0)	4.42 (0.54)	0.126 (0.030)	284 (16)
1.7×	WW	3.48 (0.58)	0.094 (0.015)	271 (0)	4.71 (0.93)	0.107 (0.024)	264 (10)
2×	WW	3.62 (1.97)	0.107 (0.041)	293 (5)	3.59 (0.44)	0.098 (0.005)	288 (9)
2.5×	WW	4.17 (0.08)	0.107 (0.013)	274 (8)	2.42 (0.23)	0.073 (0.016)	300 (0)
CF	WS	4.54 (0.51)	0.108 (0.033)	253 (14)	4.76 (0.04)	0.100 (0.004)	252 (5)
NF	WS	6.07 (0.80)	0.122 (0.011)	246 (0)	5.37 (0.43)	0.115 (0.017)	255 (5)
1.7×	WS	5.83 (1.05)	0.128 (0.017)	248 (7)	4.63 (1.94)	0.099 (0.052)	253 (9)
2×	WS	4.43 (1.06)	0.094 (0.012)	264 (2)	4.94 (0.24)	0.105 (0.013)	264 (2)
2.5×	WS	5.09 (0.88)	0.127 (0.018)	269 (7)	4.44 (0.23)	0.084 (0.013)	254 (1)
		ANOVA[d]					
O_3		0.64	0.48	0.79	0.61	0.57	0.34
Linear		0.35	0.19	0.56	0.21	0.17	0.16
Quadratic		0.94	0.46	0.90	0.40	0.56	0.52
Residual		0.49	0.61	0.09	0.99	0.88	0.65
Water regime		0.15	0.37	0.05	0.06	0.69	0.02
Interaction		0.36	0.16	0.29	0.48	0.88	0.47

[a] Values are means and standard errors (in parentheses) of four measurements taken during the last two sample periods.
[b] Ozone levels CF, NF, 1.7×, 2.0×, and 2.5× are charcoal filtered, nonfiltered, 1.7 times ambient, 2.0 times ambient, and 2.5 times ambient, respectively.
[c] Water regimes WW and WS are well-watered and water-stressed, respectively.
[d] Tabular values are probability levels associated with ANOVA. Degrees of freedom associated with the sources of variation are O_3, 4 df; O_3 contrasts, 1 df; water regime, 1 df; interaction, 4 df.

cated that $\delta^{13}C$ of foliage and stem tissue in WW loblolly pine seedlings were linearly related to O_3 exposure ($P < 0.05$), but $\delta^{13}C$ of WS loblolly pine seedlings was not significantly related to O_3. This suggests an $O_3 \times$ water stress interaction. The opposite relationship was observed in shortleaf pine foliage, in which $\delta^{13}C$ of WS seedlings was linearly related to O_3 exposure ($P < 0.01$), but $\delta^{13}C$ of WW seedlings was not significantly related to O_3.

C. Gas Exchange Characteristics

Results for A and g, which have been reported previously (Elsik *et al.*, 1992), are reviewed, to provide a basis for comparison with carbon isotope and instantaneous c_i data. Divergence in A and g were initially observed during the 13 September measurement period in both species; however, significant differences were not observed until the 13 October measurement period. The seasonal trends in A and g for each O_3 level have been illustrated elsewhere (Elsik *et al.*, 1992). Assimilation rate and g decreased linearly due to O_3 in loblolly pine during the last two measurement periods (Table III). Downward trends in A and g with O_3 exposure were observed in shortleaf pine, but were not significant (Table IV). Water-stressed seedlings of both species tended to possess higher A than WW seedlings, but this was significant in loblolly pine only during the 13 October measurement period ($P < 0.05$) and in shortleaf pine only during the 27 October measurement period ($P < 0.1$). Water stress tended to decrease g in both species, but this was significant in loblolly pine only during the 13 October measurement period ($P < 0.1$) and was not significant in shortleaf pine. There were no consistent $O_3 \times$ water stress interactions on A or g in either species.

There were no consistent $O_3 \times$ water stress interactions on c_i (Tables III and IV), so data were averaged over moisture regimes to reveal the seasonal pattern in O_3 main effects on c_i (Fig. 3). Internal CO_2 concentration tended to increase due to O_3 exposure during most measurement periods, but this was significant ($P < 0.05$) only in shortleaf pine during the 13 October measurement period (Table IV). The pattern was not consistent throughout the season, often with a decrease in c_i in the 1.7× treatment compared to CF. Water deficit caused decreased c_i during all measurement periods beginning 22 August in loblolly pine ($P < 0.001$ to $P < 0.05$) and shortleaf pine ($P < 0.01$ to $P < 0.05$).

IV. Discussion

The linear increase in $\delta^{13}C$ with O_3 exposure in both species is consistent with previous studies using stable carbon isotope composition to assess O_3 response by loblolly pine (Taylor, 1991), shortleaf pine (Boutton and Flagler, 1990), and other C_3 plant species (Greitner and Winner, 1988; Martin *et al.*, 1988; Saurer *et al.*, 1991). These data indicate a decreased c_i,

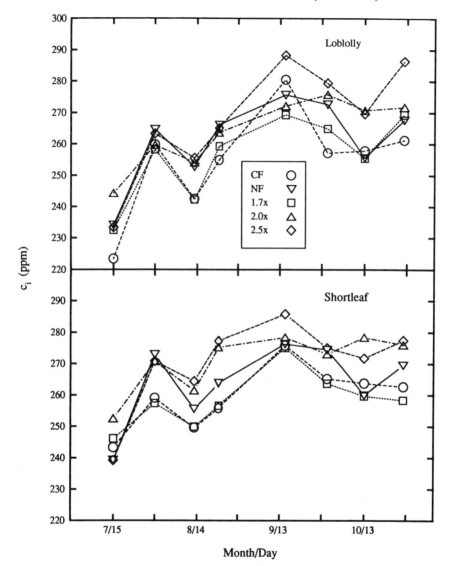

Figure 3. Seasonal pattern of instantaneous internal CO_2 concentration (c_i) of loblolly and shortleaf pine seedlings as affected by five levels of O_3 ($n = 8$).

which could result from either increased A or, more likely, decreased g. Instantaneous gas exchange measurements have indicated a linear decrease in A due to O_3 in loblolly pine. Consequently, it can be concluded that decreased c_i due to O_3 is a result of decreased g. These results are in agreement with measurements of g, which tended to decrease in response to O_3 exposure in both species, although significant in loblolly pine only. The $\delta^{13}C$ values indicate increased W with higher O_3 concentrations, in agreement with a previous study on shortleaf pine (Boutton and Flagler, 1990).

The increase in $\delta^{13}C$ in stem tissue of shortleaf pine due to water deficit indicated a water-stress-induced increase in W, in agreement with instantaneous water-use efficiency (A/E) (Elsik *et al.*, 1992). This is in accordance with previous studies on the effect of water deficit on $\delta^{13}C$ in C_3 plants (Hubick *et al.*, 1986; Martin and Thorstenson, 1988). However, the water stress did not have a consistent effect on $\delta^{13}C$ in loblolly pine, and water deficit did not affect A/E in this species (Elsik *et al.*, 1992).

Increased instantaneous c_i values attributable to O_3 suggested that the O_3-induced decrease in A was a result of biochemical processes, such as light harvesting or dark reactions of photosynthesis, rather than g, in accordance with a previous study on loblolly pine (Sasek and Richardson, 1989). This does not agree with the integrated c_i values as measured by $\delta^{13}C$. Instantaneous water-use efficiency was not significantly related to O_3 (Elsik *et al.*, 1992), also contrary to $\delta^{13}C$, which indicated increased W with elevated O_3. This variation may be related to the calculation of c_i in instantaneous gas exchange measurements. The calculation assumes that stomata are uniformly open or there is sufficient conductance between substomatal cavities, but a significant overestimation of c_i may occur if stomatal apertures vary or if lateral diffusion is limited (Laisk, 1983; Downton *et al.*, 1988; Terashima *et al.*, 1988). The homogeneity of stomatal response to O_3 has not been reported in loblolly or shortleaf pine. This may explain findings in a previous study, in which instantaneous gas exchange measurements indicated nonstomatal limitations to photosynthesis attributable to O_3 in loblolly pine (Sasek and Richardson, 1989).

In addition to possible errors in the calculation of c_i values in instantaneous gas exchange measurements, error may arise due to the need for constant environmental conditions throughout each measurement period of gas exchange. Light, humidity, and temperature were maintained relatively constant in the cuvette of the gas analyzer during each measurement, but the seedlings experienced different levels of soil moisture, humidity, and temperature throughout the day. Although the daily time frame within which measurements were made was kept to a minimum, there was still a notable change in atmospheric temperature and relative humidity during this period. Gas exchange characteristics exhibit a diurnal pattern, which may be affected by treatments. Therefore, seedlings may not be at the peak level of carbon assimilation when measurements were made. Both c_i and A/E are especially sensitive to environmental conditions, because they are dependent on A and g. Instantaneous gas exchange measurements may also have been obscured by the method in which treatments were applied, because seedlings within the same water regime but different chambers were at varying levels of moisture stress when gas exchange measurements were made.

In addition to variability in immediate environmental conditions, physiological responses to O_3 exposure may confound the relationship between

$\delta^{13}C$ and c_i. Changes in respiration due to O_3 have been reported in pine species (Barnes, 1972; McLaughlin *et al.*, 1982; Yang *et al.*, 1983) and would obscure the response of $\delta^{13}C$ to O_3 if fractionation is associated with respiration. It is assumed that isotope discrimination associated with respiration is negligible, but this has not been confirmed (O'Leary, 1988). PEP carboxylation is another process affecting carbon isotope composition. A fourfold increase in PEP carboxylase activity due to O_3 was observed in Scots pine (*Pinus sylvestris*) (Leuthy-Krause *et al.*, 1990). Since PEP carboxylation discriminates against ^{12}C (Farquhar *et al.*, 1989a), an increase in PEP carboxylase activity would cause increased $\delta^{13}C$. Fractionation during secondary metabolism could also obscure $\delta^{13}C$ response to O_3. Sharkey *et al.* (1991) reported fractionation associated with the synthesis of isoprene in red oak (*Quercus rubra*), in which the magnitude of discrimination was dependent on plant response to environmental conditions. Finally, variability in resistance to CO_2 diffusion within the leaf may modify $\delta^{13}C$ (Vitousek *et al.*, 1990). While $\delta^{13}C$ reflects c_i at the sites of carboxylation, gas exchange parameters yield c_i of the substomatal cavities (Evans *et al.*, 1986). In this study, O_3 significantly increased specific leaf area (m^2 leaf area/g foliage biomass), indicating decreased needle thickness or less densely packed cells (Elsik *et al.*, 1992). In either case, the result may be decreased internal resistance to CO_2 diffusion to the sites of carboxylation. Thus, the effect of O_3 on specific leaf area would result in increased c_i at the sites of carboxylation, consequently decreasing $\delta^{13}C$. This would conceal the effect of stomatal response to O_3 on $\delta^{13}C$, in which decreased g causes decreased c_i in the substomatal cavities, increasing $\delta^{13}C$.

There were no $O_3 \times$ water-stress interactions on instantaneous c_i or other gas exchange measurements. The relationship between watering and fertilizing schedules for each chamber with the timing of O_3 peaks may have obscured this potential interaction. Since O_3 was added in proportion to ambient, and watering and fertilizing cycles within a water stress treatment were different in each chamber, it is possible that seedlings within the same watering regime, but different chambers, were at different water-stress levels when O_3 peaks occurred. The integrated nature of stable carbon isotope composition allowed an interaction to be detected, because, unlike instantaneous gas exchange measurements, $\delta^{13}C$ is not immediately dependent on environmental variables at the time of measurement, which may or may not represent typical conditions.

The interaction suggested by regression analysis, in which the effects of O_3 on $\delta^{13}C$ of foliage and stem tissue in WW seedlings of loblolly pine were greater than those of WS seedlings, indicated that water stress provided protection from O_3 through stomatal closure. This is supported by the tendency of WS loblolly pine seedlings to possess lower g than WW loblolly pine seedlings. Similar interactions, in which the WS treatment lessened the O_3 effect, were observed in foliage biomass of loblolly pine (Elsik *et al.*, 1992). The opposite interaction was observed in foliar $\delta^{13}C$ of shortleaf pine, with a greater O_3 effect in WS seedlings than in WW seedlings. This interaction in shortleaf pine suggested that O_3 may have increased stomatal response to water stress. Instantaneous measurements of g in shortleaf pine, indicating that water stress tended to increase g at the higher O_3 levels, do not support this hypothesis. Alternatively, this apparent interaction may be a result of the high degree of variability in O_3 response of foliar $\delta^{13}C$ of WW seedlings as opposed to WS seedlings.

The results reported here and the previously reported growth measurements (Elsik *et al.*, 1992) indicate that the loblolly pine family was more sensitive to O_3 than the shortleaf pine family. Differences in O_3 sensitivity have been attributed to inherent differences in g, because O_3 uptake is limited by stomatal aperture (Reich, 1987). This is not the case in the present study. The loblolly pine family used here (GR1-8) is drought-hardy (van Buijtenen, 1966) and is expected to exhibit lower g than a drought-sensitive family. Seiler and Johnson (1988) reported that this family had lower rates of transpiration than two drought-susceptible families. In this study, instantaneous gas exchange measurements indicated that loblolly pine did indeed possess significantly lower g ($P < 0.05$) than shortleaf pine throughout the experiment. Alternatively, the difference in O_3 sensitivity may be related to mesophyll properties (Taylor *et al.*, 1982). According to Knauf and Bilan (1977), a drought-hardy loblolly pine family possessed more closely packed mesophyll than loblolly pine families from mesic seed sources. This would provide more surface area on which O_3 molecules can dissolve. The difference may also be related to the ability of the plant to compensate for O_3 damage. Loblolly pine exhibited significantly greater height growth ($P < 0.001$) throughout the season, while shortleaf pine exhibited significantly greater A ($P < 0.05$) (Elsik *et al.*, 1992). The additional carbon assimilated in shortleaf pine may have been allocated to compensatory processes. Differences in carbon allocation to roots and foliage may also have contributed to the difference in response and has been addressed elsewhere (Elsik *et al.*, 1992).

The difference between the two tissues in their response of $\delta^{13}C$ to the treatments may be related to the timing of seedling susceptibility to treatment and carbon allocation patterns. Ozone injury does not occur until a plant is unable to compensate for the cellular damage caused by O_3 (Tingey and Taylor, 1982). The stem tissue may contain a higher percentage of the carbon that had been assimilated previous to O_3 injury. Stem tissue may, therefore, provide a better estimate of integrated seedling response. The lack of significant stem $\delta^{13}C$ response to O_3 in shortleaf pine, despite the significant increase in foliar $\delta^{13}C$ due to O_3, is most likely related to the timing of O_3 injury. Loblolly pine, which showed a significant O_3 response in both tissues, and was more responsive to O_3 in growth and gas exchange characteristics (Elsik *et al.*, 1992), was unable to compensate for O_3 damage at a time preceding the onset of O_3 injury in shortleaf pine.

The increase in A due to water stress may cause one to question the effectiveness of the WS treatment. However, the aim of the WS treatment was merely to provide a mild water deficit compared to the WW treatment.

Although the WS treatment did not consistently result in significantly lower g, it did tend to decrease g. The increased A in WS seedlings may be explained by a fertilizer effect, in which nitrate was leached from the fritted clay medium more quickly in the WW treatment than in the WS treatment. This is supported by higher foliar N content and foliar chlorophyll concentration in WS seedlings than WW seedlings (Elsik, 1992).

V. Summary

Chronic environmental stress, such as O_3 exposure, may have subtle effects on physiological processes in trees, while greatly impacting growth over time. Instantaneous measurements of plant gas exchange response may not be sufficient to detect subtle changes in physiology. Therefore, an integrated measure is necessary to determine the long-term effect of physiological response to O_3. In this study, container-grown seedlings were exposed to both chronic O_3 and mild water stress in open-top field chambers throughout one growing season. Stable carbon isotope composition of foliage and stem tissue was determined in order to assess integrated gas exchange response, which may not have been detected by instantaneous measurements.

After 4 months of treatments, O_3 significantly increased $\delta^{13}C$ values of foliage and stem tissue in loblolly pine and foliage tissue in shortleaf pine, indicating decreased c_i, evidence that O_3 had a greater effect on g than on light harvesting or photosynthetic enzyme processes in loblolly and shortleaf pine. The results also indicate that O_3 exposure increased W in both species. Stable carbon isotope composition did not reflect instantaneous c_i measurements. Interactions between O_3 and water stress were not observed in gas exchange measurements, but the integrated nature of stable carbon isotope analysis permitted the detection of an interaction effect on $\delta^{13}C$. The interaction in loblolly pine suggested that water deficit provided protection from O_3 through partial stomatal closure, in agreement with instantaneous measurements of g. The interaction in shortleaf pine suggested that O_3 increased stomatal response to water deficit, but this is not supported by instantaneous measurements of g. Stable carbon isotope analysis proved to be an important tool in resolving long-term O_3 effects, while instantaneous gas exchange measurements were not sufficient to detect effects of chronic O_3 exposure on c_i and W. The application of this technique to air pollution studies can be enhanced through the investigation of O_3 effects on processes other than RuBP carboxylation that may affect $\delta^{13}C$.

Acknowledgments

This research was supported by a grant from the Texas Agricultural Experiment Station. We thank John Lock, Brad Toups, Jeff Anderson, Jimmie Exley, Andrew Midwood, Daniel Watts, and Xing Wang for their expert technical assistance.

References

Barnes, R. L. 1972. Effects of chronic exposure to ozone on photosynthesis and respiration of pines. *Environ. Pollut.* **3:** 133–138.

Becker, K., M. Saurer, A. Egger, and J. Fuhrer. 1989. Sensitivity of white clover to ambient ozone in Switzerland. *New Phytol.* **112:** 235–243.

Boutton, T. W. 1991. Stable carbon isotope ratios of natural materials. I. Sample preparation and mass spectrometric analysis, pp. 155–171. *In* D. C. Coleman and B. Fry (eds.), Carbon Isotope Techniques. Academic Press, New York.

Boutton, T. W., and R. B. Flagler. 1990. Growth and water-use efficiency of shortleaf pine as affected by ozone and acid rain. Proceedings of the 83rd Annual Meeting and Exhibition of the Air & Waste Management Association, 90-187.7. Air & Waste Management Association, Pittsburgh, PA.

Chappelka, A. H., and B. I. Chevone. 1992. Tree responses to ozone, pp. 271–324. *In* A. S. Lefohn (ed.), Surface Level Ozone Exposures and Their Effects on Vegetation. Lewis, Chelsea, MI.

Chappelka, A. H., B. I. Chevone, and J. R. Seiler. 1988. Growth and physiological responses of yellow-popular seedlings exposed to ozone and simulated acid rain. *Environ. Pollut.* **49:** 1–18.

Downton, W. J. S., B. R. Loveys, and W. J. R. Grant. 1988. Non-uniform stomatal closure induced by water stress causes putative non-stomatal inhibition of photosynthesis. *New Phytol.* **110:** 503–509.

Elsik, C. G. 1992. Growth, Physiology, and $\delta^{13}C$ of Loblolly and Shortleaf Pine as Affected by Ozone and soil Water Deficit. Thesis. Department of Forest Science, Texas A&M University, College Station, TX.

Elsik, C. G., R. B. Flagler, and T. W. Boutton. 1992. Effects of ozone and water deficit on growth and physiology of *Pinus taeda* and *Pinus echinata*, pp. 225–245. *In* R. B. Flagler (ed.), Response of Southern Commerical Forests to Air Pollution. Transactions of the Air & Waste Management Association, TR-21. Air & Waste Management Association, Pittsburgh, PA.

Evans, J. R., T. D. Sharkey, J. A. Berry, and G. D. Farquhar. 1986. Carbon isotope discrimination measured concurrently with gas exchange to investigate CO_2 diffusion in leaves of higher plants. *Aust. J. Plant Physiol.* **13:** 281–292.

Farquhar, G. D., J. R. Ehleringer, and K. T. Hubick. 1989a. Carbon isotope discrimination and photosynthesis. *Annu. Rev. Plant Physiol. Plant Mol. Biol.* **40:** 503–537.

Farquhar, G. D., K. T. Hubick, A. G. Condon, and R. A. Richards. 1989b. Carbon isotope fractionation and plant water-use efficiency, pp. 21–40. *In* P. W. Rundel, J. R. Ehleringer, and K. A. Nagy (eds.), Stable Isotopes in Ecological Research. Springer-Verlag, New York.

Freyer, H. D. 1979. On the record in tree rings. Part II. Registration of microenvironmental CO_2 and anomalous pollution effect. *Tellus* **31:** 308–312.

Ginn, S. E., J. R. Seiler, B. H. Cazell, and R. E. Kreh. 1991. Physiological and growth responses of eight-year-old loblolly pine stands to thinning. *For. Sci.* **37:** 1030–1040.

Greitner, C. S., and W. E. Winner. 1988. Increases in $\delta^{13}C$ values of radish and soybean plants caused by ozone. *New Phytol.* **108:** 489–494.

Guderian, R., D. T. Tingey, and R. Rabe. 1985. Effects of photochemical oxidants on plants, pp. 129–333. *In* R. Guderian (ed.), Air Pollution by Photochemical Oxidants—Formation, Transport, Control, and Effects on Plants. Springer-Verlag, New York.

Harkov, R., and E. Brennan. 1980. The influence of soil fertility and water stress on the ozone response of hybrid poplar trees. *Phytopathology* **70:** 991–994.

Heagle, A. S., D. E. Body, and W. W. Heck. 1973. An open-top field chamber to assess the impact of air pollution on plants. *J. Environ. Qual.* **2:** 365–368.

Hubick, K. T., G. D. Farquhar, and R. Shorter. 1986. Correlation between water-use efficiency and carbon isotope discrimination in diverse peanut (*Arachis*) germplasm. *Aust. J. Plant Physiol.* **13:** 803–816.

Keller, T., and R. Hasler. 1984. The influence of a fall fumigation with ozone on the stomatal behavior of spruce and fir. *Oecologia* **64:** 284–286.

Knauf, T. A., and M. V. Bilan. 1977. Cotyledon and primary needle variation in loblolly pine from mesic and xeric seed sources. *For. Sci.* **23:** 33–36.

Laisk, A. 1983. Calculation of leaf photosynthetic parameters considering the statistical distribution of stomatal apertures. *J. Exp. Bot.* **34:** 1627–1635.

Leuthy-Krause, B., I. Pfenninger, and W. Landolt. 1990. Effects of ozone on organic acids in needles of Norway spruce and Scots pine. *Trees* **4:** 198–204.

Lock, J. E., and R. B. Flagler. 1992. Comparison of gas exchange rates for attached and detached needle fascicles of shortleaf pine seedlings, submitted for publication.

Martin, B., and E. K. Sutherland. 1990. Air pollution in the past recorded in width and stable carbon isotope composition of annual growth rings of Douglas-fir. *Plant Cell Environ.* **13:** 839–844.

Martin, B., and Y. R. Thorstenson. 1988. Stable carbon isotope composition ($\delta^{13}C$), water use efficiency, and biomass productivity of *Lycopersicon esculentum, Lycopersicon pennellii* and the F_1 hybrid. *Plant Physiol.* **88:** 213–217.

Martin, B., A. Bytnerowicz, and Y. R. Thorstenson. 1988. Effects of air pollutants on the composition of stable carbon isotopes, $\delta^{13}C$, of leaves and wood, and on leaf injury. *Plant Physiol.* **88:** 218–223.

McLaughlin, S. B., R. K. McConathy, D. Duvick, and L. K. Mann. 1982. Effects of chronic air pollution stress on photosynthesis, carbon allocation, and growth of white pine trees. *For. Sci.* **28:** 60–70.

Moldau, H., J. Sober, and A. Sober. 1990. Differential sensitivity of stomata and mesophyll to sudden exposure of bean shoots to ozone. *Photosynthetica* **24:** 446–458.

Nobel, P. 1991. Physicochemical and Environmental Plant Physiology. Academic Press, New York.

O'Leary, M. H. 1988. Carbon isotopes in photosynthesis. *Bioscience* **38:** 328–336.

Olszyk, D. M., and T. W. Tibbitts. 1981. Stomatal response and leaf injury of *Pisium sativum* L. with SO_2 and O_3 exposures. II. Influence of moisture stress and time of exposure. *Plant Physiol.* **67:** 545–549.

Reich, P. B. 1987. Quantifying plant response to ozone: A unifying theory. *Tree Physiol.* **3:** 63–91.

Reich, P. B., and R. G. Amundson. 1985. Ambient levels of ozone reduce net photosynthesis in tree and crop species. *Science* **230:** 566–570.

Reich, P. B., and J. P. Lassoie. 1984. Effects of low level O_3 exposure on leaf diffusive conductance and water-use efficiency in hybrid poplar. *Plant Cell Environ.* **7:** 661–668.

Runeckles, V. C., and B. I. Chevone. 1992. Crop response to ozone, pp. 189–269. *In* A. S. Lefohn (ed.), Surface Level Ozone Exposures and Their Effects on Vegetation. Lewis, Chelsea, MI.

Sasek, T. W., and C. J. Richardson. 1989. Effects of chronic doses of ozone on loblolly pine: Photosynthetic characteristics in the third growing season. *For Sci.* **35:** 745–755.

Saurer, M., J. Fuhrer, and U. Siegenthaler. 1991. Influence of ozone on the stable carbon isotope composition, $\delta^{13}C$, of leaves and grain of spring wheat (*Triticum aestivum* L.). *Plant Physiol.* **97:** 313–316.

Seiler, J. R., and J. D. Johnson. 1988. Physiological and morphological responses of three half-sib families of loblolly pine to water-stress conditioning. *For. Sci.* **34:** 487–495.

Sharkey, T. D., F. Loreto, C. F. Delwiche, and I. W. Treichel. 1991. Fractionation of carbon isotopes during biogenesis of atmospheric isoprene. *Plant Physiol.* **97:** 463–466.

Sheffield, M., and N. D. Cost. 1987. Behind the decline. *J. For.* **85:** 29–33.

Taylor, G. E., D. T. Tingey, and H. C. Ratsch. 1982. Ozone flux in *Glycine max* (L.) Merr.:Sites of regulation and relationship to leaf injury. *Oecologia* **53:** 179–186.

Taylor, G., Jr. 1991. Interaction of ozone with naturally occurring stresses, pp. 103–113. *In* Influence of Ozone, Acidic Precipitation and Soil Magnesium Status on the Physiology and Growth of *Pinus taeda* L. (Loblolly Pine) under Field Conditions. Biological Sciences Center, Desert Research Institute, University of Nevada System, Reno, NV.

Terashima, I., S. C. Wong, C. B. Osmond, and G. D. Farquhar. 1988. Characterization of non-uniform photosynthesis induced by abscisic acid in leaves having different mesophyll anatomies. *Plant Cell Physiol.* **29**: 385–394.

Teskey, R. O., and T. M. Hinckley. 1986. Moisture: Effects of water stress on trees, pp. 9–33. *In* T. C. Hennessey, P. M. Dougherty, S. V. Kossuth, and J. D. Johnson (eds.), Stress Physiology and Forest Productivity. Nijhoff, The Netherlands.

Tingey, D. T., and G. E. Taylor. 1982. Variation in plant response to ozone: a conceptual model of physiological events, pp. 113–138. *In* M. H. Unsworth and D. P. Ormrod, (eds.), Effects of Gaseous Air Pollution in Agriculture and Horticulture. Butterworth Scientific, London.

van Buijtenen, J. P. 1966. Testing Loblolly Pines for Drought Resistance. Texas Forest Service Research Report 13. Texas Forest Service, College Station, TX.

Vitousek, P. M., C. B. Field, and P. A. Matson. 1990. Variation in foliar $\delta^{13}C$ in Hawaiian *Metrosideros polymorpha:* A case of internal resistance? *Oecologia* **84**: 362–370.

Yang, Y. S., J. M. Skelley, B. I. Chevone, and J. B. Birch. 1983. Effects of long-term ozone exposure on photosynthesis and dark respiration of eastern white pine. *Environ. Sci. Technol.* **17**: 371–373.

III

Agricultural Aspects of Carbon Isotope Variation

Carbon isotopes have been extensively applied to understanding water relations, especially water-use efficiency, of different crop groups. In this section, the contributions describe application of stable isotopes to understanding carbon–water relations in six different agronomic situations. Two of the contributions relate to annual crop species: peanut (Wright *et al.*) and common bean (Fu *et al.*) Two of the contributions describe patterns among perennial forage crops: crested wheatgrass (Johnson *et al.*) and crested wheatgrass, ryegrass, orchardgrass, and tall fescue (Johnson and Tieszen). Two of the contributions establish linkages between gas exchange and carbon isotopes for tree species: eucalyptus (Hubick and Gibson) and coffee (Meinzer *et al.*). In all groups, there appears to be a significant range of carbon isotope values, implying substantial variation in physiological characteristics due to the effects of both environmental and genetic differences. The extent to which carbon isotope composition can be used as a temporal-integrating measure of physiological behavior is a common theme of each of these contributions.

17

Genetic and Environmental Variation in Transpiration Efficiency and Its Correlation with Carbon Isotope Discrimination and Specific Leaf Area in Peanut

Graeme C. Wright, Kerry T. Hubick, Graham D. Farquhar, and R. C. Nageswara Rao

I. Introduction

Identification of physiological traits contributing to superior yield performance of crop plants under drought conditions has been a long-term goal of plant scientists. Transpiration efficiency (W), defined as the ratio of dry matter production to transpiration, is one such trait which can contribute to productivity when water resources are scarce. Reviews of literature on this topic over the past decade have concluded that the intraspecific variations in W are small and are likely to only be increased by either crop management (Fischer, 1979) or modifying the environment (Tanner and Sinclair, 1983). The early work of Briggs and Shantz (1914) showed significant variation in W among and within species. Their findings have subsequently been confirmed over the past few years, with numerous reports of substantial variation in W both between and within species (e.g., Farquhar and Richards, 1984; Frank et al., 1985; Hubick and Farquhar, 1989; Dingkuhn et al., 1991).

The difficulty in making accurate measurements of crop transpiration in the field has no doubt made the demonstration of species and cultivar differences in W difficult. While transpiration may be readily measured in pots in which soil evaporation can be minimized, it is much more difficult to estimate soil evaporation in the field. In the case of dry matter production, the lack of quality data on root dry matter nearly always results in W being calculated on an aboveground dry matter basis. Clearly, if differences in partitioning of dry matter to roots and shoots occur due to genetic or environmental effects, considerable error in W calculations can also arise.

Recently, a new approach has been proposed for identifying variation in W in C_3 plants, which may overcome the considerable problems involved in field measurement of total dry matter production and transpiration. It exploits theory based upon discrimination against ^{13}C by leaves (Δ) during photosynthesis (Farquhar *et al.*, 1982; Farquhar and Richards, 1984), and Δ in leaf tissue has been shown to be negatively correlated with W in numerous species (Farquhar and Richards, 1984; Hubick *et al.*, 1986; Hubick and Farquhar, 1989; see several chapters in this book). The experimental confirmation of the relationship between W and Δ supported the possibility of using Δ as a criterion to exploit variation in W in breeding programs.

Peanut is a crop of global economic significance not only for its widespread commercial production, but also as an important source of oil and protein in developing countries. In this paper we present results from a series of experiments investigating genotypic and environmental variation in W in peanut and its correlation with Δ under both controlled environment and field conditions. The relationship between pod yield and W is also investigated, as well as an assessment of whether W and Δ are heritable characters. Such information is needed before Δ could be recommended as a selection trait for improving W in peanut-breeding programs.

II. Peanut Cultivar Variation in Transpiration Efficiency and Correlation with Δ at the Whole-Plant Level

Using medium-sized pots (13 kg capacity) in a glasshouse study, Hubick *et al.* (1986) showed there was significant variation in W among seven *Arachis hypogaea* cultivars and two wild *Arachis* species, ranging from 1.41 to 2.29 g/kg. A close negative correlation ($r^2 = 0.66$) between W and Δ was also observed, as expected on the basis of theory and data presented by Farquhar and Richards (1984) (Fig. 1). Differences in photosynthetic capacity were largely responsible for W variation, as dry matter production was negatively correlated with Δ, while water use showed no such relationship with Δ. The lack of a relationship between water use and Δ may be associated with the use of small pots in this study, where plants were forced to use most of the available water and therefore ended up having the same total water use. Differing responses may occur in the field where access to soil water can be relatively unrestricted.

A recent glasshouse experiment (Wright, unpublished observations) using small pots (2 kg capacity) has confirmed that significant W variation exists among peanut cultivars. Variation in W among 10 cultivars ranged from 2.96 to 4.58 g/kg under well-watered conditions, and from 3.41 to

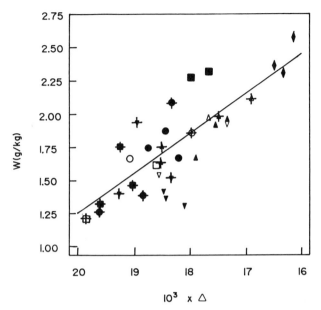

Figure 1. Transpiration efficiency (shoots plus roots) versus carbon isotope discrimination in leaves ($r = -0.81$). Open symbols represent well-watered plants and closed symbols represent plants that were droughted. Symbols represent genotypes as follows: *Arachis villosa*, ▼; *A. glabrata,* ♈; *A. hypogaea* cultivars—Chico,♣; PI314817,♦; PI259747, ●; UF78114-3, ■; VB187,♦; Florunner, ▲; Tifton-8, ♦ (redrawn from Hubick *et al.*, 1986).

4.74 g/kg under a terminal drought stress during vegetative growth (which killed all plants). A highly significant correlation between W and Δ ($r^2 = 0.74$) was observed under well-watered conditions (Fig. 2a); however, under droughted conditions no relationship was detected (Fig. 2b). Interestingly, the ranking of cultivars for W was not significantly affected by the drought treatment, as is illustrated in Fig. 3a where a highly significant correlation ($r^2 = 0.87$) between W under well-watered and droughted conditions was observed. In contrast, Δ under both watering regimes was poorly correlated (Fig. 3b), which could imply that under the severely water-stressed treatment, Δ was somehow affected such that the relationship between W and Δ broke down. Conversely, it could be interpreted that for a given Δ, W in the stressed treatment was less than in the well-watered treatment because of greater respiratory losses as a proportion of carbon gained, ϕ_c. Theory suggests that, at reduced growth rates, ϕ_c is increased (Masle *et al.*, 1990).

The experimental confirmation that variation in W exists among peanut cultivars, and that a strong relationship between W and Δ often exists under glasshouse conditions, suggests that Δ could be used as a criterion to exploit variation in W in breeding programs. There are however a number

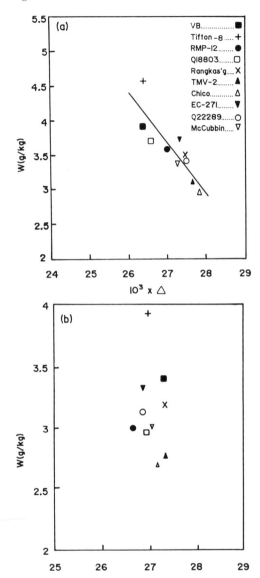

Figure 2. Transpiration efficiency (shoots plus roots) versus carbon isotope discrimination in leaves of 10 peanut cultivars grown in small pots in a glasshouse for (a) well-watered plants and (b) plants droughted from emergence to 50 days after emergence (unpublished data of Wright).

of potential sources of discrepancy between results from glasshouse plants in pots and plants grown under field conditions. For example,

1. There are difficulties in correctly apportioning water use into that lost by transpiration and that lost by evaporation. In field studies there are problems in measuring and estimating soil evaporation, in contrast

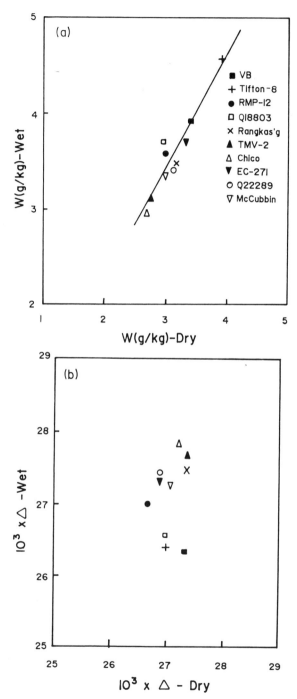

Figure 3.   Transpiration efficiency under well-watered and droughted conditions (a) and carbon isotope discrimination under well-watered and droughted conditions (b) for 10 cultivars grown in small pots in a glasshouse (unpublished data of Wright).

to pots where it can be minimized (Turner, 1986). Complications can also arise from differences between cultivars in the extent and timing of soil evaporation (Condon *et al.*, 1991).

2. There is generally a lack of data on root dry matter in field studies and W usually is based on aboveground dry matter. Differences among cultivars in apportioning of dry matter to roots and shoots may lead to erroneous comparisons of W defined on this basis. This error may be particularly large in severe drought conditions where total dry matter accumulation may be dominated by roots.

3. The aerial environment of field canopies is characterized by complex interactions involving transfer of heat and water vapor, and the interactions are different from those around isolated potted plants. Reduced W of isolated plants that occurs because of reduced stomatal conductance may not necessarily be reflected at the canopy level, if the crop boundary layer conductance is relatively small (Cowan, 1971, 1977, 1988; Jarvis and McNaughton, 1985; Farquhar *et al.*, 1989).

Definitive experiments aimed at assessing variation in W among peanut cultivars, and the correlation between W and Δ, therefore, need to be conducted in canopies under field conditions. This information is essential in order to ascertain that we are sure W variation actually exists under field conditions and that Δ can be confidently used as a selection criteria for W. Also, this assessment needs to be conducted under both well-watered and water-limited conditions; as evidence shown in this section indicates, the correlation between W and Δ may break down under severe plant water deficits.

III. Peanut Cultivar Variation in Transpiration Efficiency and Correlation with Δ in Field Canopies

A. Measurement of W under Field Conditions

Wright *et al.* (1988) described a minilysimeter facility in combination with a portable rain-out shelter (Hatfield *et al.*, 1989) which allowed an effective and low-cost assessment of plant performance in a canopy situation. It enabled isolation of the root zone, for accurate water application and measurement, and recovery of total root and shoot dry matter.

Briefly, intact soil cores were excavated using a soil-coring machine. As the core was being dug, a PVC storm water pipe (dimensions 0.30 m i.d. 0.31 m o.d. and 0.8 m deep) was slid over the core. After digging, the intact soil core and pipe were lifted aboveground using a gantry device, and a galvanized iron circular cap (1 mm thick) was screwed into the base of the pipe. A 1-mm-thick galvanized iron sheet (0.8 m deep) was then fitted around the circumference of the hole where the core had been removed, to prevent loose soil from falling back into the excavation, and the core was replaced. Plants were then established in and around the minilysimeters to simulate a field canopy. Water loss from pots was estimated by weighing with an electronic load cell (accuracy of ± 0.1 kg) mounted on a tractor-driven gantry which straddled the plot area. This facility allowed rapid and repeated measurement of minilysimeter weights and hence transpiration. Soil evaporation was estimated from the changes in water content (weight) of bare soil minilysimeters, adjusted for fractional radiation interception by the canopy (Cooper *et al.*, 1983).

B. Variation in *W* under Well-Watered and Water-Limited Conditions

Two large field experiments using the minilysimeter facility described above were conducted to determine whether cultivar differences in *W* were occurring in small field canopies. One experiment was conducted under full irrigation (Wright *et al.*, 1988), while the other imposed two levels of soil water deficit (Wright *et al.*, 1993). In both experiments *W* was measured only during the period between full canopy development (ca. 45 days after planting, DAP) and early podfilling (ca. 90 DAP). This was done to minimize the effects of soil evaporation and avoid any confounding effects arising from maturity differences among cultivars.

The results from experiments clearly indicates that significant differences in *W* existed among peanut cultivars in the field, under both water nonlimiting and limiting conditions (Table I). In general, variation in *W* among cultivars was associated with differences in dry matter accumulation rather than to differences in transpiration. This result indicates that photosynthetic capacity, rather than leaf/canopy stomatal conductance, was dominating the *W* differences among peanut cultivars.

Table I Dry Matter (Including Roots), Transpiration, *W* and Δ in Peanut Cultivars Grown in Minilysimeters in Field Canopies under Well-Watered Conditions (Wright *et al.*, 1988) and Two Levels of Water-Limited Conditions (Wright *et al.*, 1993)

Study	Cultivar	Biomass (kg)	Water use (kg)	*W* (g/kg)	Δ (× 10³)
Well watered	Tifton-8	63.1	17.0	3.71	19.7
	VB-81	46.9	16.2	2.90	20.1
	Robut 33-1	55.3	19.0	2.91	20.8
	Shulamit	51.6	16.8	3.07	20.8
	McCubbin	48.6	16.9	2.88	20.8
	Cianjur	43.4	16.3	2.66	20.9
	Rangkasbitung	41.6	16.9	2.46	20.9
	Pidie	47.3	16.6	2.85	20.6
lsd *P* = 0.05		7.0	1.5	0.3	0.55
Water limited	Tifton-8	37.5	12.2	3.07	19.4
(intermittent stress)	Shulamit	35.7	12.8	2.79	19.9
	McCubbin	36.3	13.4	2.71	20.7
	Chico	20.5	11.4	1.80	21.1
(terminal stress)	Tifton-8	31.3	10.0	3.13	18.6
	Shulamit	29.0	9.9	2.93	18.8
	McCubbin	26.8	10.0	2.68	19.4
	Chico	17.8	8.8	2.20	20.9
lsd *P* = 0.05		5.61	1.90	0.38	0.93

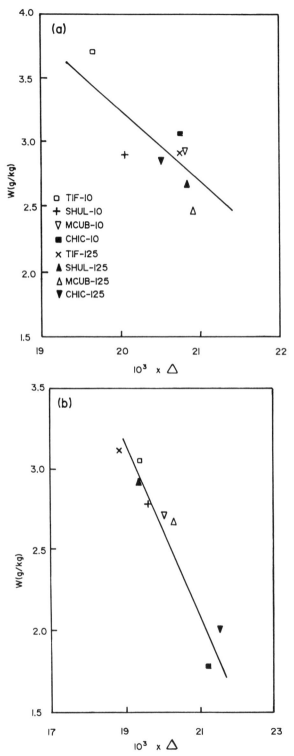

Figure 4. Relationship between transpiration efficiency and carbon isotope discrimination under well-watered (redrawn from Wright *et al.*, 1988) and droughted (redrawn from Wright *et al.*, 1993) conditions for a range of peanut cultivars growing in a field.

C. Correlation between *W* and Δ in the Field

Highly significant negative correlations were observed between Δ and *W* under both well-watered ($r^2 = 0.67$) and water-limited conditions ($r^2 = 0.92$) (Figs. 4a and 4b). These relationships for field-grown peanuts support the suitability of Δ as a selection criterion for screening for high *W*.

Changes in p_i/p_a, the ratio of internal CO_2 concentration in the leaf to ambient CO_2 concentration and Δ, can arise from changes in the balance between leaf stomatal conductance and photosynthetic capacity. Where p_i/p_a changes are due to stomatal movements, the relationship between *W* and Δ observed for well-ventilated, isolated leaves may break down in plants grown in canopies in the field because of significant canopy boundary-layer resistances to fluxes of water vapor and heat (Cowan, 1977, 1988; Farquhar *et al.*, 1989). Where p_i/p_a changes in response to variation in photosynthetic capacity, the problem associated with weak coupling between the crop canopy and atmosphere is not as important, as increased p_i/p_a and Δ arise because of decreased assimilation rate, which causes a relatively small change in the CO_2 concentration in the air above the canopy and no effect on heat and vapor transfer through the boundary layer. The observation that total dry matter production (TDM) was negatively correlated with Δ for the peanut cultivars examined in the field studies of Wright *et al.* (1988, 1993) (Fig. 5) suggests that variation in photosynthetic capacity was the predominant source of variation in p_i/p_a (and therefore Δ). Indeed, a very

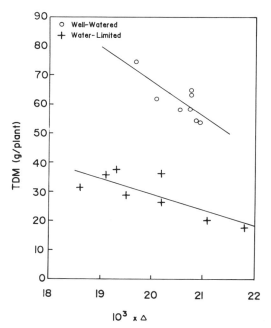

Figure 5. Total dry matter (shoots plus roots) versus carbon isotope discrimination in leaves for well-watered (Wright *et al.*, 1988) and droughted (Wright *et al.*, 1993) peanut cultivars growing in a field.

Table II Regression Equations Relating Total Dry Matter Production (TDM Includes Roots) to Δ from a Range of Glasshouse Experiments

Expt	Fitted regression	r^2	Comments	Source
1	TDM = 74.3−3.24(Δ × 10³)	0.44	Glasshouse, well-watered	Hubick et al. (1986)
	TDM = 43.6−1.95(Δ × 10³)	0.75	Glasshouse, droughted	Hubick et al. (1986)
2	TDM = 687.6−26.46(Δ × 10³)	0.84	Glasshouse, well watered, min N	Hubick (1990)
	TDM = 679.9−28.70(Δ × 10³)	0.93	Glasshouse, well watered, nodule N	Hubick (1990)
	TDM = 149.5−5.61(Δ × 10³)	0.95	Glasshouse, droughted, min N	Hubick (1990)
	TDM = 223.1−9.17(Δ × 10³)	0.97	Glasshouse, droughted, nodule N	Hubick (1990)
3	TDM = 364.0−16.46(Δ × 10³)	0.84	Glasshouse, well watered, 3 cv.	Hubick et al. (1990)
	TDM = 303.0−11.85(Δ × 10³)	0.64	Glasshouse, well watered, F2 plants	Hubick et al. (1990)

strong negative relationship between TDM and Δ has been observed in peanut cultivars over a wide range of experimental conditions (provided well-watered and droughted treatments are not mixed). Table II summarizes the statistical parameters of this relationship measured in a number of glasshouse studies and shows that extremely strong correlations (range in $r^2 = 0.44$ to 0.97) exist over a wide range of environmental conditions.

IV. Genotype X Environment Interaction and Heritability for *W* and Δ

Genotype × environment interaction for *W* appears to be small in peanut. Wright *et al.* (1988) found that although there were large differences in *W* and Δ in "above-ground" as compared to "in-ground" minilysimeters, cultivar ranking in these parameters was largely maintained across the two contrasting environments. Correlation coefficients (*r*) for *W* and Δ in in-ground versus aboveground minilysimeters were 0.91 and 0.83, respectively. *W* was strongly correlated ($r^2 = 0.74$) with Δ under well-watered and droughted environments in the pot study reported earlier (see Fig. 2a) which again indicates there is low genotype × environment interaction for *W*. Hubick *et al.* (1986, 1988) also reported that the ranking of *W* and Δ was consistent in a range of cultivars under two contrasting water regimes in glasshouse studies. Hubick (1990) showed that, although *W* and Δ varied significantly in response to watering treatment and source of nitrogen (mineral N versus nodule N), the ranking of *W* and Δ was similar under each treatment, again indicating there is low genotype × environment interaction for these parameters.

In 16 peanut cultivars grown at 10 sites with widely different rainfall patterns in subtropical and tropical areas of Queensland, Australia, there was significant genotypic variation in Δ, with no significant interaction between genotype and environment (Hubick *et al.*, 1988). A broad sense heritability (ratio of genotypic variance to the total or phenotypic variance) or repeatability of Δ in this experiment was 81%.

Inheritance of Δ was studied in plants grown in pots using crosses of cultivars with contrasting Δ and *W* (Hubick *et al.*, 1988). The F_1 progeny had Δ values similar to those of the low Δ cultivar, Tifton-8, and considerably smaller than those of Chico, the high Δ cultivar. This response suggests a degree of dominance for small Δ or large *W* in these genotypes. In the F_2 generation, the distribution of Δ exceeded the range between Tifton-8 and Chico, with two F_2 plants having smaller Δ values than those of the low Δ parent, Tifton-8 (Fig. 6). The F_2 distribution for Δ strongly suggested quantitative rather than qualitative inheritance for this trait.

The results from the study of Hubick *et al.* (1988), in combination with the evidence we present here indicating that *W* and Δ have low genotype × environment interaction, suggests that effective selection for Δ, and hence *W*, could be conducted in a restricted number of environments. Indeed,

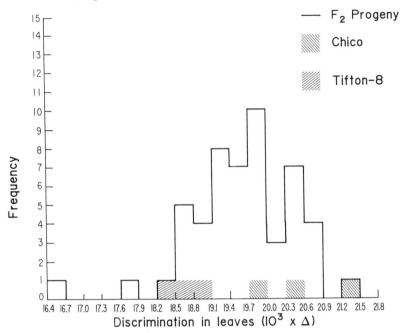

Figure 6. Frequency distribution of carbon isotope discrimination in leaves of well-watered plants of peanut cultivars Tifton-8 and Chico and their F_1 and F_2 progeny population grown together in one glasshouse environment (redrawn from Hubick *et al.*, 1988).

the results indicate selection could possibly take place in a single environment, be it well watered or water limited, and in a glasshouse or field situation.

V. Relationships between Specific Leaf Area, *W*, and Δ

It has been observed over many experiments that specific leaf area (SLA, cm^2/g, which is negatively related to leaf thickness) is closely and negatively correlated with *W* and also that SLA and Δ are positively correlated. Examples of the relationships between SLA and *W*, and SLA and Δ, measured in the minilysimeter study by Wright *et al.* (1993) are illustrated in Figs. 7a and 7b, respectively. These observations are consistent with our earlier hypothesis that cultivars with high *W* have higher photosynthetic capacity. If it is assumed that the N : C ratio does not vary among cultivars then it is possible that those cultivars with thicker leaves had more photosynthetic machinery and the potential for greater assimilation per unit of leaf area. Indeed, Nageswara Rao and Wright (unpublished observations) have shown that specific leaf nitrogen (gN/m^2) is linearly related with SLA, such that thicker leaves had higher nitrogen contents (data not shown). Similar relationships between *W* and SLA, and Δ and SLA, have been reported

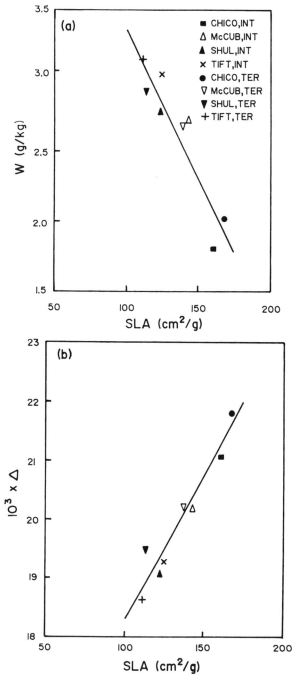

Figure 7. Transpiration efficiency versus specific leaf area (a) and carbon isotope discrimination in leaves versus specific leaf area (b) for four peanut cultivars grown under two levels of drought in the field (data derived from Wright *et al.,* 1993).

elsewhere (Wright *et al.*, 1988). A highly significant relationship between Δ and SLA was also observed for some 300 F_3 plants derived from a single cross of high and low Δ Indonesian cultivars grown in the field (Wright *et al.*, 1992). Thus, there is considerable evidence to support the hypothesis that a very strong association between Δ and SLA exists. This finding has significant implications for breeding programs, where selection for W may be practiced, as SLA is simple and inexpensive to measure, compared to the Δ measurement, which requires an isotope ratio mass spectrometer.

An experiment has recently been conducted to determine the generality of the SLA relationship with Δ by growing four cultivars with contrasting Δ in two contrasting temperature environments, under irrigated and rain-fed conditions (Nageswara Rao and Wright, unpublished data). The two sites, Kingaroy and Bundaberg, were similar except for their minimum night temperatures, in that mean minimum temperatures during the season were 16°C at Kingaroy compared to 20°C at Bundaberg. Table III shows how environment, cultivar, and watering regime all influenced the magnitude of SLA and Δ. For instance, SLA and Δ for each cultivar were significantly higher in the warmer Bundaberg environment, while water deficits associated with the rain-fed treatment tended to reduce SLA and Δ for each cultivar but not in Kingaroy. This effect was particularly apparent at Bundaberg where lower rainfall resulted in greater crop water deficits. The data clearly show that leaves of all cultivars became "thicker" in response to low temperature and water deficits, possibly due to affects on leaf expansion and translocation of assimilate from the leaf (Bagnall *et al.*, 1988). Of more interest, however, was the observation that cultivar ranking for SLA and Δ remained the same in each environment and watering regime. Analysis of variance indeed showed the main effects of location, irrigation treatments, and cultivar were highly significant ($P < 0.05$) for SLA and Δ, while the genotype $\times$ environment interactions were nonsignificant. These results are consistent with the low genotype $\times$ environment interactions for W and Δ reported earlier.

The strong correlation between Δ and SLA reported previously (Fig. 7b) was again apparent for this data set (Fig. 8) even given the interactions noted above.

Table III Specific Leaf Area (cm²/g) and Δ (‰) Measured at Maturity for Four Peanut Cultivars Grown at Two Sites (Bundaberg and Kingaroy) under Two Watering Regimes (Irrigated and Rain Fed)

Site	Treatment	Chico SLA	Chico Δ	McCubbin SLA	McCubbin Δ	Shulamit SLA	Shulamit Δ	Tifton SLA	Tifton Δ
Kingaroy	Irrigated	155.2	22.44	145.9	21.17	124.0	21.24	117.7	20.40
	Rain fed	138.9	22.15	164.8	21.45	124.9	21.35	132.7	20.77
Bundaberg	Irrigated	184.9	23.21	174.2	22.68	166.8	22.97	148.3	21.50
	Rain fed	186.3	22.29	166.2	21.50	128.5	21.39	136.8	20.84

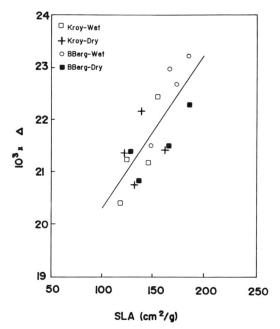

Figure 8.   Carbon isotope discrimination in leaves versus specific leaf area for four peanut cultivars grown in the field at Bundaberg and Kingaroy, Qld, Australia (unpublished data of Nageswara Rao and Wright).

Interestingly, the data from the contrasting temperature and water-stress environment form a universal relationship. Even the data presented in Fig. 7b, and other data we have measured elsewhere (e.g., Wright *et al.*, 1992), fit well onto this relationship. The physiological mechanisms involved are unknown and need further investigation. The significant application of the relationship is however obvious, in that breeders could use the inexpensively measured SLA, in lieu of Δ, to screen for high *W* among peanut germplasm within particular environments.

VI. Negative Association between *W* and Harvest Index

A number of glasshouse and field experiments have shown that while TDM at maturity was negatively correlated with *W* and Δ, pod yield was not, suggesting that selection for low Δ would substantially increase TDM while having only minimal influence on pod yield improvement (see Hubick *et al.*, 1988; Wright *et al.*, 1988, 1993). An example of the negative relationship between *W* and partitioning ratio (calculated as the proportion of total dry matter allocated to pods during the podfilling period) for seven cultivars contrasting in *W* characteristics is shown in Fig. 9. The relationship still exists when dry matter is converted to glucose equivalent (to account for

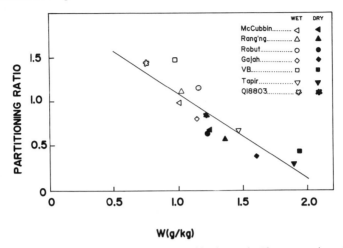

Figure 9. Transpiration efficiency versus partitioning ratio (the proportion of total biomass partitioned to pods between early podfill and maturity) for seven peanut cultivars grown under well-watered (open symbols) and droughted (closed symbols) conditions in pots in a glasshouse (unpublished data of Wright).

the higher energy costs of synthesizing oil in kernels, data not shown), indicating that the relationship is not an artifact of dry matter energy costs. It is also of interest that Hubick (1990) recently found that much of the increased dry matter produced in plants given mineral N compared to nodulated plants was allocated to leaves, not pods. Thus variation in harvest index (HI) resulted in no correlation between W, or Δ, and pod yield.

The negative association observed between Δ and partitioning in peanut may be limited to the particular set of cultivars used in our studies to date. Some preliminary genetic studies have been conducted to determine whether these traits are genetically linked, and whether the linkage could be broken through breeding. In this study (Cruickshank and Wright, unpublished data), a cross between a low W, high HI Indonesian line (Rangkasbitung) and a high W, low HI Indonesian line (Tapir) was made. Both these lines were spanish botanical types, with similar maturity of about 120 days in the Kingaroy environment.

The F_2 progeny were grown under well-watered conditions in the glasshouse, and W (via Δ) and HI were measured at maturity. Figure 10a shows that HI and Δ were negatively correlated, although considerable scatter in the relationship suggested crossing may have disturbed the association.

A number of F_3 and F_4 families derived from individual F_2 plants were selected on the basis of low Δ, high HI, or the best of both traits and grown in the field under well-watered and rain-fed conditions. Figures 10b and 10c again shown that W (via SLA measurements) and HI were negatively associated, with no obvious outliers possessing both high W and high HI. The moderate strengths of the correlation (ca. $r = -0.55$) suggest that concurrent improvement in these traits may be difficult, but should be

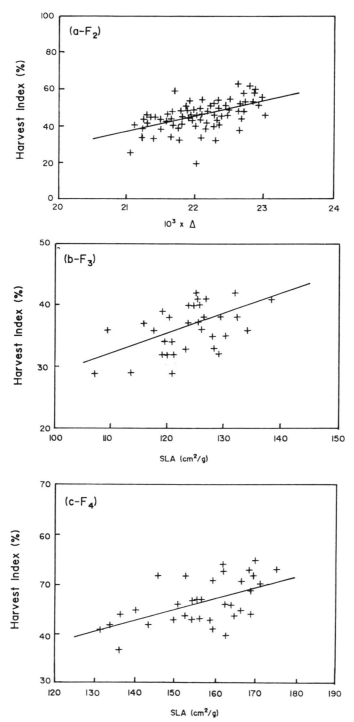

Figure 10. Relationship between (a) harvest index and carbon isotope discrimination for F_2 progeny, and harvest index and specific leaf area for F_3 progeny (b) and F_4 progeny (c), derived from a cross between Tapir and Rangkasbitung growing in a well-watered field environment (redrawn from Wright *et al.*, 1992).

possible. The extent of the negative association needs to be investigated using a wider range of peanut germplasm.

VII. Selection for *W* in Peanut-Breeding Programs

To screen large numbers of lines for *W* would be extremely difficult because of the need to accurately measure both transpiration and total biomass (including roots) under glasshouse or field condition. This is no doubt the major reason why cultivar variation in *W* in a range of species has not been widely demonstrated, or pursued as a selection trait in breeding programs. For maximum effectiveness in developing cultivars with improved *W*, selection should be conducted in the large segregating heterogeneous populations that occur at various stages of a breeding program. The results of numerous studies which have been reviewed here indicate that Δ or SLA could be used to effectively select for *W* in such large populations.

Hubick *et al.* (1986) found that Δ of all plant components were highly correlated with the Δ of leaf material. It is therefore considered that Δ in leaves should provide a reasonable guide to select peanut cultivars for improved *W*. Hall *et al.* (1993) cautioned that Δ may be different in plant material produced during stress periods, so sampling procedures would need to be developed to account for this effect. In terms of purely seeking improved *W* there may be an advantage in selecting for low Δ (or low SLA)

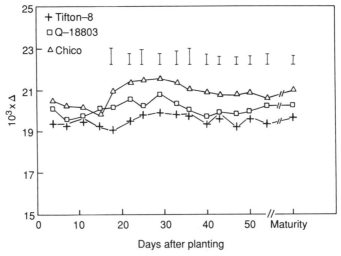

Figure 11. Change in carbon isotope discrimination in leaves and error variation with time for well-watered plants of cultivars Tifton-8, Q18803, and Chico grown in a glasshouse (redrawn from Wright *et al.*, 1992).

under well-watered conditions, so as to minimize potential drought effects on Δ.

The question of how early in a plant's life cycle Δ (or SLA) could be selected for and still represent its *W* characteristics is also pertinent in relation to selection in a breeding program. Figure 11 shows the temporal change in Δ at four daily intervals until 54 days after emergence (and also at maturity). It is clear that after about 15 days after emergence, Δ remains very constant. A similar procedure (at two weekly intervals) was also carried out for four cultivars in a field experiment under well-watered and droughted conditions (Wright *et al.*, 1991). There was no significant interaction for Δ between irrigation treatment and time of sampling for the four cultivars. Based on these observations, the stability of Δ throughout crop ontogeny indicates that selection could take place very early during crop development.

VIII. Summary

Significant variation among peanut cultivars in transpiration efficiency (*W*) under well-watered and water-limited conditions has been shown in isolated plants in the glasshouse and in small canopies in the field. There is considerable scope for *W* improvement in currently grown commercial cultivars. *W* was shown to be highly correlated with carbon isotope discrimination in leaves (Δ), and also with specific leaf area (SLA), or leaf thickness.

Genotype × environment interaction for *W*, Δ, and SLA was shown to be very low, while heritability of Δ was high, indicating that these traits could be used for selecting high *W* in peanut-breeding programs. The results indicate that selection could take place in a single environment, be it well watered or water limited, and in a glasshouse or field situation.

A worrying negative association between *W* and partitioning of dry matter to pods among peanut cultivars is apparent. Preliminary genetic studies aimed at assessing whether the association is due to a genetic linkage indicated that the negative association persisted up to the F_4 generation in a cross of two contrasting Indonesian peanut cultivars. Further research aimed at identifying cultivars with high levels of both *W* and harvest index is warranted.

Based on our observations to date, it would seem that selection for low Δ or SLA in peanut may be appropriate in certain water-limited cropping systems, for instance in developing countries, where both pod yield for human consumption and vegetative yield for animal fodder need to be maximized. In cropping systems where pod yield is of primary concern, breeders will need to be aware of the potential negative association between harvest index and Δ in any breeding program incorporating Δ as a selection trait.

Acknowledgments

Much of the information reported in this paper resulted from research supported by the Australian Centre for International Agricultural Research (Projects 8419, 8550, 8834). Their financial support is gratefully acknowledged. This paper was submitted as journal article No. 754 by the International Crops Research Institute for the Semi-Arid Tropics (ICRISAT), India.

References

Bagnall, D. J., R. W. King, and G. D. Farquhar. 1988. Temperature dependent feedback inhibition of photosynthesis in peanut. *Planta* **175:** 348–354.

Briggs, L. J., and H. L. Shantz. 1914. Relative water requirement of plants. *J. Agric. Sci.* **3:** 1–64.

Condon, A., F. Dunin, R. A. Richards. T. Denmead, R. Leuning, D. De Pury, C. Wong, and G. D. Farquhar. 1991. Does a genotypic difference in stomatal conductance result in a difference in transpiration efficiency at the paddock scale? Abstract 42. *In* 31st Annual General Meeting of Australian Society of Plant Physiologists. Canberra, Australia.

Cooper, P. J. M., J. D. H. Keatinge, and G. Hughes. 1983. Crop evapotranspiration—A technique for calculation of its components by field measurements. *Field Crops Res.* **7:** 299–312.

Cowan, I. R. 1971. The relative role of stomata in transpiration and assimilation. *Planta* **97:** 325–336.

Cowan, I. R. 1977. Water use in higher plants, pp. 71–107. *In* A. K. McIntyre (ed.), Water, Planets, Plants and People. Australian Academy of Science, Canberra, Australia.

Cowan, I. R. 1988. Stomatal physiology and gas exchange in the field, pp. 160–172. *In* W. L. Steffan, and O. T. Denmead (eds.), Flow and Transport in the Natural Environment: Advances and Application. Springer-Verlag, New York.

Dingkuhn, M., G. D. Farquhar, S. K. De Datta, and J. C. O'Toole. 1991. Discrimination of ^{13}C among upland rices having different water use efficiencies. *Aust. J. Agric. Res.* **42:** 1123–1131.

Farquhar, G. D., and R. A. Richards. 1984. Isotopic composition of plant carbon correlates with water-use efficiency of wheat cultivars. *Aust. J. Plant Physiol.* **11:** 539–552.

Farquhar, G. D., M. H. O'Leary, and J. A. Berry. 1982. On the relationship between carbon isotope discrimination and intercellular carbon dioxide concentration in leaves. *Aust. J. Plant Physiol.* **9:** 121–137.

Farquhar, G. D., K. T. Hubick, A. G. Condon, and R. A. Richards. 1989. Carbon isotope fractionation and plant water-use efficiency, pp. 21–40. *In* P. W. Rundel, J. R. Ehleringer. and K. A. Nagy (eds.), Applications of Stable Isotope Ratios to Ecological Research. Springer-Verlag, New York.

Fischer, R. A. 1979. Growth and water limitation to dryland wheat yield in Australia: A physiological framework. *J. Aust. Inst. Agric. Sci.* **45:** 83–94.

Frank, A. B., J. D. Berdahl, and R. E. Barker. 1985. Morphological development and water-use in clonal lines of four forage grasses. *Crop Sci.* **25:** 339–344.

Hall, A. E., G. D. Farquhar, K. T. Hubick, A. Condon, R. A. Richards, and G. C. Wright. 1993. Carbon isotope discrimination and plant breeding. *Plant Breed. Rev.* **11,** in press.

Hatfield, P. M., G. C. Wright, and W. R. Tapsall. 1989. A large retractable, low cost and re-locatable rainout shelter design. *Exp. Agric.* **26:** 57–62.

Hubick, K. T. 1990. Effects of nitrogen source and water limitation on growth, transpiration efficiency and carbon-isotope discrimination in peanut cultivars. *Aust. J. Plant Physiol.* **17:** 413–430.

Hubick, K. T., and G. D. Farquhar. 1989. Carbon isotope discrimination and the ratio of carbon gained to water lost in barley cultivars. *Plant Cell Environ.* **12:** 795–804.

Hubick, K. T., G. D. Farquhar, and R. Shorter. 1986. Correlation between water-use efficiency and carbon isotope discrimination in diverse peanut (*Arachis*) germplasm. *Aust. J. Plant Physiol.* **13:** 803–816.

Hubick, K. T., R. Shorter, and G. D. Farquhar. 1988. Heritability and genotype × environment interactions of carbon isotope discrimination and transpiration efficiency in peanut (*Arachis hypogaea L*). *Aust. J. Plant Physiol.* **15:** 799–813.

Jarvis, P. G., and K. G. McNaughton. 1985. Stomatal control of transpiration scaling up from leaf to region, pp. 1–49. *In* A. MacFayden and E. D. Ford (eds.), Advances in Ecological Research. Academic Press, London.

Masle, J., G. D. Farquhar, and R. M. Gifford. 1990. Growth and carbon economy of wheat seedlings as affected by soil resistance to penetration and ambient partial pressure CO_2. *Aust. J. Plant Physiol.* **17:** 465–487.

Tanner, C. B., and T. R. Sinclair. 1983. Efficient water use in crop production: Research or re-search, pp. 1–28. *In* H. M. Taylor, W. R. Jordan, and T. R. Sinclair (eds.), Limitations to Efficient Water Use in Crop Production. American Society of Agronomy, Crop Science Society of America, Soil Science Society of America, Madison, WI.

Turner, N. C. 1986. Crop water deficits: A decade of progress. *Adv. Agron.* **39:** 1–51.

Wright, G. C., K. T. Hubick, and G. D. Farquhar. 1988. Discrimination in carbon isotopes of leaves correlates with water-use efficiency of field-grown peanut cultivars. *Aust. J. Plant Physiol.* **15:** 815–825.

Wright, G. C., K. T. Hubick, and G. D. Farquhar. 1991. Physiological analysis of peanut cultivar response to timing and duration of drought stress. *Aust. J. Agric. Res.* **42:** 453–470.

Wright, G. C., R. C. Nageswara Rao, and G. D. Farquhar. 1993. Variation in water-use efficiency and carbon isotope discrimination in peanut cultivars under drought conditions in the field. *Crop. Sci.* **33,** in press.

Wright, G. C., T. A. Sarwanto, A. Rahmianna, and D. Syarifuddin. 1992. Investigation of drought tolerance traits conferring adaptation to drought stress in peanut, pp. 74–84. *In* G. C. Wright and K. J. Middleton (eds.), Peanut Improvement: A Case Study in Indonesia. ACIAR Preceedings No. 40, Canberra, Australia.

18

Genotypic and Environmental Variation for Carbon Isotope Discrimination in Crested Wheatgrass, a Perennial Forage Grass

Douglas A. Johnson, Kay H. Asay, and John J. Read

I. Introduction

Plants growing in semiarid environments regularly are exposed to water-limiting conditions and must have adaptations that allow them to escape drought, or avoid or tolerate dehydration (Fischer and Turner, 1978; Schulze, 1988). As a result, plant-breeding programs for agronomically important crop species for these areas should be able to identify breeding lines that exhibit superior adaptation to drought. Because of limited water for plant growth, one important characteristic to consider is the efficient use of water or transpiration efficiency (W, amount of dry matter produced per unit of transpiration).

Numerous soil, climatic, and plant factors influence W (Stanhill, 1986). Chemical and physical characteristics of the soil can influence W through the content, movement, and availability of soil water; impedance to root penetration; water infiltration rate; soil nutrient status; and soil surface features. Important climatic factors include amount and distribution of rainfall and the vapor pressure deficit of the air. Plant responses such as stomatal characteristics and photosynthetic capacity also influence W.

In plant-breeding programs, selection for specific leaf characteristics and plant responses generally has not led to enhanced W. This is because various leaf characteristics are time consuming to measure and do not reflect the component effects of other leaf attributes that may override or mask the measured characters. In addition, physiologically based measurements of plant responses often are instantaneous values that may not necessarily provide an indication of their integrated response through time. Direct measurements of W have not been practical within a plant-breeding program because pot-weighing techniques are laborious and are not capable of handling the large numbers of breeding lines that must be evaluated.

Farquhar *et al.* (1982) and Farquhar and Richards (1984) theorized that variations in carbon isotope discrimination (Δ) in C_3 plants depended on leaf intercellular CO_2 concentration (c_i). Farquhar and Richards (1984) reported that A/E in wheat (*Triticum aestivum* L.) genotypes was correlated with Δ and suggested that Δ could be useful in improving W in breeding programs for C_3 species. Numerous studies subsequently have documented the negative relationship between Δ and W in various C_3 species (Farquhar *et al.*, 1989). Farquhar *et al.* (1988) pointed out that this association between Δ and W is not causal, but occurs because changes in W and Δ are independently linked through changes in c_i. Consequently, measurements of Δ may be an indirect method for estimating W that could be useful in evaluating genotypes in a plant-breeding program.

Crested wheatgrass [*Agropyron desertorum* (Fischer ex Link) Schultes] is a C_3 forage grass used extensively for revegetating deteriorated, semiarid rangelands in western North America. Crested wheatgrass is a persistent, drought- and cold-resistant bunchgrass native to Eurasia that was introduced into the United States in the early 1900s (Rogler and Lorenz, 1983). This semiarid grass grows primarily during the spring and early summer, becomes dormant by midsummer, and, depending on rainfall, may initiate growth in the autumn. Crested wheatgrass is an important source of forage in the early spring and autumn for both domestic livestock and wildlife. Inasmuch as crested wheatgrass provides a stable forage resource in the western United States and because the lack of water limits forage production of this grass, cultivars of crested wheatgrass with enhanced W could increase the efficiency of forage production on rangelands of the western United States.

Bleak and Keller (1973) determined W of five 'Nordan' crested wheatgrass clones in a glasshouse and reported that it might be possible to improve W through selection. Frank *et al.* (1987) and Frank and Karn (1988) found significant variation for W among three accessions of crested wheatgrass (two cultivars and one experimental strain). Barker *et al.* (1989) also showed that clones within these same three accessions varied significantly in W and that broad-sense heritabilities for W were 73%. Although based on calculations from only four clones, their results suggested that W could be improved by selection, particularly if more effective screening techniques could be developed for evaluating W. Using Δ to estimate W may be such a technique for crested wheatgrass and other C_3 species.

Johnson *et al.* (1990) and Read *et al.* (1991, 1992) measured Δ in clones of crested wheatgrass and determined how Δ was related to W and various leaf gas exchange characteristics, both in glasshouse and field environments. This chapter reviews work with crested wheatgrass and discusses the association of Δ with W, the effect of environment on Δ, the genotypic variation present for Δ, the stability of the Δ response, and the association between Δ and forage yield.

II. Association of Δ with *W*

Johnson *et al.* (1990) reported results with crested wheatgrass plants grown in pots in a glasshouse. Pots were watered every 3 to 4 days to two water levels. Transpiration efficiency was negatively associated with Δ of shoots across water levels ($r = -0.73$**, $df = 44$) (Fig. 1a). Negative correlations were observed within the droughted ($r = -0.83$**, $df = 22$), but not the

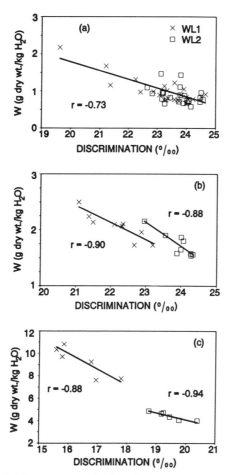

Figure 1.   Relationship between transpiration efficiency (*W*, g dry wt/kg H_2O) and carbon isotope discrimination (‰) for crested wheatgrass grown under two water levels from seed in pots in the glasshouse (a), from clonal propagules in pots in the glasshouse (b), and from clonal propagules in pots in the field (c). From Johnson *et al.* (1990) and Read *et al.* (1991, 1992).

well-watered ($r = -0.37$, $df = 20$) treatment. Values of W ranged from 0.8 to 2.3 g dry matter/kg H_2O, and Δ varied from 19.6 to 24.7‰. Another glasshouse pot study involved nine vegetatively propagated clones of crested wheatgrass with a known range of Δ responses (Read *et al.*, 1991). Plants in this study also were watered to two water levels every 3 to 4 days. Values of W in this study ranged from 1.5 to 2.5 g dry matter/kg H_2O, whereas Δ varied from 21.1 to 24.3‰. This smaller range of values for both W and Δ in the second glasshouse study was probably because the differential between watering treatments (-0.04 and -0.10 MPa soil water potential for the well-watered and droughted treatments, respectively) was smaller than that in the first glasshouse study (-0.03 and -0.30 MPa soil water potential, respectively). Results from this second study also showed that Δ was negatively correlated with W across water levels ($r = -0.89$, $df = 16$) (Fig. 1b). Correlations between Δ and W were significant for both the well-watered ($r = -0.88**$, $df = 7$) and droughted ($r = -0.90**$, $df = 7$) treatments.

In a field study of six of the same vegetatively propagated clones of crested wheatgrass included in the second glasshouse study, pots were embedded in the soil and watered with different amounts of water to two soil water potential levels (-0.04 and -0.07 MPa) every 2 to 3 days (Read *et al.*, 1992). A negative relationship between Δ and W was also found across water levels in this study ($r = -0.98**$, $df = 10$) (Fig. 1c). The relationship between Δ and W was significant within both the well-watered ($r = -0.94**$, $df = 4$) and droughted ($-0.88*$, $df = 4$) levels. Consequently, these results suggest that selection for low Δ should be a promising breeding tool for increasing W in crested wheatgrass.

Intrinsic water-use efficiency (A/g) (net photosynthetic rate/stomatal conductance) as measured periodically throughout the experiments by leaf gas exchange techniques also was associated closely with Δ of shoots. Read *et al.* (1991) found that Δ was negatively correlated with A/g across two water levels in the glasshouse ($r = -0.95**$, $df = 16$) (Fig. 2a). Negative correlations also were obtained between Δ and A/g within the well-watered ($r = -0.89**$, $df = 7$) and droughted ($r = -0.79*$, $df = 7$) treatments. Read *et al.* (1992) also found a highly significant negative association between Δ and A/g across water levels ($r = -0.93**$, $df = 10$) (Fig. 2b) and also within both their well-watered ($r = -0.92**$, $df = 4$) and droughted ($r = -0.91**$, $df = 4$) treatments in their field pot study. In another field study, Read *et al.* (1992) found that Δ and A/g were negatively correlated at two sites ($r = -0.87**$, $df = 13$; $r = -0.77**$, $df = 13$) (Fig. 2c).

In the glasshouse study, g and Δ were positively related under well-watered ($r = 0.74*$, $df = 7$) and droughted ($r = 0.92**$, $df = 7$) conditions (Read *et al.*, 1991). Similarly high correlations between g and Δ were found in the field for both the well-watered ($r = 0.90*$, $df = 4$) and droughted ($r = 0.96**$, $df = 4$) clones (Read *et al.*, 1992). Thus, differences among clones in Δ apparently are largely associated with changes in g rather than changes in

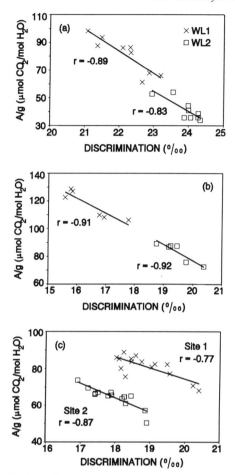

Figure 2. Relationship between intrinsic water-use efficiency (A/g, μmol CO_2/mol H_2O) and carbon isotope discrimination (‰) for clonal propagules of crested wheatgrass grown under two water levels in pots in the glasshouse (a) and the field (b) and at two sites (Evans Farm, Site 1; Deer Pens, Site 2) in northern Utah (c). From Read *et al.* (1991, 1992).

carbon assimilation. Because these relationships are evident within a water level, differences in g and Δ are attributable to genotypic effects.

III. Effect of Environment on Δ

In the glasshouse pot experiments of Read *et al.* (1991), mean Δ decreased significantly from 23.9‰ for the well-watered treatment to 22.2‰ for the drought treatment in nine clones of crested wheatgrass (Fig. 3a). Values of Δ also decreased with drought in a field study of 29 vegetatively propa-

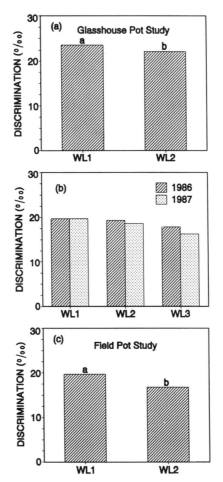

Figure 3. Relationship between carbon isotope discrimination (‰) and water level (WL1, wet; WL2, moderate; and WL3, dry) for clonal propagules of crested wheatgrass grown under two water levels in pots in the glasshouse (a), for 2 years under three water levels located along a water application gradient in a field rain-out shelter facility (b), and under two water levels in pots in the field (c). From Johnson *et al.* (1990) and Read *et al.* (1991, 1992).

gated clones of crested wheatgrass sampled at three water levels along a gradient of water application (Fig. 3b, Table I) (Johnson *et al.*, 1990). The difference in Δ between the high and low water treatments was only 1.4‰ the first year, but as drought along the gradient increased, the difference expanded to 2.8‰ the second year of the study. In the field pot study of Read *et al.* (1992), 6 clones of crested wheatgrass had an average Δ of 19.5‰ under well-watered conditions and 16.5‰ under drought conditions (Fig. 3c).

Table I Summary of Parameters Related to Δ (‰) Determined for Crested Wheatgrass Grown at Three Water Levels (WL1, Wet; WL2, Moderate; and WL 3, Dry) Located along a Water Application Gradient in the Field[a]

Parameter	16 June 1986				20 July 1987				Over 2 years			
	WL1	WL2	WL3	Comb.	WL1	WL2	WL3	Comb.	WL1	WL2	WL3	Comb.
Maximum	21.0	21.0	19.6	20.6	20.8	19.8	18.0	19.3	20.6	20.0	18.2	19.5
Minimum	18.4	17.6	16.1	17.5	18.2	17.6	15.2	17.0	18.9	17.9	16.1	17.7
Mean	19.7	19.3	17.9	19.0	19.7	18.6	16.3	18.2	19.7	18.9	17.1	18.6
SE	0.4	0.5	0.7	0.5	0.3	0.3	0.4	0.2	0.3	0.3	0.4	0.3
Gen. Var.[b]	0.26**	0.25*	0.18	0.22*	0.24**	0.14**	0.18**	0.15**	0.11*	0.19**	0.08	0.11*
Heritability[c]	0.63	0.46	0.29	0.49	0.74	0.60	0.59	0.83	0.43	0.67	0.27	0.47

[a]Values at individual water levels represent determinations from 29 clones with two replications ($n = 58$), whereas combined values for individual harvests are compiled over the three water levels ($n = 174$). Adapted from Johnson et al. (1990).

[b]Gen. Var. = variance component arising from clonal lines (σ_c^2).

[c]Heritability = broad-sense heritability computed on a mean basis.

*,**Mean squares from which the variance components (σ_c^2) were computed were significant at $P \leq 0.05$ and 0.01, respectively.

IV. Genetic Variation in Δ

In the most comprehensive studies to date of genetic variation for Δ in crested wheatgrass, Johnson *et al.* (1990) evaluated Δ in leaves of 29 clones of crested wheatgrass across three water levels (WL1, wet; WL2, moderate; and WL3, dry) sampled during a peak-season harvest in each of 2 years. Significant differences in Δ among clonal lines were found in each of the three water levels for both years, except for WL3 in 1986 (Table I). For the individual harvests and water levels where significant differences among clones were found, broad-sense heritabilities ranged from 0.46 to 0.74. In the combined analysis across the 2 years and three water levels, similar trends in the genetic variation were found, and the broad-sense heritability approached 0.50. The magnitude of narrow-sense heritability values found in our more recent studies (Read *et al.*, 1993) indicates that genetic variation among clonal lines will be recovered in their progenies. These results suggest that ample opportunity exists for selection within crested wheatgrass populations for Δ and that selection would effectively alter Δ in succeeding generations.

V. Stability of Δ Response

Information about genotype by environment interactions or the consistency of genetic differences across environments is essential to an effective breeding program. To obtain such background data, the responses of Δ values of crested wheatgrass clones were evaluated under six environmental regimes (Table II). The clone by environment interactions (C × E) were not significant in three of the six instances, and simple correlation coefficients were always positive and usually significant. The consistency of the Δ value response across environments is also evident in the comparisons of Δ

Table II Significance Level of *F* Tests for Clone by Environment Interactions (C × E) and Simple Correlation Coefficients (*r*) between Clonal Means for Δ across Environments for Clones of Crested Wheatgrass in Five Field Experiments and One Greenhouse Experiment

Study	Environmental factor	C × E df	C × E Significance	Correlation coefficients (*r*)
Rain-out shelter (1986)	Water(3)	56		0.81**, 0.75**, 0.83**
Rain-out shelter (1987)	Water(3)	56	*	0.46**, 0.38*, 0.49**
Rain-out shelter (1988)	Water(3)	16		0.80**, 0.68*, 0.61
Competition (1989)	Sites(2)	14		0.40
Greenhouse pots (1990)	Water(2)	8	**	0.29
Field pots (1990)	Water(2)	5	**	0.80*

*,**Significant at $P \leq 0.05$ and 0.01, respectively.

Table III Correlation Matrix Depicting Relationships among Δ Values for Six Clones of Crested Wheatgrass That Were Common to Seven Experiments

Experiment	Experiment[a]					
	EVANS	DRPENS	FD W	GH W	ROS-86	ROS-87
DRPENS	0.86*					
FD W	0.88*	0.66				
GH W	0.91*	0.67	0.94**			
ROS-86	0.80	0.87*	0.82*	0.69		
ROS-87	0.73	0.83*	0.69	0.73	0.78	
ROS-88	0.90*	0.75	0.99**	0.91*	0.90*	0.76

[a]EVANS and DRPENS were two field competition experiments; FD W was a field pot study; GH W was a greenhouse pot study; and ROS-86, ROS-87, and ROS-88 were field experiments conducted under a rain-out shelter facility in 1986, 1987, and 1988, respectively.

*,**Indicates significance at $P \leq 0.05$ and 0.01, respectively ($df = 4$).

values of six clones of crested wheatgrass obtained in six field experiments and one glasshouse experiment (Table III). Simple correlation coefficients (r) were relatively high and ranged from 0.67 to 0.99; however, because of only 4 degrees of freedom, only about half of these correlations were significant. These results and those in Table II indicate that, in general, the rankings of the crested wheatgrass clones were consistent across environments. This consistency is confirmed by the field studies of Johnson and Bassett (1991) for crested wheatgrass and three other cool-season perennial forage grasses. Nevertheless, caution should be used in extrapolating the results from one environment to another. Consequently, as with most characteristics important in a breeding program, genotypes should be evaluated across the range of environments for which the clones are being developed.

VI. Relationship between Δ and Yield

For forage species, forage yield is of primary importance in identifying superior breeding lines. Consequently, the relationship between forage yield and Δ is of concern. Across three water levels for peak-season forage harvests in a rain-out shelter experiment, simple correlation coefficients (r) between forage yield and Δ were 0.59** during 1986 and 0.80** ($df = 85$) during 1987 (Figs. 4a and 4b) (Johnson *et al.*, 1990). This suggests that selection for increased W (low Δ) may lead to decreases in forage yield. However, this relationship is apparent only when data were combined across the three water levels. When the association was evaluated within a water level, r values were not significant and ranged from 0.07 to 0.33 for 1986 and 0.09 to 0.44 for 1987. This was similar to the pattern observed in the field pot study of Read *et al.* (1992) where shoot dry weight increased with Δ ($r = 0.83**$, $df = 10$) across the two water levels (Fig. 4c), but was not

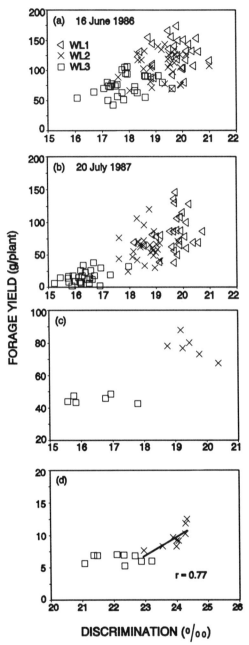

Figure 4. Relationship between forage yield (g/plant) and carbon isotope discrimination (‰) for clonal propagules of crested wheatgrass grown under three water levels located along a water application gradient in a field rain-out shelter facility in 1986 (a) and 1987 (b), under two water levels in a field pot study (c), and under two water levels in a glasshouse pot study (d). Note scale differences for c and d. From Johnson *et al.* (1990) and Read *et al.* (1991, 1992).

correlated within either the well-watered ($r = -0.71$, $df = 4$) or droughted ($r = -0.10$, $df = 4$) treatments. In the greenhouse pot study of Read *et al.* (1991), shoot dry weight of crested wheatgrass increased with Δ in the well-watered treatment ($r = 0.77*$, $df = 7$), but not in the droughted treatment ($r = -0.14$, $df = 7$) (Fig. 4d). Although we might logically expect selection for low Δ to be accompanied by a decrease in total plant biomass in water-limited environments, our results to date have not shown a consistent relationship with forage yield. A negative relationship between genetic variation for Δ and total plant biomass under a uniform water-limited environment is essential for Δ to be a useful selection criterion to improve *W*.

VII. Summary

Considerable research has focused on whether Δ could be used to increase *W* in crested wheatgrass, a perennial forage grass grown widely on semi-arid rangelands in western North America. Field and glasshouse research to date has documented a negative association between Δ and *W* in crested wheatgrass. In addition, studies in both field and glasshouse environments showed that Δ decreased with increasing drought. This work also has shown that significant genetic variability for leaf Δ existed among 29 clones of crested wheatgrass when leaves were sampled during two peak-season forage harvests and that broad-sense heritability for Δ approached 0.50. These data, combined with recently obtained estimates of narrow-sense heritability for Δ, suggest that excellent opportunities exist to genetically alter Δ in crested wheatgrass. The positive correlation between Δ values in clonal lines of crested wheatgrass across a number of experiments indicates that genetic differences for Δ are consistent across the environments that we studied. Additional research is needed to document the association between forage yield and Δ and the influence of water level on this relationship. Results to date suggest that selection for low Δ appears to be a promising screening tool for improving *W* in crested wheatgrass.

References

Barker, R. E., A. B. Frank, and J. D. Berdahl. 1989. Cultivar and clonal differences for water-use efficiency and yield in four forage grasses. *Crop Sci.* **29:** 58–61.

Bleak, A. T., and W. Keller. 1973. Water requirement, yield, and tolerance to clipping of some cool-season, semiarid range grasses. *Crop Sci.* **13:** 367–370.

Farquhar, G. D., and R. A. Richards. 1984. Isotopic composition of plant carbon correlates with water-use efficiency of wheat genotypes. *Aust. J. Plant Physiol.* **11:** 539–552.

Farquhar, G. D., M. H. O'Leary, and J. A. Berry. 1982. On the relationship between carbon isotope discrimination and the intercellular carbon dioxide concentration in leaves. *Aust. J. Plant Physiol.* **9:** 121–137.

Farquhar, G. D., K. T. Hubick, A. G. Condon, and R. A. Richards. 1988. Carbon isotope fractionation and plant water-use efficiency, pp. 21–40. *In* P. W. Rundel, J. R. Ehleringer, and K. A. Nagy (eds.), Ecological Studies. 68. Stable Isotopes in Ecological Research. Springer-Verlag, New York.

Farquhar, G. D., J. R. Ehleringer, and K. T. Hubick. 1989. Carbon isotope discrimination and photosynthesis. *Annu. Rev. Plant Physiol. Plant Mol. Biol.* **40**: 503–537.

Fischer, R. A., and N. C. Turner. 1978. Plant productivity in the arid and semiarid zones. *Annu. Rev. Plant Physiol.* **29**: 277–317.

Frank, A. B., and J. F. Karn. 1988. Growth, water-use efficiency, and digestibility of crested, intermediate, and western wheatgrass. *Agron. J.* **80**: 677–680.

Frank, A. B., R. E. Barker, and J. D. Berdahl. 1987. Water-use efficiency of grasses grown under controlled and field conditions. *Agron. J.* **79**: 541–544.

Johnson, D. A., K. H. Asay, L. L. Tieszen, J. R. Ehleringer, and P. G. Jefferson. 1990. Carbon isotope discrimination: Potential in screening cool-season grasses for water-limited environments. *Crop Sci.* **30**: 338–343.

Johnson, R. C., and L. M. Bassett. 1991. Carbon isotope discrimination and water use efficiency in four cool-season grasses. *Crop Sci.* **31**: 157–162.

Read, J. J., D. A. Johnson, K. H. Asay, and L. L. Tieszen. 1991. Carbon isotope discrimination, gas exchange, and water-use efficiency in crested wheatgrass clones. *Crop Sci.* **31**: 1203–1208.

Read, J. J., D. A. Johnson, K. H. Asay, and L. L. Tieszen. 1992. Carbon isotope discrimination: Relationship to yield, gas exchange, and water-use efficiency in field-grown crested wheatgrass. *Crop Sci.* **32**: 168–175.

Read, J. J., K. H. Asay, and D. A. Johnson. 1993. Divergent selection for carbon isotope discrimination in crested wheatgrass. *Can. J. Plant Sci.* **73**, in press.

Rogler, G. A., and R. J. Lorenz. 1983. Crested wheatgrass—Early history in the United States. *J. Range Manage.* **36**: 91–93.

Schulze, E. D. 1988. Adaptation mechanisms of noncultivated arid-zone plants: Useful lessons for agriculture? pp. 159–177. *In* F. R. Bidinger and C. Johansen (eds.), Drought Research Priorities for the Dryland Tropics. International Crop Research Institute for the Semi-Arid Tropics, Pantancheru, India.

Stanhill, G. 1986. Water use efficiency. *Adv. Agron.* **39**: 53–85.

19

Carbon Isotope Discrimination, Water Relations, and Gas Exchange in Temperate Grass Species and Accessions

Richard C. Johnson and Larry L. Tieszen

I. Introduction

The Western Regional Plant Introduction Station (Pulman, WA) maintains an active germplasm collection available for distribution to researchers worldwide. Evaluation of germplasm from diverse sources within the collection has revealed genetic variation among accessions for carbon isotope discrimination (Δ) in four temperate grass species. These species are *Agropyron desertorum* (Fishcher ex Link) Schultes (standard crested wheatgrass), *Lolium perenne* L. (perennial ryegrass), *Dactylis glomerata* L. (orchardgrass), and *Festuca arundinacea* Schreb. (tall fescue) (Johnson and Bassett, 1991). These grasses are widely used in agriculture and are all heterogenous, perennial species. Crested wheatgrass was introduced from Eurasia ca. 1900 and has become an established rangeland plant of the intermountain western United States. Perennial ryegrass, orchardgrass, and tall fescue were all introduced to the United States from Europe and are established forage crops. In addition to utilization as forage, perennial ryegrass and tall fescue are also used for turf grass. Tall fescue alone is grown for both forage and turf on an estimated 14 million ha in the United States, especially in a transition zone between the warmer southeast and the colder north extending from the central United States to the Atlantic coast (Sleper, 1988).

Following initial field evaluation for Δ, estimates of water-use efficiency in whole plants were made on high and low Δ accessions within perennial ryegrass, orchardgrass, and tall fescue. Differing tall fescue accessions were then examined for photosynthetic gas exchange characteristics to determine the relationship between c_i/c_a and Δ. And finally, accessions of the tall fescue cultivar 'Kentucky 31' are being used to develop high and low Δ germplasm lines in populations with and without the *Acremonium* fungal endophyte. In this chapter we discuss research showing Δ as a promising tool for developing germplasm to improve water-use efficiency in temperate grass species.

II. Field Evaluation of Δ in Temperate Grass Species and Accessions

A. Materials and Methods

1. Four Temperate Grass Species   Seeds of seven accessions from each of the four C_3 grass species described in Section I above were obtained from the Western Regional Plant Introduction Station (Pulman, WA). The species chosen represented forage grasses adapted to climates ranging from arid to humid (Heath *et al.*, 1978), and the accessions within each species represent germplasm obtained from diverse origins.

An identical experiment was established in three environments by transplanting seedlings during the spring of 1988 as described by Johnson and Bassett (1991). An irrigated experiment and a dryland experiment were established at the Central Ferry Research Station and an additional dryland experiment was established at Pullman, Washington, on 28 April 1988. Central Ferry, located in the Snake River canyon, is a drought-prone environment with irrigation available, facilitating comparative irrigated and drought-stress experiments. Within each of the three experiments at different locations (environments) plots consisting of 10 plants each were randomized in complete blocks with three replications. A block consisted of four species each represented by seven accessions for a total of 28 plots.

Periodic estimates of leaf-water potential (ψ) were made at midday (1130 to 1300 h) on plants in each environment on a randomly selected accession of each species. For each plot sampled, two water potential samples were obtained using individually calibrated "leaf-cutter"-type thermocouple psychrometers as described by Johnson *et al.* (1986).

The central 3 to 4 cm from recently fully emerged leaves were obtained for Δ analysis from the 10 plants within each plot. Leaf samples from each plant were bulked for a given plot, dried at 70°C to constant weight, and ground. Samples for Δ determinations were obtained on 16 August 1988 in the Central Ferry plots and 23 August 1988 in the Pullman plots. Analysis for isotopic carbon composition ($\delta^{13}C$), the ratio $^{13}C/^{12}C$ relative to the PeeDee belemnite standard, was completed on each sample using an isotope ratio mass spectrometer as described by Read *et al.* (1991).

2. Tall Fescue Accessions with Contrasting Δ   Plots of PI 428522 and PI 231561 from the experiments described above were retained at Central

Ferry, Washington, and used in 1990 to sample individual plants for forage production, Δ, and water relations in irrigated and dryland environments as described by Johnson (1993). These accessions were chosen because they differed in Δ in the 1988 field studies in all three environments. Leaf sheaths of PI 438522 and PI 231561 were examined for the presence of viable *Acremonium* endophytic fungi according to the methods given by Wilson *et al.* (1991) and none were detected. (Fungal endophytes are associated with fescue toxicosis in grazing animals and have been suggested as a factor in both biotic and abiotic stress resistance, see Section V.)

Each accession within each environment originally consisted of 10 plants spaced 0.5 m apart. All 10 plants in each plot survived for the 1990 study except for those in one plot of PI 231522 in the dryland environment. In that plot, 7 plants died over winter in 1989–1990. Samples for Δ analysis were collected on individual plants of both accessions in irrigated and dryland environments on 7 June and 9 August 1990.

Water relations data were not taken on 7 June 1990 because soil was moist in both irrigated and dryland plots due to recent rain. On 9 August 1990 water relations data were taken on the same day Δ samples were taken. Field samples for determining leaf ψ, solute potential (ψ_s), and turgor pressure (ψ_p) were taken using individually calibrated leaf-cutter thermocouple psychrometers. One psychrometer sample was taken on an upper, fully emerged leaf of each plant and wet bulb depression was read after 3 h in a 20°C water bath. The psychrometers were then frozen overnight at -20°C and measured again after 3 h equilibration in a 20°C water bath. Values of ψ_s were determined on freezed/thawed tissue and ψ_p by the difference between ψ and ψ_s.

B. Results and Discussion

1. Four Temperate Grass Species Differences in leaf ψ between irrigated and dryland environments became clearly apparent 18 July 1988. From that date until samples for Δ were taken (16 August 1988) leaves from irrigated plots averaged about -1.8 MPa and leaves from the dryland plots about -2.8 MPa. The dryland plots at Central Ferry and Pullman, although separated by about 80 km and 500 m elevation, had comparable ψ values.

Mean Δ of the Central Ferry irrigated plants was 17.6‰ and did not differ from the Central Ferry dryland plots at 17.5‰. We had expected, as others have observed (Hall *et al.*, 1992; Meinzer *et al.*, 1990; Read *et al.*, 1991), a shift downward in Δ as a result of drought. This did not occur even though there was a clear reduction in ψ of about 1.0 MPa in the dryland compared to the irrigated environment at Central Ferry. The reason for this is not clear but is discussed in more detail in Section II, B, 2 below.

Analyses of variance for Δ showed highly significant effects associated with environment, species, and accessions within species. The environment $\times$ species interaction was significant for Δ, but the environment $\times$ accession (within-species) interaction was not significant. In other words,

Table I   Mean Shoot Dry Weight and Carbon Isotope Discrimination (Δ) for Grasses Grown under Irrigated and Dryland Conditions at Central Ferry, Washington, 1988[a]

Species	Environment	Shoot dry wt (g plant^{-1})	Resistance (%)	Δ (‰)
Agropyron desertorum	Dryland	32.5*	76	18.6*
	Irrigated	42.8		18.9
	Means	37.6a**		18.8a
Dactylis glomerata	Dryland	40.5*	58	17.2*
	Irrigated	70.2		16.9
	Means	55.3b		17.1c
Festuca arundinacea	Dryland	66.5*	59	16.9
	Irrigated	112.9		17.1
	Means	89.7c		17.0c
Lolium perenne	Dryland	51.1*	50	17.2
	Irrigated	101.9		17.3
	Means	76.5d		17.3b

[a]Each species was represented by seven accessions within the irrigated and dryland environments. Drought resistance (%) was calculated as the dryland dry matter value divided by the irrigated value. (From Johnson and Bassett, 1991, used by permission.)

*Significant difference between irrigated and dryland means within a species at the 0.05 probability level using Fisher's LSD.

**Within columns and among species, means followed by the same letter are not significantly different at the 0.05 probability level using Fisher's LSD.

species differences in Δ depended on environment, but the relative differences in Δ among accessions were generally consistent across the three environments.

Even though low Δ accessions were identified the relationship of Δ to drought resistance was not clear. For example, crested wheatgrass would be considered the species in this study that is best adapted to semiarid rangeland conditions. If drought resistance is defined as dry matter productivity in the dryland environment at Central Ferry environment divided by dry matter productivity in the irrigated environment at Central Ferry, crested wheatgrass had the highest drought resistance (Table I). Yet crested wheatgrass also had higher Δ than other species at Central Ferry, suggesting lower water-use efficiency compared to that of other species. Read and Farquhar (1991) found a negative relationship between Δ and precipitation for 22 species of *Nothofagus*. They hypothesized that species of drier climates may have mechanisms for maintaining water uptake and high g_s during periods of mild water deficit, leading to Δ generally higher than that of species adapted to wetter climates. On the other hand, if drought resistance is defined simply as dry matter productivity under dryland conditions for the grass species in Table I, higher productivity was associated with lower Δ and presumably higher water-use efficiency.

2. Tall Fescue Accessions with Contrasting Δ   Consistent with results in 1988 (Johnson and Bassett, 1991), lower Δ was observed for tall fescue PI 438522 than for PI 231561 in 1990 indicating higher water-use efficiency for 438522 (Table II). Water potential values were less negative for irrigated compared to the dryland environments as expected, but no difference between environments for Δ_m (the mean of Δ samples taken on 7 June and 9 August 1990) was observed.

In the current study and in previous work described in Section II, B, 1 with these accessions, field drought associated with the dryland environment did not result in lower Δ. The absence of lower Δ in response to lower ψ could have resulted from a general balance between stomatal and nonstomatal factors so that c_i/c_a and Δ of plants in both irrigated and dryland environments remained nearly constant. Or there may have been an initial decline in c_i/c_a at mild stress followed by an increase with more severe stress as observed by Johnson *et al.* (1987) in *Triticum* species. This could result in equal Δ and average c_i/c_a between irrigated and dryland environments. The effect of lower ψ on growth of dryland plants was not dramatic; forage yields were about 15% lower in the dryland than in the irrigated environment. It seems likely, therefore, that a balance between stomatal and nonstomatal factors was maintained in the field environment so that Δ did not decline with lower ψ.

For ψ_s there was a strong accession effect; PI 438522 had more negative ψ than PI 231561, especially in the dryland environment. In the absence of a difference in ψ between accessions together with the more negative ψ_s of PI 438522, one would expect higher ψ_p for PI 438522. Mean ψ_p was 0.17 MPa higher for PI 438522 than for PI 231561, but this was not significant (Table II). Yet this tendency for PI 438522, the low Δ accession, to have higher ψ_p is consistent with previous results (Johnson and Bassett, 1991).

Linear correlations between Δ sampled on 7 June 1990 and Δ sampled 9 August 1990 were relatively weak although significant in the irrigated ($r = 0.38^*$, $n = 30$) and dryland ($r = 0.49^{**}$, $n = 30$) environments for PI 438522. For PI 231561 the correlation between sampling times was significant in the dryland environment ($r = 0.60^{**}$, $n = 23$) but not in the irri-

Table II Leaf Carbon Isotope Discrimination Sampled on 7 June 1990 (Δ_1), 9 August 1990 (Δ_2), and the Mean of the Two Dates (Δ_m) along with Leaf Water Potential (ψ), Solute Potential (ψ_s), and Turgor Pressure (ψ_p) Measured on 9 August 1990[a]

	Δ_1 (‰)	Δ_2 (‰)	Δ_m (‰)	ψ (MPa)	ψ_s (MPa)	ψ_p (MPa)
Environment						
Irrigated	19.1	18.8	18.9	-0.89^{**}	-2.04^{**}	1.14*
Dryland	18.9	18.9	18.9	-1.99	-2.70	0.43
Accession						
PI 438522	18.6*	17.9**	18.3**	-1.50	-2.51^{**}	1.01
PI 231561	19.5	19.8	19.6	-1.38	-2.22	0.84

[a]Samples were taken from two tall fescue accessions growing in irrigated and dryland environments. (From Johnson, 1993, used by permission.)

*,**Indicates significant difference between environmental means or accession means at $P = 0.05$ and $P = 0.01$, respectively.

Table III Linear Correlation Coefficients (r) for Mean Carbon Isotope Discrimination (Δ_m), Seasonal Forage Yield (FY), Water Potential (ψ), Solute Potential (ψ_s), and Turgor Pressure (ψ_p) of Individual Plants of Tall Fescue PI 438522 and PI 231522 Grown under Dryland Conditions ($n = 53$)

	Δ_m	FY	ψ	ψ_s	ψ_p
Δ_m	—				
FY	0.28*	—			
ψ	0.26	0.58**	—		
ψ_s	0.57**	0.43**	0.69**	—	
ψ_p	0.01	0.52**	0.90**	0.30*	—

*,**Significant correlation coefficient at $P = 0.05$ and $P = 0.01$, respectively.

gated environment ($r = 0.11$, $n = 30$). The results show inconsistencies in Δ between sampling dates and suggest that it may be advisable to obtain more than one Δ to effectively identify parental plants for germplasm enhancement in forage grass species.

Except for a weak positive correlation between ψ_p and forage yield ($r = 0.29*$, $n = 60$), no significant correlation coefficients were observed among Δ_m, forage yield, ψ, ψ_s, and ψ_p in the irrigated environment (data not shown). In the dryland environment, however, all correlation coefficients among the above factors were significant and positive except Δ_m versus ψ and ψ_p (Table III). Individual plants with higher ψ, ψ_s, and ψ_p in the dryland environment tended to have higher forage yield, and this is consistent with the idea that plants with a higher water status should produce more. Dehydration stress often reduces Δ and increases water-use efficiency, but the lack of correlation between Δm and ψ and between Δ_m and ψ_p suggests that variation in Δ was not consistently explained by simple dehydration effects.

III. Water-Use Efficiency in Whole Plants and Δ

A. Materials and Methods

One accession each of orchardgrass, tall fescue, and perennial ryegrass showing high and low leaf Δ in the field experiment described in Section II, A, 1 above were grown in pots in a greenhouse experiment to determine whole-plant water-use efficiency. Pots were filled with 1800 g of dry sand and saturated with water (18% by weight). The sand was contained within a plastic bag inside each pot similar to the procedures described by Read *et al.* (1991). Three plants were established in each pot and grown under greenhouse conditions. Every second or third day, the weight of water loss per pot was recorded and replenished with 0.25-strength Hoagland's solution.

A pot with no plants (blank pot) was added to each block of both well-watered and drought-stressed treatments to estimate pot evaporation. After 50 days of growth, the drought-stress treatment was imposed by replen-

Table IV Dry Weight and Water Use Efficiency (W_{et}, Dry Matter/Pot Evapotranspiration) for Three Grass Species[a]

Species and accession	Field Δ	Water regime	GH Δ	Shoot dry wt (g plant⁻¹)	Total dry wt (g plant⁻¹)	Shoot W_{et} (g kg⁻¹)	Whole-plant W_{et} (g kg⁻¹)
Dactylis glomerata							
325287	Low[b]	I	14.6**	3.05**	4.51**	2.15**	3.18**
384848	High	I	15.7	2.02	2.87	1.59	2.25
325287	Low	D	14.7**	1.58**	2.04**	1.49**	1.93**
384848	High	D	15.6	0.95	1.31	1.04	1.43
Festuca arundinacea							
438522	Low	I	14.8**	3.05**	4.51**	2.15**	3.18**
231561	High	I	15.7	2.42	3.47	1.76	2.51
438522	Low	D	14.1**	1.78*	2.28	1.84**	2.34
231561	High	D	15.2	1.29	1.94	1.44	2.15
Lolium perenne							
303041	Low	I	14.2**	4.09**	5.01**	2.81**	3.43**
231575	High	I	14.6	3.24	3.95	2.40	2.92
303041	Low	D	14.5	2.21**	2.70*	2.21**	2.69**
231575	High	D	14.3	1.64	2.02	1.70	2.10

[a]Accessions within species were selected for high or low carbon isotope discrimination values (Δ, ‰) from field data and then grown in a greenhouse (GH) under well-irrigated (I) and drought-stressed (D) conditions. (From Johnson, 1991, used by permission.)

[b]Low and high Δ (‰) averaged across field environments was 15.8 and 17.4 for *D. glomerata*, 15.9 and 18.0 for *F. arundinacea*, and 16.5 and 17.6 for *L. perenne*.

*,**Significant differences at $P = 0.05$ and $P = 0.01$, respectively, between accessions within a species and water regime for factors related to high or low field Δ values.

ishing designated pots to only 8% water content at each watering. The experiment was terminated 98 days after seedlings were transplanted to pots. Calculations of water-use efficiency as dry matter/pot evapotranspiration (W_{et}) and dry matter/transpiration (W) were made. Values of W were calculated by subtracting pot evapotranspiration from estimates of pot evaporation from blank pots. For shoot W_{et} and W, all plant material except roots was used in calculating water-use efficiency ratios. For whole-plant W_{et} and W, total plant dry weight was used.

B. Results and Discussion

Accessions selected for low Δ in the field always had high W_{et} based on shoot dry weight in the greenhouse pot experiments (Table IV). Whole-plant W_{et} also averaged higher values for accessions selected for low Δ in all treatments except the drought-stressed tall fescue treatment. In all cases except the perennial ryegrass drought-stress treatment, accession differences in Δ for leaf tissue from greenhouse plants were consistent with accession differences observed in the field (Table IV).

Drought-stressed greenhouse plants, unlike field plants, tended to have lower Δ than well-watered plants (Table IV), but the difference between water treatments for Δ was not significant. Values of W were about two times higher than those for W_{et} and appeared unusually high. It appeared that blank pots, for reasons that are not clear, may have overestimated evaporation in drought-stressed treatments. This notwithstanding, accessions with low Δ had generally higher W_{et} (Table IV) and W (data not shown) than accessions with high Δ, and this was consistent with theoretical expectations that Δ is inversely related to W (Farquhar *et al.*, 1989).

Since accessions selected for lower Δ in the field had higher W in greenhouse experiments, and a significant portion of the variation in Δ appears to be under genetic control (Ehleringer, 1988; Martin and Thorstenson, 1988; Hall *et al.*, 1990; Hubick *et al.*, 1988; Read *et al.*, 1991), this link between field Δ and W increases the confidence with which field selections for W can be made through Δ.

IV. Gas Exchange and Δ in Tall Fescue Genotypes

A. Materials and Methods

Based on 1990 determinations of Δ, three tall fescue plant genotypes of PI 438522 with high mean Δ and three with low mean Δ were selected from the irrigated environment established in 1988 (Section II above). Portions of tillers and roots were removed from field plots on 5 July 1991 as described by Johnson (1993). Four vegetative ramets of approximately equal size were obtained from each plant and grown under greenhouse conditions in separate pots with a soil volume of 2.4 liters. The ramets, representing three high and three low Δ genotypes, were randomized into complete blocks with four replications.

On 27 and 28 July 1991 well-watered plant treatments were transported to the laboratory for measurements of photosynthetic gas exchange characteristics. Gas exchange characteristics were measured on a recently fully emerged leaf from each plant using a steady-state system described by Johnson *et al.* (1987). Standard measurement conditions were 1800 μmol m^{-2} s^{-1} photon flux density, 210 ml O$_2$ liter^{-1}, 25°C, and an ambient [CO$_2$] of 350 μl liter^{-1}. After steady-state conditions were achieved on a given leaf, which usually took about 15 min, the leaf was removed from the gas exchange chamber. Estimates of ψ and leaf area were made immediately after removing the leaf from the chamber and the leaf was dried and ground for Δ analysis. Transpiration (E), A (CO$_2$ exchange rate), stomatal conductance to CO$_2$ (g_s), and c_i were calculated on a per unit leaf area basis according to von Caemmerer and Farquhar (1981). Water-use efficiency based on gas exchange factors was calculated as A/E and as A/g_s.

After completion of gas exchange measurements under well-watered conditions on 28 July 1991, the plants were returned to the greenhouse and trimmed to about 50 mm from the soil surface. These plants were then subjected to water deficits by treatment with 8000 mw polyethylene glycol (PEG). This was done to obtain as uniform a soil water deficit as possible among plant material growing in pots. Solutions of 10% (w/v) PEG were prepared in 0.25-strength Hoagland's solution and applied to pots every other day until solution drained from the base of the pots. Water potential of the 10% PEG solution was measured with thermocouple psychrometers and determined to be -0.38 MPa. Gas exchange factors and leaf water potential were measured on PEG-stressed plants as described above on 17 and 18 September 1991.

B. Results and Discussion

As a class, plants selected for low Δ had higher water-use efficiency when expressed in terms of gas exchange factors than plants selected for high Δ (Table V). This was observed under both well-watered and PEG-stressed conditions and when water-use efficiency was expressed as either the A/E ratio or as the A/g_s ratio. In addition, the low Δ class also had lower c_i/c_a and c_i than the high Δ class as expected by theory (Farquhar *et al.*, 1989). Stomatal conductance and A were lower in the low Δ class under well-watered conditions but not under stress. Unlike results with field drought, PEG-stressed plants had lower Δ than well-watered plants (Table V).

Values of Δ were positively correlated with c_i/c_a and negatively correlated with the A/g_s ratio when well-watered and stressed plants were analyzed together (Fig. 1). When analyzed separately, however, only Δ versus c_i/c_a ($r = 0.90^*$, $n = 6$) and Δ versus A/g_s ($r = -0.90^*$, $n = 6$) from well-watered plants were correlated significantly.

Table V Net Carbon Exchange Rate (A), Stomatal Conductance to CO_2 (g_s), Water-Use Efficiency as the A/Transpiration (A/E) and as A/g_s, Internal CO_2 Concentration/Ambient CO_2 Concentration (c_i/c_a), and Carbon Isotope Discrimination (Δ) for Plants Selected for High and Low Δ from Field Plots[a]

Δ Class	A (μmol m^{-2} s^{-1})	g_s (mol m^{-2} s^{-1})	A/E (mmol mol^{-1})	A/g_s (μmol mol^{-1})	c_i/c_a	c_i (μmol mol^{-1})	Δ (‰)
			Well-watered[b]				
High	32.7*	0.41**	3.84*	81.4**	0.748**	260**	21.9*
Low	29.7	0.33	4.06	91.5	0.716	248	21.1
			PEG-stressed[b]				
High	12.7	0.11	4.29*	118.1*	0.643**	227**	20.1**
Low	13.3	0.10	4.64	129.8	0.609	215	19.2

[a]Values were obtained on leaves of tall fescue grown under greenhouse conditions and measured under steady-state conditions in the laboratory. Each Δ class was represented by three high and low Δ genotypes with four replicate clones from PI 438522. (From Johnson, 1993, used by permission.)
[b]Leaves from well-watered plants had mean water potential values of −0.51 MPa and PEG-stressed leaves had mean values of −1.94 MPa.
*,**Within a water level, means are significantly different at $P = 0.05$ and $P = 0.01$, respectively, as determined by preplanned single degree of freedom comparisons.

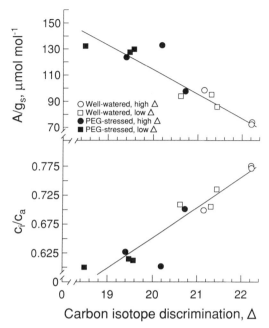

Figure 1. Internal leaf CO_2 concentration/ambient CO_2 concentration (c_i/c_a) and leaf water-use efficiency (A/stomatal conductance, A/g_s) as a function of carbon isotope discrimination (Δ) in well-watered plants and plants stressed with a -0.38 MPa polyethylene glycol solution. Linear equations are $A/g_s = -18.2\Delta + 479$ $(r = -0.93**)$ and $c_i/c_a = 0.052\Delta - 0.388$ $(r = 0.93**)$. Each value represents the mean of four replications per genotype. (From Johnson, 1993, used by permission.)

Linear correlations between Δ and both A and g_s were positive and highly significant when data from the well-watered and the PEG-stress experiments were combined (Fig. 2). For Δ versus A under well-watered conditions the linear correlation was significant $(r = 0.82*, n = 6)$ but under PEG-stress this correlation was not significant $(r = 0.61, n = 6)$. Likewise, the linear correlation between g_s and Δ for well-watered plants was highly significant $(r = 0.98**, n = 6)$ and the same correlation for PEG-stressed plants was not significant $(r = 0.50, n = 6)$. The linear correlations between A and g_s for the combined data were strong (Fig. 3), and it also was significant for well-watered $(r = 0.89*, n = 6)$ and PEG-stressed $(r = 0.90*, n = 6)$ plants.

The above data and the relationships between c_i/c_a and Δ and A/g_s and Δ (Fig. 1) show it is possible to relate instantaneous gas exchange estimates of water-use efficiency in tall fescue genotypes with Δ, an integrated estimate of water-use efficiency. Hubick *et al.* (1988) in *Arachis hypogaea* L., Condon *et al.* (1990) in *Triticum aestivum* L., Meinzer *et al.* (1990) in *Coffea arabica* L., and Read *et al.* (1991) in crested wheatgrass clones also found gas exchange factors related to Δ in a way consistent with theory. Nevertheless, instantaneous values of c_i/c_a or A/g_s obtained in gas exchange measurements do not

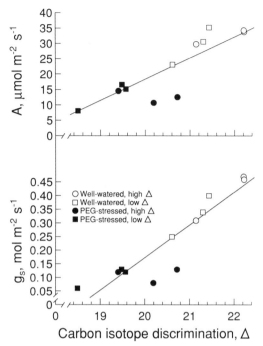

Figure 2. Stomatal conductance (g_s) and CO_2 exchange rate (A) as a function of carbon isotope discrimination (Δ) in well-watered plants and plants stressed with a -0.38 solution of polyethylene glycol. Linear equations are $A = 7.65\Delta - 135$ ($r = 0.88$**) and $g_s = 0.119\Delta - 2.21$ ($r = 0.92$**). Each value represents the mean of four replications per genotype. (From Johnson, 1993, used by permission.)

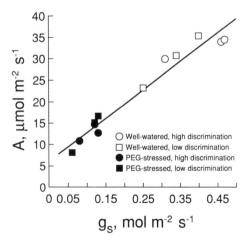

Figure 3. Relationship between CO_2 exchange rate (A) and stomatal conductance (g_s) in well-watered plants and plants stressed with a -0.38 MPa solution of polyethylene glycol. The linear equation is $A = 65.7g_s + 6.4$. Each value represents the mean of four replications per genotype.

always correlate with Δ (Hall *et al.*, 1992), and this is perhaps not surprising given that they represent only a small fraction of the time integrated by Δ.

If, theoretically, the diffusion barrier is removed, c_i should equal c_a and c_i/c_a would equal one. Under these conditions fractionation associated with carboxylation processes would determine Δ. The fractionation associated with carboxylation, mainly through RuBisCO, has been estimated as 27‰ (Farquhar *et al.*, 1982). When c_i/c_a is set to one in the regression equation describing Δ and c_i/c_a (Fig. 1), the predicted Δ is 26.7‰, very close to the 27‰ obtained by Farquhar *et al.* (1982).

The differences between high and low Δ classes (Table V) would have been much greater except for inconsistent Δ values from a single genotype. That plant was originally placed in the high Δ class in 1990 but subsequent Δ determinations in the field and greenhouse in 1991 showed it had c_i/c_a and Δ values more closely associated with the low Δ class (see high Δ class points at 21.2 and 19.4 for well-watered and PEG-stressed, respectively, Fig. 1). It is not clear why this occurred, but Read *et al.* (1991) also report some inconsistencies between original and subsequent Δ rankings. In their study, g_s, c_i, and A/E in Δ selection classes did not differ as expected under well-watered conditions, although they did differ as expected under drought stress.

Although the response of individual plants may vary, the positive correlation of Δ with both A and g_s across well-watered and PEG treatments shows the tendency for water-use efficiency and A to be inversely related. It also suggests an important role for reduced stomatal conductance as a factor leading to lower c_i/c_a and Δ in this study. This is because reduced g_s restricts the diffusion of CO_2 into leaves, and if the capacity for photosynthesis remains constant, both c_i and A are reduced. But g_s and photosynthetic leaf capacity may be colimiting in such a way that Δ could correlate with both g_s and internal leaf photosynthetic capacity as observed by Condon *et al.* (1990) with wheat. In any case, the A versus g_s relationship was strong (Fig. 3) suggesting a general coupling between A and g_s. The mechanism by which individual plants maintain lower Δ and higher water-use efficiency, whether stomatal or nonstomatal, may affect potential productivity of plants in various environments and the utility of using Δ to select plants for water-use efficiency in crop canopies (Farquhar *et al.*, 1989). It is therefore an important consideration for future research.

V. Germplasm Enhancement in Tall Fescue

A. Materials and Methods

Populations of the tall fescue cultivar Kentucky 31 infected and uninfected with the endophytic fungus *Acremonium coenophialum* were established in 1991 at Central Ferry, Washington. The presence of the endophyte in tall fescue has been associated with enhanced arthropod resistance, stand persistence, and drought resistance, but also can cause toxicity to grazing animals (Siegel *et al.*, 1987). Kentucky 31 was chosen for germplasm enhancement for two reasons. First, it is the most widely distributed tall fescue cultivar in the United States and it is grown in areas subject to environmental extremes. And second, both endophyte-free and -infected seed of Kentucky 31 collected at the Suiter Farm in Kentucky, the site where the original Kentucky 31 was collected in 1931, were available. In April 1991, individual plants were transplanted in blocks 6.5 m wide and 7.5 m long with plants spaced about 0.4 m apart. A group of 55 interior plants in an endophyte-free block and 54 plants in an endophyte-infected

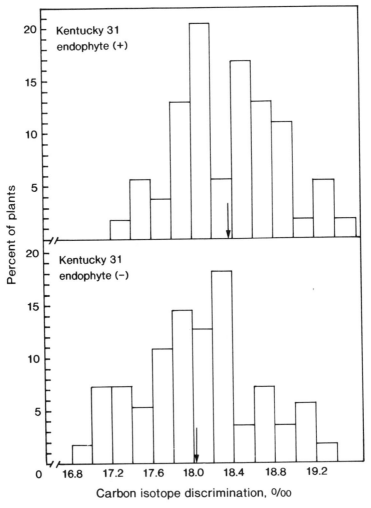

Figure 4. Frequency distribution of populations of Kentucky 31 tall fescue genotypes with and without the *Acremonium* endophytic fungus. Population size was 54 for the plants with endophyte (+) and 55 for those without endophyte (−). The arrow indicates the mean for each population.

block were sampled for Δ at two dates. The first sampling date was on 26 June 1991 when soil was moist and plants actively growing. The second sampling date was on 1 August 1991 when plants were drought stressed. Water potential, ψ_s, and ψ_p of each plant was sampled at both sampling times as described in Section II, A, 2.

B. Results and Discussion

Correlation coefficients between Δ for the two sampling dates were highly significant for both endophyte-free and -infected populations. As with other tall fescue accessions (Section II, B, 2), however, less than half the total variation was explained by correlating Δ at the two sampling dates. Also consistent with results above, Δ values were not generally reduced by the development of drought stress. In both populations Δ values were normally distributed (Fig. 4). The range of Δ values in both populations was 2.3‰, which is sufficient to proceed with selections for high and low Δ parental material. Even though the mean Δ for the endophyte-free population was about 0.4‰ less than that for the endophyte-infected population, comparisons would be premature because the populations were not replicated. The ψ values from the endophyte-free population were consistently lower than the infected population, but this could have been the result of field block position or physiological factors attributable to endophytes. Eight high and eight low Δ plants have been selected from each population as parental material for germplasm enhancement. These selections will be grown for seed under isolation and an additional cycle of selection completed. Endophyte and Δ effects on agronomic characteristics will then be tested in replicated experiments with high and low Δ germplasm.

VI. Summary

Genetic variation for Δ in crested wheatgrass, perennial ryegrass, orchardgrass, and tall fescue accessions has been demonstrated. In experiments with whole plants of perennial ryegrass, orchardgrasss, and tall fescue growing in pots, accessions with high and low Δ had low and high water-use efficiency consistent with theoretical expectations. Also consistent with expectations, a group of tall fescue clones selected for low Δ had lower c_i/c_a than those selected for high Δ. These results indicate that Δ provides an integrated estimate of water-use efficiency that can be determined on a relatively large number of samples and is therefore a promising tool for developing germplasm lines with high and lower water-use efficiency. High and low Δ parents have been selected for germplasm enhancement in populations of Kentucky 31 tall fescue with and without the *Acremonium* endophyte. Once developed, this germplasm will be useful for determining when high or low Δ is most desirable for productivity in contrasting environments and for testing the importance of Δ in grass swards, where crop canopy factors may be important determinants of water-use efficiency.

References

Caemmerer, S. von, and G. D. Farquhar. 1981. Some relations between the biochemistry of photosynthesis and gas exchange of leaves. *Planta* **153:** 376–387.

Condon, A. G., G. D. Farquhar, and R. A. Richards. 1990. Genotypic variation in carbon isotope discrimination and transpiration efficiency in wheat: Leaf gas exchange and whole plant studies. *Aust. J. Plant Physiol.* **17:** 9–22.

Ehleringer, J. R. 1988. Correlations between carbon isotope ratio, water-use efficiency and yield, pp. 165–191. *In* J. W. White, G. Hoogenboom, F. Ibarra, and S. P. Singh (eds.), Research on Drought Tolerance in Common Bean. CIAT, Cali, Colombia.

Farquhar, G. D., M. C. Ball, S. von Caemmerer, and Z. Roksandic. 1982. Effect of salinity and humidity on $\delta^{13}C$ value of halophytes—evidence for diffusional isotope fractionation determined by the ratio of internal/atmospheric partial pressure of CO_2 under different environmental conditions. *Oecologia* **52:** 121–124.

Farquhar, G. D., J. R. Ehleringer, and K. T. Hubick. 1989. Carbon isotope discrimination and photosynthesis. *Annu. Rev. Plant Physiol. Plant Mol. Biol.* **40:** 503–537.

Hall, A. E., R. G. Mutters, K. T. Hubick, and G. D. Farquhar. 1990. Genotypic differences in carbon isotope discrimination by cowpea under wet and dry field conditions. *Crop Sci.* **30:** 300–315.

Hall, A. E., R. G. Mutters, K. T. Hubick, and G. D. Farquhar. 1992. Genotypic and drought-induced differences in carbon isotope discrimination and gas exchange of cowpea. *Crop Sci.* **32:** 1–6.

Heath, M. E., D. S. Metcalf, and R. F. Barnes. 1978. Forages: The science of grassland agriculture, 3rd Ed. The Iowa State Univ. Press, Ames, IA.

Hubick, K. T., R. Shorter, and G. D. Farquhar. 1988. Heritability and genotype × environment interaction of carbon isotope discrimination and transpiration efficiency in peanut. *Aust. J. Plant Physiol.* **15:** 799–813.

Johnson, R. C. 1993. Carbon isotope discrimination, water relations, and photosynthesis in tall fescue. *Crop Sci.* **33:** 169–174. In press.

Johnson, R. C., and L. M. Bassett. 1991. Carbon isotope discrimination and water use efficiency in four cool-season grasses. *Crop Sci.* **31:** 157–162.

Johnson, R. C., H. T. Nguyen, R. W. McNew, and D. M. Ferris. 1986. Sampling error for leaf water potential measurements in wheat. *Crop Sci.* **26:** 380–383.

Johnson, R. C., D. W. Mornhinweg, J. J. Heitholt, and D. M. Ferris. 1987. Leaf photosynthesis and conductance of selected *Triticum* species at different water potentials. *Plant Physiol.* **83:**1014–1017.

Martin, B., and Y. R. Thorstenson. 1988. Stable isotope composition ($\delta^{13}C$), water use efficiency, and biomass productivity of *Lycopersicon esculentum, Lycopersicon pennellii*, and the F_1 hybrid. *Plant Physiol.* **88:** 213–217.

Meinzer, F. C., G. Goldstein, and D. A. Grantz. 1990. Carbon isotope discrimination in coffee genotypes grown under limited water supply. *Plant Physiol.* **92:** 130–135.

Read, J., and G. Farquhar. 1991. Comparative studies in *Nothofagus* [Fagaceae]. I. Leaf carbon isotope discrimination. *Funct. Ecol.* **5:** 684–695.

Read, J. J., D. A. Johnson, K. H. Asay, and L. L. Tieszen. 1991. Carbon isotope discrimination, gas exchange, and water-use efficiency in crested wheatgrass clones. *Crop Sci.* **31:** 1203–1208.

Siegel, M. R., G. C. M. Latch, and M. C. Johnson. 1987. Fungal endophytes of grasses. *Annu. Rev. Phytopathol* **25:** 293–315.

Sleper, D. A. 1988. Cool-season grasses for humid areas, pp. 58–62. *In* K. H. Asay (ed.), Status Report Crop Advisory Committee Forage and Turf Grasses. USDA-ARS, Logan, UT.

Wilson, A. D., S. L. Clement, and W. J. Kaiser. 1991. Survey and detection of endophytic fungi in *Lolium* germ plasm by direct staining and aphid assays. *Plant Dis.* **75:** 169–173.

20

Environmental and Developmental Effects on Carbon Isotope Discrimination by Two Species of *Phaseolus*

Qingnong A. Fu, Thomas W. Boutton, James R. Ehleringer, and Richard B. Flagler

I. Introduction

Isotopic discrimination (Δ) against $^{13}CO_2$ during photosynthesis in C_3 plants is positively correlated with the long-term, integrated ratio of intercellular to atmospheric CO_2 concentration (c_i/c_a), a parameter which reflects the balance between consumption of CO_2 by photosynthetic activity and supply of CO_2 through stomatal diffusion (Farquhar *et al.*, 1982, 1989). This relationship has been shown to fit the equation

$$\Delta = a + (b - a)(c_i/c_a), \tag{1}$$

where a is the isotopic fractionation due to the slower diffusion of $^{13}CO_2$ versus $^{12}CO_2$ in air (4.4‰), and b is the net isotopic fractionation associated with carboxylation activities (~27‰) (Farquhar *et al.*, 1989). As a consequence of an independent relationship between c_i/c_a and water-use efficiency (the ratio of photosynthetic carbon gain (A) to transpirational water loss (E)), Δ is negatively correlated with A/E (Farquhar and Richards, 1984; Farquhar *et al.*, 1989). This relationship between Δ and A/E has been verified under controlled laboratory conditions and in the field (Farquhar and Richards, 1984; Hubick *et al.*, 1986; Farquhar *et al.*, 1989) and has been used extensively to estimate W, long-term A/E in crops and native plants. Because c_i/c_a in C_3 plants is responsive to environmental conditions (e.g., irradiance, soil moisture, soil nutrient status, salinity, gaseous air pollutants), Δ can function as a useful integrator of plant physiological response to environment.

In this study, we have investigated the effects of environment and plant developmental status on carbon isotope discrimination by two species of *Phaseolus*. *Phaseolus vulgaris* L. (common bean) is the most important of the five cultivated species in the genus *Phaseolus*. However, this species is very drought-sensitive, and yields are often reduced by even mild water stress (Haterlein, 1983; Castonguay and Markhart, 1991). Common bean follows the theoretically expected relationship between Δ and c_i/c_a (Seeman and Critchley, 1985; Ehleringer *et al.*, 1991), and there is considerable genetic variability with respect to Δ within this species (Ehleringer, 1990; Ehleringer *et al.*, 1990, 1991; White *et al.*, 1990).

By contrast, *P. acutifolius* A. Gray (tepary bean) is native to the southwestern United States and northern Mexico, is adapted to heat and drought stress, and outyields *P. vulgaris* under hot, dry conditions (Petersen and Davis, 1982; Thomas *et al.*, 1983; Pratt and Nabhan, 1988). Although cultivation of tepary bean is presently limited to subsistence farming in the southwestern United States and northern Mexico, this species may represent an important source of variability for genetic improvement of common bean (Schinkel and Gepts, 1989); furthermore, tepary bean could play a role in the development of low-input, sustainable agricultural systems in regions affected by drought and/or salinity.

In order to better understand the ecophysiological characteristics of these two species, we compared carbon isotope discrimination by *P. vulgaris* and *P. acutifolius* in response to environmental variation, plant developmental status, and their interactions.

II. Methods and Materials

P. vulgaris L. var. "Cahone" and *P. acutifolius* A. Gray var. "Sonora" were grown in the Biology Experimental Garden on the University of Utah campus in Salt Lake City during the summer of 1989. Soil in the experimental garden belonged to the Parleys Series (fine-silty, mixed, mesic Calcic Argixeroll). Seeds of both species were germinated in vermiculite in small pots outdoors. Half of the seeds from each species were germinated approximately 5 weeks earlier (Group 1) than those of the other group (Group 2) in order to have plants at two different stages of development. Seedlings were transplanted into the field at the primary leaf stage, at a row spacing of 100 cm and a distance between plants within rows of 50 cm. All plants were watered daily with a drip irrigation system.

Four days after transferring Group 2 seedlings to the field, all plants from both groups were each fertilized with 5 g of multipurpose N : P : K fertilizer (16 : 16 : 8). At late seedling stage, half of the plants in Groups 1 and 2 were assigned to a high nitrogen (HN) treatment and each plant received an additional 10 g of nitrogen (46 : 0 : 0) fertilizer, while the remainder of the plants were assigned to a low nitrogen (LN) treatment and were not fertilized for the remainder of the study. One week after installation of the nitrogen treatments, water treatments were initiated. Half of the plants in each nitrogen treatment continued to receive daily drip irrigation as before (HW), while the other half of the plants (LW) received no further irrigation for the remainder of the study. There was no rainfall during the study period. Each of the above treatment combinations was replicated twice.

Ten days after initiating the water treatments, leaves were collected for stable carbon isotope analysis from plants of both species in Group 1 (fruiting stage) and Group 2 (vegetative stage). From two plants in each replicate, two terminal leaflets from each of three leaf age classes were sampled randomly from the tops of the plant canopies. The leaf age classes were (1) newly emerged leaves (main vein length less than half the length of the main vein in a fully mature leaf); (2) expanding leaves (main vein length greater than half the length of main vein in fully mature leaf); and (3) mature leaves.

Leaves were dried at 75°C, ground to a fine powder, and combusted to CO_2 using a sealed-tube technique (Boutton, 1991). The CO_2 was isolated and purified cryogenically, and its isotopic composition determined relative to the international PDB standard on a dual-inlet, triple collector gas isotope ratio mass spectrometer (VG Micromass 903; VG Isogas, Middlewich, UK). $\delta^{13}C_{PDB}$ values were determined with an overall precision (machine error plus sample preparation error) of <0.15‰ (±1 SD). Δ values were calculated according to the equation

$$\Delta = \frac{\delta_a - \delta_p}{1 + \delta_p},\tag{2}$$

where δ_a is the $\delta^{13}C_{PDB}$ value of atmospheric CO_2 (−8‰; Mook *et al.*, 1983), and δ_p is the $\delta^{13}C_{PDB}$ value of the plant sample.

Each treatment combination was replicated twice, and each replicate comprised measurements made on two different plants. Data were analyzed by analysis of variance to test for differences due to the main effects of water, nitrogen, species, developmental stage, and leaf age. In addition, all two-, three-, and four-way interactions between treatment effects were evaluated.

III. Results and Discussion

A. Response of Δ to Environmental and Developmental Variation

Mean Δ values for *P. vulgaris* had a range of 3.7‰ (16.8 to 20.5‰) across all treatment combinations (Table I), equivalent to an overall difference in c_i among treatments of approximately 56 μl liter^{-1}. A comparable range of Δ values was found in *P. vulgaris* plants exposed to salinity (Seeman and Critchley, 1985). Mean Δ values of *P. acutifolius* had a range of 2.7‰ (16.3 to 19.0‰) across all treatment combinations (Table I), indicating a differ-

Table I Mean Foliar Δ Values (‰) for Both *Phaseolus vulgaris* and *Phaseolus acutifolius*[a]

	Low water				High water			
	Low nitrogen		High nitrogen		Low nitrogen		High nitrogen	
	Vegetative	Fruiting	Vegetative	Fruiting	Vegetative	Fruiting	Vegetative	Fruiting
Leaf age-class 1								
P. vulgaris	19.3(0.8)	17.5(0.1)	17.7(0.0)	16.8(0.5)	19.9(0.1)	19.3(0.2)	19.6(0.1)	18.8(0.4)
P. acutifolius	17.2(0.1)	16.6(0.4)	16.3(0.3)	16.5(0.1)	18.5(0.3)	17.6(0.5)	18.7(0.0)	18.2(0.5)
Leaf age-class 2								
P. vulgaris	20.1(0.8)	19.0(0.5)	19.2(0.0)	18.0(0.2)	19.7(0.3)	19.3(0.0)	19.8(0.2)	18.9(0.2)
P. acutifolius	17.3(0.2)	16.9(0.3)	17.3(0.3)	16.7(0.1)	18.4(0.9)	18.7(0.3)	18.5(0.2)	17.8(0.8)
Leaf age-class 3								
P. vulgaris	20.4(0.5)	20.5(0.1)	19.1(0.2)	18.9(0.5)	19.9(0.0)	19.9(0.1)	19.7(0.0)	18.7(0.1)
P. acutifolius	18.4(0.2)	18.4(0.5)	17.8(0.1)	18.0(0.3)	19.0(0.0)	18.6(0.9)	18.9(0.0)	18.1(0.8)

[a] Parentheses indicate SEM.

ence in c_i among treatments of approximately 38 μl liter^{-1}. These data demonstrate that both species have a high degree of plasticity in terms of the response of c_i/c_a to soil water and nitrogen status during different stages of plant development.

B. Interspecific Variation in Δ

Leaf Δ values were significantly lower ($P < 0.001$) for *P. acutifolius* (17.8 ± 0.1‰; mean ± SE) than for *P. vulgaris* (19.2 ± 0.1‰) in all experimental treatments (Tables I and II, Fig. 1). The average difference between species across all treatments was 1.3‰, corresponding to a difference in c_i of 22 μl liter^{-1}. These data indicate greater long-term A/E for *P. acutifolius* relative to *P. vulgaris*, suggesting that high A/E may be one mechanism underlying the drought resistance of this species. Since comparisons between these species were made in identical environmental and developmental circumstances, differences in Δ values should be a consequence of genetic variation.

C. Effect of Soil Moisture Availability

Plants from both species grown at low water availability had significantly lower Δ values ($P < 0.01$) than those grown at high water availability (Tables I and II, Fig. 2), indicating higher A/E for the low water plants. Similarly, White *et al.* (1990) found that 10 genotypes of *P. vulgaris* grown under rain-fed conditions had lower Δ values than the same genotypes grown under irrigation. Many additional studies on both native and crop species have documented lower carbon isotope discrimination by plants growing under conditions of low soil water availability (Winter, 1981; Farquhar and Richards, 1984; Hubick and Farquhar, 1987; Ehleringer and Cooper, 1988).

A significant water × species interaction ($P < 0.05$; Table II) indicated that the difference in Δ values between high and low water treatments was greater in *P. acutifolius* than in *P. vulgaris*. Mean Δ values of *P. vulgaris* in

Table II   Results of Analysis of Variance (ANOVA)[a]

Source of variation	Degrees of freedom	Mean square	*F* value	Probability
Nitrogen	1	6.0	17.8	0.05
H$_2$O	1	18.2	90.7	0.01
Species	1	41.7	196.3	0.001
Developmental stage	1	6.9	21.1	0.001
Leaf age	2	7.5	22.9	0.001
H$_2$O × species	1	1.9	8.7	0.05
Nitrogen × species	1	1.9	8.9	0.05
H$_2$O × leaf age	2	4.1	12.4	0.001

[a] Only significant interactions are given.

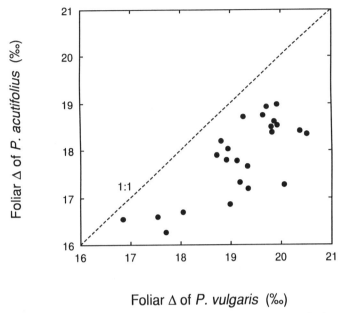

Figure 1. Relationship between foliar Δ values of *Phaseolus vulgaris* and *Phaseolus acutifolius* grown in the field under similar treatments. Values represent the means of each treatment combination. The dashed line represents 1:1.

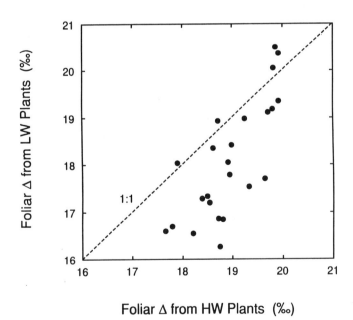

Figure 2. Relationship between foliar Δ values of *Phaseolus vulgaris* and *Phaseolus acutifolius* from high water (HW) and low water (LW) treatments. Values represent the means of each treatment combination. The dashed line represents 1:1.

the high water treatment were 19.5 ± 0.1‰, while those in the low water treatment were 18.9 ± 0.2‰. By contrast, *P. acutifolius* plants in the high water treatment had mean Δ values of 18.4 ± 0.1‰, and those in the low water treatment were 17.3 ± 0.2‰. These data indicate that *P. acutifolius* operates at lower c_i than *P. vulgaris* under both well-watered and water-limited conditions and that *P. acutifolius* can greatly reduce c_i and increase A/E in response to water limitation.

Similarly, Markhart (1985) found that stomatal conductances of *P. acuti-folius* decreased earlier and to a greater extent than those of *P. vulgaris* during the imposition of water stress. While *P. vulgaris* maintained rela-tively high stomatal conductance at leaf-water potentials as low as -1.8 MPa, stomata of *P. acutifolius* were effectively closed at leaf-water potentials of -1.0 MPa (Markhart, 1985). As result, *P. acutifolius* is able to maintain relatively high leaf-water potentials under conditions of soil water limita-tion, thereby postponing dehydration and maintaining cell volume more effectively than *P. vulgaris* (Markhart, 1985; Castonguay and Markhart, 1991).

D. Effect of Soil Nitrogen Status

Leaf Δ values were significantly higher ($P < 0.05$) for plants grown at low soil nitrogen levels (18.8 ± 0.2‰) than for those grown at high soil nitrogen levels (18.3 ± 0.2‰) (Tables I and II, Fig. 3). These results suggest lower c_i

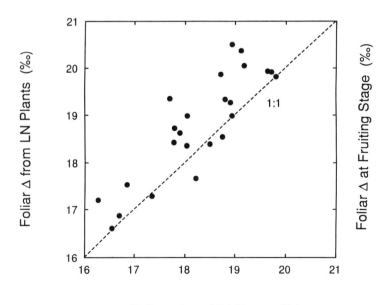

Figure 3.   Relationship between foliar Δ values of *Phaseolus vulgaris* and *Phaseolus acutifo-lius* from high nitrogen (HN) and low nitrogen (LN) treatments. Values represent the means of each treatment combination. The dashed line represents 1:1.

and higher A/E for plants grown at high soil nitrogen availability. A significant nitrogen × species interaction ($P < 0.05$, Table II) revealed that Δ values of *P. acutifolius* were relatively unresponsive to nitrogen treatment; however, *P. vulgaris* plants in the low nitrogen treatment had Δ values (19.6 ± 0.2‰) that averaged 0.8‰ higher than those of plants in the high nitrogen treatment (18.8 ± 0.2‰).

Data evaluating potential relationships between Δ and mineral nutrition are limited. White *et al.* (1990) found no effect of leaf nitrogen concentration on Δ in 10 genotypes of *P. vulgaris* growing under rain-fed and irrigated conditions in two locations in Columbia. Similarly, Hubick (1990) reported that nitrogen treatments had no effect on Δ in genotypes of peanut (*Arachis hypogaea* L.). Results on *P. acutifolius* in the present study conform to this pattern of no effect of nitrogen on Δ.

However, some studies demonstrate that mineral nutrition can in fact influence Δ. For example, Bender and Berge (1979) found that *Phleum pratense* L. plants grown at optimum temperature and fertilized with nitrogen and potassium generally had lower Δ values than unfertilized plants. Fu and Ehleringer (1992) demonstrated that for both container- and field-grown *P. vulgaris*, plants in high fertilizer treatments had significantly lower Δ values than those grown at low fertilizer levels, in agreement with results from the present study. Additional supporting evidence comes from gas exchange studies on *Helianthus annuus* (Fredeen *et al.*, 1991) and *Larrea tridentata* (Lajtha and Whitford, 1989), which both demonstrated higher A/E at higher concentrations of leaf nitrogen.

In the absence of gas exchange data, it is difficult to explain the mechanism by which nitrogen enhancement has resulted in higher A/E in this study. However, there is a well-documented positive correlation between photosynthetic capacity and leaf nitrogen concentration (Field and Mooney, 1986; Evans, 1989). For any given value of stomatal conductance, leaves with high nitrogen concentrations and high photosynthetic capacities should have lower c_i values and higher A/E than leaves with lower nitrogen concentrations and photosynthetic capacities. Therefore, the lower Δ values (higher A/E's) observed in the high nitrogen plants in this study may be a consequence of a relatively high ratio of photosynthesis to leaf conductance.

E. Effect of Plant Developmental Stage

For both species, leaf Δ values were significantly higher ($P < 0.001$; Table II) for plants in the vegetative stage than for those in the fruiting stage (Table I, Fig. 4), suggesting higher A/E during fruiting. This change in A/E during development may be a consequence of paraheliotropic leaf movements, which result in a higher photosynthetic photon flux density (PPFD) incident on leaves during the vegetative stage. Leaves at the tops of the canopies of both species have been shown to orient more obliquely toward direct solar radiation during the fruiting stage than during the vegetative stage (Fu and Ehleringer, 1991). The consequence of this reduction in

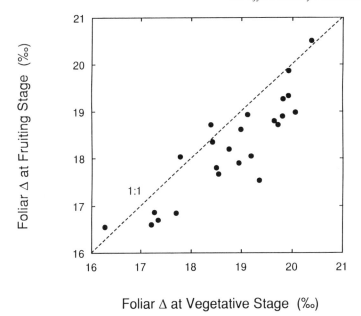

Figure 4.   Relationship between foliar Δ values of *Phaseolus vulgaris* and *Phaseolus acutifolius* from vegetative and fruiting stages of development. Values represent the means of each treatment combination. The dashed line represents 1:1.

PPFD incident on leaves during the fruiting stage may be a decrease in stomatal conductance. Because of the higher resistance to CO_2 diffusion relative to water diffusion in leaves (Nobel, 1991), stomatal closure results in a proportionally greater decrease in transpiration than in CO_2 fixation (Raschke, 1979, Cowan, 1982) and results in higher A/E. Therefore, leaves receiving less PPFD as a result of paraheliotropic leaf movement during the fruiting stage may have higher A/E than leaves during the vegetative stage. Conversely, higher PPFD incident on leaves during the vegetative growth stage should result in higher stomatal conductances and higher c_i, thereby maximizing carbon gain and allowing young plants to grow and establish rapidly.

F. Effect of Leaf Age

Leaf Δ values increased significantly ($P < 0.001$, Table II) as leaf age increased in both species, especially for the low water treatment (Table I, Fig. 5). Within treatments, variation in Δ due to leaf age was substantial. For example, in the low water–low nitrogen treatment, Δ values for *P. vulgaris* ranged from approximately 18.4‰ in the new leaves to 20.4‰ in mature leaves (Fig. 5). The large difference in Δ values between leaves of different ages within treatments emphasizes the importance of sampling leaves of similar developmental status when conducting comparative ecophysiological studies using Δ as an index.

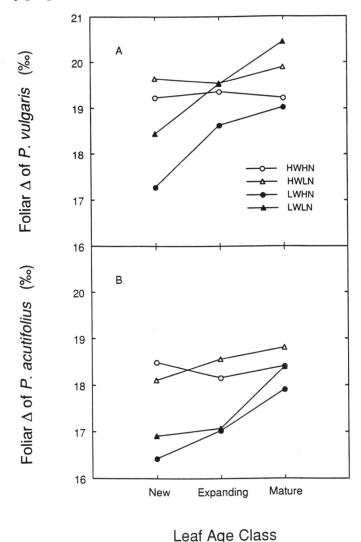

Figure 5. Relationship between foliar Δ values and leaf age-class for *Phaseolus vulgaris* (A) and *Phaseolus acutifolius* (B). LWLN, low water, low nitrogen; LWHN, low water, high nitrogen; HWLN, high water, low nitrogen; HWHN, high water, high nitrogen.

Comparable data have not been published for any crop species. However, Ducatti *et al.* (1991) documented that older leaves from several Amazonian tree species had Δ values that were approximately 0.7‰ higher than those of younger leaves. Similarly, we found that 1-year-old live oak (*Quercus virginiana*) leaves differed from those that were only 2–3 months old by approximately 3‰ (Boutton *et al.*, unpublished data). These changes in Δ may reflect changes in stomatal function and/or photosynthetic capacity associated with leaf aging. Alternatively, environmental conditions prevailing during growth and development of "old" leaves may differ significantly from those prevailing during growth and development of "new" leaves, resulting in different c_i/c_a and Δ values.

Because the effect of leaf age on Δ was more pronounced in plants grown at low water availability relative to those grown at high water availability (Fig. 5), the water × leaf-age interaction was significant ($P < 0.001$, Table II). Leaves of different ages from plants in the well-watered treatment would have all experienced optimum soil water status during growth and development. By contrast, in the low water treatment, older leaves developed prior to or soon after the cessation of irrigation, while new leaves developed later during a period of reduced water availability. Thus, lower Δ values in the younger leaves may reflect an increase in A/E in response to a decrease in soil moisture status as the growing season progressed. The near constancy of Δ values with respect to leaf age in the well-watered treatments supports an environmental rather than developmental explanation for the decrease in Δ in newer leaves in the low water treatments.

IV. Summary

P. vulgaris L. and *P. acutifolius* Gray were grown in the field at two levels of soil moisture and two levels of soil nitrogen to evaluate environmental and developmental effects on carbon isotope discrimination (Δ) and water-use efficiency (A/E). For both species, Δ was measured on three age-classes of leaves sampled from each of two developmental stages (vegetative and fruiting) growing simultaneously. Leaf Δ values were significantly lower for *P. acutifolius* than for *P. vulgaris* in all experimental treatment combinations. This suggests greater long-term A/E in *P. acutifolius*, which is adapted to warmer, drier environments. Since comparisons between these species were made under identical environmental and developmental circumstances, differences in Δ values should be a consequence of genetic variation. Plants from both species grown at low water availability had significantly lower Δ values than those grown at high water availability. A significant water × species interaction occurred because the difference in Δ values between high and low water treatments was greater in *P. acutifolius* than in *P. vulgaris*. Leaves from plants of both species grown at low soil nitrogen levels had significantly higher Δ values than those grown at high soil nitrogen levels, suggesting that plants in high nitrogen treatments had higher A/E. This result agrees with the positive correlation between leaf nitrogen content and A/E demonstrated previously by others. A significant nitrogen × species interaction revealed that *P. vulgaris* was more responsive to soil nitrogen status than *P. acutifolius*. For both species, leaf Δ values were significantly higher from plants in the vegetative stage than from those in the fruiting stage, suggesting higher A/E during fruiting. This change in A/E during development may be a consequence of stronger paraheliotropic leaf movements, which result in higher photon flux density incident on leaves during the vegetative stage. Leaf Δ values increased significantly as leaf age increased in both species. This effect was more pronounced in plants grown at low water availability compared to those grown at high water availability, indicating that it was probably due to the changing environment and not to effects of leaf development.

Acknowledgments

This research was supported by a grant from the Texas Advanced Technology Research Program (TATRP 11-078, Earth Sciences) and by the Texas Agricultural Experiment Station (Project H-6945). Anthony E. Hall provided a helpful review of the manuscript. Omer C. Jenkins (Department of Statistics, Texas A&M University) provided assistance with statistical analyses. We thank Xing Wang, Irene Perry, and Daniel Watts for their expert technical assistance.

References

Bender, M. M., and A. J. Berge. 1979. Influence of N and K fertilization and growth temperature on $^{13}C/^{12}C$ ratios of timothy (*Phleum pratense* L.). *Oecologia* **44:** 117–118.

Boutton, T. W. 1991. Stable carbon isotope ratios of natural materials. I. Sample preparation and mass spectrometric analysis, pp. 155–171. *In* D. C. Coleman and B. Fry (eds.), Carbon Isotope Techniques. Academic Press, New York.

Castonguay, Y., and A. H. Markhart III. 1991. Saturated rates of photosynthesis in water-stressed leaves of common bean and tepary bean. *Crop Sci.* **31:** 1605–1611.

Cowan, I. R. 1982. Regulation of water use in relation to carbon gain in higher plants, pp. 589–613. *In* O. L. Lange, P. S. Nobel, C. B. Osmond, and H. Ziegler (eds.), Physiological Plant Ecology. IIB. Water Relations and Carbon Assimilation. Springer-Verlag, New York.

Ducatti, C., E. Salati, and D. Martins. 1991. Measurement of the natural variation of $^{13}C:^{12}C$ ratio in leaves at Reserva Ducke Forest, central Amazonia. *For. Ecol. Manage.* **38:** 201–210.

Ehleringer, J. R. 1990. Correlations between carbon isotope discrimination and leaf conductance to water vapor in common beans. *Plant Physiol.* **93:** 1422–1425.

Ehleringer, J. R., and T. A. Cooper. 1988. Correlations between carbon isotope ratio and microhabitat in desert plants. *Oecologia* **76:** 562–566.

Ehleringer, J. R., J. W. White, D. A. Johnson, and M. Brick. 1990. Carbon isotope discrimination, photosynthetic gas exchange, and transpiration efficiency in beans and range grasses. *Acta Oecol.* **11:** 611–625.

Ehleringer, J. R., S. Klassen, C. Clayton, D. Sherrill, M. Fuller-Holbrook, Q. Fu, and T. A. Cooper. 1991. Carbon isotope discrimination and transpiration efficiency in common bean. *Crop Sci.* **31:** 1611–1615.

Evans, J. R. 1989. Photosynthesis and nitrogen relationships in leaves of C_3 plants. *Oecologia* **78:** 9–19.

Farquhar, G. D., and R. A. Richards. 1984. Isotopic composition of plant carbon correlates with water-use efficiency of wheat genotypes. *Aust. J. Plant Physiol.* **11:** 539–552.

Farquhar, G. D., M. H. O'Leary, and J. A. Berry. 1982. On the relationship between carbon isotope discrimination and intercellular carbon dioxide concentration in leaves. *Aust. J. Plant Physiol.* **9:** 121–137.

Farquhar, G. D., J. R. Ehleringer, and K. T. Hubick. 1989. Carbon isotope discrimination and photosynthesis. *Annu. Rev. Plant Physiol. Plant Mol. Biol.* **40:** 503–537.

Field, C. B., and H. A. Mooney. 1986. The photosynthesis-nitrogen relationship in wild plants. pp. 25–55. *In* T. J. Givnish (ed.), On the Economy of Plant Form and Function. Cambridge Univ. Press, Cambridge.

Fredeen, A. L., J. A. Gamon, and C. B. Field. 1991. Responses of photosynthesis and carbohydrate-partitioning to limitations in nitrogen and water availability in field-grown sunflower. *Plant Cell Environ.* **14:** 963–970.

Fu, Q. A., and J. R. Ehleringer. 1991. Regulation of paraheliotropic leaf movements in *Phaseolus* species during canopy development. *Am. J. Bot. (Suppl.)* **78:** 130–131.

Fu, Q. A., and J. R. Ehleringer. 1992. Paraheliotropic leaf movements in common bean under different soil nutrient levels. *Crop Sci.* **32:** 1192–1196.

Haterlein, A. J. 1983. Bean, pp. 157–185. *In* I. D. Teare and M. M. Peet (eds.), Crop Water Relations. Wiley, New York.

Hubick, K. T. 1990. Effects of nitrogen source and water limitation on growth, transpiration efficiency, and carbon isotope discrimination in peanut cultivars. *Aust. J. Plant Physiol.* **17:** 413–430.

Hubick, K. T., and G. D. Farquhar. 1987. Carbon isotope discrimination—Selecting for water-use efficiency. *Aust. Cotton Grower* **8:** 66–68.

Hubick, K. T., G. D. Farquhar, and R. Shorter. 1986. Correlation between water-use efficiency and carbon isotope discrimination in diverse peanut (*Arachis*) germplasm. *Aust. J. Plant Physiol.* **13:** 803–816.

Lajtha, K., and W. G. Whitford. 1989. The effect of water and nitrogen amendments on photosynthesis, leaf demography, and resource-use efficiency in *Larrea tridentata*, a desert evergreen shrub. *Oecologia* **80:** 341–348.

Markhart, A. H., III. 1985. Comparative water relations of *Phaseolus vulgaris* L. and *Phaseolus acutifolius* Gray. *Plant Physiol.* **77:** 113–117.

Mook, W. G., M. Koopmans, A. F. Carter, and C. D. Keeling. 1983. Seasonal, latitudinal, and secular variations in the abundance and isotopic ratios of atmospheric carbon dioxide. I. Results from land stations. *J. Geophys. Res.* **88:** 10915–10933.

Nobel, P. S. 1991. Physicochemical and Environmental Plant Physiology. Academic Press, New York.

Petersen, A. C., and D. W. Davis. 1982. Yield response of *Phaseolus vulgaris* L. and *Phaseolus acutifolius* A. Gray. *Annu. Rep. Bean Improv. Coop* **25:** 53–54.

Pratt, R. C., and G. P. Nabhan. 1988. Evolution and diversity of *Phaseolus acutifolius* Gray genetic resources, pp. 409–440. *In* P. Gepts (ed.), Genetic Resources of *Phaseolus* Beans. Kluwer Academic, Dordrecht, Netherlands.

Raschke, K. 1979. Movements of stomata, pp. 383–441. *In* W. Haupt and M. E. Feinleib (eds.), Physiology of Movements. Encyclopedia of Plant Physiology, new series, Vol. VII. Springer-Verlag, New York.

Schinkel, C., and P. Gepts. 1989. Allozyme variability in the tepary bean, *Phaseolus acutifolius* A. Gray. *Plant Breed.* **102:** 182–195.

Seeman, J. R., and C. Critchley. 1985. Effects of salt stress on the growth, ion content, stomatal behavior, and photosynthetic capacity of a salt-sensitive species, *Phaseolus vulgaris* L. *Planta* **164:** 151–162.

Thomas, C. V., R. M. Manshardt, and J. G. Waines. 1983. Teparies as a source of useful traits for improving common beans. *Desert Plants* **5:** 43–48.

White, J. W., J. A. Castillo, and J. Ehleringer. 1990. Associations between productivity, root growth, and carbon isotope discrimination in *Phaseolus vulgaris* under water deficit. *Aust. J. Plant Physiol.* **17:** 189–198.

Winter, K. 1981. CO_2 and water vapour exchange, malate content, and $\delta^{13}C$ value in *Cicer arietinum* grown under two water regimes. *Z. Pflanzenphysiol.* **101:** 421–430.

21

Diversity in the Relationship between Carbon Isotope Discrimination and Transpiration Efficiency when Water Is Limited

Kerry T. Hubick and Ann Gibson

I. Introduction

Plants have adapted to many different terrestrial environments. This usually means that they have evolved systems to acquire water for growth and reproduction despite the fickleness of nature in providing an adequate supply. Maximizing transpiration efficiency (W), that is maximizing dry matter gained (DM) for water lost is one means of making the best use of available water.

$$W = \text{DM/water.} \qquad (1)$$

However, species evolve in environments which vary in many parameters besides the water regime. Adaptations which maximize W do not develop independently of adaptations associated with constraints placed on the plant by light; temperature; day length; soil structure, volume and fertility; vapor pressure deficit; disease; herbivory; and competition for resources. Thus the extent to which increasing W will be a successful strategy depends on the other factors affecting evolution.

The relationship between carbon isotope discrimination (Δ) and W is useful because Δ is easier to measure than W, but the interpretation of variation in Δ requires critical analysis. We discuss, below, some of the important considerations in the interpretation of Δ when water availability varies for a number of herbaceous and woody species. We also consider the significance of Δ in plant-breeding programs aimed at increasing W and the significance of Δ in increasing understanding of the ecology of wild plants in water-limited environments.

II. Factors Affecting W and Δ

A. The Relationship between W and Δ

Transpiration efficiency depends on the gas exchange characteristics of leaves. The DM and the water lost are affected by the physical and biological environments, both of which vary temporally and spatially. Transpiration efficiency is related to Δ because of the independent linkages of W and Δ to the gas exchange characteristics of leaves. The relationship between W and Δ is

$$W = \frac{p_a(1 - \phi_c)(b - d - \Delta)}{1.6v(b - a)(1 + \phi_w)}. \tag{2}$$

The ambient partial pressure of carbon dioxide is p_a; the carbon lost by the nonphotosynthetic organs in the light and by the whole plant at night is ϕ_c; b is the inherent discrimination resulting from carboxylation; d is the inherent discrimination due to dissolution and diffusion of carbon dioxide to the site of carboxylation; a is the inherent discrimination that occurs during diffusion into the substomatal space; and ϕ_w is the loss of water not associated with uptake of carbon dioxide. The value of v, the vapor pressure difference, is often the greatest variable affecting W (Sinclair *et al.*, 1983) and can vary significantly over a season.

In the long term, loss of carbon, either by respiration or shedding of dry matter, can affect W. Information regarding the magnitude of respiratory discrimination is rare (Farquhar *et al.*, 1989).

B. Interpretation of a Relationship between W and Δ in Peanut Cultivars

1. Robustness of the Relationship   Numerous examples of a linear negative association between W and Δ have been observed in many species and cultivars within species (Farquhar *et al.*, 1989). An example of the negative correlation between W and Δ was observed in four peanut cultivars subjected to four different treatments (Hubick, 1990; Fig. 1). While the relationship was robust overall there was residual variation which was attributed to variation in W resulting from carbon losses in respiration associated with nodulation. The variation could also have been in other factors such as the value of d, but this meant the variation in d would have to be consistent over treatments. Either respiratory losses or values of d would affect W and Δ independently. In the former, W would be affected through variation in DM, whereas in the latter Δ would be affected. Further sources of residual variation may be from measurement errors and from significant differences in genotype × environment interactions for the variables in Eq. (2). Examples of these could be variation in leaf temperature causing differ-

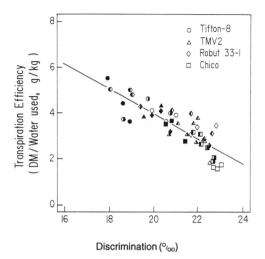

Figure 1.   Relationship between whole-plant transpiration efficiency, *W*, and carbon isotope discrimination, Δ, in leaves of four cultivars of peanut. Cultivars were either watered well (open right side of symbol) or grown on stored water (closed right side of symbols). For each water treatment half the plants were given mineral nitrogen (closed left side of symbols) and half were nodulated (open left side of symbols). (From Hubick, 1990.)

ences in the value of v or variation in cuticular resistance causing variation in ϕ_w.

2. Effect of Nutrition   Nutrition, particularly nitrogen, can affect the biochemical machinery for photosynthetic CO_2 assimilation. If CO_2 uptake is affected more than leaf conductance then *W* should reflect a nutrition effect.

Briggs and Shantz showed that fertility affected *W* in their classical work on the water requirement (inverse of *W*) of plants in 1914. Others have shown that Δ is affected by soil fertility as reviewed by Farquhar *et al.* (1989). Variation due to the source of nitrogen for peanut cultivars is shown in Fig. 1. The majority of the well-watered, nodulated plants (open symbols) in each cultivar have a greater Δ value than water-limited plants supplied with mineral nitrogen (closed symbols). Over all watering treatments nodulated plants had a value of Δ about 1‰ greater than that of non-nodulated plants. Depending on the value of v, this could give a decrease in *W* of up to 20%.

No relationship is expected between Δ and carbon allocation. The differences in the harvest index (seed wt/total shoot wt) of the plants with a high pod yield, i.e., nodulated plants, exceeded that of plants given mineral nitrogen. This contributed to the lack of relationship between Δ and yield noted for peanut (Hubick *et al.*, 1988; Wright *et al.*, 1988; Hubick, 1990). Thus there may be at least two effects of nutrition on *W*. In the first case, reduced soil fertility, particularly *N*, may affect the photosynthetic mechanism differently from leaf conductance and reduce *W* while increasing Δ

(Bender and Berge, 1979; Hubick, 1990). In the second case, fertility may affect W via effects on carbon partitioning. In the latter, Δ may not be affected or may change in an unrelated manner (Hubick, 1990).

There may also be interactions between water availability and nutrition on W and Δ. In the experiment with peanut, water limitation decreased dry matter production overall but the loss occurred more in leaf production than in pod production. These changes meant that yield of dry matter was correlated well with W but not with pod yield. Thus, although Δ would have allowed estimates of differences in W it would have given no information about yield, which is usually the variable of most agronomic importance.

C. Carbon Isotope Discrimination and the Interpretation of the Relationship between W and Δ

1. Differences between Plant Parts   The variation of Δ of plant DM results from fractionation in chemical and physical processes (O'Leary *et al.*, 1981). Differences in chemical composition among plant parts can be large. In an experiment with cultivars of barley, cotton, peanut, and cowpea under water stress the differences in Δ between stems and leaves were as great as 7‰ (Hubick, unpublished; Fig. 2). Differences in well-watered peanut plants were 3‰. In all cases stems were more enriched in ^{13}C than leaves and this effect increased with water limitation. These differences may reflect storage of assimilate in the stems during water stress. Sugars and starch may be enriched in ^{13}C relative to bulk tissue (Brugnoli *et al.*, 1988).

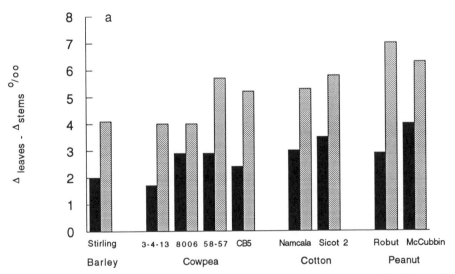

Figure 2.   Variation in carbon isotope discrimination, Δ, in plant parts of cultivars of barley, cotton, peanut, and cowpea. δ is the difference in Δ between leaves and stems (a), between leaves and roots (b), and between leaves and reproductive structures (c). Closed bars, well-watered plants; open bars, water-limited plants. (*Figure continues.*)

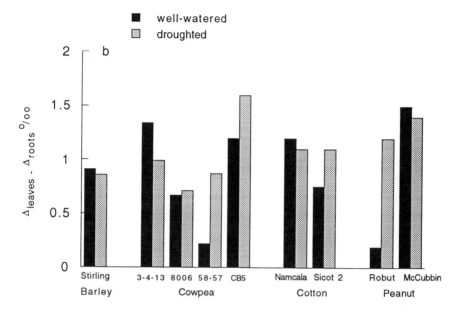

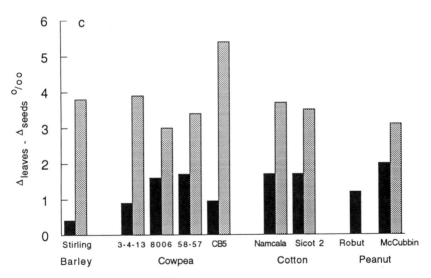

Figure 2. *(Continued)*

Similarly, seeds were always enriched in ^{13}C relative to leaves and this difference was enhanced by water limitation. In this case, the seeds developed when water stress was most severe. It would be expected that the Δ of the later developing organs would be less than the leaves which developed before the water stress was imposed. These differences indicate both a developmental change in Δ as stress is imposed and perhaps a physiological change as different compounds are produced.

It is likely that inferences about W from Δ in a well-planned and -analyzed field trial designed to compare genotypes will need to take account of a total variation of about 1‰ in the mean Δ of cultivars of some species. Leaf Δ is usually measured in such trials so confusion about differences in Δ arising from variation in chemical composition may not be important. However, it should be noted.

Little is known about fractionations when carbon is lost by respiration from whole plants. Some of the above effects may have been due to respiration effects.

2. Differences over Time   The W can vary as the plant completes its life cycle even in environments where the conditions around the plants are fairly uniform. This genetically induced variation may be due to changes in respiration during reproductive phases affecting DM accumulation and thus affecting W. Other variations also occur, such as the general decrease of W with age that has been observed in many pot experiments in glasshouses at Canberra.

Temporal variation in leaf Δ can be in the order of 1–2‰ with young leaves on a plant tending to have larger values of Δ than older leaves. This change is enhanced by water limitation during leaf development. These effects occur in addition to those induced by stress noted above. Similar variation in Δ between leaves and heads of well-watered barley was noted by Hubick and Farquhar (1989).

III. The Use of Δ to Predict W in Plant Breeding

If Δ is to be used as selection criterion for modification of plant characters to improve performance in a defined environment, W must be heritable

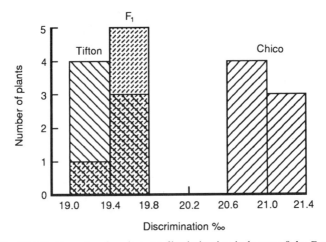

Figure 3.   Distribution of carbon isotope discrimination in leaves of the F_1 progeny of a cross between two peanut cultivars, Tifton-8 (large W, small Δ) and Chico (small W, large Δ).

and genetically correlated with Δ. The utility of Δ will depend on the robustness of the correlation between Δ of a selected plant part and whole-plant W or crop W.

Differences in Δ may arise through differences in photosynthetic capacity, leaf conductance, or both. Either a large photosynthetic capacity or a small leaf conductance would give rise to a small Δ and, hopefully, the W is negatively correlated with it. Unfortunately for plant breeders, all species do not show similar inheritance for Δ. Figures 3 and 4 show how the inheritance of Δ can differ.

In peanut (Fig. 3) Δ in the F_1 progeny indicates that inheritance of Δ tends to be dominant for small Δ, as in the Tifton parent (Hubick *et al.*,

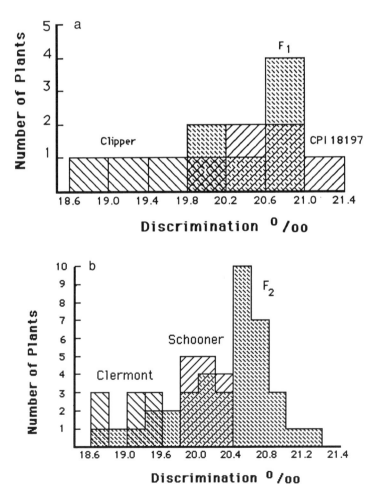

Figure 4. (a) Distribution of carbon isotope discrimination in leaves of the F_1 progeny of a cross between two barley cultivars, Clipper (large W, small Δ) and CPI 18197 (small W, large Δ). (b) Distribution of carbon isotope discrimination in leaves of the F_2 progeny of a cross between two barley cultivars, Clermont and Schooner.

1988). The same pattern has been observed in other peanut crosses (Cruickshank and Wright, unpublished data). However in barley the inheritance of Δ may be reversed compared to that of peanut as shown in Fig. 4a where the F_1 approach the large Δ parent, CP1 18197 (Hubick, Sparrow, and Farquhar, unpublished). The F_2 progeny of another barley cross showed a distribution of Δ skewed toward the large Δ parent (Fig. 4b). Also, many progeny had Δ's greater than that of the large Δ parent. Martin and Thorstenson (1988) found that the F_1 of a cross between two species of tomato was intermediate between the values of the parents. Apparently the inheritance of Δ varies among different species. This inheritance must be understood if Δ is to be successfully used as a selection criterion in breeding.

IV. Relationships between *W* and Δ in Diverse C_3 Species

Experiments with a variety of species indicate that further knowledge is needed before Δ can be used reliably as a predictor of *W*, especially in situations of water limitation.

A. *Eucalyptus camaldulensis*

Eucalyptus camaldulensis occurs on the banks and flood plains along all of the permanent and intermittent water courses of inland Australia. It is also one of the most widely used species in plantations globally. However, little is known of the physiological bases for its adaptation to a wide range of water regimes or for the success of trees from certain seed sources (provenances) in plantations where rainfall is limited.

It may be that differences in *W* contribute to differences in productivity among the provenances. Experiments using seedlings from three northern provenances have provided interesting results for the broader interpretation of the potential for using Δ as an indication of characters that are useful ecologically and in production forestry.

The provenances used were Tennant Creek, Northern Territory, in the semiarid grassland of Central Australia (rainfall <400 mm per annum and unreliable), Katherine, Northern Territory, in the dry tropics (reliable summer monsoon >1000 mm), and Petford, Queensland, in the humid tropics (reliable summer rainfall 700 mm) where early morning dew and fogs lessen the severity of the winter dry season.

Whole-plant *W* varied consistently among pot-grown seedlings of the three provenances (Hubick and Gibson, unpublished data; Fig. 5). In general Petford seedlings had the greatest values of *W* and Tennant Creek seedlings the least, while Katherine seedlings were intermediate. The differences in *W* were maintained whether plants were well watered or water limited and when temperatures were maintained at tropical levels, such as 30/20°C day/night, or more temperate conditions (18/12°C), although the dry matter gain was significantly greater in the warmer temperatures de-

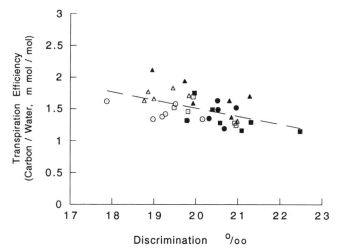

Figure 5. Plot of whole-plant transpiration efficiency against carbon isotope discrimination of leaves of well-watered (closed symbols) and water-limited (open symbols) plants of three provenances of *Eucalyptus camaldulensis*. Petford provenance (△), Katherine provenance (○), and Tennant Creek provenance (□).

spite similar light availability and growth period. The consistency of W over this wide range of conditions suggests that there is a genetic basis for the differences between provenances.

Differences in Δ between the three provenances were also significant and showed the negative correlation with W expected from theory (Fig. 5). The indications are that Δ will be useful in predicting the provenances with the greatest W in this species. It is particularly interesting that the Petford seedlings, which had the greatest W, also grew the most. Trees from this provenance generally perform best in plantations in the tropics overseas and screening seedlings for Δ may enable provenances to be selected for specific sites, obviating the need for long-term trials.

Paradoxically, Petford seedlings which had the highest W and lowest Δ come from the wettest environment while Tennant Creek seedlings with the lowest W and highest Δ come from the driest environment. The significance of W and Δ in these provenances can be interpreted by considering the range of morphological and anatomical differences between them in addition to their physiological differences and relating these plant differences to differences in the environments where they grow.

Petford seedlings have a wide range in p_i/p_a (Fig. 6; from Gibson *et al.*, 1991). The range is greater than in Katherine or Tennant Creek seedlings (Gibson *et al.*, 1991) and the higher overall growth rates and lower Δ of Petford seedlings may reflect a superior ability to utilize water when it is available and to reduce water loss in dry conditions. This strategy is appropriate in its environment where moist early mornings give way to dry days throughout the winter dry season. At Tennant Creek the severity of

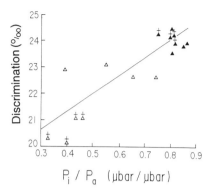

Figure 6. Relationship between carbon isotope discrimination and the ratio of internal partial pressure of carbon dioxide to ambient carbon dioxide, p_i/p_a, in leaves of the Petford provenance of *Eucalyptus camaldulensis* on well-watered (closed symbols) and water-limited (open symbols) plants with (+ above symbol) and without applied abscisic acid. (From Gibson *et al.*, 1991.)

drought due to low and unreliable rainfall and high vpd is increased by high wind speeds. The seedlings appear to have adopted an opportunistic strategy with high Δ and low W when water is available, as in the glasshouse experiments, accompanied by the production of thick-leaved, sclerophyllous plants that will withstand the rigorous dry periods and yet achieve high assimilation rates per unit leaf area when favorable conditions occur.

B. *Banksia* spp

Banksia is an Australian genus of shrubs and small trees that occurs on a wide range of soils including those that are so infertile, wet, or dry, that other woody species do not grow. A survey of carbon isotope discrimination in leaves from three species of *Banksia* growing at sites with similar climatic conditions but soils that had different water contents indicated that *B. robur*, growing in wet and boggy soils, had the smallest Δ, 21.1‰ (Table I). *Banksia marginata*, in the driest soils, had a Δ of 22.5‰ and *B. paludosa*, on soils with intermediate moisture content, had a Δ of 21.9‰. The soil water content at the *B. robur* site was eight times greater than that at the *B. marginata* site (Table I).

When cuttings from the three species were grown under well-watered conditions in the glasshouse whole-plant W of *B. robur* from the wet site tended to be greater than that of the other species (Table I). Whole-plant W increased in water-limited cuttings of all species and *B. robur* again had the greatest W. This result is consistent with the low Δ found in leaf material from the field. Thus the *Banksia* from the wettest site had the lowest Δ and largest W, the unexpected result that was noted above for *E. camaldulensis* and in Australian populations of *Nothofagus* growing in a range of environments (Read and Farquhar, 1991).

Thus Δ alone is not useful in interpreting the relationship of wild Austra-

Table I   Transpiration Efficiency and Carbon Isotope Discrimination
of Three *Banksia* Species[a]

Species	Field location		Glasshouse experiment whole plant W (g DM/kg water used)	
	Soil water content (g/g)	Leaf Δ ‰	Well watered	Water limited
B. marginata	0.04	22.5	2.72	3.14
B. paludosa	0.06	21.9	2.76	3.00
B. robur	0.31	21.1	3.02	3.34

[a]Carbon isotope discrimination determined at three field locations with different soil moisture contents and transpiration efficiency, W, of cuttings from these plants grown in well-watered and water-limited glasshouse conditions. Values of W are means of four pots with one cutting per pot. (Unpublished data from Frank Fox.)

lian species to their environments and factors other than water status must be more important for their adaptation.

C. Herbaceous Crop Species

A full understanding of the importance of W and Δ in both the ecological and agricultural contexts requires experiments in which competition for water in herbaceous species is evaluated, as well as effects of physical limitations such as soil compaction, volume, and fertility.

In many cases vpd is usually the largest determinant of W and it may be possible to scale W by vpd by the calculation of k in some cases (Hubick and Farquhar, 1989).

An experiment in which the effects of water limitation on W and Δ was investigated in four species under the same vpd conditions showed that all species do not respond in the same way. Cultivars of peanut, barley, cotton, and cowpea with previously determined differences in W and Δ were selected for the experiment and most had demonstrated a significant negative correlation between W and Δ.

Plants were grown in pots of two sizes, either 7 or 42 liters and either well watered or allowed only the water stored in the soil at field capacity. All the plants grown on stored soil water were in large pots and half of the well-watered plants, evenly distributed over all cultivars, were in small pots. Pot size affected the total growth of the well-watered plants. Those in large pots grew more than, and the Δ was different from, those in small pots even though they were given similar amounts of nutrients.

In the same environment well-watered barley plants were significantly larger than cowpeas, cotton, and peanuts, which did not differ in dry matter production. Growing plants on stored water reduced dry matter production by about 70% and barley again accumulated the most dry matter. This occurred despite a much greater reduction in leaf relative water content (RWC) in barley than in the other species. By 68 days from sowing

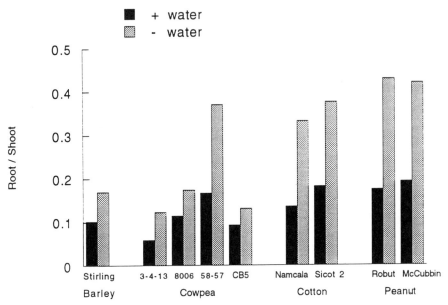

Figure 7. Root : shoot ratio of cultivars of four different species in two watering treatments, well watered (open bars) and water limited (closed bars).

the RWC of water-limited barley was only 0.54 while the RWC other species was little different from that of the well-watered plants. By 94 days from sowing, cotton and peanut plants showed a reduction in RWC but the reduction in cowpea was minimal.

Water limitation caused a change in carbon allocation from shoots to roots. Root : shoot ratio was always greater in water-limited plants. This was particularly clear in cotton and peanut cultivars (Fig. 7). Such a response has been noted in previous pot experiments with herbaceous species (Hubick *et al.*, 1986; Hubick and Farquhar, 1989; Hubick, 1990) and also tends to occur in tree species, including *E. camaldulensis*. This is consistent with an increase in carbon allocation to water-harvesting organs in response to

Table II Variation in Transpiration Efficiency, *W* (g DM/kg Water Used) at Maturity and Carbon Isotope Discrimination of the Lint in Four Cultivars of Cotton Grown under Well-Watered and Water-Limited Conditions

Cultivar	Whole plant *W* (g DM/kg water used)		Discrimination in lint (‰)	
	Well watered	Water limited	Well watered	Water limited
PSP 251 671-61	2.37	2.26	18.5	17.4
Coker 383	2.23	2.41	19.0	16.6
Sicot 2	2.34	2.24	19.2	18.6
Namcala	2.74	2.41	18.4	16.8

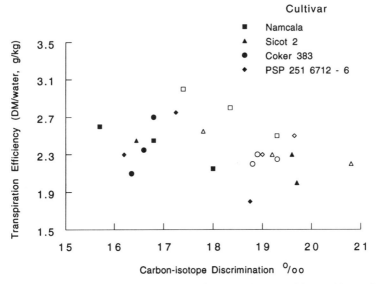

Figure 8. Plot of whole-plant transpiration efficiency of plants of four cultivars of cotton against carbon isotope discrimination in the leaves. Plants were grown to maturity and either watered well (open symbols) or water limited (closed symbols) by being given approximately half of the water of the well-watered controls.

water limitation. Where genetic variation occurs in root:shoot ratios it can be important in the interpretation of the relationship between W and Δ in the field where it is not always feasible to harvest roots (Wright *et al.*, 1991).

In another experiment with four cotton cultivars grown in 42-liter pots to maturity, water limitation caused a reduction in Δ (Table II). The effect on W was complex. The two variables were negatively correlated as expected, but the relationship was different for the two treatments (Hubick and Farquhar, unpublished data; Fig. 8) much as in the case of the banksias. For well-watered plants the regression was $W = 6.53 - 0.22 \, (\Delta)$, $r^2 = 0.50$, and for water-limited plants it was $W = 4.44 - 0.12 \, (\Delta)$, $r^2 = 0.33$. The regression constant was less in the water-limited plants which is consistent with an increase in ϕ_c (Eq. (2)). However, the water-limited plants had a greater slope, not lesser, which is not easily explained in terms of Eq. (2) and may be due to the samples for isotope analysis having been taken from the lint which was produced when water stress was most severe.

V. Conclusion

Transpiration efficiency and Δ appear to be negatively correlated for C_3 species that have diverse forms and natural distributions. This correlation holds promise for breeders who wish to exploit the relationship. However, results from an increasing range of species indicate that the relationship can be complex. The breeder will need to understand the nature of the relationship between W and Δ for the species and target environment and its overall significance for growth in addition to the inheritance of the particular traits that are to be modified.

In addition to the information concerning leaf and whole-plant W that may be contained in measurements of Δ, insight into temporal and spatial variation in carbon uptake and water loss may be gained through measurements of Δ in plant parts over time in water-limited plants and comparing them with the variations in Δ that occur among plant parts in well-watered plants. These differences may be useful for interpreting ontogenetic changes in physiology and carbon translocation and distribution (Farquhar *et al.*, 1989). Knowledge of these changes may be useful in understanding the effect of water limitation on physiological changes other than transpiration efficiency.

Transpiration efficiency is only one of many traits which together may confer an ability to survive in water-limited natural environments. Other characters may be more important than increased W. This seems to be the case for the *E. camaldulensis* provenances and *Banksia* spp which inhabit the driest environments in the species' ranges. Measuring Δ may, however, be a useful guide as to the relative importance of physiological and other adaptations for a particular species in its environment.

Acknowledgments

We thank Frank Fox for permission to use the data from his *Banksia* experiments.

References

Bender, M. N., and A. J. Berge. 1979. Influences of N and K fertilization and growth temperature on $^{13}C/^{12}C$ ratios of timothy (*Phleum pratense* L.). *Oecologia* **44**: 117–118.

Briggs, L. J., and H. L. Shantz. 1914. Relative water requirement of plants. *J. Agric. Res.* **3**: 1–64.

Brugnoli, E., K. T. Hubick, S. von Caemmerer, and S. C. Wong. 1988. Correlation between carbon isotope discrimination in leaf starch and sugars of C3 plants and the ratio of intercellular and atmospheric partial pressures of CO_2. *Plant Physiol.* **88**: 1418–1424.

Farquhar, G. D., J. R. Ehleringer, and K. T. Hubick. 1989. Carbon isotope discrimination and photosynthesis. *Annu. Rev. Plant Physiol. Plant Mol. Biol* **40**: 503–537.

Gibson, A., K. T. Hubick, and E. P. Bachelard. 1991. The effects of water stress on the morphology and gas-exchange characteristics of *Eucalyptus camaldulensis* seedlings. *Aust. J. Plant Physiol.* **18**: 153–163.

Hubick, K. T. 1990. Effects of nitrogen and water availability on growth, transpiration efficiency and carbon isotope discrimination in peanut cultivars. *Aust. J. Plant Physiol.* **17**: 413–430.

Hubick, K. T., and G. D. Farquhar. 1989. Carbon isotope discrimination and the ratio of carbon gained to water lost in cultivars of barley. *Plant Cell Environ.* **12**: 795–804.

Hubick, K. T., G. D. Farquhar, and R. Shorter. 1986. Correlation between water use efficiency and carbon isotope discrimination in diverse peanut (*Arachis*) germplasm. *Aust. J. Plant Physiol.* **13:** 803–816.

Hubick, K. T., R. Shorter, and G. D. Farquhar. 1988. Heritability and genotype × environment interactions in carbon isotope discrimination and transpiration efficiency of peanut (*Arachis hypogea* L.). *Aust. J. Plant Physiol.* **15:** 799–813.

Martin, B., and Y. R. Thorstenson. 1988. Stable carbon isotope composition (δ ^{13}C), water use efficiency and biomass productivity of *Lycopersicon esculentum, Lycopersicum pennellii* and the F_1 hybrid. *Plant Physiol.* **88:** 213–217.

Mathews, R. B., D. Harris, R. C. Nageswara Rao, J. H. Williams, and K. D. R. Wadia. 1988. The physiological basis for yield differences between four genotypes of ground nut (*Arachis hypogea* L.) in response to drought. I. Dry matter production and water use. *Exp. Agric.* **24:** 191–202.

O'Leary, M. H., C. B. Osmond, and J. A. M. Holtum. 1981. Carbon isotope fractionation in CAM plants. *In* Photosynthesis. IV. George Akoyunoglou (ed.), Regulation of Carbon Metabolism. Balaban International Science Series, Philadelphia.

Ong, C. K. 1984. The influence of temperature and water deficits on the partitioning of dry matter in groundnut (*Arachis hypogea* L.). *J. Exp. Bot.* **154:** 746–755.

Read, J., and G. Farquhar. 1991. Comparative studies in *Nothofagus* (Fagaceae). I. Leaf carbon isotope discrimination. *Func. Ecol.* **5:** 684–695.

Sinclair, T. R., C. B. Tanner, and J. M. Bennett. 1983. Water use efficiency in crop production. *Biol. Sci.* **34:** 36–40.

Virgona, J. M., K. T. Hubick, H. M. Rawson, and G. D. Farquhar. 1990. Genotypic variation in transpiration efficiency, carbon isotope discrimination and dry matter partitioning during early growth in sunflower. *Aust. J. Plant Physiol.* **17:** 207–214.

Wright, G. C., K. T. Hubick, and G. D. Farquhar. 1988. Discrimination in carbon isotopes of leaves correlates with water use efficiency of field grown peanut cultivars. *Aust. J. Plant Physiol.* **15:** 815–825.

Wright, G. C., K. T. Hubick, and G. D. Farquhar. 1991. Physiological analysis of peanut cultivar response to timing and duration of drought stress. *Aust. J. Agric. Res.* **42:** 453–470.

22

Carbon Isotope Discrimination and Gas Exchange in Coffee during Adjustment to Different Soil Moisture Regimes

Frederick C. Meinzer, Guillermo Goldstein, and David A. Grantz

I. Introduction

In C_3 plants the magnitude of carbon isotope discrimination (Δ) and the resulting carbon isotope composition of leaf tissue are determined largely by the ratio of intercellular-to-atmospheric partial pressure of CO_2 (p_i/p_a) which prevails when the tissue carbon is assimilated (Farquhar et al., 1982). The ratio p_i/p_a is also directly related to the ratio of the instantaneous rates of CO_2 assimilation and leaf conductance, a measure of intrinsic water-use efficiency (W). These relationships are being increasingly exploited to establish correlations between genotypic variations in Δ and integrated W (total biomass/total transpiration) for both greenhouse- and field-grown plants (e.g., Farquhar and Richards, 1984; Hubick et al., 1986; Martin and Thorstenson, 1988; Hubick and Farquhar, 1989).

Leaf tissue Δ is an integrated measure of internal plant physiological and external environmental properties that influence photosynthetic gas exchange. Thus, variations in Δ within and among genotypes may prove useful for assessment and prediction of other aspects of plant performance in addition to W. For example, foliar Δ values have been used as an integrated measure of photosynthetic responses to changes in environmental variables such as soil water availability (Hubick et al., 1988), atmospheric humidity (Winter et al., 1982), salinity (Farquhar et al., 1982; Guy et al., 1986), and irradiance (Zimmerman and Ehleringer, 1990).

Similar foliar Δ values may be attained by contrasting mechanisms each having different ecological consequences. For example, intrinsic W may be enhanced by reduced stomatal conductance which would increase the stomatal limitation of photosynthesis and lower p_i/p_a. This apparently occurs as Δ varies genotypically in wheat (Condon *et al.*, 1987) and common bean (Ehleringer, 1990). An alternative means of enhancing intrinsic W is through variations in photosynthetic capacity and therefore p_i/p_a at similar levels of stomatal conductance. Behavior consistent with this mechanism has been observed in diverse peanut genotypes (Wright *et al.*, 1988). The former mechanism thus represents a trade-off between enhanced W and reduced rates of carbon assimilation, and the latter mechanism constitutes a trade-off between maintenance of high rates of carbon assimilation at the expense of unrestricted water use. These contrasting mechanisms for regulating intrinsic W indicate that adequate gas exchange data must be acquired before patterns of adjustment in foliar Δ values can be fully interpreted.

A conventional approach used in many studies has been to relate instantaneous measurements of gas exchange to Δ values of leaves whose carbon was assimilated during an unknown period prior to the gas exchange measurements. Evergreen woody species having leaves with relatively long lifespans present a special problem in relating current physiological performance to leaf Δ values. In coffee, for example, leaves may be retained for more than 1 year. The carbon isotope composition of leaves of different ages would thus be expected to reflect the combined influence of predictable phenological rhythms in gas exchange characteristics and of unpredictable variations in environmental variables such as moisture availability. The current behavior of a leaf may thus bear limited resemblance to the behavior of that leaf or of the rest of the plant during the period in which the leaf's cellulose was assimilated.

Members of the genus *Coffea* evolved as evergreen understorey shrubs in African tropical forests at elevations between 1700 and 2000 m. Annual rainfall is 1500–2000 mm in its native habitats and exhibits a pronounced seasonal distribution in which several consecutive months are nearly rainless. This probably has contributed to the substantial drought resistance shown by coffee (Kumar and Tieszen, 1980b; Meinzer *et al.*, 1990b) and to its requirement for a period of reduced water availability to trigger phenological events such as floral bud release (Alvim, 1960; Crisosto *et al.*, 1992). Coffee exhibits features typical of shade-adapted plants. Photosynthesis saturates at photon flux densities below 300 μmol m^{-2} s^{-1} and maximum photosynthetic rate is about 10 μmol CO$_2$ m^{-2} s^{-1}. Exposure to full sunlight increases rates of water use and the potential for development of soil water deficits. Despite their adaptations to shade, commercial coffee varieties grown in the sun often outyield shade-grown plants (Mitchell, 1988).

The studies described in this chapter were carried out with coffee (*Coffea arabica* L.) plants growing under full sunlight in the field or in a greenhouse where maximum photosynthetic photon flux density (PPFD) exceeded 75% of full sunlight. We examined relationships among plant–water relations, leaf gas exchange characteristics, and Δ for coffee plants growing under different soil moisture regimes. We were particularly interested in evaluating the utility of Δ as an integrated measure of genotypic and phenotypic variation in response to limited water availability. The results of these studies have implications concerning appropriate uses of carbon isotope discrimination and gas exchange measurements to assess integrated and current responses to a change in environment in evergreen woody species having leaves with relatively long life spans.

II. Responses to Progressive Soil Drying in the Field

Estimates of genotypic variation in Δ are often useful in interpreting phenotypic plasticity in the response of Δ to fluctuations in soil moisture and other environmental factors. A 2.1‰ genotypic variation in Δ values was observed among 2.5-year-old seedling populations of 14 coffee cultivars grown in the same field under a single irrigation regime as part of a variety evaluation trial (Meinzer *et al.*, 1991). The variation in foliar Δ values among cultivars was the same at 8 months after planting and the ranking of the cultivars according to their foliar Δ values was similar, but not identical, at the two sampling dates. In a later study, genotypic variation in Δ among 15 well-irrigated, greenhouse-grown individuals of the cultivar Guatemalan was found to be approximately 3.2‰ (Meinzer *et al.*, 1992). Genotypic variation in Δ in these well-irrigated coffee plants (corresponding to a 48% variation in intrinsic W) was substantially greater than that reported for well-irrigated wheat (Condon *et al.*, 1990), common bean (White *et al.*, 1990), and cowpea (Hall *et al.*, 1990) genotypes, but comparable to that reported for peanut genotypes (Hubick *et al.*, 1986). The magnitude of genetic variation in Δ observed in coffee is consistent with the relatively high 10 to 40% natural cross-pollination reported for *C. arabica* (Carvalho, 1988).

When Δ, photosynthetic gas exchange, and plant–water relations characteristics were monitored in five coffee cultivars during 1 month of progressive soil drying in the field, considerable variation among cultivars was observed in the rate at which leaf-water deficits developed (Meinzer *et al.*, 1990a). For example, the 1-month drought caused predawn and midday leaf-water potential (ψ_L) of cv Mokka to decrease by about 1.0 MPa relative to irrigated plants, while ψ_L for individuals of the cultivars Guatemalan and Catuai exhibited a difference of only 0.13 MPa prior to dawn and about 0.2 to 0.4 MPa at midday (Table I). Pressure–volume relationships obtained from leaves indicated that increased leaf-water deficits in droughted coffee plants led to reductions in bulk leaf elasticity, osmotic potential, and the ψ_L at which turgor loss occurred (Meinzer *et al.*, 1990b). Nevertheless, adjustments in osmotic potential, evaluated at zero turgor, were not sufficient to prevent loss or near loss of turgor in Mokka, Yellow Caturra, and Catuai at the lowest values of midday ψ_L attained.

Drought-induced changes in assimilation (A) and stomatal conductance

Table I   Leaf-Water Potential (ψ_L) of Field-Grown Coffee Plants Irrigated
Weekly (I) and Droughted for 1 Month (D)

Cultivar	Predawn ψ_L (MPa)		Midday ψ_L (MPa)	
	I	D	I	D
Catuai	−0.25	−0.38	−1.10	−1.53
Guatemalan	−0.23	−0.36	−1.01	−1.22
Mokka	−0.21	−1.23	−1.07	−2.02
San Ramon	−0.09	−0.37	−1.03	−1.34
Yellow Caturra	−0.22	−0.44	−1.09	−1.66

(g) were largely independent of bulk leaf turgor except at very low values of turgor. In droughted Mokka leaves, for example, average midday A and g at zero turgor were approximately 50% of the average maximum values recorded for irrigated Mokka plants. When variation in A and g were examined as a function of relative symplastic volume derived from pressure–volume analyses, there appeared to be an optimum relative protoplast volume associated with maximum rates of gas exchange, above and below which A and g declined. These findings are in agreement with those of other studies in which photosynthesis–protoplast volume relationships have been examined (Kaiser, 1982; Sen Gupta and Berkowitz, 1987; Santakumari and Berkowitz, 1990) and indicate the need to consider the role of both volume and turgor regulation in the maintenance of gas exchange and other processes in droughted plants (Munns, 1988). The responses observed over the range of leaf-water deficits experienced by the five coffee cultivars studied provided no evidence of genotypic differences in ability to modify the protoplast volume–ψ_L relationship via osmostic and elastic adjustment, nor for differential tolerance of photosynthesis to protoplast volume reduction.

It thus appears that a major component of differential adaptation to drought among these coffee cultivars may consist of more complex suites of behavioral traits associated with postponement of dehydration through reduced rates of water use or greater efficiency of extraction of soil water. Nevertheless, Δ was a useful probe of these differences in performance under limited water supply. The 1-month drought caused Δ of expanding leaves to decline noticeably in all five cultivars (Table II). The drop in Δ associated with a declining water supply ranged from 4.1‰ in Mokka to 1.9% in Guatemalan. Mature leaves, whose nonmobile, structural carbon was assimilated prior to the drought showed much smaller differences in Δ between irrigation treatments, with Δ apparently increasing slightly rather than decreasing in response to drought in two of the cultivars (Table II).

The pattern of change in Δ of mature leaves may reflect differences among cultivars in the response of dry matter partitioning to drought. Drought-induced changes in average dry weight per leaf were highly correlated with the magnitude and direction of changes in Δ (Fig. 1). For the

Table II Carbon Isotope Discrimination in Leaves of Field-Grown Coffee Plants Irrigated Weekly (I) and Droughted for 1 Month (D)

Cultivar	Expanding leaves (‰)			Mature leaves (‰)		
	I	D	I–D	I	D	I–D
Catuai	18.5	16.3	2.2	18.2	18.4	−0.2
Guatemalan	18.5	16.6	1.9	18.7	19.0	−0.3
Mokka	20.1	16.0	4.1	19.3	18.0	1.3
San Ramon	20.3	18.0	2.3	19.4	18.8	0.6
Yellow Caturra	19.4	16.1	3.3	18.9	18.6	0.3

two cultivars in which a substantial increase in leaf dry weight occurred, Δ also decreased substantially. Apparently, carbon assimilated at lower p_i/p_a in droughted plants had been partitioned primarily to storage in mature leaves rather than to growth. This hypothesis is supported by the predictions of a simple model. In this model Δ values for mature leaves of droughted plants (Δ_{MD}) were estimated as

$$\Delta_{MD} = \Delta_{MI}DW + \Delta_{ED}(1 - DW),$$

where Δ_{MD} is the average of Δ values of mature leaves of irrigated plants (Δ_{MI}) and Δ of expanding leaves of droughted plants (Δ_{ED}) weighted by the fractions of total leaf dry weight derived from preexisting assimilate (DW) and assimilate imported during the drought (1 − DW), respectively. Δ values predicted with this model (not shown) were within 0.2‰ of the measured values for mature leaves of droughted Mokka and San Ramon

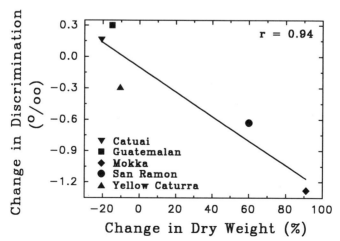

Figure 1. Relationship between the change in carbon isotope discrimination and the change in dry weight per leaf (in fully expanded leaves) for five field-grown coffee cultivars from which irrigation was withheld for 1 month.

plants shown in Table II. The remaining three cultivars exhibited a decrease in dry weight of mature leaves in response to drought (Fig. 1) suggesting a loss of carbon through increased respiration or export to other parts of the plant. The cause of the increase in Δ observed in the cultivars Catuai and Guatemalan is not apparent. Loss of leaf carbon through increased respiration should have caused the remaining leaf dry matter to become more enriched rather than depleted in ^{13}C.

Examination of adjustments in leaf gas exchange characteristics associated with the changes in Δ described above for expanding leaves revealed that a single linear function could be used to describe the relationship between A and g for all cultivars and levels of water availability (Meinzer *et al.*, 1990a). This suggests that the degree of coupling between g and A is highly conserved among coffee genotypes. The relative variation in g was larger than that of A over the range of adjustment observed, and the relationship between g and A had a positive, nonzero intercept. Intrinsic W (A/g) thus increased, and p_i/p_a decreased, with decreasing g and A. This was consistent with genotypic and drought-induced variation in Δ (Table II). In fact, a linear relationship was obtained between A/g and Δ of expanding leaves of irrigated and droughted plants (Fig. 2). A/g was used to confirm the relationship between gas exchange components and Δ in this study, since A/g, and not instantaneous W (A/transpiration) determines p_i/p_a and therefore Δ and since single leaf measurements of gas exchange provide a valid measure of g but biased estimates of transpiration (Meinzer and Grantz, 1989). Three days of field gas exchange measurements during the final 2 weeks of the drought were sufficient to characterize the relationship between gas exchange and Δ. Similar weather patterns during the 3 days

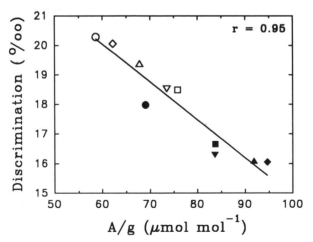

Figure 2. Relationship between carbon isotope discrimination and intrinsic water-use efficiency (A/g) for five coffee cultivars growing in the field. Open symbols represent plants irrigated weekly and closed symbols represent plants from which irrigation was withheld for 1 month. Symbols are as defined in Fig. 1. Adapted from Meinzer *et al.* (1990a).

on which gas exchange was measured, and use of expanding leaf material containing carbon assimilated after imposition of the drought for isotopic analyses, contributed to the success of this simple protocol.

III. Dynamics of Adjustment to Altered Soil Moisture Regimes

Plants growing under reduced soil moisture are known to produce leaves with lower Δ values (Martin and Thorstenson, 1988; Hubick and Farquhar, 1989; Hall *et al.*, 1990). However, there is little information available concerning the dynamics of adjustment in Δ and gas exchange following a change in water availability. Results from a recent study of coffee (Meinzer *et al.*, 1992) suggest that considerable time may be required for these adjustments to be completed in relatively slow-growing woody species. Photosynthetic gas exchange, Δ, and growth were monitored in greenhouse-grown coffee plants for 120 days after the irrigation frequency was decreased from twice daily to either twice weekly or weekly. One group of plants continued to receive irrigation twice daily throughout the 120-day period. Foliar Δ determined in the youngest fully expanded leaf decreased under all soil moisture regimes throughout the experiment (Fig. 3). At 120-days Δ had declined by 4.5‰ in the plants irrigated weekly. The nearly linear decline in Δ throughout the experiment in plants irrigated weekly and twice weekly indicated that more than 120 days were required for complete adjustment in Δ following a change in soil moisture regime. A 1‰ decline in Δ was observed even in plants that continued to receive irrigation twice daily throughout the experiment (Fig. 3). The steady production of

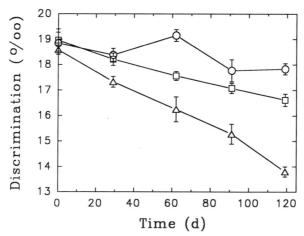

Figure 3. Carbon isotope discrimination in the youngest fully expanded leaf of green-house-grown coffee plants (cv Guatemalan) irrigated twice daily (O), twice weekly (□), and weekly (△). All plants were irrigated twice daily until Day 0. Points are means (±SE) of determinations from five plants. From Meinzer *et al.* (1992).

new leaves with progressively lower Δ values resulted in a pattern of sharply increasing Δ with increasing leaf age when leaves from different positions along the branches were sampled at 120 days (Meinzer *et al.,* 1992).

Maximum values of *A* and *g* and the relationship between *A* and *g* in these plants varied both with a change in soil moisture regime and with time elapsed since the beginning of the experiment, independently of a change in soil moisture regime (Fig. 4). For example, near the beginning of the experiment a single, nearly linear function described the relationship between *A* and *g* for plants growing under all three moisture regimes (Fig. 4; 6–14 days). Approximately 30 days later, a single, but curvilinear, function described the relationship between *A* and *g* for all plants. Maximum

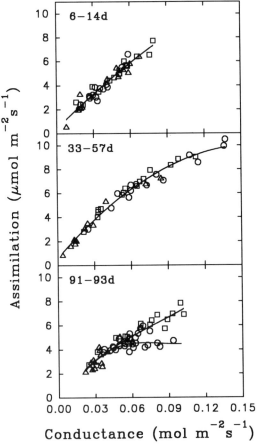

Figure 4. Relationship between assimilation and stomatal conductance of greenhouse-grown coffee plants at different time intervals after imposing the irrigation regimes given in Fig. 3. Points are the two highest values measured for each plant on each of 3 days during the periods indicated. Symbols are as defined in Fig. 3. From Meinzer *et al.* (1992).

Table III   Total Leaf Area, Gas Exchange Characteristics, and Carbon Isotope Discrimination (Δ) in Greenhouse-Grown Coffee Plants (cv Guatemalan) Growing under Three Different Soil Moisture Regimes for 120 Days[a]

Irrigation frequency	Leaf area (m²)	A (μmol m^{-2} s^{-1})	g (mol m^{-2} s^{-1})	E (mmol m^{-2} s^{-1})	A/E (μmol mmol^{-1})	Δ (‰)
Twice daily	2.24	4.50	0.060	0.43	10.5	17.8
Twice weekly	1.25	4.22	0.045	0.70	6.5	16.7
Weekly	0.53	2.22	0.023	0.89	2.5	13.8

[a]Data are average daily values of assimilation (A), stomatal conductance (g), transpiration from sap flow measurements (E), and instantaneous water-use efficiency (A/E). Δ was measured in the youngest fully expanded leaves.

values of A and g were considerably higher than those at the beginning of the experiment. Near the end of the experiment (Fig. 4; 91–93 days), two functions were required to describe the relationship between A and g because A was nearly independent of g in plants irrigated twice daily. These patterns of A and g suggest that at least two mechanisms contributed to the steadily declining Δ observed in plants irrigated weekly and twice weekly (Fig. 3). When the relationship between A and g is curvilinear (33–57 days; Fig. 4), stomatal restriction of gas exchange along the steep portion of the curve increases A/g and therefore reduces p_i/p_a and Δ. A second mechanism leading to a reduction in p_i/p_a and Δ is a higher photosynthetic rate at a given level of g, which was observed near the end of the experiment in plants irrigated twice weekly. Average values of A and g during this period (Table III) suggested that the latter mechanism may have predominated in causing reduced Δ in plants irrigated twice weekly. Assimilation rates were similar in plants irrigated twice daily and twice weekly while g was 25% lower in the latter.

IV. Sensing of Soil Water Availability

Traditionally, sensing of declining soil water availability by the shoot has been attributed to reductions in water uptake which lead to increased leaf-water deficits. Although this mechanism undoubtedly operates in many circumstances, there is increasing evidence that shoot-water status is often better in plants exposed to drying soil than in plants irrigated regularly (e.g., Bates and Hall, 1981; Turner *et al.*, 1985). In these cases leaf-water status is conserved by stomatal closure and slower leaf growth which reduce water loss. These and other modifications of shoot physiology in advance of reductions in leaf-water status have been linked to generation of a chemical signal in roots growing in drying soil which is transported to the shoot in the transpiration stream (reviewed by Davies *et al.*, 1990; Davies and Zhang, 1991).

Control of stomatal aperture and shoot gas exchange by substances transported in the xylem sap from roots growing in drying soil has been termed nonhydraulic signaling. This distinguishes it from hydraulic transmission of changes in soil water availability from the roots to the leaves via changes in xylem sap tension. Nonhydraulic signaling thus represents a feedforward means of stomatal regulation of leaf-water status, while hydraulic signaling represents a feedback system. Nevertheless, the distinction between the two may obscure the fact that changes in stomatal and root hydraulic properties are often coordinated, permitting stomatal conductance and hydraulic conductance to decline in parallel during soil drying. This coupling of vapor-phase with liquid-phase conductance is precisely what serves to maintain leaf-water status nearly constant as the soil dries (Davies and Meinzer, 1990). These mechanisms for sensing soil water availability are not necessarily mutually exclusive and it is probable that both operate in many species, especially if soil drying advances to the stage at which maintenance of constant leaf-water status becomes thermodynamically impossible.

Studies with field- and greenhouse-grown coffee plants have provided evidence for regulation of shoot gas exchange during soil drying by both chemical signals originating in the roots and by hydraulically mediated changes in leaf-water status. In the field study of five coffee cultivars described in Section II above, intrinsic W increased with decreasing absolute symplastic volume (total protoplast volume per unit leaf dry weight; Fig. 5). The correlation between symplastic volume and intrinsic W was strong both when intrinsic W was expressed as A/g derived from instantaneous gas exchange measurements and when it was expressed as Δ values of expanding leaves, an independent, integrated measure of leaf gas exchange properties, principally p_i/p_a. This is apparently an example of a direct hydraulic influence on leaf gas exchange mediated by changes in protoplast volume. Nevertheless, variations in symplast volume and in ψ_L were poorly correlated (cf. Table I and Fig. 5), especially in irrigated coffee plants which exhibited no significant differences in midday ψ_L, but accounted for nearly 50% of the total variation in symplastic volume and intrinsic W observed. This suggests that either ψ_L was not an appropriate indicator of hydraulic influences on leaf-water status in this situation, or that intrinsic W and protoplast volume may both covary in response to fluctuations in a third, unrecognized factor.

Results of experiments conducted with greenhouse-grown coffee plants having dual root systems suggest that root signals also play a role in regulation of shoot gas exchange in coffee during soil drying. In these experiments plants with two functional root systems in separate pots were prepared by grafting an accessory root system to a main shoot with its original intact root system. This permitted experimental separation of leaf/shoot water status from soil/root water status by allowing the soil in one pot to dry while keeping the other well irrigated. Withholding water from either the accessory or the main root system reduced leaf gas exchange compared

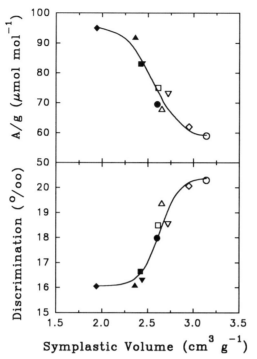

Figure 5. Relationship between intrinsic water-use efficiency and leaf absolute symplastic volume for five coffee cultivars growing in the field and either irrigated weekly (open symbols) or subjected to a 1-month drought (closed symbols). Intrinsic W was expressed as A/g determined from leaf gas exchange measurements (top) and as carbon isotope discrimination of expanding leaves (bottom). Increasing discrimination is an index of increasing p_i/p_a and therefore decreasing intrinsic W. Absolute symplastic volume per unit leaf dry weight was calculated from pressure–volume analyses and values of *in situ* leaf-water potential as described by Meinzer *et al.* (1990b). Symbols for cultivars are as defined in Fig. 1.

with that of plants in which both root systems were well irrigated even though diurnal fluctuations in ψ_L were similar in all plants, except those in which both root systems were droughted (Table IV). This relatively rapid soil drying in the greenhouse appeared to have a smaller effect on A/g than did the slower soil drying in the field even though g was inhibited to

Table IV Effect of Drought on Leaf-Water Potential (ψ_L) and Gas Exchange in Greenhouse-Grown Coffee Plants (cv Guatemalan) with Dual Root Systems

Treatment	ψ_L (MPa) Predawn	Midday	A (μmol m^{-2} s^{-1})	g (mol m^{-2} s^{-1})	A/g (μmol mol^{-1})
Both irrigated	−0.05	−2.08	5.06	0.076	67
Accessory droughted	−0.05	−2.00	4.31	0.064	67
Main droughted	−0.05	−1.92	1.81	0.021	86
Both droughted	−4.50	—	0.04	0.006	7

approximately the same extent in field-grown plants and in dual root system plants with the main root system droughted (cf. Table IV and Fig. 2). This suggests that regulation of g and photosynthetic capacity by root signals may occur over different time scales, or that a combination of chemical signals from the roots and declining leaf-water status is required to achieve maximal adjustment in intrinsic W in response to drought.

There is increasing evidence that intrinsic W may be a function of the hydraulic capacity of the roots and soil to supply the leaves with water (e.g., Küppers, 1984; White *et al.*, 1990). A linear relationship between Δ and plant hydraulic efficiency was obtained for five coffee cultivars growing at different levels of soil moisture availability in the field (Fig. 6). Plant hydraulic efficiency was defined as the ratio of stomatal conductance to the difference between predawn and midday ψ_L. This is analogous to an apparent hydraulic conductance, since stomatal conductance multiplied by the prevailing leaf-to-air vapor pressure difference yields transpiration, and the difference between predawn and midday ψ_L approximates the driving force for water flux at midday. This index of soil water availability and efficiency of water supply was more effective in accounting for variation in Δ than any other traditional measure of leaf-water status such as ψ_L, turgor, or leaf-water content (Meinzer *et al.*, 1990a). In coffee plants growing under different soil moisture regimes in the greenhouse Δ was largely independent of root hydraulic conductance when soil moisture was abundant, but declined sharply with root hydraulic conductance under drier soil

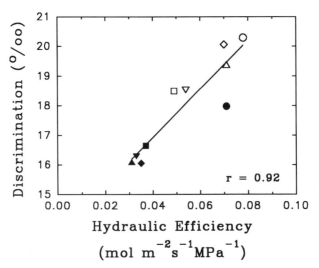

Figure 6. Relationship between carbon isotope discrimination and hydraulic efficiency for five coffee cultivars growing in the field and either irrigated weekly (open symbols) or subjected to a 1-month drought (closed symbols). Hydraulic efficiency is defined as the ratio of stomatal conductance to the difference between predawn and midday leaf-water potential. Symbols for cultivars are as defined in Fig. 1. Adapted from Meinzer *et al.* (1990a).

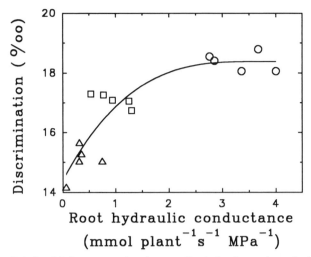

Figure 7. Relationship between carbon isotope discrimination and root hydraulic conductance in coffee plants (cv Guatemalan) growing under three different soil moisture regimes in the greenhouse. Values of discrimination are averages of samples collected from single plants at 60, 90, and 120 days after imposing new irrigation regimes. All plants were irrigated at 120 days, and decapitalized, and their root systems were pressurized to determine hydraulic conductance. Symbols for irrigation regimes are as defined in Fig. 3. From Meinzer *et al.* (1992).

moisture regimes (Fig. 7). Whether these responses are a function of strictly hydraulic signals as defined above, rather than chemical signals from the roots, is not known. Root hydraulic conductance may be linked to the production of chemical signals in the roots via the direct effect of reduced hydraulic conductance on root-water potential when transpirational water movement is occurring. These transpiration-induced root-water deficits could enhance production and export of chemical signals such as abscisic acid (Davies *et al.*, 1990). Altered hydraulic conductance both within roots and in the soil outside the roots may thus be a primary signal controlling shoot gas exchange in relation to soil water availability.

V. Importance of Spatial and Temporal Scales

In evergreen species with continuous production of leaves, the current gas exchange characteristics of a particular leaf may deviate considerably from those of leaves which may already have abscised but which were the source of the leaf's cellulose. Since foliar Δ values are determined largely by the isotopic composition of leaf cellulose, this may cause difficulties in relating the current gas exchange behavior of a leaf to its Δ value. This potential problem of the inherent difference in time scale for gas exchange and Δ measurements becomes a spatial sampling problem when a large range of leaf ages is available for sampling. Thus when irrigation frequency was abruptly reduced from twice daily to twice weekly, progressive adjustment in Δ values of expanding coffee leaves (Fig. 3) and stability of Δ in preexisting leaves caused pronounced variation in Δ among leaves at different nodes along a branch even though their current gas exchange behavior was similar (Meinzer *et al.*, 1992). In an earlier study with droughted coffee plants good agreement was obtained between measured Δ values and those predicted from p_i/p_a observations because Δ values were measured in expanding leaves containing carbon assimilated by older leaves in which p_i/p_a had been determined (Meinzer *et al.*, 1990a).

An additional problem associated with scale is the relationship between gas exchange on a unit leaf area basis and that on an entire plant basis. Changes in leaf area per plant may alter total assimilation per plant while assimilation on a unit leaf area basis remains constant. For example, at the end of 120 days the total leaf area of coffee plants irrigated twice weekly was only one-half that of plants irrigated twice daily although initially their total leaf areas were similar and their assimilation rates had been nearly equal throughout the experiment (Table III). In addition to producing fewer leaves, plants irrigated twice weekly also had higher rates of leaf senescence and produced considerably smaller leaves than plants irrigated twice daily. This pattern suggests that a major mode of adjustment to reduced soil moisture availability in coffee is the maintenance of nearly constant photosynthetic activity on a unit leaf area basis through a reduction in the rate of increase in total leaf area per plant. In long-lived woody species such as coffee, homeostasis of properties on a unit leaf area basis through reduced rates of leaf expansion may be a common form of adjustment to limited availability of water, nutrients, and other resources (Pereira, 1990). Interpretation of short-term single-leaf gas exchange measurements in species exhibiting this form of phenotypic adjustment to resource limitation may be problematic if the total leaf area of the individual is not taken into account. Even if taken into account, size differences among individuals with similar single-leaf gas exchange rates could not be correctly attributed to differences in age or resource availability unless additional information were available.

The micrometerological constraints involved in scaling water-use efficiency estimates from the leaf to the canopy level have been outlined by Farquhar *et al.* (1988, 1989). It is important to recognize that Δ values reflect p_i/p_a rather than the ratio of the actual fluxes of carbon dioxide and water vapor at the leaf level. For isolated individuals of the same species growing in pots or in the field, the decoupling influence of the unstirred air boundary layer (Jarvis and McNaughton, 1985) is expected to be minimal, and scaling directly from foliar Δ values may be justified. Nevertheless, even under these conditions, variations in canopy morphology, leaf size, and orientation and in stomatal conductance may affect leaf energy balance, the leaf-to-air vapor pressure gradient, and boundary layer conductance. Thus, even among genotypes growing in the same environment, the relative ranking with respect to Δ may not reflect the actual ranking in terms of A/E (Table III, Fig. 3). This pattern contrasts with that observed by Wright *et al.* (1988) who reported the expected negative correlation between W and foliar Δ for field-grown peanut genotypes.

The unexpected positive correlation between Δ and instantaneous W of coffee plants grown under different soil moisture regimes may have resulted from adjustments in crown morphology and leaf size which enhanced E relative to A with decreasing water availability. Large differences in total leaf area indicated that the extent of self-shading was much greater in the crowns of plants irrigated twice daily than in those irrigated twice weekly or weekly. In coffee, the resulting increase in prevailing irradiance per unit leaf area in water-limited plants enhanced E relative to A, since A is saturated at low light levels (Kumar and Tieszen, 1980a). Thus, increasing irradiance would be expected to increase leaf temperature and therefore the leaf-to-air vapor pressure difference without increasing A substantially.

VI. Discrimination as a Predictor of Performance

As suggested at the beginning of this chapter, Δ may be useful for evaluating and predicting other aspects of plant performance in addition to W because Δ integrates the response of photosynthetic gas exchange to the environment and to other internal plant physiological properties. For example, coffee cultivars, with higher initial foliar Δ values when soil moisture is adequate, deplete soil water more rapidly and experience symptoms of physiological stress more rapidly when water is withheld. This can be seen in the relationship between Δ values observed under irrigated conditions and the changes in both Δ and midday ψ_L caused by withholding irrigation (Fig. 8). Coffee cultivars with higher foliar Δ values become water limited more rapidly when irrigation is withheld because of their higher levels of g and therefore higher rates of water use. Thus, Δ of well-irrigated coffee plants can be used to predict genotype performance under water-limited conditions. Nevertheless, these predictions can be invalidated by potentially confounding factors such as an atypical crown morphology and resultant atypical canopy boundary layer properties. These may prevent a particular cultivar from having a transpiration rate similar to that of other cultivars at a given level of g (Meinzer *et al.*, 1990a). San Ramon, a dwarf cultivar with extremely shortened internodes and a dense, hemispherical crown, was excluded from the analysis shown in Fig. 8 for this reason. Lower boundary layer conductance reduced transpiration in this cultivar, limiting the change in midday ψ_L to only 0.3 MPa even though its initial Δ and g values were the highest of the five cultivars studied.

The results described above imply that coffee cultivars with lower Δ values under irrigated conditions may perform better than those with higher Δ values when they are subjected to water-limited conditions. The reverse may be true under a regime of frequent, ample irrigation. Growth

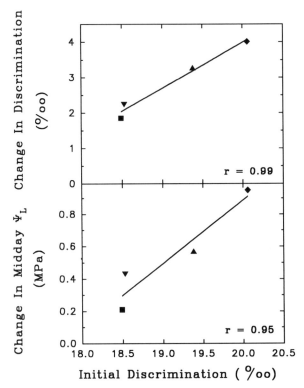

Figure 8.   Changes in carbon isotope discrimination and in midday leaf-water potential (ψ_L) in field-grown coffee during a 1-month drought, as functions of initial values of carbon isotope discrimination when plants were irrigated weekly. Symbols for cultivars are as defined in Fig. 1. Adapted from Meinzer *et al.* (1990a).

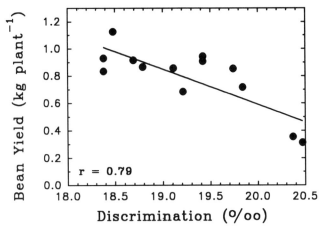

Figure 9.   Relationship between yield of green coffee beans and carbon isotope discrimination in mature leaves of 14 coffee cultivars growing in the field. The plants were irrigated weekly. Adapted from Meinzer *et al.* (1991).

may be more rapid in cultivars with higher Δ values and therefore higher rates of leaf gas exchange. Yield of green coffee beans was negatively correlated with Δ among 2.5-year-old plants of 14 cultivars irrigated at weekly intervals under field conditions (Fig. 9). This positive correlation between yield and intrinsic W was unexpected because cultivars with the lowest intrinsic W had the highest rates of photosynthesis (Meinzer *et al.*, 1990a). These results suggest that the irrigation regime employed may have limited photosynthetic gas exchange in cultivars with lower intrinsic W but not in those with higher intrinsic W or that the linkage between yield and photosynthetic rates is weak in coffee.

VII. Summary

The results obtained with coffee indicate that foliar Δ can be used as an integrated measure of several aspects of plant performance and response to specific environmental conditions once suitable background physiological data have been gathered. With regard to W it should be emphasized that foliar Δ values represent only intrinsic W (A/g) and therefore only the potential ratio of A and E rather than the actual ratio of these fluxes. Environmentally induced adjustments in leaf and canopy characteristics that alter conductances and driving forces for water vapor loss more than those for CO_2 uptake may cause realized W (A/E) to differ from that predicted with foliar Δ values.

In comparing current leaf gas exchange with Δ values in species having leaves with relatively long life spans, consideration should be given to the conditions that prevailed when a particular leaf's carbon was actually assimilated and to the time scales governing adjustments in p_i/p_a and subsequent changes in foliar Δ. Similar considerations apply to measurements at different spatial scales. Spot measurements of gas exchange in single leaves may not reveal differences in total carbon assimilation at the whole-plant level resulting from variations in plant size. The causes of size-related differences in total plant carbon assimilation cannot be correctly partitioned between age and environmental influences unless something is known of a plant's previous growth and environmental history.

Regulation of gas exchange and Δ in coffee in response to variations in soil water availability appears to be mediated both by leaf-water status and by chemical signals originating within the roots. It is noteworthy that intrinsic W in coffee is negatively correlated with the efficiency of water supply to the leaves. Hydraulic efficiency rather than the absolute amount of water available in the root zone may thus represent an important ecological variable governing genetic and environmental variation in W in coffee. This complex variable integrates internal physiological and anatomical constraints on efficiency of water transport with external availability and transport of soil water.

References

Alvim, P. 1960. Moisture stress as a requirement for flowering of coffee. *Science* **132:** 354.

Bates, L. M., and A. E. Hall. 1981. Stomatal closure with soil water depletion not associated with changes in bulk leaf water status. *Oecologia* **50:** 62–65.

Carvalho, A. 1988. Principles and practice of coffee plant breeding for productivity and quality factors: *Coffea arabica*, pp. 129–165. *In* R. J. Clarke and R. Macrae (eds.), Coffee. 4. Agronomy. Elsevier Applied Science, London/New York.

Condon, A. G., R. A. Richards, and G. D. Farquhar. 1987. Carbon isotope discrimination is positively correlated with grain yield and dry matter production in field-grown wheat. *Crop Sci.* **27:** 996–1001.

Condon, A. G., G. D. Farquhar, and R. A. Richards. 1990. Genotypic variation in carbon isotope discrimination and transpiration efficiency in wheat: Leaf gas exchange and whole plant studies. *Aust. J. Plant Physiol.* **17:** 19–22.

Crisosto, C. H., D. A. Grantz, and F. C. Meinzer. 1992. Effects of water deficit on flower opening in coffee (*Coffea arabica* L.). *Tree Physiol.* **10:** 127–139.

Davies, W. J., and F. C. Meinzer. 1990. Stomatal responses of plants in drying soil. *Biochem. Physiol. Pflanz.* **186:** 357–366.

Davies, W. J., and Zhang. 1991. Root signals and the regulation of growth and development of plants in drying soil. *Annu. Rev. Plant Physiol. Mol. Biol.* **42:** 55–76.

Davies, W. J., T. A. Mansfield, and A. M. Hetherington. 1990. Sensing of soil water status and the regulation of plant growth and development. *Plant Cell Environ.* **13:** 709–719.

Ehleringer, J. R. 1990. Correlations between carbon isotope discrimination and leaf conductance to water vapor in common bean. *Plant Physiol.* **93:** 1422–1425.

Farquhar, G. D., and R. A. Richards. 1984. Isotopic composition of plant carbon correlates with water-use efficiency of wheat genotypes. *Aust. J. Plant Physiol.* **11:** 539–552.

Farquhar, G. D., M. C. Ball, S. von Caemmerer, and Z. Roksandic. 1982. Effect of salinity and humidity on $\delta^{13}C$ value of halophytes—evidence for diffusional isotope fractionation determined by the ratio of intercellular/atmospheric CO_2 under different environmental conditions. *Oecologia* **52:** 121–124.

Farquhar, G. D., K. T. Hubick, A. G. Condon, and R. A. Richards. 1988. Carbon isotope fractionation and plant water-use efficiency, pp. 21–40. *In* P. W. Rundel, J. R. Ehleringer, and K. A. Nagy (eds.), Stable Isotopes in Ecological Research. Springer-Verlag, New York.

Farquhar, G. D., J. R. Ehleringer, and K. T. Hubick. 1989. Carbon isotope discrimination and photosynthesis. *Annu. Rev. Plant Physiol. Mol. Biol.* **40:** 503–537.

Guy, R. D., D. M. Reid, and H. R. Krouse. 1986. Factors affecting $^{13}C/^{12}C$ ratios of inland halophytes. I. Controlled studies on growth and isotopic composition of *Puccinellia nuttalliana*. *Can. J. Bot.* **64:** 2693–2699.

Hall, A. E., R. G. Mutters, K. T. Hubick, and G. D. Farquhar. 1990. Genotypic differences in carbon isotope discrimination by cowpea under wet and dry field conditions. *Crop Sci.* **30:** 300–305.

Hubick, K. T., and G. D. Farquhar. 1989. Carbon isotope discrimination and the ratio of carbon gained to water lost in barley cultivars. *Plant Cell Environ.* **12:** 795–804.

Hubick, K. T., G. D. Farquhar, and R. Shorter. 1986. Correlation between water-use efficiency and carbon isotope discrimination in diverse peanut (*Arachis*) germplasm. *Aust. J. Plant Physiol.* **13:** 803–816.

Hubick, K. T., R. Shorter, and G. D. Farquhar. 1988. Heritability and genotype x environment interactions of carbon isotope discrimination and transpiration efficiency in peanut (*Arachis hypogaea* L.). *Aust. J. Plant Physiol.* **15:** 799–813.

Jarvis, P. G., and K. G. McNaughton. 1985. Stomatal control of transpiration: Scaling up from leaf to region. *Adv. Ecol. Res.* **15:** 1–49.

Kaiser, W. M. 1982. Correlation between changes in photosynthetic activity and changes in total protoplast volume in leaf tissue from hygro-, meso-, and zerophytes under osmotic stress. *Planta* **154:** 538–545.

Kumar, D., and L. L. Tieszen. 1980a. Photosynthesis in *Coffea arabica* L. I. Effects of light and temperature. *Exp. Agric.* **16**: 13–19.

Kumar, D., and L. L. Tieszen. 1980b. Photosynthesis in *Coffea arabica* L. II. Effects of water stress. *Exp. Agric.* **16**: 21–27.

Küppers, M. 1984. Carbon relations and competition between woody species in a central European hedgerow. II. Stomatal responses, water use, and hydraulic conductivity in the root/leaf pathway. *Oecologia* **64**: 344–354.

Martin, B., and Y. R. Thorstenson. 1988. Stable carbon isotope composition ($\delta^{13}C$), water use efficiency, and biomass productivity of *Lycopersicon esculentum, Lycopersicon pennellii*, and the F_1 hybrid. *Plant Physiol.* **88**: 213–217.

Meinzer, F. C., and D. A. Grantz. 1989. Stomatal control of transpiration from a developing sugarcane canopy. *Plant Cell Environ.* **12**: 635–642.

Meinzer, F. C., G. Goldstein, and D. A. Grantz. 1990a. Carbon isotope discrimination in coffee genotypes grown under limited water supply. *Plant Physiol.* **92**: 130–135.

Meinzer, F. C., D. A. Grantz, G. Goldstein, and N. Z. Saliendra. 1990b. Leaf water relations and maintenance of gas exchange in coffee cultivars grown in drying soil. *Plant Physiol.* **94**: 1781–1787.

Meinzer, F. C., J. L. Ingamells, and C. Crisosto. 1991. Carbon isotope discrimination correlates with bean yield of diverse coffee seedling populations. *HortScience* **26**: 1413–1414.

Meinzer, F. C., N. Z. Saliendra, and C. H. Crisosto. 1992. Carbon isotope discrimination and gas exchange in *Coffea arabica* during adjustment to different soil moisture regimes. *Aust. J. Plant Physiol.* **19**: 171–184.

Mitchell, H. W. 1988. Cultivation and harvesting of the arabica coffee tree, pp. 43–90. *In* R. J. Clarke, and R. Macrae (eds.), Coffee. 4. Agronomy. Elsevier Applied Science, London/New York.

Munns, R. 1988. Why measure osmotic adjustment? *Aust. J. Plant Physiol.* **15**: 717–726.

Pereira, J. S. 1990. Whole-plant regulation and productivity in forest trees, pp. 237–250. *In* W. J. Davies and B. Jeffcoat (eds.), Important of Root to Shoot Communication in Responses to Environmental Stress. British Society for Plant Growth Regulation Monograph No. 21, Bristol, UK.

Santakumari, M., and G. A. Berkowitz. 1990. Correlation between maintenance of photosynthesis and *in situ* protoplast volume at low water potentials in droughted wheat. *Plant Physiol.* **92**: 733–739.

Sen Gupta, A., and G. A. Berkowitz. 1987. Osmotic adjustment, symplast volume, and nonstomatally mediated water stress inhibition of photosynthesis in wheat. *Plant Physiol.* **85**: 1040–1047.

Turner, N. C., E.-D. Schulze, and T. Gollan. 1985. The responses of stomata and leaf gas exchange to vapour pressure deficits and soil water content. II. In the mesophytic herbaceous species *Helianthus annuus*. *Oecologia* **65**: 348–355.

White, J. W., J. A. Castillo, and J. R. Ehleringer. 1990. Associations between root growth and carbon isotope discrimination in *Phaseolus vulgaris* under water deficit. *Aust. J. Plant Physiol.* **17**: 189–198.

Winter, K., J. A. M. Holtum, G. E. Edwards, and M. H. O'Leary. 1982. Effect of low relative humidity on $\delta^{13}C$ value in two C_3 grasses and in *Panicum milioides*, a C_3-C_4 intermediate species. *J. Exp. Bot.* **33**: 88–91.

Wright, G. C., K. T. Hubick, and G. D. Farquhar. 1988. Discrimination in carbon isotopes of leaves correlates with water-use efficiency of field-grown peanut cultivars. *Aust. J. Plant Physiol.* **15**: 815–825.

Zimmerman, J. K., and J. R. Ehleringer. 1990. Carbon isotope ratios are correlated with irradiance levels in the Panamanian orchid *Catasetum viridiflavum*. *Oecologia* **83**: 247–249.

IV

Genetics and Isotopic Variation

The seven contributions in this section focus on genetic variation in carbon isotope composition of crop species and on the prospects that carbon isotope composition can be used as a tool for directing plant-breeding programs to develop crops better adapted to water-limited environments. As in the previous section, a range of crop species are examined: cowpea (Hall *et al.*), common bean (White), barley (Acevedo), and wheat (Ehdaie *et al.*, Condon and Richards, Richards and Condon). Each of these contributions focuses on classical approaches to plant breeding and on the use of carbon isotopes as a tool for evaluating and selecting genetic material with improved water-use efficiency. Masle *et al.* examine an alternative approach, namely linking genetic variation in carbon isotope composition with variation in restriction fragment length polymorphism. In a critical and thoughtful synthesis, Richards and Condon lay out the challenges that lie ahead in the application of carbon isotopes to directing plant-breeding programs.

23

Implications for Plant Breeding of Genotypic and Drought-Induced Differences in Water-Use Efficiency, Carbon Isotope Discrimination, and Gas Exchange

Anthony E. Hall, Abdelbagi M. Ismail, and Cristina M. Menendez

I. Introduction

Breeding for increased water-use efficiency (W = the ratio of total biomass production to transpiration) or potential productivity has been limited by the lack of screening criteria and methods that could be used to select desirable genotypes from large populations under field conditions. At the leaf level, W should be strongly influenced by the ratio of CO_2 assimilation rate (A) to transpiration rate. The transpiration rate depends on the diffusive conductance to water vapor (g_h) and the vapor pressure difference between leaf and air. The diffusive conductance to CO_2 (g) is proportional to g_h and includes boundary layer and stomatal components. Theory (Farquhar et al., 1982; Hubick et al., 1986) has demonstrated that C_3 plants should exhibit an association between the extent of their discrimination against ^{13}C compared with ^{12}C during CO_2 fixation (Δ), and their leaf intrinsic gas exchange efficiency (A/g) as indicated in Eq. (1). A more complete theory is presented by Farquhar et al. in Chapter 5 of this volume.

$$(A/g)_1/(A/g)_2 = (b - d - \Delta_1)/(b - d - \Delta_2) \tag{1}$$

Subscripts 1 and 2 denote two plants growing in the same aerial environment and having either different genotypes or the same genotype but growing in different soil environments. The inherent discriminations associated with carboxylation (b) and other metabolic processes (d) are assumed to be constant. When the carbon sampled for Δ has accumulated over several days, $(b - d - \Delta)$ is associated with assimilation-weighted values of A/g over this period as indicated in Eq. (2) where i denotes individual gas exchange fluxes measured on n occasions and $\overline{A}$ is the arithmetic mean of the A_i values.

$$\frac{\sum_{i=1}^{n} (A_i/g_i) \times (A_i/\overline{A} \times n)_1}{\sum_{i=1}^{n} (A_i/g_i) \times (A_i/\overline{A} \times n)_2} \simeq \frac{(b - d - \Delta_1)}{(b - d - \Delta_2)}. \tag{2}$$

A positive association is expected between relative A/g and relative W, at the leaf level, if the vapor pressure differences between leaf and air are not substantially different requiring that leaf temperatures are similar, which may not be valid in some cases (Ehleringer *et al.*, 1991). In many cases, however, W could exhibit a negative association with Δ as indicated in

$$W_1/W_2 \simeq (b - d - \Delta_1)/(b - d - \Delta_2). \tag{3}$$

W is usually calculated from the ratio of seasonal biomass production to seasonal transpiration and this is equivalent to a biomass-weighted value of W. Studies described in this chapter with plants in pots demonstrated that relative W determined from seasonal measurements of biomass production and transpiration was associated with Δ values as described by Eq. (3).

The extent to which Δ may be useful as a selection criterion in plant breeding depends upon the consistency of genotypic ranking for Δ (which may be influenced by tissue sampling methods), physiological traits associated with differences in Δ, inheritance of Δ, heritability of Δ, and genetic correlations with other undesirable and desirable traits. Information on these matters is described in relation to strategies for using selection for Δ in plant breeding, mainly based on our research with the grain legume, cowpea (*Vigna unguiculata* (L.) Walp.).

II. Choosing Tissue to Be Sampled for C-Isotope Composition and Breeding Nursery Environments

Ideally, the tissue sampled for determining C-isotope composition should contain carbon assimilated during the stage of growth when photosynthesis has the greatest impact on economic yield. For cereals (Passioura, 1972) and grain legumes (Ziska and Hall, 1983) grain yield is strongly dependent upon water available, and presumably photosynthesis, after anthesis. Consequently, in these crops, grains may provide an effective source of carbon for sampling for Δ. There are several reasons why using grains for Δ samples may not be advisable. Studies by Hall *et al.* (1990) demonstrated that heritability for C-isotope discrimination (based on evaluating plot means of accessions) was substantially greater (+117%) for samples from leaves than from grain. The low heritability for grain was due to lower genetic variance and higher environmental and G × E variances compared with leaves. When comparing genotypes, samples should be taken that contain carbon that was assimilated during the same time period, for Eq. (3) to be valid. For leaves of cowpea, this can be achieved by sampling leaves during the same time period and at similar nodal positions on the main stem. The same approach can be used with grains, except when comparing accessions with substantial differences in time of first flowering when grains may not be available, on different accessions, which contain carbon assimilated during the same time period. Another problem with the use of grains is that they can have much lower Δ values (e.g., 16.5‰ compared with 19.3‰ for subtending leaves of cowpea) indicating that additional metabolic processes influence the Δ of grain compared with that of leaves (Hall *et al.*, 1990). These processes may have been responsible for inconsistencies that were observed in genotypic rankings for Δ based upon leaf and grain samples (Hall *et al.*, 1990). Consequently, for cowpea, leaf samples appear to be more effective than grain samples for detecting genotypic differences in C-isotope discrimination. In studies with barley, highly significant correlations were obtained between Δ from stem, peduncle, leaf, or grains when they were sampled at the same time, but not when they were sampled at different times (Acevedo, Chapter 26, this volume) indicating that care is needed in the choice of tissues for determining Δ.

Date of sampling can influence ability to detect genotypic differences in Δ. Under well-watered conditions in field studies at Riverside, California, leaves sampled from the 5th node at 38 days after germination from crosses with two sets of parents had 71 and 100% higher heritabilities (broad-sense values based on analysis of F_2, F_1, and parental plants) than Δ from leaves sampled from the 10th node at 68 days after germination. Genetic variations increased with later sampling for both crosses (43 and 53%) but environmental variances increased even more. It is likely that there is an optimum date during the season for sampling leaves to maximize ability to detect genotypic differences in Δ and, for cowpea, this optimum date is relatively early in the season.

Environmental conditions in breeding nurseries can strongly influence ability to detect genotypic differences in quantitative traits that depend on several genes. Surprisingly, genetic and environmental variances, and heritabilities (based on evaluating plot means of accessions), were similar for 17 cowpea accessions grown under well-watered conditions compared with extreme drought in a split-block field experiment (Hall *et al.*, 1990). Heritabilities for grain yield and several other quantitative traits are usually much lower for plants under extreme drought, compared with those under well-watered conditions. Apparently, when screening accessions to detect genotypic differences in Δ, a range of hydrologic conditions could be effective, and it would be advisable to use nursery conditions that are similar to the commercial production conditions for which improved cultivars are being developed. It is likely that factors such as plant nutrition (Fu and Ehleringer, 1992), plant spacing, and disease or pest attacks can influence Δ values, and any variability in the exposure of accessions to these factors could cause difficulties in interpreting the results of screening.

III. Consistency of Genotypic Ranking for C-Isotope Discrimination

Consistency of genotypic ranking is essential for breeding to be effective in modifying a particular quantitative trait. Genotypic ranking for Δ values, based upon leaf samples, has been remarkably consistent in Riverside, California, field environments with no significant genotype $\times$ environment interactions (Hall *et al.*, 1992). Correlations between mid-season and late-season values ($r = +0.86$ with $n = 17$) and values obtained under well-watered conditions and extreme drought ($r = +0.85$ with $n = 17$), or one year versus another year ($r = +0.96$ with $n = 12$) were very highly significant. In contrast, correlations between genotypic mean values for the same sets of cowpea genotypes grown at Riverside, California, and Lubbock, Texas, were much lower over 2 years ($r = +0.59$ with $n = 8$, and $r = +0.53$ with $n = 6$) and were not significant. Comparison of Δ values for a similar set of cowpea accessions grown in locations with greater environmental differences (Riverside, CA, which is subtropical versus Bambey, Senegal, which is tropical) gave no correlation ($r = -0.11$ with $n = 9$). The variation in consistency of genotypic ranking for C-isotope discrimination is similar to that of complex quantitative agronomic traits, such as grain yield. Ranking of grain yield is consistent over years for advanced breeding lines and cultivars of cowpea grown at Riverside, California, and some of these genotypes might perform well at Lubbock, Texas, but genotypic ranking for yield is usually completely different at Bambey, Senegal, compared with that at Riverside, California. Apparently, breeding cowpea for differences in Δ values might be effective in particular crop production zones but the performance, with respect to Δ, may not be transferable to radically different production zones.

Some studies with other crops have not detected significant genotype $\times$ environment interactions. Sixteen peanut genotypes were evaluated in 10 contrasting subtropical and tropical environments in Queensland, and the genotype $\times$ environment interaction was highly significant for kernel yield but not significant for Δ (Hubick *et al.*, 1988; Wright *et al.*, Chapter 17, in this volume). Studies with 21 or 23 genotypes of barley over 2 years in two environments in the United Kingdom, and three locations in Syria, gave positive and significant correlations among Δ for 20 of 25 comparisons of genotypic values for different pairs of environments (Craufurd *et al.*, 1991). Relatively consistent rankings were observed for six clones of crested wheatgrass grown in six field experiments and one glasshouse experiment and under different watering regimes (Johnson *et al.*, Chapter

18, this volume). White (Chapter 25, this volume) reported strong positive correlations for values of Δ from 10 genotypes grown under rain-fed conditions in two different locations in Colombia with different soil types. In contrast, he also observed a genotype × water-regime interaction for Δ with eight genotypes of common bean grown under five different water regimes and, in another experiment, a lack of consistency of genotypic ranking for Δ for plants grown under rain-fed conditions compared with those grown with complete irrigation.

The extent of genotype × environment interactions for Δ should be established for different crop species and environments. The extent of these interactions can determine the choice of breeding nursery environments and the success of selection programs. In addition, specific types of interactions may enhance adaptation by increasing phenotypic plasticity. For example, genotypes that rank low for Δ in water-limited environments but high for Δ in wet environments may have improved adaptation in variable rain-fed environments. This hypothesis is based on the assumptions that low Δ is negatively associated with high W and yield in dry environments, whereas high Δ is positively associated with yield in wet environments, which may or may not be valid as is discussed later. White (Chapter 25, this volume) has proposed that drought-induced changes in Δ may be more relevant for evaluating genotypes than absolute values of Δ. However, as was pointed out by White, statistical resolution of changes in Δ is substantially less than for absolute Δ values (Hall *et al.*, 1990).

IV. Physiological Basis of Plant Differences in C-Isotope Discrimination

Genotypic differences in Δ will only be useful if they are associated with useful differences in physiological traits. Theory indicates that low Δ could be associated with high A/g and high W (Farquhar *et al.*, 1982; Farquhar and Richards, 1984; Hubick *et al.*, 1986). Empirical evidence indicates that high Δ might be associated with potential productivity in wheat (Condon *et al.*, 1987), and some cowpea cultivars and advanced breeding lines that have high yields under irrigation in California also have high Δ under these conditions. A possible physiological explanation for this association is that Δ might be positively correlated with leaf conductance under well-watered conditions (Condon *et al.*, 1987; Ehleringer, 1990), but this hypothesis was not supported for cowpea by data obtained with contrasting accessions (Hall *et al.*, 1992).

Extended drought causes cowpea to have substantially lower Δ compared with those under well-watered conditions (Hall *et al.*, 1990). Afternoon gas exchange measurements on plants under field conditions demonstrated that substantial increases in A/g, which were mainly due to decreases in stomatal conductance to water vapor, could explain these drought effects on Δ (Hall *et al.*, 1992) in a manner expected based upon the theory of

Farquhar *et al.* (1982). Genotypic differences in Δ were also present in these same studies, but, they were not consistent with the significant genotypic differences in A/g that were observed (Hall *et al.*, 1992). Detailed gas exchange studies have been conducted with a putative isogenic pair of cowpea lines that differ in Δ, and similar results were obtained. Drought-induced differences in Δ were consistent with differences in A/g measured in the afternoon, whereas the genotypic differences in Δ were not associated with differences in A/g expected based on theory (Kirchhoff *et al.*, 1989). A possible explanation for the inconsistency with theory of the genotypic differences in Δ and A/g is that the Δ values represent the effects of processes integrated over time, whereas A/g values reflect fluxes measured at particular times. Subsequent studies demonstrated that geotypic differences in A/g measured in the early morning were more strongly associated with Δ, as expected based on theory, than afternoon values of A/g (Ismail, 1992). A major conclusion of these studies is that gas exchange measurements were much less effective in detecting genotypic differences in physiological processes than were measurements of Δ. Genotypic rankings for A/g, A, or g_h were inconsistent between the mid-season and late-season measurements, from year to year or between plants under well-watered conditions and drought; in contrast, for the same plants, genotypic rankings for Δ were highly consistent (Hall *et al.*, 1992).

Seasonal W measured on cowpea plants grown in pots under field conditions was negatively correlated with Δ for both genotypic and drought-induced effects, and the regression line was consistent with theory (Fig. 1;

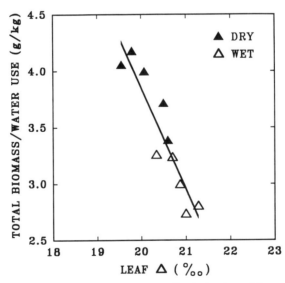

Figure 1. Water-use efficiency and C-isotope discrimination of five cowpea genotypes grown under wet or dry pot conditions in the field. $W = 0.90 \ (24.3 - \Delta \times 10^3)$ with $r^2 = 0.86$ (Ismail and Hall, 1992).

from Ismail and Hall, 1992). The linear regression for this relationship predicted a value of $(b - d)$ of 24.3‰, which is similar to values of $b = 27‰$ and $d = 3‰$ suggested by Hubick *et al.* (1986) and values for $(b - d)$ obtained with peanut of 24 and 25‰ (refer to Figs. 1 and 4 of Wright *et al.*, Chapter 17, this volume). The total genotypic variation in W of cowpea was 19 and 23% under wet and dry conditions, respectively, and was not associated with differences in plant vigor (Ismail and Hall, 1992). Drought increased W by 29% due to biomass being reduced less than water use (Ismail and Hall, 1992), which is consistent with observations on the same genotypes indicating strong drought-induced reductions in stomatal conductance (Hall *et al.*, 1992).

V. Inheritance of C-Isotope Discrimination and Water-Use Efficiency

Comparison of F_1 hybrids from reciprocal crosses and parents of cowpea grown in pots under field conditions demonstrated that differences among parents in Δ and W involve nuclear, rather than maternal, inheritance (Table I; from Ismail and Hall, 1993). These studies also were consistent with partial dominance for low Δ and high W under wet or dry pot conditions (Table I), and were confirmed by another study with four hybrids and four parents grown in pots under wet and dry field conditions (Table II;

Table I   The C-Isotope Discrimination (Δ) and Water-Use Efficiency (W) of Cowpea Accessions and F_1 Hybrids Grown under Wet and Dry Treatments in Pots under Field Conditions in 1990

Genotypes	Δ (‰)			W (g/kg)		
	Wet	Dry	Mean	Wet	Dry	Mean
UCR 237A (U)	19.6	18.6	19.1	3.92	4.13	4.03
TVx 309-1G (T)	19.9	18.6	19.2	4.04	4.15	4.09
F_1 hybrids						
U × C	19.8	18.4	19.1	3.53	4.21	3.87
C × U	19.7	18.9	19.3	3.72	4.15	3.94
T × P	19.8	19.0	19.4	3.81	4.23	4.02
P × T	20.2	18.9	19.5	3.68	4.15	3.91
CB46 (C)	20.3	19.7	20.0	2.95	2.51	2.73
Prima (P)	20.7	20.4	20.6	2.79	2.54	2.67
Mean	20.0	19.1		3.56	3.76	
Significance						
Genotypes (G)		***			***	
Treatments (T)		***			NS	
G × T		NS			NS	
$LSD^{0.05}$ (G)		0.5			0.50	

Source: Ismail and Hall (1993).
***Significant at 0.001% level.

Table II   The C-Isotope Discrimination (Δ) and Water-Use Efficiency (*W*) of Cowpea Accessions and F₁ Hybrids Grown under Wet and Dry Treatments in Pots under Field Conditions in 1991

Genotypes	Δ (‰)			*W* (g/kg)		
	Wet	Dry	Mean	Wet	Dry	Mean
UCR 237A (U)	19.8	18.5	19.1	4.27	5.39	4.83
TVx 309-1G (T)	19.7	18.4	19.1	4.46	5.54	5.00
F₁ hybrids						
U × C	20.2	18.4	19.3	4.29	5.58	4.93
U × P	19.8	18.7	19.3	4.46	5.32	4.89
T × C	20.0	18.7	19.4	4.34	5.59	4.96
T × P	20.0	18.6	19.3	4.44	5.50	4.97
CB46 (C)	20.4	18.8	19.6	3.99	5.13	4.56
Prima (P)	20.8	19.9	20.3	4.15	5.01	4.58
Mean	20.1	18.8		4.30	5.38	
Significance						
Genotypes (G)		***			**	
Treatments (T)		***			***	
G × T		NS			NS	
LSD^{0.05} (G)		0.3			0.29	

Source: Ismail and Hall (1993).
,*Significant at 0.01% and 0.001% levels, respectively.

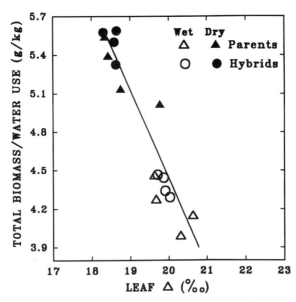

Figure 2.   Water-use efficiency and C-isotope discrimination of four cowpea parents and four F₁ hybrids grown under wet or dry pot conditions in the field. $W = 0.69 (26.5 - \Delta \times 10^3)$ with $r^2 = 0.88$ (Ismail and Hall, 1993).

from Ismail and Hall, 1993). Seasonal W was negatively correlated with $Δ$ for both genotypic and drought-induced effects, and the regression line was consistent with theory (Fig. 2; from Ismail and Hall, 1993). Studies with a larger set of reciprocal crosses under well-watered natural soil conditions confirmed that inheritance of $Δ$ is nuclear rather than maternal (Table III; from Ismail and Hall, 1993). One of the pairs of reciprocal crosses did exhibit a significant difference in $Δ$, but the cross with the highest $Δ$, T × C, had the low $Δ$ parent, TVx 309-1G, as the female in the cross indicating that the effect was not maternal. A surprising feature of the experiment under natural soil conditions was that $Δ$ appeared to be partially dominant for high $Δ$ (Table III) which is opposite from results of studies with some of the same hybrids in a similar aerial environment, but under pot conditions (Tables I and II). We hypothesized that the change in ranking of the F_1's, compared with the parents, was due to the later sampling in the natural soil condition experiment, compared with the pot experiment. However, data from two field experiments with two sets of parents and hybrids (Table IV) did not support this hypothesis. The devia-

Table III The C-Isotope Discrimination ($Δ$) of Cowpea Accessions and F_1 Hybrids Grown under Wet, Natural Soil Field Conditions in 1990

	Genotypes	$Δ$ (‰)
Parents		
	TVx 309-1G (T)	20.4
	UCR 237A (U)	20.4
	Vita 7 (V)	20.4
F_1 hybrids		
	U × C	21.7
	C × U	21.5
	T × P	21.6
	P × T	21.7
	U × P	21.1
	P × U	21.2
	T × C	22.1
	C × T	21.5
	V × P	21.8
	P × V	21.6
	V × C	21.3
	C × V	21.3
Parents		
	Prima (P)	21.8
	CB46 (C)	22.2
	LSD$_{0.05}$	0.6

Source: Ismail and Hall (1993).

Table IV Does Time of Sampling Influence Conclusions Concerning the Inheritance of C-Isotope Discrimination?

Parent 1 × parent 2	DAG[a]	F_1(‰)	P_1(‰)	P_2(‰)	$\dfrac{P_2 - P_1}{2}$ (‰)	$F_1 - \left(\dfrac{P_1 + P_2}{2}\right)$ (‰)
TVx 309-1G × CB46						
	38	20.97	20.29	21.34	0.52	+0.16
	69	20.71	20.58	21.00	0.21	−0.08
	38	20.80	20.42	21.07	0.32	+0.06
	68	20.75	19.99	21.14	0.58	+0.18
UCR 237A × CB46						
	38	21.08	20.97	21.34	0.18	−0.08
	69	20.34	19.24	21.00	0.88	+0.22
	38	21.01	20.66	21.04	0.20	+0.16
	68	20.15	19.21	21.16	0.98	−0.04

[a]Days after germination.

Table V Does the Soil Environment Influence Conclusions Concerning the Inheritance of C-Isotope Discrimination?

Parent 1 × parent 2	Soil environment[a]	F_1(‰)	P_1(‰)	P_2(‰)	$\dfrac{P_2 - P_1}{2}$ (‰)	$F_1 - \left(\dfrac{P_1 + P_2}{2}\right)$ (‰)
1990						
UCR 237A × CB46	Dry pot	18.68	18.56	19.70	0.57	−0.45
	Wet pot	19.72	19.63	20.30	0.34	−0.24
	Wet soil	21.58	20.39	22.23	0.92	+0.27
TVx 309-1G × Prima	Dry pot	18.96	18.61	20.45	0.92	−0.57
	Wet pot	20.00	19.86	20.66	0.40	−0.26
	Wet soil	21.63	20.37	21.76	0.70	+0.56
1991						
UCR 237A × CB46	Dry pot	18.36	18.48	18.82	0.17	−0.29
	Wet pot	20.16	19.78	20.45	0.34	+0.04
	Wet soil	21.01	20.66	21.04	0.20	+0.16
TVx 309-1G × Prima	Dry pot	18.65	18.40	19.88	0.74	−0.49
	Wet pot	19.99	19.74	20.79	0.52	−0.28
	Wet soil	21.09	20.48	21.58	0.55	+0.06
UCR 237A × Prima	Dry pot	18.69	18.48	19.88	0.70	−0.49
	Wet pot	19.82	19.78	20.79	0.50	−0.46
	Wet soil	21.10	20.54	21.38	0.42	+0.14
TVx 309-1G × CB46	Dry pot	18.71	18.40	18.82	0.21	+0.10
	Wet pot	20.03	19.74	20.45	0.36	−0.06
	Wet soil	20.80	20.42	21.07	0.32	+0.06

[a]Note comparisons are being made across two different experiments (pot and natural soil) conducted in a similar field aerial environment. In 1991, samples for Δ were taken at a similar early date in both experiments, whereas in 1990, samples were taken 1 month earlier in the pot than the natural soil experiment.

tion of the F_1 from the mid-parent mean did not change substantially or consistently with date of sampling (Table IV).

We tested an alternative hypothesis that the difference in Δ ranking of the F_1's, compared with the parents was due to differences in rooting environment. The test is not rigorous because we had to use data from different experiments which were not designed to test the hypothesis. These studies with plants in different soil environments but similar aerial environments over 2 years did support the hypothesis. With five of six crosses, Δ of the F_1's were substantially less than the mid-parent means under dry pot conditions, but greater than the mid-parent means under wet natural soil conditions (Table V). Under wet soil conditions, Δ of the F_1's was more similar to Δ of P_2, whereas under dry pot conditions, Δ of the F_1's was more similar to Δ of P_1. Data for W, which were only obtained in the pot experiments, are consistent with the hypothesis that the rooting environment is influencing dominance relations for W (Table VI) as well as for Δ. We speculate that the physiological basis for this effect is that the hybrids are experiencing stronger root signals than the parents, in response to dry and restricted rooting conditions. The Δ of the F_1's decreased much more than that of either parent in response to drought and restricted rooting, due possibly to enhanced levels of root signals causing greater stomatal closure of the F_1's. A phytohormonal basis for heterosis has also been proposed for maize (Rood *et al.*, 1988). The ecological significance of the cowpea data is that they indicate possible broader adaptation of the F_1's compared with either parent, assuming that low Δ and high W are advanta-

Table VI   Does the Soil Environment Influence Conclusions Concerning the Inheritance of Water-Use Efficiency?

Parent 1 × parent 2	Soil environment	F_1(g/kg)	P_1(g/kg)	P_2(g/kg)	$\frac{P_2 - P_1}{2}$ (g/kg)	$F_1 - \left(\frac{P_1 + P_2}{2}\right)$ (g/kg)
1990						
UCR 237A × CB46	Dry pot	4.18	4.13	2.51	0.81	+0.86
	Wet pot	3.63	3.92	2.95	0.48	+0.20
TVx 309-1G × Prima	Dry pot	4.19	4.15	2.54	0.80	+0.84
	Wet pot	3.74	4.04	2.79	0.62	+0.32
1991						
UCR 237A × CB46	Dry pot	5.58	5.39	5.13	0.13	+0.32
	Wet pot	4.29	4.27	3.99	0.14	+0.16
TVx 309-1G × Prima	Dry pot	5.50	5.54	5.01	0.26	+0.22
	Wet pot	4.44	4.46	4.15	0.16	+0.14
UCR 237A × Prima	Dry pot	5.32	5.39	5.01	0.19	+0.12
	Wet pot	4.46	4.27	4.15	0.06	+0.25
TVx 309-1G × CB46	Dry pot	5.59	5.54	5.13	0.20	+0.26
	Wet pot	4.34	4.46	3.99	0.24	+0.12

[a]Data are from experiments conducted with plants in pots under field conditions.

geous in dry environments, and high Δ is advantageous in wet environments. These data (Tables V and VI) also have implications for plant breeding. The adaptive G × E may only be present in F_1 hybrids, since we have not detected it in the cultivars and accessions that we have studied which are expected to be highly homozygous. Selecting hybrids with high Δ under wet conditions but low Δ under dry conditions may be an effective breeding strategy for crop species and cultural conditions where hybrid cultivars are desirable. Unfortunately, for cowpea, there are no other obvious justifications for developing hybrid cultivars, and breeding programs emphasize development of pure-line cultivars (cowpea usually is highly self-pollinated). Environmentally induced changes in dominance relations will also complicate interpretations of studies of inheritance and may be responsible for the contrasting dominance relations reported for Δ and W in wheat by Ehdaie *et al.* (Chapter 27, this volume). It should be emphasized, however, that the hypothesis proposed concerning effects of rooting environment on the ranking of parents and hybrids for Δ is speculative and has not been tested in a rigorous manner.

VI. Heritability of C-Isotope Discrimination and Genetic Correlations with Other Traits

Progress in breeding will be influenced by the heritability of Δ and any genetic correlations with other desirable and undesirable traits. Associations between low Δ and late flowering have been reported for cowpea (Hall *et al.*, 1990), peanut (Hubick *et al.*, 1986), barley (Craufurd *et al.*, 1991), and bean (White, Chapter 25, this volume), and late flowering would not be desirable in many dry environments. Studies were conducted of two crosses among parents having different Δ and earliness with moderate F_2 populations (110 and 116), F_1's, and parents grown under wide spacing in well-watered field conditions at Riverside, California. Broad-sense heritabilities were high for days to flowering (0.72 and 0.80) which was expected since this trait can be effectively manipulated by selection in the F_2 generation. In contrast, broad-sense heritabilities for Δ were low or moderate (0.27 and 0.58), indicating that family selection in advanced generations may be more effective than single plant selection in the F_2. The genetic correlations between days to flowering and Δ were negative and low to moderate (-0.14 and -0.46). This indicates that in breeding for dry environments it may be advisable to use large populations in selecting for earliness in the F_2 generation, since this may indirectly result in increases in Δ, and then select for low Δ among early families in advanced generations.

We have observed weak positive associations between Δ and harvest index (HI = the ratio of grain yield to the total shoot biomass) in cowpea, and a similar association has been reported for peanut (Wright *et al.*, Chapter 25, this volume). Any genetic correlations of this type would reduce progress in breeding for increased grain yield in dry environments through selection for low Δ, but may be advantageous in breeding to increase yield potential, through selection for high Δ. Studies were conducted of two crosses among parents having different Δ, modest differences in HI, and similar earliness, with moderate F_2 populations (144 and 148), F_1's, and parents under wide spacing in well-watered field conditions at Riverside, California. Broad-sense heritabilities were low or moderate for both HI (0.36 and 0.54), and Δ (0.30 and 0.50). The genetic correlations between HI and Δ were low and positive (0.13 and 0.36). These data indicate that selection for HI may result in modest indirect selection for high Δ.

VII. Breeding Strategies for C-Isotope Discrimination and Progress in Breeding

Theory and empirical studies have indicated that low Δ confers high W, which could improve adaptation to water-limited environments, providing it is combined with traits that enable cultivars to use most of the available water, and effectively partition carbohydrates to grain (Hall, 1990). In contrast, empirical evidence indicates associations between high Δ and high yield potential or possibly general adaptation to specific local environments. In breeding hybrid cultivars, selection for hybrids with high Δ under optimal conditions but low Δ under water-limited conditions might be effective. In breeding pure-line cultivars, it is not obvious whether, for particular commercial production zones, one should select for lower or higher Δ, and it is possible that intermediate levels are needed, with the specific level that is most adaptive varying depending upon the environmental conditions of the production zone (Hall, 1981). Consequently, given present limited understanding, it would appear advisable to breed some lines with lower Δ and others with higher Δ to test hypotheses concerning the contributions of different levels of Δ to adaptation in different field environments.

We have initiated a breeding project to develop cowpea lines with improved adaptation to water-limited environments by combining low Δ with early flowering, moderately high seasonal water use, and high HI. The F_2 populations were developed from crosses between a parent with low Δ and early flowering (TVx 309-1G) and two parents with low Δ but late flowering (UCR 237A and Vita 7). Seven thousand of these F_2 plants were subjected to water-limited conditions, being grown on stored soil moisture, but wide spacing. Plants were selected (281) which had adequate earliness, and their shoot biomass and HI were measured. Out of the 281 plants, we chose 44 which had high HI and moderate-to-high shoot biomass (as an indirect indication of deep rooting and substantial water extraction from the soil profile). These 44 F_2-derived lines were advanced two generations in a glasshouse and then sown as F_5 lines under dry, stored soil moisture conditions together with the three parents, and three lines with high Δ and high yield potential (CB46, H8-9, and H14-10) as controls. The experimental design involved four completely randomized blocks.

The three parents had about the lowest Δ (UCR 237A, Vita 7, and TVx 309-1G had values of 19.4, 19.5, and 19.8‰, respectively), whereas the three controls had the highest Δ (H14-10, CB46, and H8-9 had values of 21.0, 20.9, and 20.7‰, respectively). Of the 44 F_5 lines, we chose 29 lines that were early. These F_5 lines had intermediate Δ values ranging from 19.6 to 20.7‰ (Fig. 3) and an average of 20.2‰. The three control lines with high Δ also had the highest HI with values of 51 to 53%, whereas the low Δ parent, TVx 309-1G, had a value of 43% and the HIs of the other two low Δ parents were close to zero. The HI values of the F_5 lines ranged from 34 to 50% (Fig. 3) with an average value of 44%. Apparently, the F_2 selection for high HI (and earliness) had been effective but had also resulted in a substantial increase in Δ of 0.6‰ (mean of F_5—parental mean). Presumably this was a consequence of the genetic correlations described earlier. Studies with peanut, in which F_2 plants were selected based upon low Δ, high HI, or the best combination of both, indicated similar positive correlations between Δ and HI, and the authors concluded that concurrent improvement of W and HI may be difficult but possible in peanut (Wright *et al.*, Chapter 17, this volume).

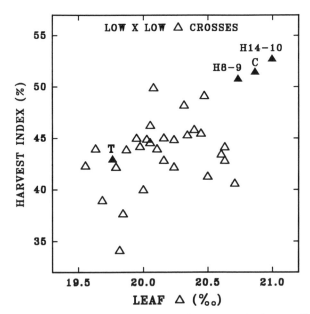

Figure 3.   Harvest index (grain yield/total shoot biomass) and C-isotope discrimination of 29 F_5 progeny (Δ) from crosses between TVx 309-1G (T), and UCR 237A and Vita 7 (data not shown) grown in four randomized complete blocks under dry stored soil moisture conditions. Variety CB46 (C) and two advanced lines (H8-9 and H14-10) were included as controls with high C-isotope discrimination.

The critical question, at this stage, is whether selecting the F_5 cowpea lines for low Δ would contribute to breeding for water-limited environments. Seven of the twenty-nine lines which had the lowest Δ also had moderately high shoot biomass (Fig. 4) and grain yield, but the line with the highest grain yield had intermediate Δ (Fig. 5). Studies are needed with more crosses and larger populations to determine whether selection for low or intermediate Δ values would result in improved adaptation to water-limited environments.

We have initiated another breeding project to combine high Δ and high yield potential. We developed an F_2 population from a cross involving two parents with high Δ (CB46 and Prima). Prima also has heat tolerance during pod set (Patel and Hall, 1990) and empirical breeding has indicated that incorporating this heat tolerance can enhance yield potential (Hall, 1992). The F_2 population was screened under extremely hot, well-irrigated, field conditions (Hall and Patel, 1990), and 20 plants were selected which exhibited abundant pod set. These 20 F_2-derived lines were advanced one generation in a glasshouse and then sown as F_4 lines under well-irrigated field conditions with optimal temperatures, together with the two parents, a low Δ control (TVx 309-1G), and two control genotypes with heat tolerance (1393 and 160). The experimental design involved four completely randomized blocks.

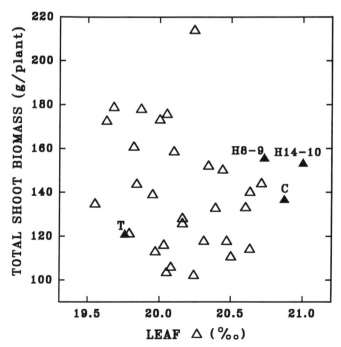

Figure 4. Total shoot biomass and C-isotope discrimination of 29 F_5 progeny (Δ), a parent (T), and three controls (C, H8-9, and H14-10) as described in the legend to Fig. 3.

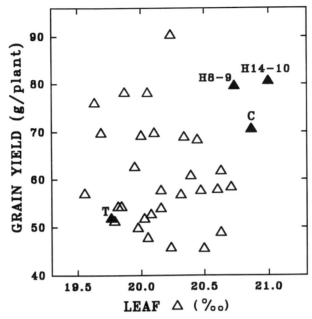

Figure 5. Grain yield and C-isotope discrimination of 29 F_5 progeny (Δ), a parent (T), and three controls (C, H8-9, and H14-10) as described in the legend to Fig. 3.

The three controls had the lowest Δ (TVx 309-1G, 160, and 1393 had values of 21.1, 21.2, and 21.4‰, respectively), and the two parents, CB46 and Prima, had values of 21.5 and 21.9‰, respectively. One of the F_4 lines was eliminated because the plants were very small. The Δ of the remaining F_4's ranged from 21.5 to 22.1 (Fig. 6) with an average value of 21.8‰, which was similar to the mid-parent mean. The heat-tolerant controls had about the highest HIs (51 and 49% for 1393 and 160, respectively) while the low Δ control, TVx 309-1G, had about the lowest HI of 38%. The HI of the F_4's ranged from 36 to 50% (Fig. 6) with a mean of 44%, which was less than the mean of the parents (48 and 45% for CB46 and Prima, respectively). Apparently, selection for abundant pod set under hot conditions in the F_2 generation did not result in F_4's with as high an average HI under more optimal temperatures as the parents but high Δ was maintained in the F_4's.

The critical question, at this stage, is whether selecting F_4 lines for high Δ would contribute to breeding for yield potential. Two of the nineteen lines had the highest grain yields but their Δ values were intermediate (Fig. 7). Studies are needed with more crosses and larger populations to determine whether selecting for high or intermediate Δ values would result in increased grain yields in favorable environments.

Attempting to determine the value of Δ in selection by comparing Δ with yield, in principle, will only be valid if genotypes are used that have similar genetic backgrounds, except for genes influencing Δ. Positive associations

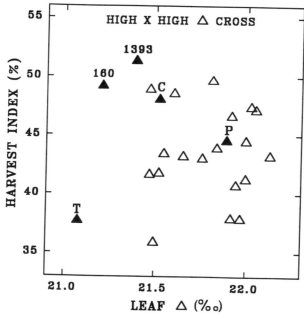

Figure 6.   Harvest index (grain yield/total shoot biomass) and C-isotope discrimination of 19 F_4 progeny (Δ) from crosses between CB46 (C) and Prima (P) grown in four randomized complete blocks under well-irrigated conditions. Controls included TVx 309-1G (T) for low C-isotope discrimination and 1393 and 160 for high yield potential through heat tolerance.

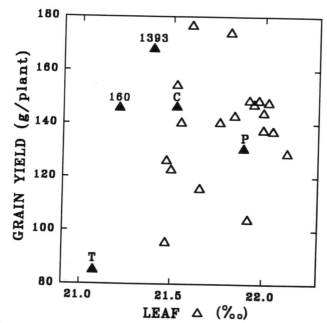

Figure 7.   Grain yield and C-isotope discrimination of 19 F_4 progeny (Δ), two parents (C and P), and controls for low C-isotope discrimination (T) and high yield potential (1393 and 160) as described in the legend to Fig. 6.

between grain yield and Δ reported for barley in dry environments (Craufurd et al., 1991) could have been due to earliness having a strong physiological influence on grain yield, and genetic correlations between earliness and Δ, and HI and Δ as were described for cowpea. Another possible explanation is that the later maturing barley genotypes suffered more drought, such that Δ in grain would have been low due to drought-induced effects, the grain yield would have been low due to the drought stress associated with later flowering. Other possible explanations are discussed by Acevedo (Chapter 26, this volume). The main conclusion, however, is that breeders should first select for appropriate levels of earliness which can be done with F_2 plants because it usually is highly heritable and easy to score. After which it is then appropriate to ask whether selection among lines with similar earliness for low, intermediate, or high Δ would result in improvements in yield. Interactions with other ideotype traits may influence correlations between Δ and grain yield. For example, genotypes with deeper rooting could have greater grain yields in water-limited environments where they are able to obtain more water, and Δ could be higher than for shallower rooting genotypes that suffer more stress. In contrast, genotypes with similar rooting depth but lower Δ may have higher grain yields due to higher W providing that the additional biomass is partitioned to grain. Interactive effects of this type could explain the inconsistent associations between grain yield and Δ reported for common bean by White (Chapter 25, this volume). Also, Johnson et al. (Chapter 18, this volume) reported a lack of consistent association between Δ and forage yield for crested wheatgrass.

A breeding strategy has been designed for enhancing yield potential of grain legumes, such as cowpea. It is based upon the hypothesis that cultivars with higher HI may have higher yield potential in the environments of the next century due to their greater responsiveness to the elevated atmospheric concentrations of CO_2 that are likely to be present (Hall, 1990). This hypothesis assumes that during thousands of years of evolution under low atmospheric CO_2 concentrations, plant species have developed a relatively low ratio of reproductive sink-to-photosynthetic source, and that adaptation to higher atmospheric CO_2 concentrations requires that plants have a higher ratio, and, consequently, higher HI. The objective is to combine high HI with high Δ, simply because high Δ has been associated with high yield potential, although as was stated earlier, intermediate Δ may be more effective. Parents with different genetic origin but high HI, high Δ, and high yield potential would be intercrossed. Selection for desirable earliness and seed characteristics and high HI would be conducted in the F_2 generation with plants at wide spacing under well-watered conditions and optimal temperatures. In advanced generations, family selection for agronomic characters and high Δ would be conducted, after which the grain yields of selected lines would be evaluated in comparison with the highest yielding controls that are available.

Breeding strategies for enhancing the adaptation of grain legumes, such as cowpea, to dry environments are more complex since they should consider the seasonal patterns of drought occurring in commercial production environments. For example, selection during the F_2 generation could be conducted with widely spaced plants under dry conditions and focus on highly heritable characters that confer breadth to adaptation such as the combination of early flowering and delayed leaf senescence (Gwathmey and Hall, 1992). It may also be appropriate to select during the F_2 generation for moderately high HI, and for high shoot biomass production under stored soil moisture conditions to indirectly select for deep rooting. It should be noted that selecting for both high HI and high shoot biomass production is equivalent to selecting for high grain production. In advanced generations, family selection for Δ would be conducted under water-limited conditions, but it is not clear at this time whether the selection should be for low or intermediate values.

VIII. Summary

Guidelines have been developed for breeding cowpeas with differences in Δ, W, adaptation to drought, and yield potential. Leaf samples were more effective than seed samples for detecting genotypic differences in Δ. An earlier sampling of leaves from lower nodes may be more effective in detecting genotypic differences because broad-sense heritabilities were higher, even though genetic variances were greater for later samples. Ability to detect genotypic differences in Δ was similar under well-watered or extremely dry field conditions. Apparently, when screening accessions for genotypic differences in Δ, a range of hydrologic conditions could be effective, and it would be advisable to use nursery conditions that are similar to the commercial production conditions for which improved cultivars are being developed.

Ranking of accessions for Δ was remarkably consistent when the same genotypes were grown over different drought conditions, years, and dates of sampling, but at the same location (Riverside, CA). In contrast, ranking varied for Δ among some accessions grown at Lubbock, Texas, compared with those grown at Riverside, California, and rankings were completely different for accessions grown at Bambey, Senegal, compared with those grown at Riverside, California. This variation in consistency of genotypic ranking for Δ is similar to that of complex quantitative agronomic traits, such as grain yield. Apparently, breeding for differences in Δ values might be effective in particular crop production zones but the performance, with respect to Δ, may not be transferrable to radically different production zones.

Inheritance of Δ and W were shown to be nuclear rather than maternal. Surprisingly, F_1 hybrids exhibited partial dominance for low Δ in dry environments with restricted rooting due to pots, values more similar to mid-parent means in wet environments with restricted rooting, and partial dominance for high Δ in wet, natural soil conditions. Broad-sense heritabilities for Δ determined for F_2 populations, F_1's, and parents from four crosses grown under well-watered field conditions were low to moderate (0.27, 0.30, 0.50, and 0.58), compared with values for HI (0.36 and 0.54) and days to flowering (0.72 and 0.80). Genetic correlations between Δ and days to flowering were negative and low to moderate (−0.14 and −0.46), and genetic correlations between Δ and HI were low and positive (0.13 and 0.36). Apparently, when breeding to combine different levels of Δ and traits influencing time of flowering and HI, it would be most effective to apply selection pressure for earliness and possibly HI during early (F_2) generations with large populations and select for Δ among families of advanced generations.

Selection studies have been initiated with cowpea which demonstrated that F_2 selection under drought for earliness and high HI was effective but that it also resulted in a substantial increase in Δ as predicted by the genetic correlations. From these studies it was not clear whether family selection for low or intermediate Δ values would be most effective in increasing yield under drought. Cowpea breeding lines with high yield potential, developed by empirical procedures, also have high Δ and high HI. Studies have been initiated to attempt to increase yield potential by selecting under well-watered conditions to combine high HI and high Δ but the results, to date, are inconclusive.

Acknowledgments

Cowpea research reported in this paper was supported by the Bean/Cowpea CRSP, USAID Grant No. DAN-1310-G-SS-6008-00, and the Southwest Consortium, New Mexico State University, USDA Subagreement No. 88-34186-3340. The opinions and recommendations are those of the authors and not necessarily those of USAID. We appreciate the assistance of Dan Krieg in carrying out the experiments in Lubbock, Texas, and Samba Thiaw in carrying out the experiment in Bambey, Senegal. We appreciate the assistance of Graham Farquhar and colleagues who conducted the measurements of tissue carbon isotope composition.

References

Condon, A. G., R. A. Richards, and G. D. Farquhar. 1987. Carbon isotope discrimination is positively correlated with grain yield and dry matter production in field-grown wheat. *Crop Sci.* **27**: 996–1001.

Craufurd, P. Q., R. B. Austin, E. Acevedo, and M. A. Hall. 1991. Carbon isotope discrimination and grain yield in barley. *Field Crops Res.* **27**: 301–313.

Ehleringer, J. R. 1990. Correlations between carbon isotope discrimination and leaf conductance to water vapor in common beans. *Plant Physiol.* **93**: 1422–1425.

Ehleringer, J. R., S. Klassen, C. Clayton, D. Sherrill, M. Fuller-Holbrook, Q. Fu, and T. A. Cooper. 1991. Carbon isotope discrimination and transpiration efficiency in common bean. *Crop Sci.* **31**: 1611–1615.

Farquhar, G. D., and R. A. Richards. 1984. Isotopic composition of plant carbon correlates with water-use efficiency of wheat genotypes. *Aust. J. Plant Physiol.* **11**: 539–552.

Farquhar, G. D., M. H. O'Leary, and J. A. Berry. 1982. On the relationship between carbon isotope discrimination and the intercellular carbon dioxide concentration in leaves. *Aust. J. Plant Physiol.* **9:** 121–137.

Fu, Q. A., and J. R. Ehleringer. 1992. Paraheliotropic leaf movements in common bean under different soil nutrient levels. *Crop Sci.* **32:** 1192–1196.

Gwathmey, C. O., and A. E. Hall. 1992. Adaptation to midseason drought of cowpea genotypes with contrasting senescence traits. *Crop Sci.* **32:** 773–778.

Hall, A. E. 1981. Adaptation of annual plants to drought in relation to improvements in cultivars. *HortScience* **16:** 37–38.

Hall, A. E. 1990. Plant adaptation to hot and dry stresses in relation to horticultural plant breeding. Plenary Lecture, XXIII International Horticultural Congress, Firenze, Italy, August 27–September 1, 1990.

Hall, A. E. 1992. Breeding for heat tolerance, Vol. 10, Chapter 5, pp. 129–168. *In* Plant Breeding Reviews. Wiley, New York.

Hall, A. E., and P. N. Patel. 1990. Breeding heat-tolerant cowpeas, pp. 41–46. *In* J. C. Miller, J. P. Miller, and R. L. Fery (eds.), Cowpea Research—A U.S. Perspective. Texas Agricultural Experiment Station, College Station, Texas.

Hall, A. E., R. G. Mutters, K. T. Hubick, and G. D. Farquhar. 1990. Genotypic differences in carbon isotope discrimination by cowpea under wet and dry field conditions. *Crop Sci.* **30:** 300–305.

Hall, A. E., R. G. Mutters, and G. D. Farquhar. 1992. Genotypic and drought-induced differences in carbon isotope discrimination and gas exchange of cowpea. *Crop Sci.* **32:** 1–6.

Hubick, K. T., G. D. Farquhar, and R. Shorter. 1986. Correlation between water-use efficiency and carbon isotope discrimination in diverse peanut (*Arachis*) germplasms. *Aust. J. Plant Physiol.* **13:** 803–816.

Hubick, K. T., R. Shorter, and G. D. Farquhar. 1988. Heritability and genotype × environment interactions of carbon isotope discrimination and transpiration efficiency in peanut (*Arachis hypogaea* L.). *Aust. J. Plant Physiol.* **15:** 799–813.

Ismail, A. M. 1992. Water-use efficiency, photosynthesis, transpiration and carbon isotope discrimination of contrasting cowpea genotypes. pp. 77. Ph.D. Dissertation, University of California, Riverside.

Ismail, A. M., and A. E. Hall. 1992. Correlation between water-use efficiency and carbon isotope discrimination in diverse cowpea genotypes and isogenic lines. *Crop Sci.* **32:** 7–12.

Ismail, A. M., and A. E. Hall. 1993. Inheritance of carbon isotope discrimination and water-use efficiency in cowpea. *Crop. Sci.* 33.

Kirchhoff, W. R., A. E. Hall, and W. W. Thomson. 1989. Gas exchange, carbon isotope discrimination, and chloroplast ultrastructure of a chlorophyll-deficient mutant of cowpea. *Crop Sci.* **29:** 109–115.

Passioura, J. B. 1972. The effect of root geometry on the yield of wheat growing on stored water. *Aust. J. Agric. Res.* **23:** 745–752.

Patel, P. N., and A. E. Hall. 1990. Genotypic variation and classification of cowpea for reproductive responses to high temperature under long photoperiods. *Crop. Sci.* **30:** 614–621.

Rood, S. P., R. I. Buzzell, L. N. Mander, D. Pearce, and R. P. Pharis. 1988. Gibberellins: A Phytohormonal basis for heterosis in maize. *Science* **241:** 1216–1218.

Ziska, L. H., and A. E. Hall. 1983. Seed yields and water use of cowpeas (*Vigna unguiculata* (L.) Walp.), subjected to planned-water-deficit irrigation. *Irrig. Sci.* **3:** 237–245.

24

Analysis of Restriction Fragment Length Polymorphisms Associated with Variation of Carbon Isotope Discrimination among Ecotypes of *Arabidopsis thaliana*

Josette Masle, Jeong Sheop Shin, and Graham D. Farquhar

I. Introduction

Genetic variation in transpiration efficiency (*W*) has been shown in numerous crop species (see other chapters in this volume and Farquhar *et al.*, 1989). This raises two questions:

1. How can this genetic variation be exploited in terms of plant breeding or genetic engineering to produce plant material with the desired transpiration efficiency?
2. What are the bases of this variation at a mechanistic level?

These two questions relate to the identification, and introduction in a new genetic background, of the gene(s) controlling transpiration efficiency. Of course the introduction of desired genes in a desired genetic background does not necessarily require that the genes be identified (see in this volume current selection programs for *W*). Many useful genes have been successfully introduced in plants using classic breeding methods, i.e., by (a) identification of suitable parents as donor and receptor, (b) crossing of these parents followed by several backcrosses to eliminate possible undesirable traits associated with the trait of interest, and (c) by selecting plants with the desired *W*. However, the process is slow and allows limited understanding of the genetic control of the trait, especially if it is a quantitative trait dependent on several individual components.

We decided to undertake the molecular analysis of plant transpiration efficiency, an approach which in the long-term can fulfill two goals—gene isolation and plant transformation—but in the short term aims primarily at identifying the genetic control of *W*.

II. Analysis of Naturally Occurring DNA Polymorphisms versus Analysis by Mutagenesis

Two routes can a priori be envisaged to conduct the molecular analysis of genetic variation in transpiration efficiency. One route is to search for DNA polymorphisms associated with variation of transpiration efficiency among lines or ecotypes of the species of interest, and once such polymorphisms have been identified proceed to map and clone the genes involved. An alternative is to screen a population of artificially mutagenized seedlings for mutations affecting transpiration efficiency and, if successful, proceed to gene mapping and complementation.

At first sight the second approach is more straightforward. Indeed when dealing with natural genetic variation as opposed to variation induced by mutation one has to expect other genetic differences than those responsible for the trait of interest. However, until recently, the only methods with proven records in generating mutants were chemical mutagenesis (usually using ethyl methanesulfonate, EMS) and mutagenesis by irradiation. Both yield single point mutations. Now, it is likely that a whole suite of genes, and probably more than one main gene, are involved in the control of transpiration efficiency. This means that single point mutations might not induce any detectable change in the phenotype. And if they do, it is most likely to be in the direction of dramatically reduced *W* via alterations of essential functions, i.e., in a way totally irrelevant to the nature of naturally occurring genetic variation. It is obvious, for example, that mutations affecting the pathway of ABA biosynthesis or the activity of fundamental enzymes of the Calvin cycle would greatly affect *W*.

In 1989, Martin *et al.* published very encouraging results regarding the genetic control of *W*, based on analysis of DNA polymorphisms between two species of tomato. They showed that 70% of the variation in carbon isotope discrimination (Δ) between *Lycopersicon esculentum* ("drought-sensitive" cultivated tomato with high Δ) and *L. penellii* ("drought tolerant" wild tomato with low Δ) was associated with only three DNA fragments which they mapped to three different chromosomes. They had previously shown that the variation in Δ among the two species and their F_1 progeny was strongly correlated with season-long transpiration efficiency (Martin and Thorstenson, 1988). One of the polymorphic markers that were identified is linked with more than one gene since it was characterized by both additive and nonadditive gene actions. The two other markers only showed linear gene action; it is still possible, however, that they are also associated with more than one gene (Martin *et al.*, 1989).

On the basis of these results and of the above considerations we decided to analyze the DNA polymorphisms associated with variation in *W* between genotypes *within* a species. We chose *Arabidopsis thaliana* as the species to study. *A. thaliana* is a small weed of the crucifera family, with a very short life cycle (ca. 6 to 8 weeks under controlled conditions). Moreover, it has a much smaller genome (0.7 to 1×10^8 nucleotide pairs), with fewer repetitive sequences of DNA, than other higher plants. Because of these unique features, *A. thaliana* has been widely adopted as a model plant for genetic studies.

III. Principle of DNA Polymorphism Analysis

The classical technique for detecting DNA polymorphisms is the analysis of restriction fragment length polymorphisms (RFLPs). This technique is shown schematically in Fig. 1. Briefly, RFLPs detect differences in the length of specific DNA fragments after digestion of genomic DNA with sequence-specific restriction endonucleases (Botstein *et al.*, 1980). Several thousand so-called restriction fragments are generated by endonuclease digestion. If the, say, 6-bp recognition sequence specific to a given enzyme was distributed at random on the genome, one would expect a cut every 4096 bp (i.e., 4^6). In fact fragment lengths vary between a few hundred and 20,000 bp. Polymorphic DNA fragments are detected by Southern blot analysis using radioactively or enzymatically labeled probes. The DNA sequences used as hybridization probes are either genomic DNA or cDNA (complementary DNA) clones, whose lengths vary from a few hundred to a few thousand base pairs. They are generally selected so as to contain only low copy DNA (Chang *et al.*, 1988; Nam *et al.*, 1989). Polymorphisms are simply the results of changes in the position of the restriction sites, which are due to base deletions or insertions or to single base changes.

RFLP markers are typically codominant (i.e., one band is visible in each of the two genotypes compared), allowing the scoring of progeny as being homozygous or heterozygous (in contrast to "present vs null" markers where one band is visible in one genotype but none in the other), and segregate in a mendelian fashion (e.g., Hulbert *et al.*, 1988). This is why it was proposed about 10 years ago that they could be used as genetic markers of traits of interest (Botstein *et al.*, 1980).

By carrying out the above analysis on parental lines and on their F_2 or usually F_3 segregrating progeny, one looks at the correlation between the segregation pattern of the polymorphic fragments and the segregation pattern(s) of the trait(s) of interest. The correlation is quantified using multiple regression analysis and maximum likelihood tests. Computerized programs have been developed specifically for that purpose (Lander and Green, 1987; Lander *et al.*, 1987). The threshold used to decide that an RFLP marker and the gene of interest are linked is arbitrary; the criterion used is known as the "lod score," which is expressed as $\log_{10}$ of the ratio of

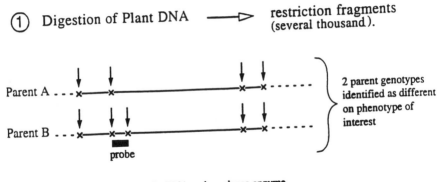

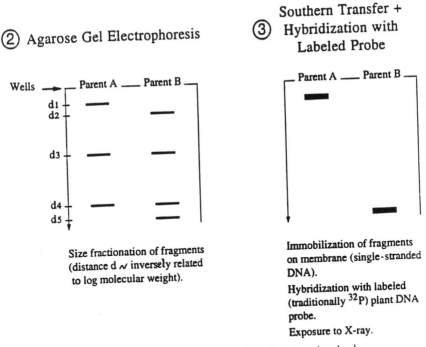

Figure 1.   Principle of RFLP analysis; key steps involved.

the probability of linkage to the probability of no linkage (Botstein *et al.*, 1980). A lod score of about 3 is often used, corresponding to a probability of 0.001 that the observed data could be generated by an unlinked marker/ gene. Gillet (1991) provides a more complete discussion of the genetic analysis and interpretation of restricted fragments patterns.

IV. Analysis of DNA Polymorphisms Associated with Variation of Carbon Isotope Discrimination in *Arabidopsis thaliana*

From the above brief description of the technique, it is clear that the chances of identifying polymorphic RFLP DNA markers associated with the phenotypic variation of interest are greater if (a) many probes (RFLP markers) are used to screen the template genome; (b) these markers are evenly spread on the different linkage groups constituting the genome and do not contain highly repeated sequences of DNA, which would complicate the segregation analysis; and (c) the probed genome also contains a small proportion of repeated sequences of DNA.

A. thaliana meets all these requirements beautifully (Pruitt and Meyerowitz, 1986). With about 100 Mbp its nuclear genome is the smallest known among higher plants (Leutweiler *et al.*, 1984) with an estimated distance between repetitive sequences of DNA of 125 kb compared to 1.4 kb in tobacco, for example, or 0.3 kb in pea. More than 200 RFLP markers are now available, which, together with phenotypic markers, provide a high-density map of the five *Arabidopsis* chromosomes (Chang *et al.*, 1988; Nam *et al.*, 1989). These same characteristics also enhance the chances of cloning the gene(s) of interest by "chromosome walking" along the polymorphic segments.

A. Genetic Variation of Carbon Isotope Discrimination among *Arabidopsis thaliana* Ecotypes

Forty ecotypes of *A. thaliana* were screened for variation in carbon isotope discrimination (Δ). Carbon isotope discrimination was chosen as a criterion rather than transpiration efficiency directly because a good and stable correlation had been found between the two characteristics across a broad range of species (Farquhar *et al.*, 1989) and because Δ is so much easier to measure. Despite its short life cycle, *A. thaliana* is a species characterized by slow establishment, with the roots being fragile for a long time and therefore hard to separate from soil by washing. This means that it is especially difficult to accurately measure whole-plant transpiration efficiency of soil-grown plants of this species.

Plants were grown in a naturally lit glasshouse, providing an irradiance of 600 to 1000 μE m^{-2} s^{-1}, at high air humidity and temperatures of about 23°C day/18°C night. There was a wide spread of Δ values (Fig. 2), with no obvious relation to geographic origin. A second screen done later in the year, under lower temperatures and irradiances, gave systematically greater values, covering a somewhat smaller range. Most importantly, however, the genetic ranking was usually maintained ($\Delta_2 = 0.51\Delta_1 + 11.93$, $r = +0.83$, $n = 19$, where Δ_1 and Δ_2 refer to the Δ values obtained in the first and second run, respectively). Three ecotypes (circled on the histogram in Fig. 2) were chosen at each end of the spectrum of carbon isotope discrimination values to be analyzed for DNA polymorphisms.

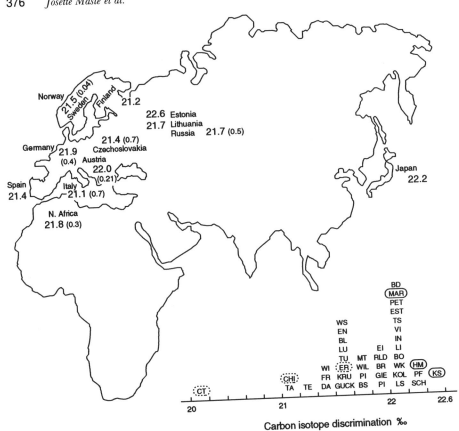

Figure 2. Distribution of leaf carbon isotope discrimination measured on glasshouse-grown *Arabidopsis thaliana* ecotypes (ecotypes identified by 2 to 4 letters). Circled are the three low Δ and three high Δ ecotypes (dotted and solid circle, respectively) which were selected for RFLP analysis. The map shows the mean Δ value for ecotypes originating from the same country. Δ values were calculated from leaf carbon isotope composition, taking −8‰ for the isotopic composition of carbon in CO_2 in air.

B. Analysis of DNA Polymorphisms among *Arabidopsis thaliana* Ecotypes Differing in Carbon Isotope Discrimination

Twenty-five *Arabidopsis* DNA probes (cloned in λ phage) were initially used to probe genomic DNA of the six selected ecotypes. Most of these probes were anonymous low copy number clones of *Arabidopsis* genomic DNA; a few were cloned genes. These 25 probes (kindly provided to us by E. Meyerowitz's group) were chosen among the molecular markers mapped by Chang *et al.* (1988) on the *Arabidopsis* genome (with a mean insert size of 12.5 kb) so as to be evenly distributed over the five *Arabidopsis* linkage groups. Genomic DNA was digested with two restriction enzymes in the first instance, *Bgl*II and *Eco*RI. These are two endonucleases with a 6-bp recognition sequence, of high A + T nucleotides content, which therefore are more likely to detect RFLPs in the *Arabidopsis* genome which has a 58% content of A + Ts. With each restriction enzyme, 19 of the 25 probes tested revealed at least one polymorphism between at least one low Δ line and one high Δ line (Fig. 3).

All nine pairs of low–high Δ ecotypes were polymorphic, to a varying degree (see Fig. 3), with two pairs, CHI and HM, and CT and HM, being highly polymorphic (polymorphisms detected by two-thirds of the probes). Most DNA markers revealed only one polymorphic locus between a given pair of ecotypes; several probes, however, revealed several polymorphic loci. The polymorphic markers were distributed throughout the genome (Fig. 4).

Based on these promising results, the most different of the six ecotypes analyzed above were crossed to generate F_1 plants, and the F_2 and F_3 progenies from the cross of the most polymorphic pair of ecotypes, CHI and HM, were grown for analysis of the RFLPs and phenotypic segregation patterns.

Restriction Enzyme	Bgl II	EcoRI
X-ray illegible	4	3
Number of probes detecting at least 1 polymorphic locus between at least 1 low Δ and 1 high Δ line	19	19
No polymorphisms detected	2	3
Total Probes tested	25	25

Number of polymorphic loci detected by 19 probes (using either Bgl II or EcoRI enzyme) between the following pairs of ecotypes:

CHI	/	HM	15*
CHI	/	KS	9
CHI	/	MAR	9
CT	/	HM	14
CT	/	KS	11
CT	/	MAR	6
ER	/	HM	9
ER	/	KS	9
ER	/	MAR	10

↑ Low Δ ecotypes   ↑ High Δ ecotypes

Figure 3. Results of the initial RFLP analysis conducted on six extreme ecotypes (circled in Fig. 2). *Ecotypes whose F_2 and F_3 progeny were chosen to be analyzed first.

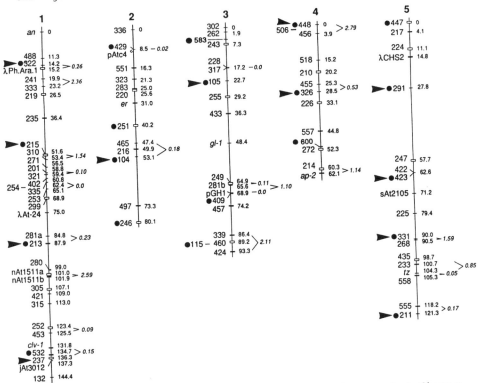

Figure 4. Genetic position on the *Arabidopsis thaliana* RFLP linkage map (as in Chang *et al.*, 1988) of the polymorphic DNA markers identified through the initial RFLP analysis summarized in Fig. 3, with two restriction enzymes (●) *Bg*III and (▶) *Eco*RI.

C. Interactions between Genetic and Nongenetic Factors on Carbon Isotope Discrimination in *Arabidopsis thaliana*

Since similar atmospheric conditions could not be reproduced with confidence for successive glasshouse runs, the six parental lines were systematically grown in each run together with their progeny. This revealed a major, unexpected problem, which has temporarily impeded any attempt at correlating phenotypic and molecular data. The 60 F_2 plants initially grown showed, as expected, a wide spread of Δ values over a 2‰ range. The carbon isotope discriminations measured on the parental lines, however, all collapsed together (see run marked by a thin arrow in Fig. 5). Such phenotypic instability was completely unexpected as earlier published genetic studies with other species showed that discrimination was a highly heritable trait (e.g., Condon *et al.*, 1987; Hubick *et al.*, 1988; and also Richards and Condon, Chapter 29, this volume). All values were low and shifted toward the expected values for the low Δ lines, which suggested that plants might have experienced some stress. However, all six parents also exhibited similar isotope compositions in another run (marked by a thick arrow in Fig. 5),

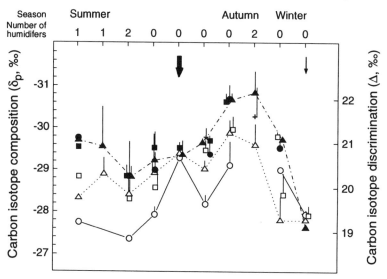

Figure 5. Variation of leaf carbon isotopic composition and corresponding isotope discrimination between different glasshouse runs for six *Arabidopsis thaliana* ecotypes, identified by ○ (CT), △ (CHI), □ (ER), ▲ (HM), ● (KS), ■ (MAR), as in Fig. 2, and also * (Ld, Landsberg) and + Col, Columbia). Vertical bars represent standard errors of the mean. N.B. Runs are ordered by season, i.e., successive runs are not shown in successive order. The arrows refer to two runs where Δ differences between ecotypes were very small (see text).

in a range of much higher isotopic discriminations. And in several subsequent runs the two groups of lines originally selected as low or high Δ lines overlapped. It may be seen from Fig. 5 that some ecotypes were phenotypically more unstable than others. What may be the explanations for such observations? *A. thaliana* is not a singular species in terms of the relationship between carbon isotope discrimination and gas exchange properties. Figure 6 shows that at the leaf level, Δ is linearly related to the ratio p_i/p_a of intercellular to ambient partial pressure of CO_2 as predicted from theory and seen in other species. A set of physiological experiments where carbon isotope discrimination and gas exchange properties were analyzed under a range of well-defined growth conditions showed that:

1. *Arabidopsis* ecotypes differ in the sensitivity of carbon isotopic discrimination to light and atmospheric humidity. This is illustrated in Fig. 7. Some ecotypes were consistently low or high Δ lines (e.g., CT and Landsberg, respectively), across a range of air vapor pressure deficits (vpd), irradiances, and light spectra. Others, like ER or Columbia are low Δ lines under high irradiances, but high Δ lines under low irradiances. Differences in sensitivity to light were confirmed in the glasshouse by relating pot-to-pot variation in Δ to spatial variation in light level within the glasshouse. For some ecotypes (e.g., KS), ranking of Δ depended on the irradiance–vpd combination. Overall, genetic differ-

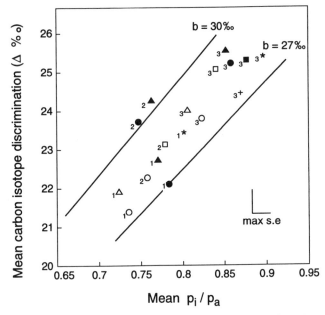

Figure 6. Relationship between leaf carbon isotope discrimination (6- to 8-week-old plants) and the ratio p_i/p_a of intercellular to ambient partial pressure of CO_2 measured by leaf gas exchange just before plants were harvested in a standard gas exchange system for several *Arabidopsis* ecotypes grown in controlled chambers. Digits beside symbols, which are the same as those in Fig. 5, refer to the growth and gas exchange measurement conditions: 13 mbar air vapor pressure deficit (vpd), 520 μE m^{-2} s^{-1} irradiance (1) or 320 μE m^{-2} s^{-1} (2); 3 mbar vpd and 320 μE m^{-2} s^{-1} (3); in all cases leaf temperature was 23°C. The lines represent simplified theoretical relationships of the form $\Delta = a + (b - a)p_i/p_{a'}$, where a (= 4.4 × 10^{-3}) and b (= 27 or 30 × 10^{-3}) denote fractionation during diffusion and carboxylation, respectively (Farquhar *et al.*, 1982).

ences in Δ were usually the greatest under low light and low vpd conditions, reaching then a maximum of about 2‰.

2. These differences in sensitivity of Δ to environmental conditions may reflect ecotypic differences in the nature of the variation in Δ (stomatal conductance versus mesophyll capacity for photosynthesis). For example, within the most polymorphic pair of low/high Δ lines (as identified in our original screens) which was chosen for subsequent RFLP analysis, the difference in isotopic discrimination is mostly due to a difference in mesophyll capacity for photosynthesis (Fig. 8). Another pair of low/high Δ lines (KS and ER) seems to differ in both stomatal conductance and photosynthetic capacity (data not shown).

3. There are substantial seed effects on Δ in *Arabidopsis*. Table I compares several ecotypes for leaf carbon isotope discrimination of plants grown together but originating from three different lots of seeds harvested in three different glasshouse runs (and therefore stored for different durations before sowing). Except for one ecotype (CT)

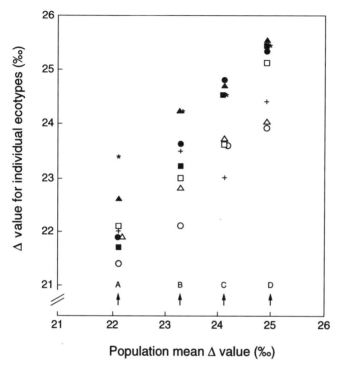

Figure 7. Effect of light (I) and air vapor pressure deficit (vpd) on genetic variation in leaf carbon isotope discrimination. The individual Δ values measured for eight *Arabidopsis* ecotypes (same symbols as in Fig. 5) under each of four different I × vpd combinations are plotted against the population mean Δ value in each environment (represented by a capital letter and an arrow on the x-axis): (A) high vpd and high I (13 mb and 520 μE m^{-2} s^{-1}, respectively); (B) high vpd and low I (320 μE m^{-2} s^{-1}); (C) low vpd (4 mb) and high I; (D) low vpd and low I.

Table I Seed Effects on Leaf Carbon Isotope Discrimination (Δ, ‰) in Six *Arabidopsis* Ecotypes (Same as Circled in Fig. 2)[a]

Seed lot	Ecotype					
	CT	ER	CHI	HM	KS	MAR
1	20.4ab	20.9c	20.8c	21.4d	21.2d	21.6e
2	20.6a	21.3d	21.1d	22.0e	21.8e	22.5f
3	20.3b	21.4d	21.8e	22.2e	21.9e	22.3f

[a]For each ecotype, plants originating from three different lots of seeds are compared in a single planting; the three lots of seeds were produced in three successive glasshouse runs and then stored in a desiccator at 3°C for 17, 12, and 5 months for lot 1, 2, and 3, respectively, until sowing. Values followed by the same letter are not significantly different ($P = 0.05$) within columns and lines.

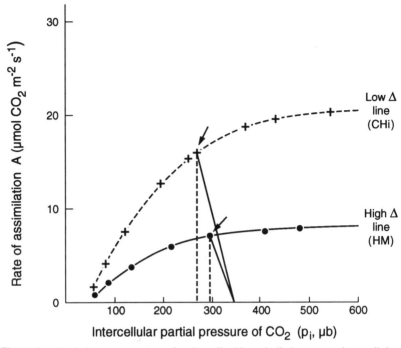

Figure 8. Typical response curve of carbon dioxide assimilation rate to intercellular partial pressure of CO_2 (p_i) for one pair of low/high Δ lines (CHI and HM, respectively). The arrows indicate the operating point on each curve and the dotted vertical lines show the corresponding p_i. The slope of the line between the operating point and p_a on the x-axis equals the stomatal conductance for CO_2.

plants from seed lot 1 showed significantly lower Δ values than those from the two other lots of seeds. It is known that seeds progressively loose viability during storage (Roberts and Ellis, 1982). However, in the present case, all seeds had been stored for a rather short time, under cool and dry conditions under which loss of viability should be slow (Roberts and Ellis, 1980). It seems therefore more likely that the seed effects shown in Table I reflect differences in environmental conditions during ripening such as in irradiance and temperature. Both the amount and the biochemical composition of seed reserves (e.g., protein type) are known to be sensitive to conditions during seed development on the mother plant and to influence enzymatic activity in the germinating seed and early seedling growth.

The fact that the six ecotypes compared in Table I were differentially affected is consistent with other reports. Roberts and Ellis (1977), for example, found a sevenfold difference in longevity between barley cultivars under identical storage conditions. It has been shown that decrease in seed viability manifests itself through decreased rates of germination and of seedling growth, through mechanisms still poorly understood. Our data show that these effects carry through to much later stages (plants in Table I were 6 to 9 weeks old and were beginning to flower), even in a species where seed reserves represent an infinitesimal contribution to seedling growth (an *Arabidopsis* seed weighs 20 to 30 μg) and affect metabolic functions involved in photosynthesis and transpiration. Seed effects were responsible for a variation of up to 1‰ in Δ, equivalent to as much as 50% of the maximum genetic difference in Δ between the most extreme parents. They therefore potentially represent an important confounding factor in the RFLP analysis of genetic variation in Δ. The practical consequence is that all the seeds necessary for segregation analysis should be generated at once, in an homogeneous controlled environment, and all F_3 families and parental lines should be assessed from seeds of same age.

The above observations show the complexity of the physiological processes involved in determining plant carbon isotope fractionation and transpiration efficiency. We believe that the effects that we identified in *A. thaliana* are likely to apply to other species and that there are other effects to be aware of when conducting genetic studies or breeding programs on Δ, such as those of root environment (Masle and Farquhar, 1988). These effects need not necessarily be understood before attempting to analyze the molecular bases of genetic variation in Δ. As a first step they only need be identified in order to control the appropriate environmental parameters when assessing the phenotype and to set these parameters so as to get maximum expression of genetic variation in Δ.

V. Prospects

Provided that genetic differences in carbon isotope discrimination, or transpiration efficiency, may be reliably assessed, i.e., strong interactions with environmental parameters are not present, RFLP analysis is a promising route with respect to the isolation of the genes involved and also to gaining insight into the mechanistic links between Δ and associated traits.

A. Genetic Analysis of the Link between Carbon Isotope Discrimination and Associated Traits

It is clear that RFLP analysis of carbon isotopic discrimination could be extended to other associated traits with only the cost of more extensive phenotypic screening, the same molecular data being used for all traits. Several such traits, often species specific, are mentioned in other chapters of this volume, such as the ratio of leaf dry weight to leaf area for species where photosynthetic capacity is the dominant source of variation in Δ (Hubick *et al.*, 1988; and see Wright, Chapter 17, this volume). Recent data (Masle *et al.*, 1992) point to a genetic association between carbon isotope discrimination and leaf mineral concentration. This association was found in a range of monocot- and dicotyledonous species of the C_3 photosynthetic pathway (wheat, *Hordeum spontaneum*, rice, sunflower, tobacco) and also in sorghum, a C₄ species. The association is also present in *A. thaliana*, among the ecotypes which were selected for RFLP analysis of Δ (Fig. 9). In *Arabidopsis*, as in most other species examined, potassium is one of the minerals contributing the most to the relationship. Other elements showing a high correlation with Δ are calcium and sulfur. The underlying mechanisms are as yet unclear. Genetic variations in transpiration or CO_2 assimilation rate per se (passive nutrient uptake) are only a minor component. We have evidence, in both tobacco and wheat, that the association between Δ (or W) and mineral content occurs in part through genetic variation in leaf-water content (the genotypes with lowest W, highest Δ, having leaves with greater fresh weight to dry weight ratio). Comparison of the DNA polymorphisms associated with genetic variation in Δ and in physiological traits which covary with Δ will be a powerful way to distinguish correlations from functional links and to provide tools for analyzing the mechanisms involved.

B. Prospects of Gene Isolation

There are three main steps in map-based gene cloning: (a) identifying markers that are tightly linked to a target gene, (b) walking to the gene using various genomic libraries constructed in λ vectors or YACs (yeast artificial chromosomes), and (c) complementing the recessive phenotype by transformation

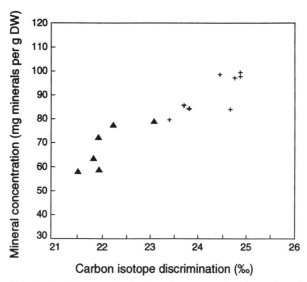

Figure 9. Relationship between leaf mineral concentration (mg minerals/g dry wt) and leaf carbon isotope discrimination for six *Arabidopsis* ecotypes (same as in Fig. 5 excluding Col and Ld), grown in controlled chambers under 3 mb vpd, 520 μE m^{-2} s^{-1} (+), or 13 mb and 520 μE m^{-2} s^{-1} (▲), and 23°C day/20°C night. Mineral concentration is the sum of concentrations in individual elements (major elements, excluding N, down to trace elements) measured by x-ray fluorescence as in Masle *et al.* (1992).

The RFLP analysis presented earlier is obviously relevant to the first step. It will provide the identification of DNA fragments linked with genes controlling Δ (i.e., which cosegregate with these genes with a high frequency). From the results obtained for *inter*specific variation in *W* in tomato (Martin *et al.*, 1989), we are hopeful that these fragments will be very few. Once RFLP markers associated with Δ or *W* have been identified, it may be necessary to flank the gene(s) of interest by closer markers in order to decrease the walking distance to these genes (step b above). A new polymorphism assay (RAPDs) devised by Williams *et al.* (1990) may then prove very useful. "Template" genomic DNA is amplified by a polymerase chain reaction with random primers that do not require any information on the sequence of the target gene; the primers need contain a minimum of nine bases, only, with a minimum G + C content of 40% (Williams *et al.*, 1990). Hundreds of fragments are generated very quickly; the polymorphic fragments are simply visualized by ethidium bromide staining of a gel [i.e., gel transfer on a membrane, probe labeling, and hybridization (necessary for the detection of polymorphic RFLPs) are not needed].

With the advent of such a technique and the use of large-size DNA clones (several YACs libraries are available in *Arabidopsis*) gene cloning in *Arabidopsis* today appears much more realistic than only 2 years ago. The first successes in cloning a gene by chromosome walking have apparently just been achieved. For a polygenic trait like transpiration efficiency, the major difficulty will lie in the complementation step. The rate of success of plant transformation techniques is *Arabidopsis* is now reasonable, but unless only one or two major genes are involved in controlling *W*, it will be extremely difficult to demonstrate reversion of the phenotype. Obviously, the lower the environmental noise, the greater the chances of success. This puts even more emphasis on the necessity of gaining a better understanding of the physiology of Δ and of identifying experimental means of stabilizing the expression of genetic variation in this trait. In fact therein possibly lies the greatest challenge.

References

Botstein, D., R. L. White, M. Skolnick, and R. W. Davis. 1980. Construction of a genetic linkage map in man using restriction fragment length polymorphisms. *Am. J. Hum. Genet.* **32**: 314–331.

Chang, C., J. L. Bowman, A. W. DeJohn, E. S. Lander, and E. M. Meyerowitz. 1988. Restriction fragment length polymorphism linkage map for *Arabidopsis thaliana. Proc. Natl. Acad. Sci. USA* **85**: 6856–6860.

Condon, A. G., R. A. Richards, and G. D. Farquhar. 1987. Carbon isotope discrimination is positively correlated with grain yield and dry-matter production in field-grown wheat. *Crop Sci.* **30**: 300–305.

Farquhar, G. D., M. H. O'Leary, and J. A. Berry. 1982. On the relationship between carbon isotope discrimination and the intercellular carbon dioxide concentration in leaves. *Aust. J. Plant Physiol.* **9**: 121–137.

Farquhar, G. D., J. R. Ehleringer, and K. T. Hubick. 1989. Carbon isotope discrimination and photosynthesis. *Annu. Rev. Plant Phyiol. Plant Mol. Biol.* **40:** 503–537.

Gillet, E. M. 1991. Genetic analysis of nuclear DNA restriction fragment patterns. *Genome* **34:** 693–703.

Hubick, K. T., R. Shorter, and G. D. Farquhar. 1988. Heritability and genotype environment interactions of carbon isotope discrimination and transpiration efficiency in peanut (*Arachis hypogea L.*). *Aust. J. Plant Physiol.* **15:** 799–813.

Hulbert, S. H., T. W. Ilott, E. J. Legg, S. E. Lincoln, E. S. Lander, and R. W. Michelmore. 1988. Genetic analysis of the fungus, *Bromia lactucae*, using restriction fragment length polymorphisms. *Genetics* **120:** 947–958.

Lander, E. S., and P. Green. 1987. Construction of multilocus genetic linkage maps in humans. *Proc. Natl. Acad. Sci. USA* **84:** 2363–2367.

Lander, E. S., P. Green, J. Abrahamson, A. Barlow, M. Daly, S. Lincoln, and L. Newberg. 1987. Mapmaker: An interactive computer package for constructing primary genetic linkage maps of experimental and natural populations. *Genomics* **1:** 174–181.

Leutwiler, L. S., B. R. Hough-Evans, and E. M. Meyerowitz. 1984. The DNA of *Arabidopsis thaliana*. *Mol. Gen. Genet.* **194:** 15–23.

Martin, B., and Y. R. Thorstenson. 1988. Stable carbon isotope composition ($\delta^{13}C$), water use efficiency, and biomass productivity of *Lycopersicon esculentum*, *Lycopersicon pennellii*, and the F_1 hybrid. *Plant Physiol.* **88:** 213–217.

Martin, B., J. Nienhuis, G. King, and A. Schaefer. 1989. Restriction fragment length polymorphisms associated with water use efficiency in tomato. *Science* **243:** 1725–1728.

Masle, J., and G. D. Farquhar. 1988. Effects of soil strength on the relations of water use efficiency and growth to carbon isotope discrimination in wheat seedlings. *Plant Physiol.* **86:** 32–38.

Masle, J., G. D. Farquhar, and S. C. Wong. 1992. Transpiration efficiency and plant mineral content are related among genotypes of a range of species. *Aust. J. Plant Physiol.* **19:** 709–721.

Nam, H-G., J. Giraudat, B. den Boer, F. Moonan, W. D. B. Loos, B. M. Hauge, and H. M. Goodman. 1989. Restriction length polymorphisms linkage map of *Arabidopsis thaliana*. *Plant Cell* **1:** 699–705.

Pruitt, R. E., and E. M. Meyerowitz. 1986. Characterization of the genome of *Arabidopsis thaliana*. *J. Mol. Biol.* **187:** 169–183.

Roberts, E. H., and R. H. Ellis. 1977. Prediction of seed longevity at sub-zero temperatures and genetic resources conservation. *Nature* **268:** 431–433.

Roberts, E. H., and R. H. Ellis. 1980. The influence of temperature and moisture on seed viability period in barley (*Hordeum distichum L.*). *Ann. Bot.* **45:** 31–37.

Roberts, E. H., and R. H. Ellis. 1982. Physiological, ultrastructural and metabolic aspects of seed viability, pp. 465–485. *In* A. A. Kahn (ed.), The Physiology and Biochemistry of Seed Development, Dormancy and Germination. Elsevier Biochemical Press, Amsterdam/New York.

Williams, J. G. K., A. R. Kubelik, K. J. Livak, J. A. Rafalski, and S. V. Tingey, 1990. DNA polymorphisms amplified by arbitrary primers are useful as genetic markers. *Nucleic Acids Res.* **18:** 6531–6535.

25

Implications of Carbon Isotope Discrimination Studies for Breeding Common Bean under Water Deficits

Jeffrey W. White

I. Introduction and Review

Common bean (*Phaseolus vulgaris* L.) is frequently grown under rain-fed conditions where yields are limited by water deficits. In Latin America, over 60% of bean production areas suffer moderate to severe water deficits; a similar situation is thought to occur in Africa. Although common bean is considered drought-sensitive, it is often found in low rainfall areas where few other crops are grown. This reflects the very short growth cycles of many bean cultivars. Some cultivars require as little as 60 days from planting to maturity, and this trait reduces the total water requirement of the crop to levels well below that of many species considered more drought-adapted.

Bean germplasm and breeding lines evaluated under water deficits show considerable variation in adaptation as measured by seed yield (Laing *et al.*, 1983). Studies on adaptation mechanisms indicate that a major component is related to differences in root growth, leading to desiccation postponement through greater extraction of soil moisture (Sponchiado *et al.*, 1989). To assess the relative importance of root and shoot traits in determining yields under deficits, adapted and sensitive bean cultivars have been grafted, interchanging roots and shoots. In a series of trials, a strong effect of the genotype of the root was always detected, while shoot effects were only found in studies where large numbers of shoots were tested on rootstocks of a single cultivar (White and Castillo, 1989, 1992). Early maturity provides an escape mechanism, but its utility varies greatly with environment and requires sacrificing high yields when precipitation is adequate (White and Izquierdo, 1991).

Strategies for breeding beans under water deficits were reviewed recently (White and Singh, 1991). Selection for yield under drought gave as much as a 16% gain from one generation to the next and resulted in bred lines with yields 60% greater than tolerant checks after four seasons of selection (CIAT, 1992, and J. Kornegay, unpublished observations).

While these results suggest direct selection for yield under drought is practical, expected progress is slow enough that there is ample justification to search for additional selection criteria or, at a minimum, to identify novel sources of variation among parental lines. Promising results from studies using carbon isotope discrimination (Δ) as an indirect indicator of water-use efficiency in other species (e.g., Farquhar and Richards, 1984; Hubick *et al.*, 1988) have led to investigation of variation in Δ values in common bean germplasm. To evaluate the implications of such research for bean breeding, this chapter examines variation in Δ among diverse bean germplasm, inheritance of Δ values, and its associations with growth, yield, and physiological traits. A final section considers how research on Δ might guide bean-breeding efforts in the future.

II. Variation in Carbon Isotope Discrimination in Bean Germplasm

Two surveys of variation in Δ values over large sets of bean germplasm have been conducted in the field under rain-fed conditions. At Cortez, Colorado, values of Δ in leaves varied from 16.7 to 19.7‰ over 99 genotypes (Ehleringer *et al.*, 1990). At Palmira, Colombia, 90 genotypes gave leaf Δ values from 17.3 to 20.4‰ (unpublished data). In both trials, early flowering or maturing genotypes showed greater discrimination (Fig. 1). In the trial in Colombia, tepary bean (*P. acutifolius* Gray) and lima bean (*P. lunatus* L.) accessions were also evaluated (Fig. 1). Various other traits measured in this trial were correlated with leaf Δ, but it was considered surprising that seed weight was not (Table I) since this trait frequently shows a negative association with growth and seed yield in common bean (White and Gonzalez, 1990).

Values of Δ were correlated in a comparison of 10 genotypes grown under rain-fed conditions at Palmira and Quilichao, Colombia (White *et al.*, 1990; Fig. 2). These sites have similar weather patterns but contrast in soil types. Palmira has a deep, fertile mollisol, while Quilichao has a shallow, acidic oxisol. No significant correlation was found between Δ for the rain-fed treatment at Palmira and a well-irrigated treatment at the same site. A genotype by water-regime interaction was also found in a comparison of 8 genotypes grown under five water regimes at Palmira (Fig. 3; data for only 4 genotypes are shown). The comparison of BAT 1224 and A 170 is particularly instructive. BAT 1224 showed high values of Δ which increased with the amount of water applied, while A 170 had low and relatively constant values throughout.

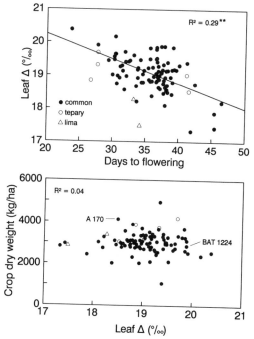

Figure 1. Relation between leaf Δ measured at 56 days and days to flowering (top) and crop dry weight at 56 days (bottom) for germplasm of common, lima, and tepary beans grown under water deficits at Palmira, Colombia.

Table I Correlations between Various Traits and Leaf Δ, Days to Maturity, and Seed Weight for 90 Bean Genotypes Grown under Water Deficits at Palmira, Colombia

	Correlation with		
Trait	Leaf Δ	Days to maturity	Seed weight
Seed weight	−0.11	−0.10	1.00
Days to flower	−0.54**	0.87**	−0.39**
Days to maturity	−0.58**	1.00	−0.10
Specific leaf area	−0.25**	0.13	0.26*
Leaf nitrogen conc.	−0.33**	0.69**	−0.39**
Leaf phosphorus conc.	−0.46**	0.76**	−0.23*
Leaf potassium conc.	−0.19	0.49**	0.02
Crop dry weight			
At 56 days	0.02	−0.08	−0.31**
At maturity	0.14	0.14	−0.45**
Seed yield	0.07	−0.08	−0.40**

*,** Significant at the $P = 0.05$ and $P = 0.01$ levels, respectively.

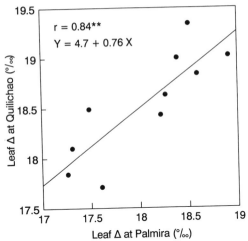

Figure 2.   Comparison of leaf Δ in 10 cultivars grown under water deficits at Palmira and Quilichao, Colombia (White *et al.*, 1990).

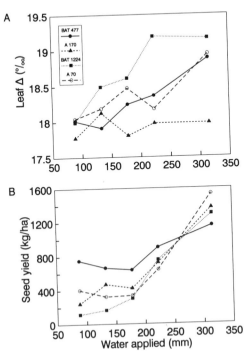

Figure 3.   Comparisons of leaf Δ and seed yield for four bean cultivars in relation to water applied through five irrigation regimes at Palmira, Colombia. The regimes were preplanting irrigation only, irrigated to preflowering, irrigated to flowering, preplanting plus a recovery irrigation at 63 days after planting, and fully irrigated. (A) Leaf Δ. (B) Seed yield.

III. Inheritance of Carbon Isotope Discrimination

Heritability of leaf Δ was examined among eight parents adapted to water deficits either in the semiarid highlands of Mexico or at the Palmira station of CIAT in Colombia, along with an additional drought-sensitive line, BAT 1224 (CIAT, 1992; White *et al.*, 1993). The parents were crossed in a diallel without reciprocals, and F_2 and F_3 populations were evaluated for leaf Δ, seed yield, and other traits in separate seasons at Palmira and Quilichao. Leaf samples for Δ and leaf N and K concentrations were taken at approximately 50 days after planting (onset of seed filling), and consisted of 30 central leaflets from fully expanded leaves of different plants. At Quilichao, only rain-fed conditions were used, but at Palmira, both rain-fed and irrigated trials were conducted.

Values of leaf Δ for parents ranged from 19.2 to 20.2‰ at Quilichao, 17.6 to 18.7‰ for the rain-fed trial at Palmira, and 18.3 to 19.5‰ for the irrigated trial. Realized heritability (estimated by parent–offspring regression) of leaf Δ was lower than that for many other traits, including leaf N and K concentrations, and even seed yield, which is usually thought to have a moderate to low heritability (Table II). Correlations of leaf Δ with seed

Table II   Realized Heritability (h^2), Correlation with Seed Yield, and Gain in Yield from Selection (from F_2 to F_3 Generations) for Various Traits at Quilichao (Q), Palmira—Rain-Fed (P_R), and Palmira—Irrigated (P_I)

Trait	Trial	h_2	Correlation with yield F_2	Correlation with yield F_3	Gain in yield from selection (%)[a]
Seed yield	Q	0.28	—	—	2.9
	P_R	0.57**	—	—	15.7
	P_I	0.50**	—	—	3.9
Δ	Q	0.10	−0.02	−0.38**	−2.0
	P_R	0.00	−0.15	0.45**	7.2
	P_I	0.12	0.04	0.19	0.6
Leaf OD[b]	Q	0.33**	−0.22	0.07	1.2
	P_R	0.11**	−0.51**	−0.37*	12.0
	P_I	0.13	−0.43**	−0.57**	2.9
Leaf N (%)	Q	0.39**	0.38*	0.20	4.4
	P_R	0.44**	0.32*	0.67**	4.3
	P_I	0.22**	0.42**	0.37**	3.1
Leaf K (%)	Q	0.04	0.27	0.20	2.0
	P_R	0.29**	0.03	0.39**	9.4
	P_I	0.22**	0.07	−0.13	1.0
100-seed wt.	Q	0.57**	−0.11	−0.55**	2.1
	P_R	0.69**	−0.04	−0.18	−6.6
	P_I	0.70**	−0.43**	−0.63**	1.7

Source: CIAT (1992).

[a]For Δ, leaf OD, and 100-seed weight, lower values were considered desirable.

[b]Leaf OD is optical density measured with a hand-held chlorophyllometer (Hardacre *et al.*, 1984) and was based on the mean value of 30 leaflets per plot.

**Significant at the $P = 0.01$ level.

Table III   Mean Squares for General (GCA) and Specific (SCA) Combining Ability for Leaf Δ Measured in a 9 × 9 Diallel Cross Grown at Quilichao and Palmira, Colombia

| | | | | Palmira | | | |
| | | Quilichao | | Rain-fed | | Irrigated | |
Source	df	F_2	F_3	F_2	F_3	F_2	F_3
GCA	8	0.689**	0.578**	0.677**	0.210	0.639**	0.388**
SCA	35	0.186*	0.076*	0.178	0.144	0.227	0.220*
Error	88	0.116	0.059	0.141	0.270	0.187	0.133

Source: White *et al.* (1993).
*,** Significant at the $P < 0.05$ and $P < 0.01$ levels, respectively.

yield were inconsistent across trials. Realized gain in yield from selection for low Δ was low at Quilichao and under irrigated conditions at Palmira, but a 7% gain was obtained for rain-fed conditions at Palmira (Table II). Although it might be argued that these disappointing results reflect a low range of variation in Δ values of parents, an inappropriate sampling strategy, or other methodological problems, leaf N concentration had heritabilities of 0.39 and 0.44 ($P < 0.01$) under water deficit, was consistently correlated with yield and was an effective selection criterion.

General and specific combining abilities were also examined in these trials using Model 2, Method I of Griffing (1956). Mean squares for general combining ability (GCA) tended to be large in comparison with those for specific combining ability (SCA), suggesting additive gene action (Table III). BAT 1224 stood out for its high positive GCA for Δ, while San Cristobal 83 and Rio Tibagi had negative GCAs (Table IV).

Table IV   General Combining Ability Effect for Δ Obtained from a 9 × 9 Diallel Cross Grown at Two Locations in Colombia

| | | | | Palmira (‰) | | |
| | | Quilichao (‰) | | Rain-fed | | Irrigated |
Parent	F_2	F_3	F_2	F_3	F_2	F_3
BAT 477	0.133*	0.033	−0.076	0.032	0.117	0.179**
V 8025	−0.048	−0.055	−0.030	0.100	−0.052	−0.059
Rio Tibagi	−0.131*	−0.242**	−0.133*	−0.075	−0.093	0.084
S. Critobal 83	−0.134*	−0.103*	−0.211**	0.113	−0.020	−0.083
Apetito	−0.222**	−0.041	−0.062	−0.058	−0.274**	−0.097
Bayo Cr. Llano	0.003	0.122**	0.105	0.007	0.066	−0.049
Bayo Rio Grande	0.130*	−0.022	0.224**	−0.027	−0.052	−0.049
Durango 222	0.065	0.188**	0.173**	−0.126	0.163*	−0.080
BAT 1224	0.203**	0.120**	0.011	0.036	0.145*	0.155**

*,** Significantly different from 0 at $P < 0.05$ and $P < 0.01$ levels, respectively.

IV. Associations of Δ with Physiological Traits, Crop Growth, and Seed Yield

Transpiration efficiency (W = total biomass/transpiration) and leaf Δ were negatively associated among bean cultivars grown in containers in Utah (Ehleringer *et al.*, 1991), for a range of W from 0.5 to 1.1 g kg^{-1}. Variation in stomatal conductance (g) was positively associated with Δ among dry bean cultivars, but not for snap beans (Ehleringer *et al.*, 1990). Bean cultivars vary in their capacity for heliotropic leaf movements (Sato, 1988) in response to air temperature (Fu and Ehleringer, 1989), which could affect carbon isotope discrimination through leaf temperature effects. As a consequence, correlations between W and Δ may not be as strong as has been demonstrated for other species.

Leaf nitrogen, phosphorus, and potassium concentrations (dry weight basis) usually are lower in cultivars showing higher values of Δ (Table V). This might reflect enhanced efficiency of photosynthesis for leaves with higher nutrient concentrations. An additional explanation would be that genotypes subjected to greater drought stress, due possibly to shallower roots, have reduced leaf expansion and smaller cell size; thus their nutrient concentrations are higher. At the same time, stomatal controls cause an increase in A/E and a decrease in carbon isotope discrimination.

Relations between whole-plant traits and carbon isotope discrimination are less consistent. At Palmira, greater biomass and seed yield were associated with higher values of Δ in a comparison of 10 genotypes, and the relation was much stronger if BAT 1224 was considered an "outlier" (White *et al.*, 1990; Figs. 4A and 4B). In contrast, at Quilichao, no clear trend was found, and it appeared that intermediate values of Δ were associ-

Table V Correlations between Δ Values and Leaf Concentrations (Dry Weight Basis) of Nitrogen, Phosphorus, and Potassium for Various Trials in Colombia

Location and type of trial	No. of entries	Correlation of Δ with leaf concentration of		
		Nitrogen	Phosphorus	Potassium
Palmira, rain-fed	90	−0.33**	−0.46**	−0.19
Quilichao, rain-fed	10	−0.61		−0.65*
Palmira, rain-fed	10	−0.53	−0.72*	−0.74*
Palmira, irrigated	10	−0.27		−0.47
Quilichao, rain-fed F$_2$	45	−0.11		0.13
Quilichao, rain-fed F$_3$	45	−0.75**		−0.45**
Palmira, rain-fed F$_2$	45	−0.67**		−0.48**
Palmira, rain-fed F$_3$	45	0.26		0.00
Palmira, irrigated F$_2$	45	−0.44**		−0.44**
Palmira, irrigated F$_3$	45	−0.22		−0.51**

Source: White *et al.* (1990, 1993).
*,** Significant at the P = 0.05 and P = 0.01 levels, respectively.

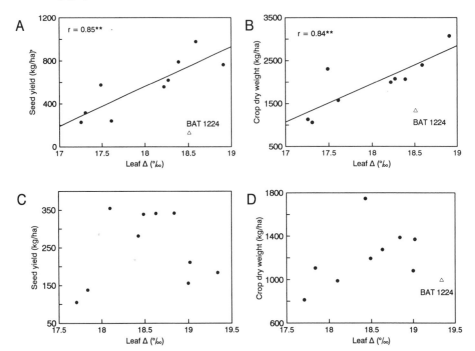

Figure 4. Comparisons of seed yield and crop dry weight at 64 days in relation to leaf Δ at Palmira and Quilichao, Colombia (White *et al.*, 1990). (A) Seed yield at Palmira, (B) crop dry weight at Palmira, (C) seed yield at Quilichao, (D) crop dry weight at Quilichao.

ated with higher yields (Figs. 4C and 4D). The germplasm survey at Palmira also indicated that extreme values of Δ might be associated with slower growth (Fig. 1), and in the inheritance study, correlations of seed yield and final crop dry weight with Δ varied greatly among the trials.

One plausible explanation for these results is that the relation between Δ and growth or yield will vary according to the pattern of water deficit the crop experiences (Ehleringer *et al.*, 1990; White *et al.*, 1990). In a deep soil with stored water, deep-rooting cultivars will maintain transpiration and photosynthesis and have high discrimination. Shallow-rooting cultivars will have reduced transpiration and photosynthesis, lower discrimination, and low yields due to restricted growth. For a shallow soil (or a deep one with little stored water), cultivars which depend upon abundant stored water may show rapid initial growth and high discrimination, but when the soil moisture is exhausted, they will become severely stressed and give low yields. A shallow-rooted cultivar with more conservative water use throughout the growth cycle will maintain a slow growth rate, have low discrimination, and have intermediate yields. Intermediate circumstances are easily envisaged where neither profligate nor miserly water use is the most advantageous, and an intermediate level of discrimination gives the greatest growth and yield. Which of these three relations holds in a given experiment will depend upon soil characteristics, weather patterns, and the cultivars being tested.

Data for root growth support this analysis in the comparison of 10 genotypes grown at Palmira and Quilichao (Fig. 5). Root length density was positively correlated with Δ at both sites. However, since Quilichao has a shallower soil with lower moisture-holding capacity, cultivars with more extensive root development, while growing well at Palmira, may have exhausted available soil moisture prematurely at Quilichao (White *et al.,* 1990). This apparent importance of genotypic differences in root growth coincides with the grafting studies which concluded that genes acting in the root system are of greater importance in determining growth and yield under water deficits than genes acting in the shoot (White and Castillo, 1989, 1992).

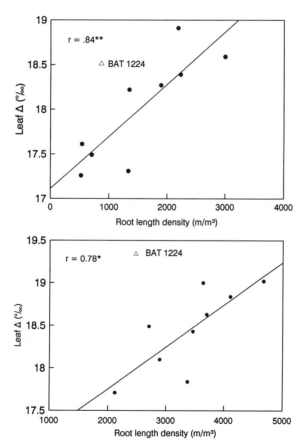

Figure 5. Relation between leaf Δ and root length density for 10 bean cultivars grown under water deficits at Palmira (top) and Quilichao (bottom), Colombia (White *et al.,* 1990). At Palmira root samples were taken at 65 days from 0.4 to 1.2 m soil depth, and at Quilichao, at 65 days from 0.15 to 0.45 m soil depth.

These analyses all assume that absolute values of Δ provide the most effective measure of plant response to water deficits. An alternative view is that changes in discrimination in relation to water deficits are more relevant than absolute levels. Thus, one should analyze change in Δ, defined as the difference between Δ under well-watered and water-deficit conditions. This is equivalent to assuming that variation in c_i has two components, a base-level c_i expressed under nonstress conditions, and a reduction in c_i due to effects of water deficits (White *et al.*, 1990). A large reduction in Δ would suggest a large increase in W. Partial justification for assuming that there is variation in a base-level c_i is provided by the significant differences in Δ typically found among bean genotypes grown under well-watered conditions.

In the 10-genotype study, a small reduction in Δ ($\Delta_{irrigated} - \Delta_{rainfed}$) at Palmira was associated with greater growth and yield in the rain-fed treatment (White *et al.*, 1990). In the comparison of 8 genotypes under five water regimes (Fig. 3), reduction in Δ calculated for each water-deficit regime still showed significant effects of regime, genotype, and their interaction, and reduction in Δ was not correlated with yield for any water regime (J. White, unpublished observations). This suggests reduction in Δ would not be more stable than Δ per se, nor would it be more closely associated with growth or yield. A further practical limitation of reduction in Δ is that its error tends to be large since it is calculated as the difference of two variables which usually have considerable measurement error themselves.

V. Implications for Breeding Common Bean

To date, results of studies on carbon isotope discrimination provide little support for the immediate use of Δ values as a selection criterion for bean yields under water deficits. The possible low heritability of Δ and its inconsistent relation with seed yield would make Δ an unreliable criterion. Nonetheless, this conclusion should be viewed with caution.

Studies of inheritance are notoriously sensitive to selection of parental materials. For the nine-parent study discussed herein, parents were not selected for extreme values of Δ. A study including A 170, which appears to have stable, low values of Δ, and BAT 1224, which consistently has higher values, could result in larger heritabilities.

The inconsistent relations between yield and carbon isotope discrimination seem to reflect differences in stress patterns. For bean production regions such as the coast of Peru and some parts of Mexico where stress patterns are fairly predictable, sufficient consistency might be found to justify use of Δ for selection. In other regions a valid, albeit conservative strategy might be to avoid both extremely low and high Δ values.

Most of the difficulties in use of Δ in bean breeding appear to stem from the fact that reduced carbon isotope discrimination in bean seems to be caused largely by processes which limit photosynthesis. Increases in A/E are achieved through reductions in E, which reduce A sufficiently to reduce growth. There is an obvious need to search for variation in Δ where differences in A predominate over differences in E. Field studies seem poorly suited for such work since under water-limited conditions, cultivar differences in root growth may increase variation in E. Research using containers offering a limited soil volume may provide more meaningful results if the concomitant restriction to root growth reduces genotypic differences in E. Studies of different shoot genotypes grafted onto a uniform rootstock should also be of interest since these also might minimize variation in Δ due to reduced E for cultivars with shallow or sparse root development.

VI. Summary

Common bean cultivars show variation in Δ values, and these differences are associated with variation in c_i and W as predicted by theory. The physiological basis of variation in Δ appears to be dominated by effects of E rather than A, and it is proposed that cultivar differences in root growth and leaf movements may be primary factors controlling E. The relatively greater importance of variation in E probably explains why cultivar rankings for Δ appear to be more sensitive to growing conditions than in other species and also may explain reported low heritabilities and variable relations of Δ with growth and seed yield. Nonetheless, certain bean genotypes (e.g., BAT 1224 and A 170) show patterns of variation in Δ which are not easily attributable to variation in E. Furthermore, large mean squares for GCA of Δ are suggestive of significant additive gene action, and the negative correlation between Δ and days to flowering remains to be explained.

References

CIAT. 1992. Annual Report Bean Program. CIAT, Cali, Colombia.

Ehleringer, J. R., J. W. White, D. A. Johnson, and M. Brick. 1990. Carbon isotope discrimination, photosynthetic gas exchange, and transpiration efficiency in beans and range grasses. *Acta Oecol.* **11**: 611–625.

Ehleringer, J. R., S. Klassen, C. Clayton, D. Sherill, M. Fuller-Holbrook, Q. A. Fu, and T. A. Cooper. 1991. Carbon isotope discrimination and transpiration efficiency in common bean. *Crop Sci.* **31**: 1611–1615.

Farquhar, G. D., and R. A. Richards. 1984. Isotopic composition of plant carbon correlates with water-use efficiency of wheat genotypes. *Aust. J. Plant Physiol.* **9**: 121–137.

Fu, Q. A., and J. R. Ehleringer. 1989. Heliotropic leaf movements in common beans controlled by air temperature. *Plant Physiol.* **91**: 1162–1167.

Griffing, B. 1956. Concept of general and specific combining ability in relation to diallel crossing systems. *Aust. J. Biol. Sci.* **9**: 463–493.

Hardacre, A. K., H. F. Nicholson, and M. L. P. Boyce. 1984. A portable photometer for the measurement of chlorophyll in intact leaves. *N. Z. J. Exp. Agric.* **12**: 357–362.

Hubick, K. T., R. Shorter, and G. D. Farquhar. 1988. Heritability and genotype × environment interactions of carbon isotope discrimination and transpiration efficiency in peanut (*Arachis hypogaea* L.). *Aust. J. Plant Physiol.* **15:** 799–813.

Laing, D. R., P. J. Kretchmer, S. Zuluaga, and P. J. Jones. 1983. Field bean, pp. 227–248. *In* Symposium on Potential Productivity of Field Crops under Different Environments. IRRI, Los Baños, Philippines.

Sato, H. 1988. Studies on leaf orientation movements in kidney beans (*Phaseolus vulgaris* L.). II. Difference among cultivars and the inheritance. *Jpn. J. Crop Sci.* **57:** 340–345.

Sponchiado, B. N., J. W. White, J. A. Castillo, and P. G. Jones. 1989. Root growth of four common bean cultivars in relation to drought tolerance in environments with contrasting soil types. *Exp. Agric.* **25:** 249–257.

White, J. W., and J. A. Castillo. 1989. Relative effect of root and shoot genotypes on yield of common bean under drought stress. *Crop Sci.* **29:** 360–362.

White, J. W., and J. A. Castillo. 1992. Evaluation of diverse shoot genotypes on selected root genotypes of common bean. *Crop Sci.* **32:** 762–765.

White, J. W., and A. Gonzalez. 1990. Characterization of the negative association between seed yield and seed size among genotypes of common bean. *Field Crops Res.* **23:** 159–175.

White, J. W., and J. Izquierdo. 1991. Physiology of yield potential and stress tolerance, pp. 287–282. *In* A. van Schoonhoven and O. Voysest (eds.), Common Beans: Research for Crop Improvement. CAB International, Wallingford, UK.

White, J. W., and S. P. Singh. 1991. Breeding for adaptation to drought, pp. 501–560. *In* A. van Schoonhoven and O. Voysest (eds.), Common Beans: Research for Crop Improvement. CAB International, Wallingford, UK.

White, J. W., J. A. Castillo, and J. Ehleringer. 1990. Associations between productivity, root growth and carbon isotope discrimination in *Phaseolus vulgaris* under water deficit. *Aust. J. Plant Physiol.* **17:** 189–198.

White, J. W., J. A. Castillo, J. R. Ehleringer, S. P. Singh, and J. Garcia. 1993. Relations of carbon isotope discrimination and other physiological traits to seed yield in common bean under rainfed conditions. *J. Agric. Sci. Cambridge*, in press.

26

Potential of Carbon Isotope Discrimination as a Selection Criterion in Barley Breeding

E. Acevedo

I. Introduction

The effectiveness of selection for yield per se is relatively low because of the large number of genes involved and the additional attributes required by new cultivars. Hence, it is difficult to find desired gene combinations in the progenies. Many economic traits like grain yield have low heritability values since they are controlled by many genes and their expression is greatly influenced by the environment (Kronstad and Moss, 1991). Therefore, the use of associated traits, such as physiological attributes, in selecting for yield is becoming more important. In determining the possible role of a physiological trait, it is useful to keep in mind the various steps in a breeding program. These include (1) identifying factors that limit yield; (2) determining if genetic variability exists and the nature of this variation for specific attributes; (3) identifying the most promising parents to hybridize; (4) applying appropriate selection pressures to isolate the desired progeny; and (5) evaluating, multiplying, and disseminating new cultivars. Physiological traits may play a role in the first four steps.

The associations between physiological mechanisms and yield potential as well as adaptation of winter cereal genotypes to abiotic stresses are being intensively studied with the aim to define breeding goals and increase efficiency (Evans, 1987; Acevedo et al., 1991b; Quisenberry, 1982; Srivastava et al., 1987). Plant breeders have only slowly adopted traits related to these mechanisms for use in selection because of (1) the difficulty in measuring the trait or mechanism in a wide range of genetic materials, (2) the lack of convincing evidence to support a relationship with increased yield or increased stability of production, possibly due to the small number of appropriate selection experiments (Ceccarelli et al., 1991; Acevedo et al.,

1991c), (3) insufficient knowledge about inheritance of the character, and (4) inadequate cross-disciplinary (physiology/breeding) communications.

There now appears to be widespread acceptance that if physiological traits are to be useful in breeding, they must comply with some minimum criteria. These include (1) a greater heritability than yield, (2) a significant correlation with yield and/or stability of yield, (3) a causal relationship with yield, (4) physiological assays that are easy to use, and (5) the possibility of screening early generations when yield per se cannot be assessed (Acevedo and Fereres, 1992). Ideally physiological techniques that are nondestructive, repeatable, and adaptable to evaluating large numbers of plants at specific stages in the growth cycle would be most useful to breeders in providing a greater degree of selection efficiency.

In this chapter, the potential of carbon isotope discrimination (Δ) as an indirect selection criterion in barley breeding for dry environments is analyzed. The analysis refers to information obtained at the International Center for Agricultural Research in the Dry Areas (ICARDA) during 1985–1990. Part of the work was done in collaboration with the Institute of Plant Science Research (IPSR), Cambridge, United Kingdom.

II. Carbon Isotope Discrimination

Carbon isotope discrimination has been proposed as an indirect selection trait to assist in genotype selection of C_3 species for drought environments (Farquhar and Richards, 1984; Austin et al.,, 1990; Richards, 1991). Both stomatal control of assimilation and the assimilative capacity of photosynthetic organs affect the discrimination against atmospheric CO_2 and contribute to determining the Δ value. From the values of inherent discrimination of $^{13}CO_2$ by diffusion and carboxylation, it can be inferred that as the stomata open wider, thus increasing the internal partial pressure of CO_2 (p_i), the discrimination by RuBisCO will be greater as compared to discrimination by diffusion, increasing the overall Δ value. The overall balance between photosynthesis and transpiration is such that increased p_i, due to the wider stomata openings and hence a greater Δ value, is associated with decreased transpiration efficiency (W). Our work shows that the relation between Δ and W is indeed negative in barley, which agrees with results reported by Hubick and Farquhar (1989) (Fig. 1).

The Δ value is an interesting parameter to look at as an indirect selection criterion from a plant-breeding perspective, since Δ is a measure of gas exchange integrated over time and hence provides a seasonal average of W. This is an important advantage for a physiological trait to be used in breeding since the instantaneous and/or short-term processes of photosynthesis (A) and transpiration (E) are too variable with weather fluctuations to be of much value as selection criteria. It can be argued that the Δ value is high in the hierarchy of processes determining biomass production in C_3 plants, a requirement for a trait to be useful in breeding at present (Passioura, 1986;

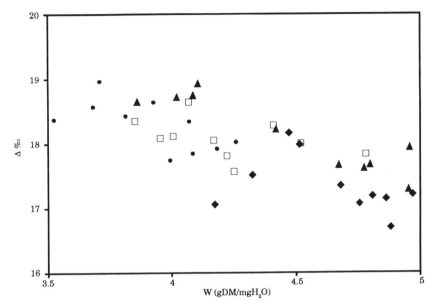

Figure 1.   Relation between Δ measured in the peduncle and transpiration efficiency (W) calculated based upon the mean growing period vapor pressure deficit (vpd = 0.53 kPa). Ten barley genotypes were grown in a greenhouse at temperatures of 25°C maximum and 5°C minimum through heading; the temperature regime was changed to 25/10 from heading to maturity. Four watering treatments in 5-kg soil (3 sand : 1 clay) pots were given immediately after planting: watering to field capacity when the pots reached one-half of the available water (FC, dots); watering to one-half available water when the pots reached the wilting point (½ FC, diamonds); ½ FC through heading and FC from heading through physiological maturity (½ FC—FC, triangles); and FC through heading followed by ½ FC to maturity (FC—½FC, squares). The surface of the pots was sealed with aluminum foil perforated to allow two plants per pot to pass through (Acevedo, unpublished).

Acevedo and Ceccarelli, 1989). At the plant level, yield under drought stress can be expressed as the product of three terms that are part of an identity: transpiration (E), transpiration efficiency (W), and harvest index (HI) (Passioura, 1977; Fischer and Turner, 1978). If these terms are largely independent, increasing any of them would increase yield. It follows that, in dry environments, genotypes with low Δ would perform better (higher W)—other factors being similar and the interactions between the three terms being nonsignificant, but the latter is highly unlikely.

A. Experimental Work Being Conducted at ICARDA on Carbon Isotope Discrimination by Barley

The experimental approach that ICARDA used to evaluate the potential of carbon isotope discrimination as a selection criterion in barley breeding for abiotically stressed environments involved studying the relationship between genotypes for Δ values and barley grain yield under stress. After a relationship between Δ and yield was shown to exist for the most stressed environments, experiments were designed to test if the relationship would hold in spaced plants, typical of F_2 generation plantings. The correlation between dense and spaced plantings indicated that spacings used in breeding nurseries would not affect genotype ranking significantly and therefore the study should be pursued further.

A set of two-row barley genotypes differing in grain Δ was planted in a line source sprinkler system (LSS) to determine if there was a genotype × drought-stress interaction. The plantings were done during the 1989 season at Breda in northern Syria, an experimental site with a Mediterranean-type climate. Although the site has a long-term average rainfall of 281 mm, only 180 mm fell during the 1989 season.

The same set of genotypes studied in the LSS was grown in a greenhouse at Tel Hadya, ICARDA's headquarters, under four moisture regimes in order to reassess the correlations between Δ and transpiration efficiency (total biomass production/transpiration). Positive correlations between Δ and grain yield under drought stress had been found in previous studies, implying a negative relationship between W and yield (e.g., Condon *et al.* 1987; Craufurd *et al.*, 1991).

Crosses were made between high and low Δ two-row barley genotypes and Δ was determined in the F_2 progeny of each of seven crosses. F_3- and F_4-derived bulks from each F_2 plant were grown to study genetic parameters and the breeding implications of early-generation (F_2) selection for Δ.

B. Carbon Isotope Discrimination and Grain Yield

A strong positive correlation ($r = 0.952$) between mean trial Δ in grain and mean grain yield was found for barley trials grown in dry environments (years and locations) as different as the dry Mediterranean types of northern Syria and trials in which drought was imposed by rain-out shelters in Cambridge in the United Kingdom (Craufurd *et al.*, 1991). High-yielding trials had higher Δ values and possibly lower Ws (Fig. 2). The genotypes were the same in all trials, hence this is essentially an environmental effect on Δ values. As drought increased in severity (lower rainfall and higher vapor pressure deficit, vpd), Δ values decreased presumably mainly as a result of increased stomatal control of transpiration. Associated with the decrease in Δ values was a corresponding decrease in grain yield. Similar drought effects on Δ and yield have been found for wheat grown in an LSS in the Yaqui Valley of Mexico by CIMMYT (Acevedo, Sayre, and Austin, unpublished). The within-trial relationship between variety Δ values and grain yield was found to vary with environment. Where barley crops were droughted, the relationship was positive, as reported by Condon *et al.* (1987) for wheat in well-watered conditions. But for wheat the positive correlation was stronger as the water-stress level increased. In mild water-stress environments, the phenotypic correlation dropped to zero and it became negative when water was nonlimiting in apparent contrast to the results of Condon *et al.* (1987). We have found a significant correlation ($r = -0.853$) between trial mean grain yield and the correlation coefficient be-

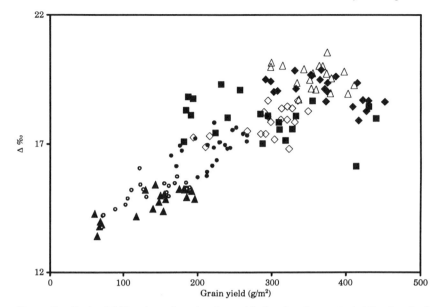

Figure 2. Grain yield has been found to be positively related to grain Δ. The data in the figure show the values of grain Δ and grain yield of 21 or 23 two-row barley genotypes grown under field conditions either in Cambridge (●, ◇, △) or in northern Syria (▲, ○, ■). The grain yield is indicative of the stress intensity.

tween variety grain yield and Δ as shown in Fig. 3 from Austin *et al.* (1990). Varieties with a high Δ value had higher yields in stressed environments with the reverse occurring in nonstressed environments. Furthermore, barley landraces from the Fertile Crescent tended to have a lower W than improved genotypes (Table II). We also noted a strong negative correlation between Δ and days to ear emergence (Craufurd *et al.*, 1991). It is possible that variations in earliness may have been responsible for differences in grain yield in the dry environments.

As expected from the previous discussion, significant correlations between Δ value and grain yield were not always found between genotypes. This correlation was weak but significant ($r = 0.316$, $P < 0.05$, $n = 45$) in a breeder's advanced barley yield trial at ICARDA in northern Syria (Acevedo and Ceccarelli, unpublished) despite significant variations in grain Δ values (ranging from 12.86 to 14.86‰) and very low seasonal rainfall (186 mm). The correlation in days to ear emergence and grain yield was nonsignificant in this trial. A barley yield trial showed no significant association between grain Δ value and grain yield nor grain Δ value and days to ear emergence despite a highly significant association ($r = -0.559$, $P < 0.001$, $n = 45$) between days to ear emergence and grain yield. The variation in Δ values across genotypes in the preliminary yield trial was highly significant, ranging from 12.88 to 14.42‰.

Many reasons have been suggested for the contrasting relationships

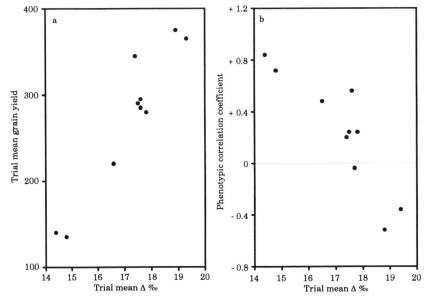

Figure 3. As the water stress decreased (a), mean trial grain yield and mean trial grain Δ increased. The within-trial phenotype correlation coefficient between grain yield and grain Δ was negatively related to trial mean Δ (b). Of the 10 trials in the figures, 6 were conducted in northern Syria and 4 in Cambridge. After Austin *et al.* (1990).

found between grain yield and genotype in dry and wet environments. Barley genotypes with early ear emergence, which are higher yielding in Mediterranean environments (Acevedo *et al.*, 1991a), can fix much of their grain carbon in a relatively drought-stress-free period and hence the value of Δ would be dominated by a relatively high p_i/p_a. The relationship between Δ and grain yield would be positive. Second, earliness is generally associated with early vigor and better ground cover and hence higher water-use efficiency at the crop level (by reducing soil surface evaporative losses) but not necessarily at the plant level (W) (Acevedo and Ceccarelli, 1989; López-Castañeda, 1992), resulting in higher yield under terminal drought stress. Low W genotypes that cover the soil quickly may have high crop water-use efficiency and high Δ values due possibly to a higher specific leaf area (SLA, m^2/g) (López-Castañeda, 1992). It is also possible that the tissue sampled may have fixed most of its carbon at a time when water stress was low, e.g., tissue produced before anthesis in terminal drought environments. At the crop level, the relationship between Δ and W may be affected because of an uncoupling of stomatal openings and transpiration caused by the leaf and crop boundary layer (Farquhar *et al.*, 1988; see chapters in this volume). High W may be associated with high Δ values in this case. The relocation of pre-anthesis assimilates to the grain may affect grain Δ values. This relocation increases with drought stress (Richards and Townley-Smith, 1987) and there seems to be variation among genotypes in relocation of pre-anthesis assimilates (Blum *et al.*, 1983; Turner and Nicholas, 1987). Grain Δ would reflect p_i/p_a of the leaves that produced the stored assimilates. If relocation of assimilates substantially contributes to grain yield, grain Δ and grain yield would be positively associated. Under nonstress conditions, grain Δ will reflect carbon gain by leaves and other photosynthetic organs during grain filling because the proportion of relocated assimilates to the grain is low. A relatively low Δ due to partial stomatal closure may be observed in long-duration varieties that have high grain yields. This would show a negative relationship between grain Δ and grain yield. If stomatal conductance is equally high in all genotypes throughout the season, variations in p_i/p_a and hence Δ values would be a reflection of variation in photosynthetic capacity. Genotypes having a lower p_i/p_a and low Δ would have a higher yield.

The relationship between genotype Δ and grain yield is complex. It is dependent on stress level and stresses are usually present in combination. Other stresses may also influence yield in dry rain-fed Mediterranean environments (Ceccarelli *et al.*, 1991) where low winter temperatures and terminal heat stress are nearly always present. This fact would also complicate relationships between Δ values and yield.

C. Carbon Isotope Discrimination and Transpiration Efficiency

Carbon isotope discrimination has been shown, in theory and experimentally, to be related to transpiration efficiency (Farquhar *et al.*, 1982; Farquhar and Richards, 1984), and variation in Δ and W across genotypes has been reported in several C_3 species by various authors (e.g., Farquhar and Richards, 1984; Hubick *et al.*, 1986; Farquhar *et al.*, 1989). It has therefore been suggested that crop physiologists should search for exploitable variation in W by using Δ. Figure 1 shows that W and Δ are negatively related in barley, a finding that was also reported by Hubick and Farquhar (1989). Many authors have reported a linear variation of dry matter (DM) with water transpired, which implies a constant W (e.g., Arkley, 1963; Tanner and Sinclair, 1983), but as much as 35% of the variation among barley genotypes has been found to be unexplained by transpiration by Hubick and Farquhar (1989). Furthermore, the conclusion of a constant W is valid only if the regression of DM on E passes through the origin. Tables I and II show the significant variation in W and Δ among barley genotypes. Stressed plants had a larger W value than did well-watered ones. The rankings of stressed and mildly stressed genotypes were similar. The carbon isotope discrimination values for barley in Hubick and Farquhar (1989) were highly and negatively correlated to W for both whole plants and plant parts (leaves, heads, stems, and roots). Our greenhouse studies are in agreement with this finding, but they also show a negative relation between W and total biomass (Fig. 4) and a highly significant positive relationship between Δ measured in the peduncle or grain at harvest and total biomass and grain yield (ranging from 0.46 to 0.60, $P < 0.001$), probably indicating that in barley the greater leaf area is more than enough to counteract the effect of

Table I Δ at Various Irrigation Levels in a Line Source Sprinkler System[a]

| | Irrigation (mm) | | | |
Variety	82	181	325	Mean
Harmal	16.42	17.32	18.58	17.43
WI2198	16.18	17.78	18.96	17.67
WI2198/WI2291	16.49	17.43	18.79	17.61
Tadmor	16.21	17.29	18.96	17.49
Roho	16.46	18.02	18.89	17.77
SLB 39-99	16.53	17.95	19.00	17.76
SLB 62-35	16.32	18.19	19.28	17.97
SLB 62-99	17.39	18.09	19.80	18.36
SLB 8-6	16.70	18.54	19.66	18.24
SBON 96	16.45	17.26	18.79	17.47
Mean	16.51	17.78	19.07	

Note. LSD0.05 for genotypes at each irrigation level, 0.58; LSD0.05 for genotype means at each irrigation level, 0.18; LSD0.05 for irrigation levels at each genotype level, 0.55; LSD0.05 for irrigation means for each genotype, 0.26. The variety × irrigation interaction is insignificant.

[a]The Δ measurements were done in peduncles collected at harvest. The experiment was conducted at Breda, northern Syria, in the 88/89 season. Total seasonal rainfall was 180 mm.

Table II Transpiration Efficiency (W, mg DM/g H_2O kPa)[a] of Selected Two-Row Barley Genotypes[b]

| | | FC | | ½ FC | |
Type	Genotype	W	Δ	W	Δ
Improved					
	Harmal	4.1	18.68	4.7	19.09
	WI2198	4.0	19.14	4.8	18.61
	WI2198/WI2269	3.9	18.93	4.7	18.68
	ROHO	3.7	19.45	4.6	18.54
	SBON 96	3.8	18.96	4.0	18.51
Landraces					
	Tadmor	3.9	20.08	4.6	19.20
	SLB 62-99	3.6	20.71	4.3	19.03
	SLB 39-99	3.5	20.18	4.3	19.48
	SLB 62-35	3.4	20.18	4.2	19.42
	SLB 8-6	3.8	20.76	4.5	18.61

Note. LSD0.05 for genotype W at the same level of watering, 0.51; LSD0.05 for watering treatment at the same level of cultivar, 0.56; LSD0.05 for Δ at the same level of watering, 0.67. The genotype × watering interaction is insignificant for W and Δ.

[a]Dry mass includes roots. Mean day time vpd: 0.53 kPa.

[b]Plants grown in pots in a greenhouse and irrigated up to field capacity (FC) or ½ FC. In the FC treatment the pots were allowed to dry until they reached one-half of the available water. In the ½ FC treatment the pots reached the witting point before irrigation.

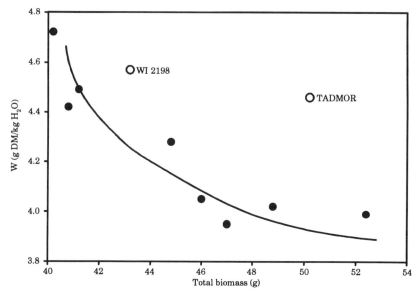

Figure 4. Mean transpiration efficiency and mean biomass production of 10 barley geno-types grown in a greenhouse under four irrigation treatments (described under Fig. 1). Each point in the figure represents the mean transpiration efficiency for a genotype.

a decreased W due to higher SLA (López-Castañeda, 1992). The significant difference in Δ values found for various plant parts by Hubick and Farquhar (1989) was attributed to chemical composition, to different contribution to carbon fixation by PEP carboxylation, or to temporal variations in p_i/p_a while carbon was being laid down in the various plant parts. Our barley work also shows such differences in Δ values between plant parts at a given time of measurement (Fig. 5). The grain tends to have lower Δ values when compared with leaf lamina and peduncle. This may be related to the time at which assimilates are laid down in the various organs. Of interest, however, is that for given sampling dates, anthesis, and maturity, the values of various organs were found to be highly correlated (Table III). Strong correlations among Δ values of different plant parts at various growth stages were observed by Craufurd *et al.* (1991). In other cases, weak correlations have occurred for genotypes sampled at the five-leaf stage, anthesis, and maturity in a Mediterranean environment (Table IV). This result may be explained by the temporal variation of plant-water status at the various growth stages.

D. Carbon Isotope Discrimination, Drought Resistance, and Stability of Yield

Breeders measure drought resistance as the ability of a genotype to yield under stress compared with standard cultivars. In low-rainfall environments, there is a high year-to-year variability of rainfall that results in a

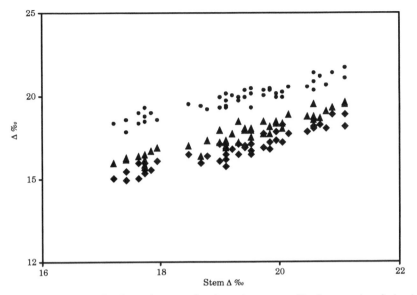

Figure 5. Δ determination at harvest of various plant parts of barley at various irrigation levels: leaf lamina (dots), peduncle (triangles), and grain (diamonds). Data from Breda, northern Syria, in the 1989 season.

high genotype × environment interaction for crops sown without irrigation. The joint regression analysis (Finlay and Wilkinson, 1963; Blum, 1988) has been used to assess both stress resistance and stability of yield across environments. In the joint regression analysis, a nursery of genotypes is grown in various environments and the yield of each genotype is regressed against the nursery mean yield of each environment. The nursery mean yield is used as an environmental index. At least two parameters

Table III Simple Correlation Coefficients for Δ of Various Plant Parts within Each of Two Sampling Times (n = 45; Line Source Sprinkler, Breda, Northern Syria)

		Stem	Peduncle	Leaf lamina
Anthesis				
	Stem	1.000	—	—
	Peduncle	0.903***	1.000	—
	Leaf lamina	0.945***	0.878***	1.000
	Heads	0.899***	0.887***	0.920***
Maturity				
	Stem	1.000	—	—
	Peduncle	0.946***	1.000	—
	Leaf lamina	0.931***	0.910***	1.000
	Grain	0.933***	0.959***	0.927***

***Significance at 0.001% level.

Table IV Correlations between Δ Measured in Various Plant Tissues at Various Growth Stages (n = 45; Line Source Sprinkler, Breda, Northern Syria)

	Five-leaf stage	Anthesis			
		Stem	Leaf lamina	Peduncle	Heads
Five leaf (whole plant)	1.000	0.195	0.154	0.210	0.172
Maturity					
Stem	0.004	0.226	0.223	0.320*	0.186
Leaf lamina	0.133	0.232	0.226	0.310*	0.188
Penducle	0.168	0.296*	0.284	0.383**	0.275
Grain	0.130	0.283	0.274	0.371*	0.234

*.**Significance at 0.05 and 0.01%, respectively.

characterize each genotype, the intercept of extrapolated genotype yield for a mean nursery yield of zero and the slope of the regression that represents the stability (or responsiveness) of the yield of the genotype as the environment changes. If the slope of the regression has a value of 1, the genotype has average (of the genotypes in the nursery) stability; a slope above 1 indicates below-average stability; a slope below 1 indicates above-average stability. In the joint regression analysis, the intercept may be

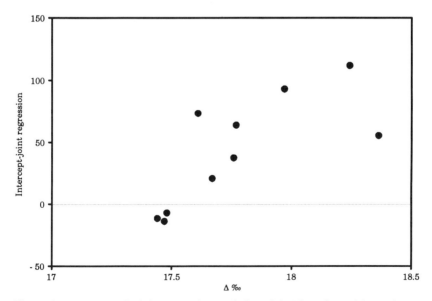

Figure 6. Intercept of a joint regression analysis and Δ. The values of Δ are the mean across irrigation levels in a line source sprinkler (Breda, northern Syria, 1988/1989). The genotypes are the same as those in Table II, grown across nine environments in northern Syria. The joint regression analysis was done using the nine Syrian environments.

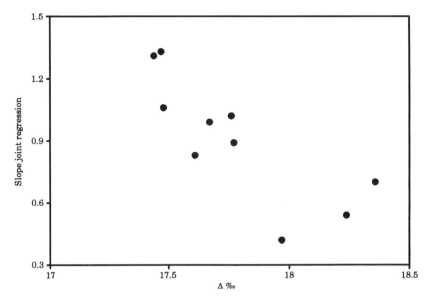

Figure 7. Slope of the joint regression analysis versus Δ. Data obtained as for Fig. 6.

interpreted as a measure of the genotype yield under severe stress (stress resistance in breeder terminology) and the slope as a measure of how stable the yield will be with changing environments. Ideally, breeders are searching for genotypes with high stress resistance as well as responsiveness to an improving environment, or stable across years if the year-to-year variability is high.

The 10 barley genotypes in Table II were grown in nine environments in northern Syria (years and locations) and the joint regression analysis was performed for each genotype. Figure 6 shows the intercept of the joint regression plotted against the nursery mean grain Δ. There is a clear tendency in Mediterranean environments for a higher grain Δ to be associated with a higher yield under drought. In addition, the barley genotypes with a high Δ had high stability of production (Fig. 7).

III. Screening for Carbon Isotope Discrimination in Early Generations

The number of crosses and segregating populations handled in a breeding program can be massive. In a typical cycle for CIMMYT's spring bread wheat-breeding program, which operates two cycles per year, from 500,000 to 750,000 segregating genotypes at the F_2 level are reduced to 10,000 to 15,000 bulks at the F_3 level (S. Rajaram, personal communication). This large reduction in the number of F_2 plants is based essentially on screening for biotic stress resistance (mainly diseases) and visually assessed morphological traits that may bear a relation to yield potential and/or abiotic stresses. These include plant height, spike size, maturity types, head number, and leaf posture. There is no objective assessment of yield potential or yield under stress until at least the F_5 generation when the grain yield can be measured. An objective method is needed to determine the potential for grain and/or yield under stress, such as drought, in earlier generations. There is a high probability that among the voluminous material discarded between the F_2 and F_5 stages of selection (98%), many high-yielding and/or stress-resistant materials are discarded as well.

Several questions need to be answered before a trait such as Δ can be recommended for use in early-generation screening. Does the Δ measured in spaced plants reflect the value of a dense plant situation? The what extent do the measurements made in early generations (F_2) reflect the correlated responses that the breeder is looking for in later generations? What fraction of the trait's variance is heritable?

A. Carbon Isotope Discrimination in Spaced versus Dense Plantings

To allow selection, F_2 plants are grown as spaced (30×10 cm) plants. The Δ value would increase its potential for screening early-generation material if the values determined in a spaced-plant situation would show similar rankings as those measured in dense plantings. To answer this question, we undertook an experiment in which we grew 24 genotypes in spaced and dense plots. The trials were grown at Breda in northern Syria and at Cambridge, United Kingdom, under drought. Table V shows that grain Δ as well as Δ values of other plant parts measured on dense plantings strongly correlated with that measured on spaced plants. These results supported the proposed use of Δ to screen in early generations.

Table V Correlation between Δ Values Measured in Various Plant Parts on Spaced and Drilled Plots[a]

	Breda[b] 1988	Breda[b] 1989	Cambridge[c] droughted 1988
Grain	0.714**	0.635**	0.661**
Peduncle at maturity	0.943**	0.810**	0.894**
Straw at maturity	0.854**	NA	0.542**
Whole plant at heading	0.511*	NA	0.668**
Peduncle at anthesis	NA[d]	0.678**	NA
Whole plant at tillering	0.740**	0.758**	0.665**

[a]The seasonal rainfall at Breda was 415 mm in 1988 and 195 mm in 1989. Total evapotranspiration at Cambridge for plants grown in a rain-out shelter was 109 mm.
[b]$N = 24$.
[c]$N = 21$.
[d]NA, not available.
*,**Significance at 0.05 and 0.01%, respectively.

B. Genetic Parameters and Correlated Yield Response to Carbon Isotope Discrimination

Knowledge of genetic correlations and the heritability of direct and indirect characters is required to obtain the expected response of a character (e.g., grain yield or W) when selection is applied to another character (e.g., Δ). Table VI gives estimated broad-sense heritabilities for grain yield and W and the genetic correlations between grain yield, W, and Δ values. The values used in the calculation were replicated plots of the 10 genotypes grown in the LSS and greenhouse experiments. The calculated heritability for W and Δ values is in the order of 81%. These values compared well with those reported by Condon et al. (1987) for wheat Δ (60 to 90%) (see other chapters in this volume). The broad-sense heritability of Δ values calculated by Hubick et al. (1988) for 16 peanut genotypes across 10 sites was 81% and Hall et al. (1990) reported a value of 76% for leaf Δ in cowpeas.

The merit of indirect selection relative to that of direct selection may be expressed as the ratio of the expected responses, CRx/Rx, where CRx is the correlated response to selection in character X when selection is done for character Y and Rx is the direct response when selection is done for character X (Falconer, 1983). Using a 10% selection intensity for the two characters and the value of the square root of the heritabilities as well as the genetic correlations given in Table VI, selecting indirectly for grain yield and W using Δ would have an expected merit of 0.82 for grain yield and 0.98 for transpiration efficiency.

With the above results in mind, we proceeded to cross two-row barley genotypes with similar phenology having low and high Δ's measured in previous trials. Seven crosses were made, the F_2's were grown in the field at Breda in the 1988/1989 season (rainfall, 180 mm), Δ was determined in the straw at harvest, and yield was measured in F_4 families (1990/1991, 290 mm rainfall).

Table VI Genetic Correlations (rg) between Δ, Grain Yield, and Transpiration Efficiency (W), and Broad-Sense Heritabilities (H) Obtained from 10 Barley Genotypes (pure lines)[a]

	Genetic correlation (rg)	Broad-sense heritability H
Grain yield	0.775 (0.760)	0.71
W	−0.910	0.82
Δ	—	0.81

[a]Values of rg were estimated in a greenhouse experiment. Those in parentheses were estimated from a line source sprinkler experiment. All heritability estimates are from the line source experiment (Breda, northern Syria, 1989).

Table VII ^{13}C Discrimination of Parents and Mean F_2 Progenies of Seven Barley Crosses with Parents Diverging in Δ^a

Cross	^{13}C discrimination (‰) P_1	P_2	$\bar{x} F_2$ (range)	Shoot biomass (g/m²) $\bar{x} F_4$ (range)	Grain yield (g/m²) $\bar{x} F_4$ (range)
SLB8-6/WI2198	18.32c*	18.88a	18.64a (16.39–20.15)	1290ab (736–1739)	404a (177–604)
SLB8-6/Roho	18.32c	18.66b	18.26c (16.28–19.31)	1306a (892–1733)	405a (146–579)
Tadmor/SLB62-99	18.59b	18.93a	18.61a (17.2–20.4)	1007d (550–1428)	256e
Tadmor/Roho	18.59b	18.66b	18.07d (16.28–19.52)	1257b (819–1642)	363b (126–513)
Tadmor/SLB62-35	18.59b	19.00a	18.36a (17.32–19.10)	1010d (656–1317)	251e (83–441)
SLB8-6/SLB62-99	18.32c	18.93a	18.66a (17.12–19.84)	1117c (653–1564)	274d (78–498)
Tadmor/WI2198	18.59b	18.88a	18.22c (16.80–20.15)	1274ab (589–1756)	315c (129–587)

[a]The mean shoot biomass and grain yield of F_4 progenies are also given.
*Cross means followed by different letters differ at 0.05%. Values of Δ for parents of a cross followed by a different letter differ at 0.05%.

Despite significant variation in F_2 Δ, F_4 grain yield, and aboveground biomass (Table VII), no subsequent correlation was found between F_2 Δ values and F_4 family yields either within or between crosses (Table VIII) (Acevedo, Ceccarelli, and Dakeel, unpublished). The relation between F_2 Δ values and days to ear emergence in the F_4 was nonsignificant. W and Δ values have not been measured in the F_4 progenies. This experiment is in progress; the results obtained so far with indirect selection, using Δ as a surrogate for grain yield, indicated however that the correlated responses

Table VIII   Phenotypic Correlation Coefficient between Straw Δ in F_2 and Grain Yield (g/m²) or Shoot Biomass (g/m²) in F_4 Families

Cross	No. (g/m²)	Grain yield (g/m²)	Shoot biomass
SLB8-6/WI2198	90	0.109	0.070
SLB8-6/ROHO	120	0.126	0.120
TADMOR/SLB62-99	51	−0.161	−0.267[a]
TADMOR/ROHO	70	−0.047	−0.122
TADMOR/SLB62-35	35	0.007	0.106
SLB8-6/SLB62-99	129	−0.155	−0.107
TADMOR/WI2198	86	−0.081	0.044

[a]Significance at 0.1% level.

were limited. The distributions of the Δ values were normal for the seven F_2 progenies indicating that Δ is a quantitative character. Out of seven crosses of genotypes differing in Δ value, the F_2 distributions showed a significant skewness to lower values only in three, indicating that some degree of dominance of low Δ may not be a general phenomenon in barley as suggested by Hubick *et al.* (1988) for peanuts.

C. Genotype × Environment Interaction of Δ

The genotype × environment (G × E) interaction of Δ in our experiments has been low and usually nonsignificant (Craufurd *et al.*, 1991). Additional evidence of low G × E comes from our greenhouse study on W of barley (Table II) and from the LSS work conducted at Breda in the 1988/1989 season (Table I). Both experiments had the same genotypes. The varieties in these experiments were chosen because they had similar flowering dates when grown in the field in northern Syria and they represented a range of Δ values measured in the previous experiments.

The interaction between irrigation levels and variety Δ in the LSS and in the pot-grown varieties was nonsignificant. Varieties at each irrigation level and varietal means across irrigation levels differed significantly in Δ values (Tables I and II). The LSS and the greenhouse created different environments. As a result, varieties differed significantly ($P < 0.001$) in their flowering date in these experiments.

A strong G × E in Δ would be expected for locally adapted barley genotypes, which are usually responsive to temperature and photoperiod. The similar genotype rankings for Δ values in the greenhouse experiment and the LSS (Tables I and II, rank correlations for means $r = 0.70$, $P < 0.05$) indicate, however, that G × E in this parameter is low despite the large dissimilarity between the two environments. This subject warrants further analysis.

IV. Summary

Carbon isotope discrimination may have the potential to assist yield-based selections in barley breeding for drought-stressed environments. Experimental evidence showed that, for barley, there is genetic variability for Δ and that the G × E is low, and the broad-sense heritability is high. In Mediterranean environments, the trait can be measured at anthesis or at the end of the growing cycle in any plant tissue as long as consistency is maintained in the sampling procedure. The Δ values for dense and spaced plantings were highly correlated, which suggests that measurements could be done in early generations, a time when yield and/or transpiration efficiency cannot be assessed. However, the efficiency of this trait for selection in early barley generations remains to be shown. There were no correlations between Δ values measured in F_2 plants of seven barley crosses and the grain yield of the F_4 families derived from each F_2 plant. Work is in progress through a divergent selection experiment to study this point further.

The experiments conducted showed relations between Δ, transpiration efficiency, and grain yield. The relation of Δ with grain yield was found to be generally strong and positive under water stress, but it was also generally found that Δ was related to earliness in flowering. Transpiration efficiency was negatively related to Δ across barley genotypes.

Carbon isotope discrimination was positively related to genotype grain yield at the intercept of a joint regression analysis, which included nine Mediterranean environments, i.e., the larger the Δ values, the higher the yield under drought stress. The Δ values were negatively related to the slope of genotype yield over nursery mean yield. The higher the discrimination, the higher the genotypic stability as well as yield under stress. The apparent paradox between higher yield under water-limited conditions and lower transpiration efficiency needs to be resolved. Factors involved may be, among others, variations in earliness of flowering and/or in specific leaf area.

Acknowledgments

The author acknowledges the continuous discussions on this subject with Dr. R. B. Austin and Dr. S. Ceccarelli. Mrs. Issam Nagi, Riad Sakkal, and Mufid Adjami of ICARDA helped in all phases of data collection. Dr. G. D. Farquhar provided much stimulus to this work. Part of this work has been financed by USAID Grant No. DHR-5542-G-SS 9085-00. Most of this work was carried out while the author was working at the International Center for Agricultural Research in the Dry Areas (ICARDA), P.O. Box 5466, Aleppo, Syria.

References

Acevedo, E., and S. Ceccarelli. 1989. Role of the physiologist breeder in a breeding program for drought resistance conditions, pp. 117–139. *In* F. W. Baker (ed.), Drought Resistance in Cereals. CAB International, Wallingford, U.K.

Acevedo E., and E. Fereres. 1993. Resistance to Abiotic Stresses. *In* M. D. Hayward, N. O. Bosemark, and I. Romagosa (eds.), Plant Breeding Principles and Prospects. Chapman & Hall, London, In press.

Acevedo, E., P. Q. Craufurd, R. B. Austin, and P. Perez-Marco. 1991a. Traits associated with high yield in barley in low rainfall environments. *J. Agric. Sci. Cambridge* **116**: 23–36.

Acevedo, E., A. P. Conesa, P. Monneveux, and J. P. Srivastava (eds.). 1991b. Physiology—Breeding of Winter Cereals for Stressed Mediterranean Environments. INRA, Paris.

Acevedo E., E. Fereres, C. Gimenez, and J. P. Srivastava (eds.). 1991c. Improvements and Management of Winter Cereals under Temperature, Drought and Salinity Stresses. INIA, Madrid.

Arkley, R. J. 1963. Relationship between plant growth and transpiration. *Hilgardia* **34**: 559–584.

Austin, R. B., P. Q. Craufurd, B. da Silveira Pinheiro, M. A. Hall, E. Acevedo, and E. C. K. Ngugi. 1990. Carbon isotope discrimination as a means of evaluating drought resistance in barley, rice and cowpeas. *Bull. Soc. Bot. Fr. 137, Actual. Bot.* **1**: 21–30.

Blum, A. 1988. Plant Breeding for Stress Environments. CRC Press, Boca Raton, FL.

Blum, A., H. Poiarkova, G. Golan, and J. Meyer. 1983. Chemical desiccation of wheat plants as a simulator of post anthesis stress: Effects on translocation and kernel growth. *Field Crops Res.* **6:** 51–58.

Ceccarelli, S., E. Acevedo, and S. Grando. 1991. Breeding for yield stability in unpredictable environments: Single traits, interaction between traits, and architecture of genotypes. *Euphytica* **56:** 169–185.

Condon, A. G., R. A. Richards, and G. D. Farquhar. 1987. Carbon isotope discrimination is positively correlated with grain yield and dry matter production in field grown wheat. *Crop Sci.* **27:** 996–1001.

Craufurd, P. Q., R. B. Austin, E. Acevedo, and M. A. Hall. 1991. Carbon isotope discrimination and grain-yield in barley. *Field Crops Res.* **27:** 301–313.

Evans, L. T. 1987. Opportunities for increasing the yield potential of wheat, pp. 79–93. *In* CIMMYT (ed.), The Future Development of Maize and Wheat in the Third World. CIMMYT, Mexico.

Falconer, D. S. 1983. Introduction to Quantitative Genetics. Longman, London.

Farquhar, G. D., and R. A. Richards. 1984. Isotopic composition of plant carbon correlates with water use efficiency of wheat genotypes. *Aust. J. Plant Physiol.* **11:** 539–552.

Farquhar, G. D., H. M. O'Leary, and J. A. Berry. 1982. On the relationship between carbon isotope discrimination and intercellular carbon dioxide concentration in leaves. *Aust. J. Plant Physiol.* **9:** 121–137.

Farquhar, G. D., K. T. Hubick, A. G. Condon, and R. A. Richards. 1988. Carbon isotope fractionation and plant water use efficiency, pp. 21–40. *In* P. W. Rundel, J. R. Ehleringer, and K. A. Nagy (eds.), Applications of Stable Isotope Ration to Ecological Research. Springer, New York.

Farquhar, G. D., J. A. Ehleringer, and K. T. Hubick. 1989. Carbon isotope discrimination and photosynthesis. *Annu. Rev. Plant Physiol. Plant Mol. Biol.* **40:** 503–537.

Finlay, K. W., and G. N. Wilkinson. 1963. The analysis of adaptation in a plant-breeding programme. *Aust. J. Agric. Res.* **14:** 742–754.

Fischer, R. A., and N. C. Turner. 1978. Plant productivity in the arid and semiarid zones. *Annu. Rev. Plant Physiol.* **29:** 277–317.

Hall, A. E., R. G. Mutters, K. T. Hubick, and G. D. Farquhar. 1990. Genotypic differences in carbon isotope discrimination by cowpea under wet and dry field conditions. *Crop Sci.* **30:** 300–305.

Hubick, K. T., and G. D. Farquhar. 1989. Carbon isotope discrimination and the ratio of carbon gained to water lost in barley cultivars. *Plant Cell Environ.* **13:** 795–804.

Hubick, K. T., G. D. Farquhar, and R. Shorter. 1986. Correlation between water use efficiency and carbon isotope discrimination in diverse peanut (Arachis) germplasm. *Aust. J. Plant Physiol.* **15:** 779–813.

Hubick, K. T., R. Shorter, and G. D. Farquhar. 1988. Heritability and genotype × environment interactions of carbon isotope discrimination and transpiration efficiency in peanuts (*Arachis hypogea* L.). *Aust. J. Plant Physiol.* **15:** 799–813.

Kronstad, W. E., and D. N. Moss. 1991. Requirements of plant breeders from physiologists to increase selection efficiency, pp. 319–335. *In* E. Acevedo, A. P. Conesa, P. Monneveux, and J. P. Srivastava (eds.), Physiology-Breeding of Winter Cereals for Stressed Mediterranean Environments. INRA, Versailles.

López-Castañeda, C. 1992. A Comparison of Growth and Water Use Efficiency in Temperate Cereal Crops, Thesis. Research School of Biological Sciences, Australian National University, Canberra, Australia.

Passioura, J. B. 1977. Grain yield, harvest index and water use of wheat. *J. Aust. Inst. Agric. Sci.* **34:** 117–120.

Passioura, J. B. 1986. Resistance to drought and salinity: Avenues for improvement. *Aust. J. Plant Physiol.* **13:** 191–201.

Quisenberry, J. E. 1982. Breeding for drought resistance and water use efficiency, pp. 193–212. *In* M. N. Christiansen and C. F. Lewis (eds.), Breeding Plants for Less Favorable Environments. Wiley, New York.

Richards, R. A. 1991. Crop improvement for temperate Australia: Future opportunities. *Field Crops Res.* **26:** 141–169.

Richards, R. A., and T. F. Townley-Smith. 1987. Variation in leaf area development and its effect on water use, yield, and harvest index of droughted wheat. *Aust. J. Agric. Res.* **38:** 938–992.

Srivastava, J. P., E. Porceddu, E. Acevedo, and S. Varma (eds.). 1987. Drought Tolerance in Winter Cereals. Wiley, New York.

Tanner, C. B., and T. R. Sinclair. 1983. Efficient water use in crop production: Research or re-research, pp. 1–28. *In* H. Taylor (ed.), Limitations to Efficient Water Use in Crop Production. ASA-CSSA-SSA, Madison, WI.

Turner, N. C., and M. E. Nicholas. 1987. Drought resistance of wheat for light-textured soils in a Mediterranean climate, pp. 203–216. *In* J. P. Srivastava, E. Porceddu, E. Acevedo and S. Varma, editors. Drought Tolerance in Winter Cereals. Wiley, New York.

27

Genetic Analyses of Transpiration Efficiency, Carbon Isotope Discrimination, and Growth Characters in Bread Wheat

Bahman Ehdaie, David Barnhart, and J. G. Waines

I. Introduction

Transpiration efficiency (W = total dry matter/water transpired) in bread wheat (*Triticum aestivum* L.) has not been improved during the past few decades, mainly because of a lack of an effective selection method applicable in breeding programs. Genetic variation for W has been reported for various crop species (Briggs and Shantz, 1913, 1914; Farquhar and Richards, 1984; Frank *et al.*, 1985; Hubick *et al.*, 1986, 1988; Martin and Thorstenson, 1988; Hubick and Farquhar, 1989; Virgona *et al.*, 1990; Ehleringer *et al.*, 1990; Ehdaie *et al.*, 1991; Ismail and Hall, 1992). However, breeding programs have not exploited this variation due to the difficulty of assessing W under field conditions.

Early studies described variation in plant carbon isotope composition among species (Farquhar *et al.*, 1989). Recently, Farquhar and his colleagues (Farquhar and Richards, 1984; Hubick *et al.*, 1986) suggested that carbon isotope composition may be a useful criterion to assess variation in W within C$_3$ crop species. Plants discriminate against the naturally occurring and heavier isotope of carbon (^{13}C) during the diffusion and fixation of CO_2 in photosynthesis. The extent of the discrimination (Δ) is a function of the ratio of the intercellular and atmospheric partial pressure of CO_2 (p_i/p_a) (Farquhar *et al.*, 1982). Based on theory, they suggested that Δ measured in plant dry matter should be positively correlated with the ratio of p_i/p_a, and negatively associated with W.

Several studies with different C_3 plant species have reported significant differences among genotypes for Δ and for W and a negative correlation between the two traits. Farquhar and Richards (1984) and Ehdaie *et al.* (1991) reported a negative correlation between W of pot-grown plants and Δ for several bread wheat genotypes grown under different levels of drought or well-watered conditions. Negative correlations between Δ and W have been reported for diverse peanut (*Arachis* spp) germplasm (Hubick *et al.*, 1986), for tomato (*Lycopersicon* spp) (Martin and Thorstenson, 1988), for barley (*Hordeum vulgare* L.) (Hubick and Farquhar, 1989), for common bean (*Phaseolus vulgaris* L.), for a range grass (*Agropyrum desertorum*) (Ehleringer *et al.*, 1990), and for cowpea (*Vigna unguiculata* L.) (Ismail and Hall, 1992). Little or no genotype × environment interactions have been reported for Δ (for exceptions see Hall *et al.*, Masle *et al.*, and White, Chapters 23, 24, and 25, respectively, this volume). The variance components for Δ associated with genotype × environment interactions were relatively much smaller than genotypic variance for Δ exhibited by several bread wheat genotypes (Ehdaie *et al.*, 1991). Furthermore, the broad-sense heritabilities reported for Δ were relatively large (Ehleringer, 1988; Hubick *et al.*, 1988; Hall *et al.*, 1990; Ehdaie *et al.*, 1991). These observations resulted in the suggestion that selection for Δ, under either well-watered or water-stressed conditions, could indirectly improve W.

The relative efficiency or merit of indirect selection relative to direct selection in either a well-watered or a water-stressed environment depends upon the magnitude of the heritabilities of transpiration efficiency and Δ, and the genetic correlation between the two characters (Falconer, 1988). The relative efficiency estimated under well-watered conditions should be compared to that estimated under water-stressed conditions in order to define the most effective selection environment for Δ.

A successful breeding program for improving transpiration efficiency in bread wheat requires knowledge of the inheritance of W, of Δ, and of associated growth characteristics, and the interrelationships among these characters. Detailed information on the inheritance of W and Δ in bread wheat is not available in the literature. The objectives of this study were to (1) determine the inheritance of W, of Δ, and of yield and growth characters in pot experiments under well-watered and water-stressed conditions using segregating generations derived from a cross between contrasting parents, (2) estimate heritability of these characters and their genetic correlations, (3) predict the relative efficiency of direct and indirect selection for Δ, and (4) determine the most effective environment to select for Δ.

II. Materials and Methods

A. Parents and Water Treatments

Two spring wheat genotypes chosen on the basis of their contrasting growth characteristics, W, and Δ were used as parents in this study. The first parent, 'Chinese Spring' (P_1), is a tall, later-maturing, landrace cultivar with relatively high W and low Δ. The second parent, 'Yecora Rojo' (P_2), is a dwarf, early-maturing, modern cultivar with low W and high Δ (Ehdaie *et al.*, 1991). Seeds from the P_1, P_2, F_1, and F_2 generations were planted in plastic pots containing a double-layered plastic bag filled with 5 kg of soil, composed of 84.2% sand, 11.8% silt, and 4.0% clay with water-holding capacity of $\approx21.5\%$ by weight. Pots were arranged in a randomized complete block design with eight replicates in an unheated, water-cooled glasshouse at the University of California, Riverside. Each replicate contained one pot of each of the parents, two pots of F_1, and four pots of F_2 generations. Two complete sets of pots were established, one under a well-watered and the other under a water-stressed treatment.

Each pot in both treatments was brought to water-holding capacity by adding 1075 ml of half-strength Hoagland's solution on 3 January 1990. Two seeds were planted in each pot on the same day. In each treatment, there were three unplanted pots to quantify evaporative water loss from the pots. Throughout the study, pots were irrigated with the same solution. Pots were weighed every 2 or 3 days and an amount of solution equal to the loss in weight was added. Sixteen days after sowing, when plants were at the two-leaf stage, one seedling was removed from each pot, leaving the most vigorous one. At this time, 60 g of medium-size perlite was added to the top of each pot, including the unplanted pots, to reduce surface evaporation.

In the well-watered treatment, hereafter referred to as the "wet" environment, pots were irrigated, as described, until the color of the main tiller turned yellow. In the water-stressed experiment, hereafter referred to as the "dry" environment, irrigation was terminated as each plant reached heading stage, i.e., when the spike of the main tiller was partially emerged from the flag leaf sheath. Genotypes received different amounts of water in both wet and dry environments due to genotypic differences in heading and maturity. Flag leaves were collected from each plant in the dry environment on 25 April 1990 and in the wet environment on 10 May 1990, when all plants reached maturity. Shoots were removed from roots at the soil surface. The pots were weighed to the nearest gram and the soil was carefully washed from the roots. Plant parts were dried at 65°C for 10 days before weighing. Season-long W was calculated for each plant as g total dry matter/kg water transpired with correction for the water loss due to evaporation from the soil surface.

The molar ratio of carbon isotopes in the flag leaves was determined using a ratio mass spectrometer (Hubick *et al.*, 1986). The flag leaves were dried at 80°C for 2 days and were separately ground in a Wiley mill to a fine powder. The carbon isotope discrimination (Δ) for each plant was calculated according to an equation (Hubick *et al.*, 1986) assuming an isotopic composition of the air relative to the standard Pee Dee belemnite of $-8\permil$ at Riverside. The other plant characteristics measured on a plant basis were plant height, number of spikes, shoot dry matter, root dry matter, total dry matter, ratio of shoot-to-root dry matter, grain yield, and transpiration efficiency (W) as the ratio of total dry matter-to-water transpired.

B. Genetic Models and Analyses

The data obtained in both wet and dry treatments on a single-plant basis were subjected to analysis of variance. Significant differences among generations for a character are an indication of genetic differences. As a result, the generation means analysis proposed by Hayman (1958) and Mather and Jinks (1971) was carried out for all characters except the ratio of shoot-to-root dry matter which did not show significant differences among generations. In an additive-dominance model, the expectations of generation means in terms of m, mean effect; a, pooled additive effects; and d, pooled dominance effects are as follows:

$$\overline{P}_1 = m + a$$
$$\overline{P}_2 = m - a$$
$$\overline{F}_1 = m + d$$
$$\overline{F}_2 = m + \tfrac{1}{2}d.$$

A weighted least-squares technique (Mather and Jinks, 1971), using all individual plot means, was used to obtain estimates of m, a, and d (Ehdaie and Ghaderi, 1978). Variation among generations was partitioned into variation due to additive and dominance effects and that due to residual or deviation from the additive-dominance model. This permitted a test of the adequacy of the model and an evaluation of the importance of the additive and dominance gene effects in their contributions to the genetic variation. The full model as described by Mather and Jinks (1971) was used for analysis of plant height, root dry matter, grain yield, and Δ, all of which showed significant generation $\times$ environment interactions. The genetic models used were appropriate as the χ^2 tests for goodness of fit (Mather and Jinks, 1971) were nonsignificant. The genetic and environmental variances and covariances were estimated, after removing block effects, using P_1, P_2, F_1, and F_2 generations (Mather and Jinks, 1971) and these estimates were used to determine the broad-sense heritabilities (h^2) for different characters and their pairwise genetic correlations.

C. Predicted Response to Selection

The predicted response to direct selection (R_W) was estimated as $R_W = ih_W\sigma_W$ (Falconer, 1988), where i is the selection intensity, h_W is the square root of the heritability of W, and σ_W is the square root of estimated genetic variance of W. Correlated response to indirect selection (CR_W) due to selection for Δ was estimated as $CR_W = ih_\Delta\sigma_W r_{W,\Delta}$ (Falconer, 1988), where h_Δ is the square root of the heritability of Δ and $r_{W,\Delta}$ is the genetic correlation between W and Δ.

The relative efficiency (RE) in % or merit of indirect selection relative to direct selection was expressed as

$$RE = \frac{CR_W}{R_W} \times 100 = r_{W,\Delta}\frac{h_\Delta}{h_W} \times 100,$$

(Falconer, 1988) assuming similar selection intensities.

III. Results and Discussion

A. General Observations

The combined ANOVA indicated highly significant differences between wet and dry environments for all the characters examined, except for root dry matter, W, and Δ (Table I). Highly significant differences were present among different generations for all the characters, except for the ratio of shoot-to-root dry matter (Table I). Plant height, root dry matter, grain yield, total water transpired, and Δ exhibited significant genotype × environment interactions.

All of the measured characters exhibited significant differences among generation means under the wet environment except the ratio of shoot-to-root dry matter (Table II). Cultivar Chinese Spring had a higher mean than Yecora Rojo for each of the growth characters, grain yield, total water transpired, and W. Mean Δ for Chinese Spring (20.0‰) was lower than that of Yecora Rojo (22.9‰) under wet treatment and the difference observed, 2.9‰ was similar to our previous observations of 2.8‰ in 1987 and 2.5‰ in 1988 (Ehdaie *et al.*, 1991). However, the difference observed between these two cultivars for W (1.0 g/kg), though significant, was less than that reported previously (1.33 g/kg). Among the measured traits, Δ exhibited the lowest overall coefficient of variation (1.4%), indicating that the environmental errors associated with its measurement were the lowest compared to other traits (Table I).

The drought treatment caused substantial stress, reducing plant height, number of spikes per plant, total dry matter, and grain yield (Table III). Root dry matter exhibited mixed reactions to drought depending on the different generations. Mean root dry matter of Chinese Spring and the F_2 generation were higher under drought than under well-watered conditions. In contrast, root dry matter of Yecora Rojo and the F_1 generation were reduced under drought. Ehdaie *et al.* (1991) reported that Chinese Spring produced higher root dry matter under dry conditions than under

Table I Summary of Combined ANOVA and Test of Significance for Plant Height (PH), Number of Spikes (NS), Total Dry Matter (TDM), Shoot Dry Matter (SDM), Root Dry Matter (RDM), Ratio of Shoot-to-Root Dry Matter (S/R), Grain Yield (GY), Total Water Supplied (TWS), Total Water Transpired (TWT), Transpiration Efficiency (W), and Carbon Isotope Discrimination of Flag Leaves (Δ) of Wheat in P_1 (Chinese Spring), P_2 (Yecora Rojo), F_1, and F_2 Generations under Well-Watered and Water-Stressed Treatments

Source	df	PH	NS	TDM	SDM	RDM	S/R	GY	TWS	TWT	W	Δ
Treatments (T)	1	**	**	**	**		**	**	**	**		
Generations (G)	3	**	**	**	**	**		**	**	**	**	**
T × G	3	**				**		**		**		**
Pooled error	42											

**Significant at the 0.01 level of probability.

Table II Generation Means and Coefficient of Variation (CV) of Plant Height (PH), Number of Spikes (NS), Total Dry Matter (TDM), Shoot Dry Matter (SDM), Root Dry Matter (RDM), Ratio of Shoot-to-Root Dry Matter (*S/R*), Grain Yield (GY), Total Water Supplied (TWS), Total Water Transpired (TWT), Transpiration Efficiency (*W*), and Carbon Isotope Discrimination of Flag Leaves (Δ) of Bread Wheat Generations under Well-Watered Conditions

Characters	$P_1{}^a$	$P_2{}^a$	F_1	F_2	LSD (0.05)	CV (%)
			Generations			
PH (cm)	122 a*	66 b	100 c	95 d	3.1	3.1
NS (No.)	17.4 a	12.5 c	15.3 b	17.3 a	1.7	10.6
TDM (g)	81.4 a	57.9 c	72.2 b	74.2 b	3.1	4.1
SDM (g)	74.6 a	53.3 c	66.0 b	67.9 b	2.6	3.8
RDM (g)	6.8 a	4.6 b	6.2 a	6.4 a	0.9	13.8
S/R (g/g)	11.4 a	11.9 a	10.9 a	11.1 a	—	14.6
GY (g)	33.3 a	29.7 b	34.6 a	33.4 a	2.0	5.8
TWS (kg)	18.3 a	15.9 d	16.8 c	17.6 b	0.4	2.2
TWT (kg)	14.7 a	12.7 c	13.6 b	14.3 a	0.5	3.2
W (g/kg)	5.6 a	4.6 c	5.3 b	5.2 b	0.2	4.2
Δ (‰)	20.0 c	22.9 a	23.0 a	22.2 b	0.3	1.4

[a]P_1, Chinese Spring; P_2, Yecora Rojo.
*Means in a row followed by a common letter do not differ significantly as tested by LSD (0.05).

Table III Generation Means and Coefficient of Variation (CV) of Plant Height (PH), Number of Spikes (NS), Total Dry Matter (TDM), Shoot Dry Matter (SDM), Root Dry Matter (RDM), Ratio of Shoot-to-Root Dry Matter (*S/R*), Grain Yield (GY), Total Water Supplied (TWS), Total Water Transpired (TWT), Transpiration Efficiency (*W*), and Carbon Isotope Discrimination of Flag Leaves (Δ) of Wheat Generations under Water-Stressed Conditions

Characters	$P_1{}^a$	$P_2{}^a$	F_1	F_2	LSD (0.05)	CV (%)
			Generations			
PH (cm)	92 a*	52 c	78 b	77 b	7.2	9.3
NS (No.)	9.4 a	5.0 c	6.9 b	6.9 b	1.5	20.5
TDM (g)	42.6 a	18.2 d	28.9 c	33.4 b	4.3	13.4
SDM (g)	33.5 a	14.5 c	23.1 b	23.9 b	3.3	13.1
RDM (g)	9.1 a	3.7 d	5.8 c	7.2 b	1.1	17.0
S/R (g/g)	3.7 a	4.1 a	4.0 a	3.9 a	—	13.4
GY (g)	0.9 a	0.4 c	0.8 ab	0.7 b	0.2	24.0
TWS (kg)	9.7 a	6.4 c	7.4 b	8.0 b	0.7	8.1
TWT (kg)	7.6 a	4.0 d	5.5 c	6.3 a	0.7	12.1
W (g/kg)	5.6 a	4.5 c	5.2 b	5.3 b	0.2	3.9
Δ (‰)	19.7 d	23.6 a	22.5 a	21.8 c	0.5	2.3

[a]P_1, Chinese Spring; P_2, Yecora Rojo.
*Means in a row followed by a common letter do not differ significantly as tested by LSD (0.05).

wet conditions. Transpiration efficiency (W) of different generations did not change from wet to dry environments (Tables I and II) because the amount of water transpired and the total dry matter produced were reduced proportionately from wet to dry environments. Mean percentage reduction in water transpired and total dry matter produced under drought were, respectively, 48 and 47% for Chinese Spring, 68 and 69% for Yecora Rojo, 59 and 60% for the F_1 generation, and 56 and 57% for the F_2 generation. The ranking of different generations for W did not change from the wet to the dry environment and, consequently, there was no significant generation × environment interaction for this character. Changes in W from well-watered to water-stressed conditions were not consistent in our previous studies with various bread wheat genotypes (Ehdaie *et al.*, 1991). Transpiration efficiency (W) for a set of diverse bread wheat cultivars increased under dry conditions relative to that under wet conditions in 1987, but decreased in 1988 (Ehdaie *et al.*, 1991). Ismail and Hall (1992) reported increased W under drought conditions compared to wet conditions for five cowpea genotypes due to biomass production being reduced less by drought than water transpired. The significant generation × environment interaction observed for $Δ$ was due to an increase and a decrease in mean $Δ$ of Yecora Rojo and the F_1 generation, respectively, from the wet to the dry environment (Tables II and III). Therefore, while W of the generations and their ranking did not change across wet and dry environments, mean $Δ$ of Yecora Rojo and the F_1 generation varied over the two water treatments. In our previous studies (Ehdaie *et al.*, 1991), Yecora Rojo also responded differently to drought with respect to W and $Δ$. In the pot experiments conducted in 1987, Yecora Rojo responded to drought by exhibiting increased W but no change in $Δ$. In the 1988 experiments, W of Yecora Rojo changed little across the irrigation treatments, but $Δ$ decreased significantly under drought. Hubick *et al.* (1988) reported that the F_1 generation derived from a peanut cross responded to severe stress with little change in $Δ$, but with a reduction in W. These results are difficult to explain although they may reflect different stress exposures with their consequent effects on plant growth. In the present studies, drought was imposed late during heading and would be expected to have small effects on $Δ$ of flag leaves and W. The coefficient of variation for $Δ$, 1.4%, was also the lowest under the dry condition compared to those of other characters (Table III). The ratios of shoot-to-root dry matter were the same for different generations in both wet and dry environments (Tables II and III). However, these ratios were much reduced under drought, mainly due to reduced shoot dry matter (Table III). Reduction in shoot dry matter under the water-stressed condition was mainly due to reduced number of tillers and spikes per plant and reduced plant height (Table III). There was a strong correlation coefficient between total dry matter and water transpired under wet ($r = 0.98*$, significant at $P = 0.05$) and under dry ($r = 1.00**$, significant at $P = 0.01$) environments which are consistent with the studies of Tanner and Sinclair (1983), Hubick *et al.* (1988), and Ismail and Hall (1992).

The amounts of water supplied to Chinese Spring and Yecora Rojo were, respectively, 18.3 and 15.9 kg in the wet treatment (Table II) and were, respectively, 9.7 and 6.4 kg in the dry treatment (Table III). These differences in water requirements between the two cultivars were mainly due to differences in dates to heading and to flowering and were also due to variation in transpiration rates. Chinese Spring reached heading 14 and 12 days later than Yecora Rojo in the wet and dry conditions, respectively. Chinese Spring was later than Yecora Rojo in anthesis by 11 and 9 days, respectively, in the wet and dry treatments. The transpiration rate (total water transpired/duration of irrigation) of Chinese Spring was 156 g/day and that of Yecora Rojo was 142 g/day in the wet environment. Under the drought condition, the transpiration rate was 116 g/day for Chinese Spring and 90 g/day for Yecora Rojo.

Grain transpiration efficiency (grain yield/water transpired) was the same for Chinese Spring (2.28 g/kg) and Yecora Rojo (2.34 g/kg) in the wet treatment. This trait was not calculated for the dry condition because very little grain was produced due to termination of irrigation at the time of heading for these plants.

B. Genetic Analyses

The additive-dominance genetic model used to characterize the variation observed among different generations for number of spikes per plant, total dry matter, shoot dry matter, and transpiration efficiency (W) was appropriate for both dry and wet environments (Tables IV and V). This model accounted for more than 95% of the variation among generations for these characters except for number of spikes per plant under the wet environment, in which it accounted for 76% of total variation (Tables IV and V). The reduction in sum of squares due to fitting additive gene effects was

Table IV Summary of ANOVA, Test of Significance, Percentage Variation Accounted for by Different Items, and Estimates of Gene Effects on Number of Spikes (NS), Total Dry Matter (TDM), Shoot Dry Matter (SDM), and Transpiration Efficiency (W) under Well-Watered Conditions for Different Wheat Generations

Source	df	NS	TDM	SDM	W
Generations	3	**	**	**	**
Genetic model	2	**	**	**	**
Additive	1	**76%	**95%	**96%	**92%
Dominance	1		* 2%	* 2%	** 8%
Deviation	1	**22%	* 3%	* 3%	
Error	21				
m^a		15.5 ± 0.5	7.0 ± 1.0	64.2 ± 0.8	5.1 ± 0.1
a		2.9 ± 0.5	11.7 ± 1.0	10.6 ± 0.8	0.5 ± 0.1
d		0.2 ± 0.8	2.6 ± 1.6	2.1 ± 1.2	0.3 ± 0.1

[a]m, a, and d are mean effect, pooled additive effects, and pooled dominance effects, respectively.
*,**Significant at the 0.05 and 0.01 level of probability, respectively.

Table V Summary of ANOVA, Test of Significance, Percentage Variation Accounted for by Different Items, and Estimates of Gene Effects on Number of spikes (NS), Total Dry Matter (TDM), Shoot Dry Matter (SDM), and Transpiration Efficiency (*W*) under Water-Stressed Conditions for Different Wheat Generations

Source	df	NS	TDM	SDM	W
Generations	3	**	**	**	**
Genetic model	2	**	**	**	**
Additive	1	**99%	**96%	**100%	**92%
Dominance	1				* 5%
Deviation	1		* 3%		
Error	21				
m^a		7.2 ± 0.3	30.9 ± 1.1	24.1 ± 0.9	5.1 ± 0.1
a		2.2 ± 0.4	12.6 ± 1.1	9.5 ± 0.9	0.5 ± 0.1
d		−0.3 ± 0.7	−1.3 ± 1.7	−0.9 ± 1.4	0.2 ± 0.1

[a] *m*, *a*, and *d* are mean effects, pooled additive effect, and pooled dominance effects, respectively.
*,**Significant at the 0.05 and 0.01 level of probability, respectively.

highly significant and accounted for 92 to 100% of total variation observed for total dry matter, shoot dry matter, and *W* under wet conditions and for *W* under dry conditions. Dominance variation accounted for only 2 to 8% of total variation for these characters. The deviation from the additive-dominance model which is a measure of epistatic gene effects was only significant for number of spikes per plant under the wet environment.

Since the generation × environment interactions (G × E) were significant for plant height, root dry matter, grain yield, and Δ, the genetic model was extended to include estimates of additive × environment (*a* × *e*) and dominance × environment (*d* × *e*) effects (Table VI). Significant additive

Table VI Summary of ANOVA, Test of Significance, Percentage Variation Accounted for by Different Items for Plant Height (PH), Root Dry Matter (RDM), Grain Yield (GY), and Carbon Isotope Discrimination of Flag Leaves (Δ) of Different Wheat Generations under Wet and Dry Treatments

Source	df	PH	RDM	GY	Δ
Generations (G)	3	**	**	**	**
Genetic model	2	**	**	**	**
Additive	1	**98%	**94%	**53%	**86%
Dominance	1	** 2%		**47%	**14%
Deviation	1		* 6%		
Environments (E)	1	**		**	
G × E	3	**	**	**	**
Additive × E	1	** 7%	**75%	** 1%	**53%
Dominance × E	1	**38%		**49%	**43%
Deviation × E	2	**55%	*25%	**50%	
Pooled error	42				

*,**Significant at the 0.05 and 0.01 level of probability, respectively.

variations were observed for these characters. The dominance variations were significant for plant height, grain yield, and Δ and they accounted, respectively, for 2, 47, and 14% of total variation exhibited by the generations. Significant epistatic gene action, as estimated by the deviation from the genetic model, was observed for root dry matter, but it accounted for only 6% of total variation in this trait. Dominance $\times$ environment and epistasis $\times$ environment interactions together accounted for more than 90% of GE observed for plant height and grain yield (Table VI). Additive $\times$ environment interactions for plant height and grain yield were significant but accounted, respectively, for only 7 and 1% of GE observed for these characters. Additive $\times$ environment interactions were also significant for root dry matter and Δ, but for these two characters they accounted for more than 50% of GE. Dominance $\times$ environment interaction was also significant for Δ (Table VI).

C. Gene Effects and Direction of Dominance

The additive effects of genes involved in the expression of number of spikes per plant, total dry matter, shoot dry matter, and transpiration efficiency (W) were at least twice as large as dominance effects under wet and dry conditions (Tables IV and V). The direction of dominance, as indicated by the sign of d, was positive and, therefore, toward a higher number of spikes per plant and higher values of total dry matter, shoot dry matter, and W under wet conditions. However, the direction of dominance was negative for these characters under dry conditions except for W which remained positive. Our observations with regard to the direction of dominance for W in bread wheat are in accordance with those reported for peanut by Hubick *et al.* (1988) and for tomato by Martin and Thorstenson (1988).

The additive gene effects were greater than dominance gene effects for plant height, root dry matter, and Δ (Table VII). The observed negative value for the additive gene effects was due to the fact that Chinese Spring with lower value for Δ was designated as P_1 in the generation means analysis. Grain yield had higher dominance gene effects than additive gene effects. The direction of dominance was toward higher values of plant

Table VII Estimates of Additive Gene Effects (a), Dominance Gene Effects (d), Environment Effects (e), and Interaction Effects (a $\times$ e and d $\times$ e) for Plant Height (PH), Root Dry Matter (RDM), Grain Yield (GY), and Carbon Isotope Discrimination of Flag Leaves (Δ) of Different Wheat Generations under Wet and Dry Treatments

Effects	PH	RDM	GY	Δ
m	83.1 ± 0.7	6.1 ± 0.2	16.1 ± 0.2	21.51 ± 0.12
a	24.2 ± 0.7	1.9 ± 0.2	1.0 ± 0.2	−1.71 ± 0.12
d	6.1 ± 1.3	0.0 ± 0.3	1.6 ± 0.3	1.21 ± 0.15
e	11.1 ± 0.7	−0.3 ± 0.2	15.5 ± 0.2	−0.04 ± 0.12
$a \times e$	4.1 ± 0.7	−0.8 ± 0.2	0.8 ± 0.2	0.25 ± 0.12
$d \times e$	−0.8 ± 1.3	0.5 ± 0.3	1.4 ± 0.3	0.31 ± 0.15

height, grain yield, and Δ (Table VII). Hubick *et al.* (1988) and Martin and Thorstenson (1988) reported, respectively, that the direction of dominance for Δ in peanut and tomato was slightly toward the parent with low Δ. The additive × environment gene effects were smaller than dominance × environment gene effects for root dry matter, grain yield, and Δ (Table VII).

Our studies indicate that W and Δ are complex traits and quantitatively inherited. The directions of dominance for W and Δ were opposite from what would be expected based on theory. Both traits are related to each other through independent relationships with the ratio of intercellular to atmospheric partial pressure of CO_2 and are affected by genetic and environmental factors. While W is a measure of transpiration efficiency throughout plant growth, carbon isotope discrimination (Δ) only reflects an integrated measure of transpiration efficiency during the assimilation of the carbon that was sampled. Therefore, gene actions involved in the expression of Δ might be overshadowed by the actions of other genes involved in the expression of W. Thus, in studies associated with the inheritance of W and Δ, one of the parents might confer genes exhibiting partial dominance for one trait, while the other parent might confer genes exhibiting partial dominance for the other trait as indicated by F_1 and F_2 means relative to means of the parents. In our study, the F_1 and F_2 means for W were closer to the mean of the parent with higher transpiration efficiency (Chinese Spring), whereas the means for Δ were closer to the other parent with higher Δ (Yecora Rojo), under both wet and dry conditions (Tables II and III), indicating that the direction of dominance was toward higher W and higher Δ, as was also evident from the genetic analyses (Tables IV, V, and VII).

D. Broad-Sense Heritabilities

Estimates of broad-sense heritabilities for plant-growth characters, grain yield, W, and Δ are presented in Table VIII. These characters exhibited

Table VIII Estimates of Broad-Sense Heritability[a] for Different Characters of Bread Wheat under Well-Watered and Water-Stressed Glasshouse Conditions Using Different Generations

Character	Well-watered	Water-stressed
Plant height	0.99	0.87
Number of spikes	0.88	0.69
Total dry matter	0.98	0.91
Shoot dry matter	0.99	0.88
Root dry matter	0.83	0.95
Grain yield	0.98	0.77
Transpiration efficiency	0.78	0.69
Carbon isotope discrimination	0.94	0.86

[a]Broad-sense heritability (H_B) = V_G/V_{F2}, where $V_G = V_{F2} - V_E$ and $V_E = \frac{1}{4}(V_{P1} + V_{P2} + 2V_{F1})$, and V_G, V_E, V_{P1}, V_{P2}, V_{F1}, and V_{F2} are variances associated with genetic effects, nongenetic effects, P_1, P_2, F_1, and F_2 generations, respectively.

higher heritabilities under well-watered conditions than under drought conditions, except for root dry matter. The range in broad-sense heritability in the wet environment was from 78% for W to 99% for shoot dry matter and plant height. In the dry environment, the range was from 69% for number of spikes per plant and W to 95% for root dry matter. Variable heritabilities across different environments, such as glasshouse and field experiments, have been reported in several studies for W and Δ (Johnson *et al.*, 1990; Ehdaie *et al.*, 1991). However, similar heritabilities for Δ under both dry and wet field conditions have been reported by Hall *et al.* (1990) and Ehdaie *et al.* (1991).

E. Genetic Correlations

Pairwise genetic correlations between different characters under wet and dry conditions are presented in Table IX. The number of significant correlations was higher under wet than under dry conditions. Transpiration efficiency (W) was positively and significantly correlated to plant height, total dry matter, shoot dry matter, root dry matter, and grain yield, but negatively correlated with Δ under both wet and dry environments. The negative genetic correlations observed between W and Δ were consistent with the theory of Farquhar *et al.* (1982). However, the negative genetic correlation between those two characters was much stronger under drought ($r = -0.77^{**}$) than under wet conditions ($r = -0.36^{+}$, significant at $P = 0.10$) (Table IX). Ehdaie *et al.* (1991) reported negative phenotypic correlations between W and Δ under both wet and dry conditions over 2 years for a set of diverse bread wheat genotypes. However, the phenotypic

Table IX Pairwise Genetic Correlations[a] between Plant Height (PH), Number of Spikes (NS), Total Dry Matter (TDM), Shoot Dry Matter (SDM), Root Dry Matter (RDM), Grain Yield (GY), Transpiration Efficiency (W), and Carbon Isotope Discrimination of Flag Leaves (Δ) in Wheat under Well-Watered (Upper Diagonal) and Water-Stressed (Lower Diagonal) Glasshouse Conditions

	PH	NS	TDM	SDM	RDM	GY	W	Δ
PH		0.32	0.50*	0.51*	0.32	-0.09	0.55**	-0.40*
NS	0.15		0.76**	0.76**	0.67**	0.36*	0.35	-0.53**
TDM	0.09	0.41*		1.00**	0.84**	0.66**	0.70**	-0.56**
SDM	0.41*	0.79**	0.32		0.80**	0.66**	0.63**	-0.59**
RDM	-0.13	0.26	0.96**	0.13		0.59**	0.72**	-0.49*
GY	0.32	0.02	-0.01	0.23	-0.11		0.49*	0.25
W	0.09	0.59**	0.39*	0.51*	0.36*	0.38*		-0.36+
Δ	-0.23	-0.39*	-0.82**	-0.25	-0.73**	0.02	-0.77**	

[a]Genetic correlation between characters x and y = $Cov_G(x,y)/\sqrt{[V_G(x)V_G(y)]}$, where $Cov_G(x,y) = Cov(x,y)$ $F_2 - Cov(x,y)E$, $Cov(x,y)E = \frac{1}{4}[Cov(x,y)P_1 + Cov(x,y)P_2 + 2 Cov(x,y)F_1]$ and $Cov_G(x,y)$, $Cov(x,y)E$, $Cov(x,y)P_1$, $Cov(x,y)P_2$, $Cov(x,y)F_1$, and $Cov(x,y)F_2$ are covariances of x and y associated with genetic effects, nongenetic effects, P_1, P_2, F_1, and F_2 generations, respectively, and $V_G(x)$ and $V_G(y)$ are genetic variances of x and y, respectively.

*,**Significant at the 0.05 and 0.01 level of probability, respectively.

correlations observed under wet conditions were higher than those obtained under dry conditions. The negative genetic correlations found in our pot experiments between Δ and plant growth characters such as shoot dry matter, root dry matter, and total dry matter (Table IX) were consistent with studies in previous years (Ehdaie *et al.*, 1991).

F. Indirect Selection

The use of indirect selection to achieve more rapid progress under selection for a correlated response than under selection for the desired character itself has been described by Falconer (1988). Predicted gain from indirect selection, which incorporates both the heritabilities of the desired character and the secondary character and their genetic correlation coefficient, could be used to quantify the relative efficiency of indirect selection compared to direct selection. The relative efficiency of indirect selection in different environments could be compared to identify the most effective selection environment. In our study, the predicted gain in W from selection applied to Δ was less than the predicted gain from direct selection in both wet and dry environments. The absolute values of the relative efficiencies of indirect to direct selection estimated were less than 1 (wet = [−0.36] [0.97/0.88] = −0.39, and dry = [−0.77][0.93/0.83] = −0.86). However, the relative efficiency of indirect selection in the dry environment was twice as large as that in the wet environment, mainly due to stronger genetic association between W and Δ in the dry environment. Since measurements of W involve pot experiments which are extremely laborious and which are not applicable to large-scale screening of germplasm or to selection practices with cultivar improvement under field conditions, Δ could be used as a selection criterion to screen breeding materials under water-stressed field conditions to improve transpiration efficiency in wheat.

IV. Breeding Implications

Direct measurement of transpiration efficiency is difficult under natural field conditions and pot studies have high labor requirements, which limits their usefulness for large breeding programs. Measurement of Δ offers advantages over both transpiration efficiency and gas exchange measurements in that Δ provides a time-integrated estimate of transpiration efficiency. It has also been demonstrated that the ranking of genotypes for Δ is more consistent across environments and more precise than gas exchange measurements (Hall *et al.*, 1992). The flag leaf or peduncle can easily be sampled from breeding materials or yield trials and dried. Analysis of carbon isotope composition of these tissue samples can be conducted during that season or delayed until the off-season after desirable breeding materials have been selected based on other characters.

In a self-fertilized species such as bread wheat, selection is usually applied in early generations after hybridization, such as in F_2 or F_3. Selection in early generations is effective if the additive effects are large and the nonadditive variance and environmental interactions are relatively small (Thakare and Qualset, 1978; Weber, 1983). Our study indicated significant and relatively large additive gene effects and variance for Δ compared to the nonadditive variance and genotype × environment interactions. Therefore, it is suggested that selection under water-limited conditions be initiated in the F_2 generation for plants with low Δ to improve transpiration efficiency in bread wheat. Selection for low Δ in a water-stressed environment might also increase total dry matter and root dry matter because there were highly significant negative correlations between each of these characters and Δ under drought conditions.

Field tests should be conducted to determine whether carbon isotope discrimination can be successfully used as a selection technique in a commercial-scale wheat-breeding program, before this method is advocated widely for improving transpiration efficiency and drought resistance (Richards and Condon, Chapter 29, this volume). A major concern in the use of carbon isotope discrimination as a selection technique for transpiration efficiency in wheat is the cost of sample analysis. One flag leaf or peduncle is sampled per plant. In a large-scale breeding program, many thousand plants are screened in segregating generations, making the cost of carbon isotope discrimination analysis very expensive. An alternative procedure would be to select first for other desirable characters such as appropriate days to maturity, grain yield, and harvest index and then screen this selected material for discrimination analysis (Hall *et al.*, Chapter 23, this volume).

V. Summary

This is the first comprehensive study of the inheritance of transpiration efficiency (W) and of carbon isotope discrimination (Δ) in bread wheat and of their genetic associations with plant-growth characters. Contrasting parents, Chinese Spring and Yecora Rojo, and their F_1 and F_2 generations were grown in replicated pot experiments in a glasshouse under well-watered and water-stressed treatments. Using generation means, a positive correlation was found between total dry matter and water transpired under well-watered and water-stressed conditions. Genetic corelations between W and Δ were negative in both wet and dry environments, as the theory predicted. However, the two characters exhibited much stronger association under drought than under well-watered conditions. W was positively and significantly correlated with shoot dry matter, root dry matter, total dry matter, and grain yield under wet and dry environments. Δ was negatively and significantly associated with plant height, number of spikes per plant, shoot dry matter, root dry matter, and total dry matter under wet conditions, and with number of spikes per plant, root dry matter, and total dry matter under drought conditions. Broad-sense heritabilities of the characters examined were relatively large, ranging from 78% for W to 99% for plant height and shoot dry matter under wet conditions. Except for root dry matter, broad-sense heritabilities were smaller under dry conditions compared to wet conditions and they ranged from 69% for W and number of spikes per plant to 95% for root dry matter.

Generation means analysis indicated significant additive and dominance variation for W under wet and dry environments with additive variance accounting for more than 90% of total variation observed among the generations. Significant additive and dominance variances, and additive × environment and dominance × environment interactions, were observed for Δ. However, for both characters, the additive effects were larger than the dominance effects. Compared to direct selection, indirect selection for W through selection for Δ was only 14% less efficient under water-stressed conditions, but 61% less efficient under well-watered conditions. Since direct measurements of W involved pot experiments which are extremely laborious and which are not applicable to large-scale selection programs associated with cultivar improvement, Δ could be used as an indirect selection criterion to advance W under water-stressed conditions. Selection for lower Δ in bread wheat could be initiated in the F_2 generation because the additive effects of genes involved in the expression of Δ were higher than the dominance effects.

Acknowledgments

We thank Ruth Shaw, A. E. Hall, and M. L. Roose for helpful comments. This research was supported by the California Agricultural Experiment Station, USDA Hatch Funds, and NRI Competitive Grants Program/USDA No. 91-37100-6614.

References

Briggs, L. J., and H. L. Shantz. 1913. The water requirements of plants. II. A review of the literature. United States Department of Agriculture Bureau of Plant Industry Bulletin **285**.

Briggs, L. J., and H. L. Shantz. 1914. Relative water requirement of plants. *J. Agric. Res. (Washington, DC)* **3**: 1–64.

Ehdaie, B., and A. Ghaderi. 1978. Inheritance of some agronomic characters in a cross of safflower. *Crop Sci.* **18**: 544–547.

Ehdaie, B., A. E. Hall, G. D. Farquhar, H. T. Nguyen, and J. G. Waines. 1991. Water-use efficiency and carbon isotope discrimination in wheat. *Crop Sci.* **31**: 1282–1288.

Ehleringer, J. R. 1988. Correlations between carbon isotope ratio, water-use efficiency and yield, pp. 165–191. *In* J. White, G. Hoogenboom, F. Ibarra, and S. P. Singh (eds.), Research and Drought Tolerance in Common Bean. Proceedings Int. Bean Drought Workshop, Cali. 19–21 Oct. 1987. International Centre for Tropical Agriculture (CIAT), Cali, Colombia.

Ehleringer, J. R., J. W. White, D. A. Johnson, and M. Brick. 1990. Carbon isotope discrimination, photosynthetic gas exchange, and transpiration efficiency in beans and range grasses. *Acta Oecol.* **11**: 611–625.

Falconer, D. S. 1988. Introduction to Quantitative Genetics. 3rd Ed. Longman, Inc., New York.

Farquhar, G. D., and R. A. Richards. 1984. Isotope composition of plant carbon correlates with water use efficiency of wheat genotypes. *Aust. J. Plant Physiol.* **11:** 539–552.

Farquhar, G. D., M. H. O'Leary, and J. A. Berry. 1982. On the relationship between carbon isotope discrimination and the intercellular carbon dioxide concentration in leaves. *Aust. J. Plant Physiol.* **9:** 121–137.

Farquhar, G. D., T. R. Ehleringer, and K. T. Hubick. 1989. Carbon isotope discrimination and photosynthesis. *Annu. Rev. Plant Physiol. Mol. Biol.* **40:** 503–537.

Frank, A. B., J. D. Berdahl, and R. E. Barker. 1985. Morphological development and water use in clonal lines of four forage grasses. *Crop Sci.* **25:** 339–344.

Hall, A. E., R. G. Mutters, K. T. Hubick, and G. D. Farquhar. 1990. Genotypic differences in carbon isotope discrimination by cowpea under wet and dry field conditions. *Crop Sci.* **30:** 300–305.

Hall, A. E., R. G. Mutters, and G. D. Farquhar. 1992. Genotypic and drought-induced differences in carbon isotope discrimination and gas exchange of cowpea. *Crop Sci.* **32:** 1–6.

Hayman, B. I. 1958. The separation of epistatic from additive and dominance variation in generation means. *Heredity* **12:** 371–390.

Hubick, K. T., and G. D. Farquhar. 1989. Carbon isotope discrimination and the ratio of carbon gained to water lost in barley cultivars. *Plant Cell Environ.* **12:** 795–804.

Hubick, K. T., G. D. Farquhar, and R. Shorter. 1986. Correlation between water-use efficiency and carbon isotope discrimination in diverse peanut (*Arachis*) germplasm. *Aust. J. Plant Physiol.* **13:** 803–816.

Hubick, K. T., R. Shorter, and G. D. Farquhar. 1988. Heritability and genotype × environment interactions of carbon isotope discrimination and transpiration efficiency in peanut (*Arachis hypogaea* L.). *Aust. J. Plant Physiol.* **15:** 799–813.

Ismail, A. M., and A. E. Hall. 1992. Correlation between water-use efficiency and carbon isotope discrimination in diverse cowpea genotypes and isogenic lines. *Crop Sci.* **32:** 7–12.

Johnson, D. A., K. H. Asay, L. L. Tieszen, T. R. Ehleringer, and P. G. Jefferson. 1990. Carbon isotope discrimination: Potential in screening cool-season grasses for water limited environments. *Crop Sci.* **30:** 338–343.

Martin, B., and Y. R. Thorstenson. 1988. Stable isotope composition ($\delta^{13}C$), water use efficiency, and biomass productivity of *Lycopersicon esculentum*, *Lycopersicon pennellii*, and the F_1 hybrid. *Plant Physiol.* **88:** 213–217.

Mather, K., and J. L. Jinks. 1971. Biometrical Genetics. The Study of Continuous Variation. Cornell Univ. Press, Ithaca, NY.

Tanner, C. B., and T. R. Sinclair. 1983. Efficient water use in crop production: Research or re-search, pp. 1–28. *In* H. M. Taylor, W. R. Jordan, and T. R. Sinclair (eds.), Limitation to Efficient Water Use in Crop Production. ASA, CSSA, and SSSA, Madison, WI.

Thakare, R. B., and C. O. Qualset. 1978. Empirical evaluation of single-plant and family selection strategies in wheat. *Crop Sci.* **18:** 115–118.

Virgona, T. M., K. T. Hubick, H. M. Rawson, G. D. Farquhar, and R. W. Downes. 1990. Genotypic variation in transpiration efficiency, carbon isotope discrimination and carbon allocation during early growth in sunflowers. *Aust. J. Plant Physiol.* **17:** 207–214.

Weber, W. E. 1983. Selection in early generations, pp. 72–81. *In* W. Lange, A. C. Zeven, and N. G. Hagenboom (eds.), Efficiency in Plant Breeding. Proc. Congress of the European Association for Research on Plant Breeding, EUCARPIA, 10th, Wageningen, 19–24 June 1983. Pudoc, Wageningen, The Netherlands.

28

Exploiting Genetic Variation in Transpiration Efficiency in Wheat: An Agronomic View

A. G. Condon and R. A. Richards

I. Introduction

From an agronomic viewpoint, interest in transpiration efficiency lies in the extent to which greater transpiration efficiency contributes to greater crop water-use efficiency and ultimately to improved crop yields. The relationship between crop transpiration efficiency, crop water-use efficiency, and crop yield is summarized in Eqs. (1) and (2). Equation (1),

$$Y = E \times W \times \text{HI}, \tag{1}$$

is in the form of the identity proposed by Passioura (1977) in which he described the grain yield of a water-limited crop (Y) as being a function of the amount of water transpired by the crop (E), the efficiency with which that transpired water is used to produce aboveground dry matter (W), and the proportion of final aboveground dry matter that is harvested as grain, i.e., the harvest index (HI). Equation (1) may be expanded to account for total water use by the crop (E_{total}), where E_{total} is the sum of transpiration from the plant canopy and the evaporation losses directly from the soil surface:

$$Y = E_{total} \times E/E_{total} \times W \times \text{HI}. \tag{2}$$

Crop water-use efficiency (W_{total}) can be seen to be the product of the second and third terms of Eq. (2), i.e., it is determined not only by crop transpiration efficiency (W) but also by that proportion of the total amount of water used by the crop that actually passes through the plants.

As noted by Passioura (1977), for Eqs. (1) and (2) to be applied most usefully there should be no strong negative interactions between any of the factors on the right-hand side of these equations. Assuming such interactions are rare, then an increase in any one of these factors should lead to greater yield. In this chapter we consider the potential to exploit genetic variation in one of these factors, transpiration efficiency, as a means of improving wheat yields in water-limited environments. In recent years such a prospect has become a distinct possibility with the realization that carbon isotope discrimination (Δ) may provide a useful indirect measure of genetic variation in transpiration efficiency of leaf gas exchange in C_3 species (Farquhar *et al.*, 1982; Farquhar and Richards, 1984). In particular, we examine the extent to which genetic variation in transpiration efficiency at the leaf level is reflected in variation in transpiration efficiency and water-use efficiency at the crop scale.

II. Genetic Variation in Transpiration Efficiency in Wheat

As predicted by theory (Farquhar *et al.*, 1982; Farquhar and Richards, 1984), a negative relationship between Δ and W has been demonstrated for a number of C_3 crop species grown in pots in the glasshouse. Wheat is no exception and in Fig. 1 we plot relationships between W of single plants and Δ for a range of wheat genotypes selected on the basis of variation in Δ. The plants were grown under both well-watered conditions (Fig. 1a) and severe water limitation in which they received no more water after sowing (Fig. 1b). There were both genotypic and environmental effects on W and on Δ.

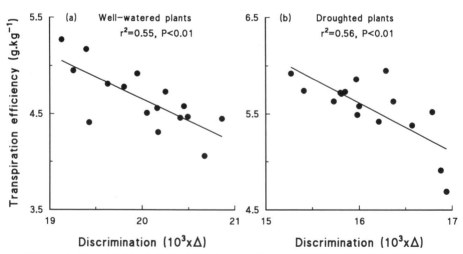

Figure 1. Relationships between transpiration efficiency of aboveground dry matter production and carbon isotope discrimination (Δ) measured in dry matter at maturity for 16 wheat genotypes grown in the glasshouse under (a) well-watered conditions (fitted regression: $Y = -0.46X + 13.76$) and (b) drought stress (fitted regression: $Y = -0.51X + 13.76$). (Adapted from Condon *et al.*, 1990.)

Transpiration efficiency was greater and Δ lower under the water-limited conditions, i.e., across environments the relationship between W and Δ was negative. More importantly in the current context, the relationships between W and Δ were also negative within environments, with substantial variation in both W and Δ being evident under both well-watered and water-stressed conditions.

These consistent relationships between plant W and Δ across treatments, the high broad-sense heritability of Δ in wheat (Condon *et al.*, 1987; Ehdaie *et al.*, 1991; Condon and Richards, 1992), and the ease with which Δ can be measured relative to W itself, all indicate that Δ could be usefully applied in breeding programs to improve the transpiration efficiency and hence yield of water-limited wheat crops.

III. Relationships between Dry Matter Production or Grain Yield and Carbon Isotope Discrimination in Field Experiments

Before Δ can be applied in such a way, a clear link must be demonstrated in the field between low Δ values and either greater yield or greater biomass production. Such a link has not yet been demonstrated in wheat. For a range of spring wheat genotypes grown in experimental plots of ca. 10 m² under relatively well-watered conditions, Condon *et al.* (1987) observed positive relationships (rather than negative relationships) between biomass production and Δ and between grain yield and Δ. If it is assumed that, in water-limited situations, E_{total} is essentially the same for all genotypes, then variation in dry matter production should provide a useful first approximation of variation in crop water-use efficiency. In further studies in southeastern Australia, the relationship between dry matter production and Δ has been examined in a wider range of environments (Condon and Richards, unpublished data). These environments varied in both the amount and distribution of growing season rainfall. Results from these experiments are summarized in Fig. 2, which shows relationships between aboveground dry matter production and Δ for 11 wheat genotypes grown at several sites over three seasons. The average yield at each site largely reflects the degree of water limitation. At Condobolin in 1987 (Fig. 2a), effective rainfall ceased in early spring and the crop canopies grew largely on a limited supply of stored soil moisture. At Moombooldool and Ariah Park in 1986 (Fig. 2c), there were regular rainfall events well into grainfilling. For these sites water limitation was not severe. For the other sites (Fig. 2b) the degree of water limitation was intermediate between these two extremes.

It should be noted that for all the relationships in Fig. 2 we have used the same values of Δ. These are mean values of Δ measured in dry matter of each genotype taken from several sites but laid down early in the season when crops were not stressed. These early season values of Δ are highly repeatable across these field environments (Condon and Richards, 1992)

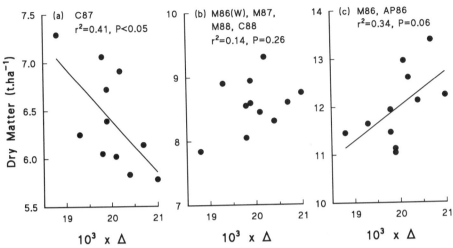

Figure 2. Relationships between aboveground dry matter production at maturity and carbon isotope discrimination (Δ) measured in early-formed dry matter for 11 wheat genotypes grown at seven field sites in southeast Australia over three seasons, 1986, 1987, and 1988. Sites are grouped on the basis of average yield level. For plots involving multiple sites, genotype yields shown are average values across sites. Values of Δ are mean values measured in dry matter of each genotype taken from several sites. Abbreviations: M, Moombooldool (New South Wales, NSW), C, Condobolin (NSW), and AP, Ariah Park (NSW). All experiments followed a pasture phase except for M86(W), which followed a wheat crop.

and represent genotypic variation in Δ which can be evaluated in a predictive sense. In other words, we might ask, what would we expect to observe, given this variation in Δ early in the season? In these experiments, what we do observe is that the relationship between dry matter production and Δ changes with the average yield level. It was only at the driest site, Condobolin in 1987, that the relationship was negative and that our expectations from leaf-level gas exchange (based on variation in Δ) were supported. At the highest yielding sites especially, the reverse seemed to apply. At intermediate yield levels there was no clear relationship between dry matter production and Δ.

The positive relationships obtained between dry matter production and Δ at higher yield levels are important for two reasons. First, from a practical viewpoint (and to the extent that they are reflected in variation in grain yield), they indicate that a substantial penalty in yield potential may result in relatively favorable environments from selecting for low Δ in breeding programs. This has important implications for the ultimate application of Δ in wheat breeding. For instance, in regions where wheat is grown principally for export and returns to the farmer are largely determined by international prices, there would seem to be little to be gained from giving farmers a yield advantage in dry years if they suffer a substantial yield limitation in those good seasons in which they make the money to survive the poor ones. Conversely, in those regions or economies where wheat is consumed directly or sold to internal markets at local prices, it may be more prudent to sacrifice yield potential in good seasons if the result is greater yield stability from season to season.

While these considerations are important for the ultimate application of Δ in breeding, the positive relationships observed between dry matter production and Δ also raise a more fundamental question: Is the relationship between transpiration efficiency and Δ still *negative* for crop canopies growing in environments where water limitation is not severe? If not, then selection for low Δ in breeding programs for variable environments, such as exist in many semiarid areas, must be questioned. If the relationship between crop transpiration efficiency and Δ is negative, even in relatively favorable environments, then low Δ may still be exploitable.

IV. Sources of Variation in Carbon Isotope Discrimination in Wheat

The problem of "scaling up" from leaf gas exchange, as represented by our measurement of Δ, to the performance of canopies in the field can be most conveniently examined by considering the impact, at the field scale, of the two factors that contribute to variation in Δ, stomatal conductance and photosynthetic capacity.

For C_3 species, carbon isotope discrimination measured in plant dry matter provides a useful integration of the assimilation-weighted ratio of p_i/p_a over the time that the dry matter was laid down (Farquhar *et al.*, 1982; Condon *et al.*, 1990). The ratio p_i/p_a reflects the balance between stomatal conductance and photosynthetic capacity. A lower value of p_i/p_a (or Δ) may result from either lower stomatal conductance and/or greater photosynthetic capacity.

In some species it seems that much of the genotypic variation in Δ can be attributed to variation in only one of these. In common bean, for example, genotypic variation in Δ seems to principally arise as a result of variation in stomatal conductance (Ehleringer, 1990). By contrast, in peanut much of the genotypic variation in Δ appears to be due to variation in photosynthetic capacity (Hubick *et al.*, 1988; Wright, 1992). In wheat, genotypic variation in Δ arises from variation in both conductance and capacity, with each contributing approximately equally.

Table I summarizes data from a study (Condon *et al.*, 1990) in which we analyzed the causes of variation in Δ among 16 wheats, mainly of Australian origin (in fact, the same wheats used in the pot studies described in Section II). For the two wheats differing most in Δ, Veery3 and M3844, the difference in Δ can be attributed to differences in both conductance (g) and photosynthetic capacity (k). However, if we compare Veery3 with the variety Sunstar, there is no difference in conductance but a substantial difference in photosynthetic capacity. Conversely, a comparison of the varieties M3844 and Cranbrook indicates that the difference in Δ is almost entirely

Table I　Variation in Carbon Isotope Discrimination (Δ) and Flag Leaf Gas Exchange Characteristics among Wheat Genotypes[a]

Genotype	$10^3 \times \Delta$	p_i/p_a	$(10^3 \times \Delta)$	g^b	k
Veery3	21.0	0.72	(20.7)	0.55	1.66
M3844	19.3	0.65	(19.1)	0.43	1.79
Sunstar	20.4	0.68	(19.8)	0.56	2.06
Cranbrook	20.3	0.69	(20.0)	0.48	1.81

[a]Gas exchange characteristics tabulated are the ratio of intercellular to atmospheric partial pressures of CO_2, p_i/p_a; leaf conductance, g; and the initial slope of the relationship between CO_2 assimilation rate and p_i, k (a measure of photosynthetic capacity). Carbon isotope discrimination values were measured in growing ears enclosed in the flag leaf sheath and (in parentheses) calculated from the p_i/p_a ratios shown using the equation $\Delta = 4.4 + 22.6 p_i/p_a$. (Adapted from Condon *et al.*, 1990.)
[b]g (mol·m^{-2}·s^{-1}); k (μmol·m^{-2}·s^{-1}·Pa^{-1}).

attributable to a difference in conductance. Among North American wheats, variation in p_i/p_a (and presumably Δ) also appears to be equally attributable to variation in stomatal conductance and photosynthetic capacity (Morgan and LeCain, 1991).

Insofar as they result in a lower value of p_i/p_a, both reduced stomatal conductance and greater photosynthetic capacity may be equally effective in increasing the transpiration efficiency of single leaves maintained under conditions of constant leaf-to-air vapor pressure difference in a well-ventilated gas exchange cuvette. Out of such a cuvette the situation is not nearly so straightforward, especially if the source of variation in Δ is stomatal conductance. This is because transpiration efficiency will be negatively related to the p_i/p_a ratio only if the leaf-to-air vapor pressure difference (v) is maintained constant. If Δ varies among genotypes as a result of variation in conductance then v is unlikely to be constant (Farquhar *et al.*, 1988).

For a genotype with relatively low conductance, leaf temperature will be greater than that of a genotype with higher conductance unless the boundary layer conductances to the diffusion of water vapor and sensible heat are very large. If these boundary layer conductances are not large, as is likely, then greater leaf temperature will cause an increase in v, driving transpiration at a faster rate than would be expected if stomatal closure had no effect on leaf temperature at all. As well, if the leaf is one of many in a canopy all behaving in a similar fashion, then the air in and above the canopy will become drier and hotter, further increasing the effective v and further offsetting the expected decrease in transpiration that might arise from a lower stomatal conductance. The magnitude of this effect will depend on leaf characteristics such as size and shape and also on canopy characteristics such as aerodynamic roughness and the area over which the canopy extends (Jones, 1976; Cowan, 1977; Jarvis and McNaughton, 1986). In summary, these effects of "scale" on the response of transpiration to a change in stomatal conductance will increase as (1) the leaf boundary layer resistances to the transfer of water vapor and heat increase, (2) canopy aerodynamic resistance increases (this will occur as the crop surface gets smoother and more extensive), (3) the ambient temperature increases, and (4) the resistance to CO_2 uptake by the canopy decreases, such as for a well-fertilized crop with a relatively large leaf area index. As stomatal control over canopy water loss decreases then so too the effect of reduced stomatal conductance on transpiration efficiency. In fact, in analyzing the potential impact of these effects, Cowan (1977, 1988) predicted that, in some circumstances, canopy transpiration efficiency may actually *decrease* as stomatal conductance is decreased.

At least for part of the growing season the conditions listed above are likely to be satisfied for many crops growing in rain-fed environments, especially in higher yielding ones. Of course, it should also be recognized that if low Δ is achieved via low conductance then dry matter production is also likely to be limited in favorable environments because of the reduced rate of photosynthesis per unit leaf area that results from the lower conductance.

Analyses of the impact of the effects of scale on transpiration efficiency, such as those of Cowan (1977, 1988) and Jarvis and McNaughton (1986), tend to paint a gloomy scenario in terms of the gains in transpiration efficiency that might be obtained by reducing stomatal conductance. However, in many rain-fed situations, including our own for wheat, the conditions listed above may apply only for a relatively short time. At the start of the season it is quite cool, the canopy is fairly rough, and soil evaporation is a large component of E_{total}. Later in the season there is almost always some water stress and leaf area index falls rapidly.

V. Testing the Effect of "Scale" on Season-Long Transpiration Efficiency of Wheat

A. Some Experiments

So, what is the impact of these scale effects on the transpiration efficiency of wheat over the whole season? To determine this we grew large canopies of two wheat varieties differing in stomatal conductance and monitored their growth and water use. The aim was to determine the extent to which the transpiration efficiency of large expanses or paddocks of wheat is influenced by a change in stomatal conductance.

The two wheat varieties grown were Matong and Quarrion. These two were chosen, first, because of the large difference in Δ between them. In well-watered plants the Δ value of Matong is typically about 2×10^{-3} greater than that of Quarrion. Second, and importantly for these experiments, this large difference in Δ was almost entirely due to a large difference in stomatal conductance. Stomatal conductance of well-watered plants is typically 40–50% greater in Matong than in Quarrion. Third, the two varieties were very similar in canopy height and flowering time. Experiments were done at two sites; Wagga Wagga, a relatively high rainfall environment, in 1989 and 1990, and Condobolin in 1990 only. At Condo-bolin seasonal rainfall tends to be less and the crop relies more on stored soil moisture. At Wagga Wagga the large expanses were 5 ha (i.e., 50,000 m²). At Condobolin they were 1.5 ha.

The data presented here summarize our measurements of dry matter production and water use by the crop canopies over the whole season. We partitioned total water use (E_{total}) into its components of soil evaporation and transpiration using the method popularized by Cooper and his col-leagues at ICARDA (Cooper *et al.*, 1983). By this method, soil evaporation from under the canopy is estimated on the basis of measurements of the change in profile water content from uncropped soil over relatively short time intervals of 2–3 weeks, and the proportion of incoming radiation that reaches the soil surface under the canopy for the same period. Radiation measurements above and below the canopy are typically taken around solar noon using a linear sensor. Measurements of soil evaporation at the Wagga Wagga site using more direct techniques have confirmed that the radiation interception technique gives accurate estimates of evaporation from the soil under the canopy in environments such as this where the soil surface is frequently rewetted (R. Leuning, O. T. Denmead, and F. X. Dunin, per-sonal communication).

B. Results

Results of this sort of analysis are shown in Fig. 3a, which depicts the time course of total crop water use for both varieties growing in 5-ha blocks at Wagga Wagga in 1989. At this site E_{total} from Matong was 24 mm more

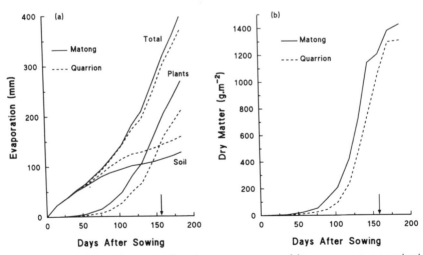

Figure 3. (a) Seasonal courses of total crop water use and its components transpiration and soil evaporation, and (b) seasonal courses of aboveground dry matter production, for two wheat varieties, Matong and Quarrion, sown in 5-ha blocks at Wagga Wagga (NSW) in 1989. The experiment was sown on May 18. Mean anthesis date is indicated by the arrow.

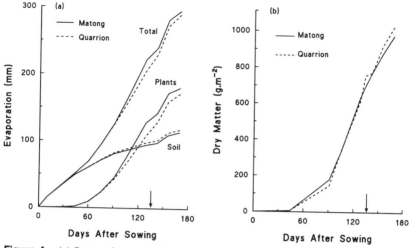

Figure 4. (a) Seasonal courses of total crop water use and its components transpiration and soil evaporation, and (b) seasonal courses of aboveground dry matter production, for two wheat varieties, Matong and Quarrion, sown in 1.5-ha blocks at Condobolin (NSW) in 1990. The experiment was sown on May 29. Mean anthesis date is indicated by the arrow.

than that from Quarrion. The difference in transpiration was much greater, 58 mm. This was because of the much greater soil evaporation from under the slower developing Quarrion canopy. This slower canopy development is reflected in the time course of dry matter production (Fig. 3b). During the early part of the growing season dry matter production was considerably slower in the low Δ genotype, Quarrion.

When the experiment was repeated in 1990 we attempted to overcome the slower early canopy growth of Quarrion by increasing the sowing rate of Quarrion relative to Matong. This had the desired effect, especially at Condobolin. At this site total soil evaporation only differed by 3 mm at the end of the season (Fig. 4a). The difference in E_{total} was also quite small, only 6 mm. The time course of dry matter production for this site is shown in Fig. 4b. Early in the season there was little difference between the two varieties, but by the end of the season Quarrion had produced significantly more dry matter.

The dry matter production and water-use data for all three plantings are summarized in Table II. At Wagga Wagga in 1989, dry matter production was greater in Matong than in Quarrion by about 10%. Total crop water use was also greater for Matong, by about 5%. Total transpiration, however, was 25% greater from the Matong canopies, with transpiration making up a considerably greater proportion of E_{total} for Matong. At the drier Condobolin site in 1990, total dry matter production was still in the order of 9.5 t/ha. Here, dry matter production was greater for Quarrion but E_{total} was less. Transpiration was also less for Quarrion, although there was little difference in the ratio of E/E_{total}, as discussed earlier. The results for the Wagga Wagga site in 1990 closely reflected those obtained at Condobolin.

Table II   Aboveground Dry Matter Production (DM), Total Crop Water
Use (E_{total}), Crop Transpiration (E), and the Ratio E/E_{total} for the Wheat
Varieties Matong and Quarrion Grown at Wagga Wagga and Condobolin,
Southeastern Australia, in 1989 and 1990

	DM ($t \cdot ha^{-1}$)	E_{total} (mm)	E (mm)	E/E_{total}
Wagga Wagga 1989				
Matong	14.27	402	273	0.68
Quarrion	13.08	378	215	0.57
Wagga Wagga 1990				
Matong	10.15	412	285	0.69
Quarrion	10.59	386	246	0.64
Condobolin 1990				
Matong	9.63	296	181	0.61
Quarrion	10.22	290	172	0.59

At Wagga Wagga in 1989, the highest yielding and wettest site, crop water-use efficiency was marginally greater in Matong (Table III). In 1990, both at Wagga Wagga and Condobolin, the advantage was with Quarrion. A summary of the results for crop transpiration efficiency (Table III) shows that at all three sites transpiration efficiency was consistently greater for the low Δ, low-conductance Quarrion than for the high Δ, high-conductance Matong. When averaged over all three experiments, transpiration efficiency was some 15% greater for Quarrion than for Matong.

But what difference in transpiration efficiency might have been expected from the difference in Δ? Early in the season, the difference in Δ was close to 2×10^{-3}. For dry matter sampled from each site at the five-leaf stage, Δ of Matong averaged 21.7×10^{-3} and for Quarrion, 19.8×10^{-3}. If we were to assume that the leaves of the two varieties were being maintained at the same temperature then for these average values of Δ we might expect transpiration efficiency for Quarrion to be potentially 35% greater than that for Matong. If we were to use grain Δ values, for which the Δ values were lower and the average difference in Δ was less (average value across

Table III   Crop Water-Use Efficiency (W_{total}) and Crop Transpiration Efficiency (W) for the Wheat Varieties Matong and Quarrion Grown at Wagga Wagga and Condobolin, Southeastern Australia, in 1989 and 1990

	W_{total} ($g \cdot m^{-2} \cdot mm^{-1}$)		W ($g \cdot m^{-2} \cdot mm^{-1}$)	
	Matong	Quarrion	Matong	Quarrion
Wagga Wagga 1989	3.55	3.46	5.23	6.08
Wagga Wagga 1990	2.47	2.74	3.56	4.30
Condobolin 1990	3.26	3.52	5.32	5.94

sites for Matong 17.4×10^{-3}, for Quarrion 16.0×10^{-3}), and make the same assumption about leaf temperature, then the potential advantage to Quarrion would be about 15%. Clearly, any evaluation of the possible effects of scale on transpiration efficiency will depend on what Δ values we take to calculate the potential difference in transpiration efficiency between the two varieties.

This problem can be overcome by measuring Δ of recently accumulated dry matter throughout the season. This was done at Wagga Wagga in 1989 (Fig. 5a). Early in the season the difference in Δ was about 2×10^{-3}. The value of Δ then fell progressively in both varieties, but faster in Matong because of more rapid soil water depletion by this variety. Approaching anthesis the difference in Δ was only about 0.5×10^{-3}. Heavy rain just after anthesis restored the initial difference but not the initial high values of Δ. This seasonal course of Δ values can be converted to a seasonal course of the values of $(1 - p_i/p_a)$ for the two varieties, as is shown in Fig. 5b. It will be recalled that if v is constant, transpiration efficiency of leaf gas exchange is directly proportional to $(1 - p_i/p_a)$. If the ratio of the values of $(1 - p_i/p_a)$ for Quarrion over those for Matong is then calculated, the time course shown in Fig. 5c is obtained. This ratio gives a measure of the potential advantage in transpiration efficiency of leaf gas exchange of Quarrion over Matong, assuming no effect of conductance on canopy temperature. Early in the season the potential advantage to Quarrion was approximately 40%. Approaching anthesis this had fallen to only about 5%. Following the relief of soil water stress it rose to about 20%. It is important to note that the potential difference in transpiration efficiency arising from the same difference in Δ between genotypes, say, 2×10^{-3}, will be greater if the mean Δ value is large, such as at the start of the season, than if the mean Δ value is smaller, as found in the grain. This is because, for the same difference in Δ, the ratio of the values of $(1 - p_i/p_a)$ will be greater at higher mean values of p_i/p_a (and Δ) than at lower mean values of p_i/p_a (and Δ) (cf. Figs. 5a–5c).

Returning to the experiment at Wagga Wagga in 1989, the potential difference in transpiration efficiency for this experiment over the whole season may be calculated by weighting each of the values of the ratio shown in Fig. 5c according to the amount of water transpired during each of the time periods shown. When this is done, the weighted average value of this ratio for the whole season is 1.19. In other words, based on measurements of Δ over the course of the season, the maximum potential advantage in crop transpiration efficiency of Quarrion over Matong was 19%. The observed advantage for Quarrion was 15%. This result suggests that the scale effects relating to canopy heating and the control of transpiration by stomata were, in fact, quite small. This is likely to be an oversimplification. Rather, these scale effects are likely to have been important, but only for that relatively short period during the growing season when the leaf area index was relatively high, say 2–4, and stomatal conductance was near its maximum in both varieties, i.e., in that period when soil water depletion was minimal. At other times of the season, the crops were too small, too stressed, or senescing.

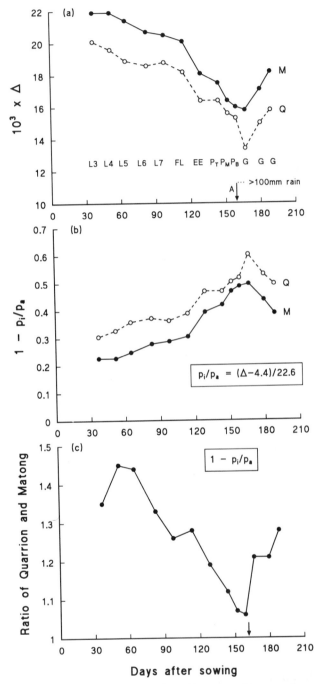

Figure 5. (a) Carbon isotope discrimination (Δ) measured in recently accumulated dry matter sampled over the course of the season at Wagga Wagga (NSW) in 1989. Plant parts sampled were recently expanded leaves 3, 4, 5, 6, and 7 and the flag leaf; the expanding ear enclosed in the flag leaf sheath; the top, middle, and bottom of the peduncle (these parts being

C. Conclusions from the "Scale" Experiments

The major conclusions that can be drawn from these scale experiments are that (1) in these relatively favorable environments there was little difference between Matong and Quarrion when crop water-use efficiency was measured over the whole season, but (2) on average, crop transpiration efficiency of the low Δ, low-conductance variety Quarrion was 15% greater than that of the high Δ, high-conductance variety Matong. A difference of 35–40% might have been expected on the basis of Δ values of early dry matter. Nevertheless, the results of these experiments confirm that there is still a substantial improvement in crop transpiration efficiency to be gained, even in favorable environments, from selecting for low conductance via low Δ.

However, for relatively unstressed environments such as those here, there are problems that need to be overcome before any advantage in transpiration efficiency arising from a difference in conductance can be effectively translated into a gain in crop water-use efficiency. In each of the plantings, total water use was less and soil evaporation greater from the low Δ variety. Together, these two factors resulted in total transpiration being less. The result was that any gain in transpiration efficiency was largely negated because less water passed through the variety with greater transpiration efficiency. It may be possible to overcome this problem agronomically, as at Condobolin in 1990. A more effective solution would be genetic manipulation of early canopy growth.

VI. The Effect of Changing Photosynthetic Capacity

Strictly, the conclusions given above only refer to this series of experiments in which we have examined the extent to which transpiration efficiency at the paddock scale is influenced by a change in stomatal conductance. To a large degree, they would also appear to be valid if we were to change Δ and transpiration efficiency by altering photosynthetic capacity—although we have yet to test this in a similar way.

Importantly, if photosynthetic capacity was altered, there should be little or no effect on canopy temperature. Therefore, there should be no scale

---

laid down in that order); and the grain at three times after anthesis. Mean anthesis date is indicated by the arrow. More than 100 mm of rain fell in a 5-day period immediately after anthesis. (b) Calculated values of $(1 - p_i/p_a)$ derived from the values of Δ plotted in (a). These values were calculated using the equation, $\Delta = 4.4 - 22.6 p_i/p_a$. This equation accurately describes the relationship between p_i/p_a measured by leaf gas exchange of source leaves and the value of Δ measured in sink dry matter (Condon *et al.*, 1990). (c) Seasonal change at Wagga Wagga in 1989 in the ratio of the values of $(1 - p_i/p_a)$ for Quarrion over those for Matong. The ratio of these values indicates the potential advantage in leaf transpiration efficiency of Quarrion over Matong assuming that leaf-to-air vapor pressure difference is a constant.

effect per se on the relationship between Δ and crop transpiration efficiency.

But changing photosynthetic capacity may influence the relationship between Δ and crop water-use efficiency. This is because, in cereals such as wheat, an increase in photosynthetic capacity is typically associated with a decrease in leaf size (Bhagsari and Brown, 1986). Specific leaf area, i.e., the leaf area per unit leaf dry weight, is reduced as the available nitrogen is concentrated such that it increases photosynthetic capacity. One outcome of this is that early canopy cover tends to be greater in wheats with low photosynthetic capacity. In environments of high winter/spring rainfall, such as those in which the scale experiments were conducted, an increase in the proportion of total water use that is lost as soil evaporation may eliminate any effect on crop water-use efficiency of an increase in photosynthetic capacity. If we wanted to lower Δ by, say, 1×10^{-3}, which might translate into a 20% potential improvement in transpiration efficiency, the data presented in Table I on the gas exchange characteristics of wheat suggest that we would need to increase photosynthetic capacity by some 25% at least. In the absence of any imaginative genetic manipulations which might allow us to increase photosynthetic capacity without reducing leaf size, this is likely to incur the cost of smaller leaves, slower canopy growth, and greater soil evaporation.

It may also, paradoxically, reduce yield potential in these rain-fed environments. Carbon gain is likely to be greater, per unit ground area, for a canopy with large, thin leaves until near-canopy-closure is reached. For many rain-fed crop canopies, near-closure is only attained for a brief period, so there is little chance for small, relatively thick-leaved canopies to make up any difference in dry matter accumulation. As well, there are strong indications that any change in the pattern of dry matter allocation responsible for a difference in photosynthetic capacity may be reflected throughout the plant. Whether variation in specific leaf area has been achieved genotypically or experimentally, some studies (Masle and Farquhar, 1988; A. G. Condon, unpublished data) indicate that low specific leaf area and greater photosynthetic capacity seem to be associated with greater allocation of carbon to nonleaf structures such as roots, reducing even further the potential for rapid carbon gain through maximum light interception.

VII. Summary

In wheat, as in other C_3 species, genetic variation in carbon isotope discrimination is reflected in variation in transpiration efficiency at both the leaf level and at the whole-plant scale. For crop canopies, however, the relationship between dry matter production and Δ varies according to the degree to which water limits crop growth. Where water limitation is severe the relationship may be negative, as might be expected from the negative relationship between Δ and transpiration efficiency of leaf gas exchange. But when crops are grown under relatively favorable conditions the relationship between dry matter production and Δ tends to be positive. In this chapter we address this apparent paradox by considering the impact of the two factors that determine crop water-use efficiency, i.e., crop transpiration efficiency and the proportion of total crop water use that actually passes through the plants.

In wheat, genetic variation in transpiration efficiency and Δ can be attributed to variation in both stomatal conductance and photosynthetic capacity. In a series of large-scale experiments we examined the effect of a change in stomatal conductance on the water-use efficiency and transpiration efficiency of wheat crops grown to maturity by comparing the growth and water use of two wheat varieties differing in stomatal conductance. Crop transpiration efficiency was 15% greater, on average, for the variety with relatively low conductance. However, this was not reflected in improved crop water-use efficiency because of the greater soil evaporation losses from under the slower developing canopy of the low-conductance, low Δ variety.

It is likely that if we were to lower Δ and improve leaf transpiration efficiency by increasing photosynthetic capacity a similar outcome would be obtained. This is because, in wheat, high photosynthetic capacity is associated with slow early canopy growth.

If Δ is to be usefully applied in breeding programs for variable environments or environments where the crop grows largely on current rainfall, then low Δ will need to be combined with more rapid early canopy cover. This could be achieved agronomically by increasing sowing rate. A more effective solution would be concurrent genetic manipulation of both Δ and early canopy growth. As discussed more fully by Richards and Condon (Chapter 29, this volume), this is just one of several challenges that will need to be overcome before carbon isotope discrimination can be successfully applied to improve crop yields.

Acknowledgments

This research was largely supported by the Wheat Research Council of Australia. A. G. Condon and R. A. Richards thank the organizers and the Australian Department of Industry, Technology and Commerce for enabling them to participate in the workshop "Perspectives of Plant Carbon and Water Relations from Stable Isotopes."

References

Bhagsari, A. S., and R. H. Brown. 1986. Leaf photosynthesis and its correlation with leaf area. *Crop Sci.* **26:** 127–132.

Condon, A. G., and R. A. Richards. 1992. Broad-sense heritability and genotype × environment interaction for carbon isotope discrimination in field-grown wheat. *Aust. J. Agric. Res.* **43,** 921–934.

Condon, A. G., R. A. Richards, and G. D. Farquhar. 1987. Carbon isotope discrimination is positively correlated with grain yield and dry matter production in field-grown wheat. *Crop Sci.* **27:** 996–1001.

Condon, A. G., G. D. Farquhar, and R. A. Richards. 1990. Genotypic variation in carbon isotope discrimination and transpiration efficiency in wheat. Leaf gas exchange and whole plant studies. *Aust. J. Plant Physiol.* **17:** 9–22.

Cooper, P. J. M., J. D. H. Keatinge, and G. Hughes. 1983. Crop evapotranspiration—a technique for calculation of its components by field measurements. *Field Crops Res.* **7:** 299–312.

Cowan, I. R. 1977. Stomatal behaviour and environment. *Adv. Bot. Res.* **4:** 117–228.

Cowan, I. R. 1988. Stomatal physiology and gas exchange in the field, pp. 160–172. *In* W. L. Steffen and O. T. Denmead (eds.), Flow and Transport in the Natural Environment: Advances and Applications. Springer-Verlag, Berlin.

Ehdaie, B., A. E. Hall, G. D. Farquhar, H. T. Nguyen, and J. G. Waines. 1991. Water-use efficiency and carbon isotope discrimination in wheat. *Crop Sci.* **31:** 1282–1288.

Ehleringer, J. R. 1990. Correlations between carbon isotope discrimination and leaf conductance to water vapor in common beans. *Plant Physiol.* **93:** 1422–1425.

Farquhar, G. D., and R. A. Richards. 1984. Isotopic composition of plant carbon correlates with water-use efficiency of wheat genotypes. *Aust. J. Plant Physiol.* **11:** 539–552.

Farquhar, G. D., K. T. Hubick, A. G. Condon, and R. A. Richards. 1988. Carbon isotope fractionation and plant water-use efficiency, pp. 21–40. *In* P. W. Rundel, J. R. Ehleringer and K. A. Nagy (eds.), Stable Isotopes in Ecological Research. Springer-Verlag, New York.

Farquhar, G. D., M. H. O'Leary, and J. A. Berry. 1982. On the relationship between carbon isotope discrimination and the intercellular carbon dioxide concentration in leaves. *Aust. J. Plant Physiol.* **9:** 121–137.

Hubick, K. T., R. Shorter, and G. D. Farquhar. 1988. Heritability and genotype × environment interactions of carbon isotope discrimination and transpiration efficiency in peanut. *Aust. J. Plant Physiol.* **15:** 799–813.

Jarvis, P. G., and K. G. McNaughton. 1986. Stomatal control of transpiration: Scaling up from leaf to region. *Adv. Ecol. Res.* **15:** 1–49.

Jones, H. G. 1976. Crop characteristics and the ratio between assimilation and transpiration. *J. Appl. Ecol.* **13:** 605–622.

Masle, J., and G. D. Farquhar. 1988. Effects of soil strength on the relation of water-use efficiency and growth to carbon isotope discrimination in wheat seedlings. *Plant Physiol.* **86:** 32–38.

Morgan, J. A., and D. R. LeCain. 1991. Leaf gas exchange and related leaf traits among 15 winter wheat genotypes. *Crop Sci.* **31:** 443–448.

Passioura, J. B. 1977. Grain yield, harvest index and water use of wheat. *J. Aust. Inst. Agric. Sci.* **13:** 191–201.

Wright, G. C., K. T. Hubick, G. D. Farquhar, and R. C. Nageswara Rao. 1992. Genetic and environmental variation in transpiration efficiency and its correlation with carbon isotope discrimination and specific leaf area in peanut. Chapter 17, this volume.

29

Challenges Ahead in Using Carbon Isotope Discrimination in Plant-Breeding Programs

R. A. Richards and A. G. Condon

I. Improving Water-Use Efficiency

Generations of plant breeders have been selecting for higher yields in dryland crops. They have generally been successful in their pursuits as the grain yield of our major crops in rain-fed areas has increased. In terms of grain yield per unit of evapotranspiration, which is the water-use efficiency for grain production (W_{GY}), plant breeders have also been successful with our major crops (Siddique *et al.*, 1990). This account is incomplete, however, as it ignores the factors responsible for yield increases. Grain yield is the product of two components, viz, biomass and harvest index (the ratio of grain weight to biomass) and almost the entire increase in grain yield in our major crops has come from an increase in harvest index (Gifford, 1986). We must conclude then that breeding has not increased W for total plant weight or biomass (W_B). If plant breeding is to continue to contribute to increased food supply for an ever-increasing population, then one of the greatest challenges contemporary and future breeders must overcome, as they approach the limits to the increase in harvest index, is to improve W for biomass production. This will be quite a formidable challenge as, so far, breeding has been unable to nudge total biomass upward (or the more usual measurement, aboveground biomass) despite enormous efforts by generations of breeders in many crops in a multitude of water-limited environments.

Why has W_B been so genetically intransigent? It must be due to one or more of the following reasons. First, there may be a lack of genetic variation in W_B on which selection can act. Second, W_B may have a very low heritability, or, possibly, a high genotype $\times$ environment interaction such that an increase in W_B is not found at all locations in all seasons. Third, an increase in W_B may be counterbalanced by negative factors that prevent its expression in the field. This may be due to genetic linkage whereby genes for high W_B are linked to genes that have a negative effect on growth and/or yield, or it could be a physiological association whereby factors that result in high W_B also result in a negative effect on yield or growth. Examples of these, which we come back to, are associations between W and slower canopy growth and/or higher canopy temperatures, both of which may negate any advantage of an improved W_B. Finally, the lack of genetic improvement in W_B may be due to the fact that no direct selection pressure has been put on improving W_B. Grain yield, which is after all the economic product and is relatively easy to select for, integrates factors that affect both biomass and harvest index and it has not been considered important to select for these components of grain yield individually. Presumably harvest index is more heritable and has been more responsive to indirect selection than biomass and this has been responsible for the yield gains. As gains in harvest index decrease, as they inevitably must, improvements in W_B may begin.

One handicap to improving W_B in the past has been the lack of a suitable selection technique. Selection for W_B in the field is often not feasible since measuring both biomass and water use is difficult, time consuming, and imprecise. A simpler procedure is to assume that variation in total water use is small and to select for aboveground biomass. This is the usual method for nongrain crops but, surprisingly, improvement in water-use efficiency after decades of selection for biomass in such crops does not seem to have been evaluated. For grain crops, progress in the selection for biomass should be possible (Hanson et al., 1985) and, providing care is taken to ensure that crop duration, phenology, and evaluation procedures do not interfere with results, it seems that this is worth attempting. Selection for W in large pots is also feasible but this is also time consuming and the results of selection may not translate to a higher W in field-grown plants. Measurement of the ratio of photosynthesis to transpiration using gas exchange equipment is even further removed from the realities of crop canopies and it is often dogged by large leaf-to-leaf and temporal variation.

Evidence for an alternative approach to improving W was presented in wheat by Farquhar and Richards (1984). This followed the development of theory that related W to the discrimination against ^{13}C relative to the more abundant ^{12}C during the fixation of carbon by photosynthesis (Farquhar et al., 1982). Thus an integrated estimate for W is the carbon isotope discrimination (Δ) of plant material. It is easier and faster to measure than total growth and water use and can be readily used to screen plants growing under identical conditions. Its measurement has opened up new opportunities for the genetic improvement of W. Pot studies, where both water use and biomass can be measured precisely, have consistently shown a negative relationship between Δ and W_B (Table I) or more correctly between Δ and transpiration efficiency as evaporation from the soil surface in these experiments is usually minimized. Clearly, in all the C_3 species so far tested and

Table I Relationship between Δ and Water-Use
Efficiency in Pot Studies

Crop	Relationship	Study
Wheat	−ve	Farquhar and Richards (1984)
Peanuts	−ve	Hubick *et al.*, (1986)
Cotton	−ve	Hubick and Farquhar (1987)
Peanut	−ve	Hubick *et al.*, (1988)
Tomato	−ve	Martin and Thorstenson (1988)
Barley	−ve	Hubick and Farquhar (1989)
Potato	−ve	Vos and Groenwold 1989
Wheat	−ve	Condon *et al.* (1990)
Sunflower	−ve	Virgona *et al.* (1990)
Wheatgrass	−ve	Johnson *et al.* (1990)
Wildrye	−ve	Johnson *et al.* (1990)
Wheatgrass	−ve	Read *et al.* (1991)
Grasses[a]	−ve	Johnson and Bassett (1991)
Wheat	−ve	Ehdaie *et al.* (1991)
Bean	−ve	Ehleringer *et al.* (1991)
Cowpea	−ve	Ismail and Hall (1992)

[a]*Dactylis, Festuca, Lolium*

over a wide range of conditions, the theory on carbon isotope discrimination matches the experimental findings superbly and provides compelling evidence for the potential use of Δ as a predictor of *W*.

II. Measurement of Carbon Isotope Discrimination

There are many features of Δ that make it very appealing as a selection criterion in breeding programs. These have to do with both its measurement and its genetic characteristics: it is not necessary to monitor water use and biomass; Δ is an integrated value over the life of the plant; only a small sample of dry matter is required for measurement; and Δ can be determined early in the life of the plant well before the seed is formed. This last point means that plants can be selected and used either for hybridization or for seed in the same cycle of growth. Furthermore, the measurement of Δ is very precise and it can be automated. Coefficients of variation are typically about 2%, which is far lower than the coefficient for grain yield of about 10% obtained in precise field experiments (Condon and Richards, unpublished). Much higher values for grain yield often occur and even higher values are usual in the measurement of either water use or biomass production. In all species studied so far genetic variation in Δ has been substantial with broad-sense heritabilities as high as 95% in several species (Condon and Richards, 1992b; Ehdaie *et al.*, 1991; Hubick *et al.*, 1988); genotype × environment interactions have correspondingly been very low (Condon *et al.*, 1987; Condon and Richards, 1992b; Hall *et al.*, 1990;

Hubick *et al.*, 1988; Read *et al.*, 1991). It should be noted, however, that these values for heritabilities are based on plot mean values, and studies with individual plants in F_2 populations give lower values (Hall *et al.*, Chapter 23, this volume).

Nevertheless, there are aspects of the measurement of Δ that have emerged that limits its use in plant breeding or provide problems in plant selection. The cost of measurement is a considerable handicap to its adoption. The present cost, which includes grinding and the determination of Δ, is about US$20 per sample. If a plant breeder were to evaluate 1000 samples each year then he/she would need to be convinced that selection for Δ would be worthwhile. A breeder may readily adopt a visual selection criterion with no guarantee that it will result in a yield improvement, as it can easily be incorporated into a breeding program and it requires few additional resources. To adopt a new test that is not cheap and that requires additional resources there must be some guarantee of success at the end; although, in rational economic terms, the cost to determine Δ may not be excessive in relation to other aspects of a crop-improvement program. Field trials are very expensive as they require a large investment in heavy machinery, labor, and time for traveling, and often the possibility of failure at any site due to biotic or edaphic factors is high. Some tests to assess important grain quality characteristics are just as costly as Δ. In addition, there may be some ways to reduce the cost of measurement of Δ when screening large numbers of genotypes. One way may be to relax the precision of the measurement of Δ. This may not only reduce the cost but also hasten the throughput of samples required in plant breeding. Increased precision may then only be required for retesting selected lines or progenies. Another possibility is the development of alternative ways to select for Δ. For example, one way may be to select for linked genetic markers such as restriction fragment length polymorphisms (RFLPs), although this also will not be cheap because Δ is under complex genetic control and a number of markers are likely to be required. Also there are other disadvantages with such techniques. Thus measuring Δ directly is still likely to be the most effective procedure.

Several surrogates for Δ have recently emerged that do not require the investment in ratio mass spectrometry. One is plant mineral content, determined by ashing plant material. Mineral content has been correlated with W in a number of different species (Masle and Farquhar, 1992). The measurement of plant mineral content instead of Δ may be particularly valuable where a ratio mass spectrometer is not available as its measurement only requires a precise balance and a muffle furnace. Another attraction is that the grinding step is not required. Specific leaf area (SLA, $cm^2 \ g^{-1}$) is another surrogate for Δ that looks promising in peanuts (Wright *et al.*, 1988). This has the attraction of being measured early in the plant's life on spaced plants or in a closed community. Nevertheless, there are some dangers in selecting for a low SLA. In some environments SLA may be positively correlated with leaf area and leaf canopy development, and hence positively correlated with yield but negatively correlated with *W*. Another danger is that selecting for low SLA instead of Δ may result in not exploiting all of the variation in Δ. Variation in Δ arises from variation in both stomatal conductance and assimilation capacity, and selection for low SLA may result in a higher assimilation capacity (the result of packing leaves with more photosynthetic machinery per unit leaf area), but it is unlikely to exploit the variation in the balance between stomatal conductance and assimilation capacity. Another possible surrogate for Δ is leaf or canopy temperature which can be readily determined using a portable infrared thermometer. We have found canopy temperature to be negatively related to Δ as is expected if stomatal conductance contributes to variation in Δ. These surrogates will never be as good as the actual measurement of Δ because heritability is unlikely to be as high and the genetic correlation between Δ and the surrogate will not be unity. However, there may be a subsidiary role for them in breeding programs in early generations when the cost of measuring Δ may prohibit large-scale screening.

The other aspect of measurement that remains a challenge has to do with sampling. Although Δ is an integrated measure of p_i/p_a (ratio of the intercellular to the ambient partial pressure of CO_2) over time, it can fluctuate from sample to sample depending on when the carbon was laid down and on the environmental conditions during that time. Any factor that alters stomatal conductance or photosynthetic capacity such as variation in soil water content, atmospheric humidity, nutrition, soil strength, or radiation, will also influence the value of Δ. Although these requirements may seem quite restrictive, conditions can be specified that make the evaluation of Δ highly repeatable with very little genotype × environment interaction. In temperate cereals for example, we have found that for plants growing in the field, sampling as early in the season as possible is best. This is when vapor pressure deficit (vpd) is still low and drought stress has not begun. Sampling plant material between stem elongation and grain formation is more unreliable (even though variation in Δ is often repeatable) because of variation in soil water, phenology, and retranslocation of earlier formed assimilates. As a general rule, sampling plants during their normal growing season before variation in soil moisture content occurs should be the most effective.

III. Genetic Challenges

The preceding recommendation imposes quite a restriction in a breeding program but one of the attractions of Δ is that it should not be dependent on assessment in the field and it may be used out of season or in a glasshouse. This would enable several generations of selection per year rather than one, as is usual, in breeding programs for yield in annual crops. However, it would not be effective if G × E for Δ in different seasons is high and results in different rankings. Published evidence indicates that, in contrast to say grain yield, G × E for Δ is very low when plants are grown during the normal field season (Condon *et al.* 1987; Hubick *et al.*, 1988; Read *et al.*, 1991; Johnson and Bassett, 1991; Hall *et al.*, 1990). For example, in our studies in southeastern Australia, although G × E values were significant, they accounted for only 5% of the total genetic variation (Condon and Richards, 1992b). In more contrasting environments, however, G × E may be quite substantial and bothersome. In wheat we have found that in some circumstances sampling vegetative material in the glasshouse can result in a different ranking of genotypes compared to that in their normal growing environment. The most likely factors responsible for this variation are high radiation or longer day lengths, although there may also be interactions with temperature or vpd and we have evidence that the bulk density of the soil in which the plants are growing may also be important. Further work is required to identify the cause of these different genotype rankings before Δ can be effectively screened out of season, and this remains an important challenge. Similar problems have emerged in sampling two contrasting ecotypes of Arabidopsis grown in controlled environments (Masle, Chapter 24, this volume).

Surprisingly little is known about the nature of genetic variation for Δ. A number of studies have reported a high broad-sense heritability, i.e., the ratio of genetic variation relative to the total genetic, G × E, and residual variation (Hubick *et al.*, 1988; Condon and Richards, 1992b; Ehdaie *et al.*, 1991; Hall *et al.*, 1990). A high broad-sense heritability shows how repeatable the measurement of Δ is in different genotypes but we still do not know whether additive genes (important if selection is to be effective) or nonadditive effects (i.e., dominance, epistasis, and maternal effects) are largely responsible for the variation in Δ. It is important that these be determined in the appropriate selection environment to assess how effective selection for Δ may be. Relatively simple genetic control has been indicated in tomato where it was found that three genetic markers accounted for 70% of the variation in Δ between two different tomato species (Martin *et al.*, 1989).

IV. The Relationship between Δ and *W* in Field-Grown Plants

Perhaps the most important immediate challenge before we use Δ in breeding programs is to determine whether Δ is also related to *W* in field-grown plants, as it is in pots. Unfortunately it is not feasible to measure the *W* of a large number of field plots of plants and to relate it to Δ, and so instead we have to rely on other measurements. Providing crops use all the available water, as they often do when water is limited, Δ should be similarly related to biomass. Table II presents published studies where both Δ and biomass have been determined in field-grown plants. The consistent negative relationship found in pot studies no longer holds and it is clear that results in the field are more complex. The expectation of a negative relationship

Table II Relationship between Δ and Biomass in Field Plots[a]

Crop	Location	Relationship	Author
Wheat	E Australia	+ve	Condon *et al.* (1987)
Peanuts	NE Australia	−ve	Wright *et al.* (1988)
Bean	Colombia	+ve,−ve	White *et al.* (1990)
Barley	Syria	+ve	Crauford *et al.* (1991)
Wheat	California	+ve	Ehdaie *et al.* (1991)
Grasses	Washington	−ve	Johnson and Bassett (1991)
Wheat	E Australia	+ve,−ve	Condon and Richards (1992a)

[a]Studies in representative field plots only are included.

between Δ and biomass is based on a number of assumptions that, if they did not hold, could markedly disturb or negate the relationship. Some of the assumptions are that genotypes use the same amount of soil water, they have the same canopy growth and hence there are no differences between them in the amount of evaporation from the soil surface, and they have the same root mass. The finding of a significant positive relationship between Δ and biomass in the first field studies (Condon *et al.*, 1987) suggests that, in wet environments at least, the above assumptions are untenable and there may in fact be a penalty in having a low value of Δ.

There are additional complications to the expected negative relationship between Δ and biomass, when crop canopies are considered, that may negate the relationship or even contribute to the positive relationships that have been observed. They arise from scaling up from a single plant in a container, usually growing in well-ventilated conditions, to a crop canopy in the field, and they are likely to be of greatest importance at the scale of a farmer's field. One complication is as follows. If genotypes with low Δ also have a low stomatal conductance as is likely in many species, then these genotypes will have a higher W than genotypes with a higher Δ when grown in a ventilated environment. However, when grown in pure stands in the field there will be a greater boundary layer resistance to water vapor and heat transfer than there is in pots and thus canopies with a lower stomatal conductance will be warmer and the air surrounding them will be both warmer and drier than that of corresponding canopies with a higher stomatal conductance. This will increase transpiration for a given stomatal conductance, and water-use efficiency should be less for a given conductance compared to that of isolated plants. This is discussed in some detail by Condon and Richards (Chapter 28, this volume) who show that, although these scale effects are significant, they are unlikely to account for the positive relationship between Δ and aboveground biomass in relatively favorable environments, as has been suggested by Cowan (1988).

We believe there are several explanations for the positive relationships reported between Δ and biomass. One has to do with the part of the plant used in the measurement of Δ. Another has to do with physiological and genetic associations with Δ that are so important that they reverse the expected negative relationship between Δ and biomass. Yield has been positively associated with Δ when Δ is measured after the commencement of reproductive growth in environments with a terminal drought. The highest yielding genotypes in these environments are usually early flowering (Crauford *et al.*, 1991) because they escape the drought. However, grain was used for the determination of Δ in the barley study by Crauford *et al.* (1991), and as the grain of early-flowering lines would be formed when drought and vpd would be lower than in the later flowering lines, then early-flowering lines should have a higher Δ than the later lines. This could account for the positive relationship between Δ and grain yield.

An association has been observed between Δ and early canopy growth (Fig. 1) which may also account for the positive relationship between Δ and biomass in some environments. For crops that receive most of their rainfall between sowing and flowering, the advantages of rapidly forming a canopy of leaves may exceed any penalty this canopy may have on the ratio of assimilation to transpiration. One advantage would arise from the saving in soil evaporation by a fast-developing canopy shading the soil surface, with the saving then being available for extra transpiration. A further advantage would be in additional growth during the early part of the season when vpd is low and seasonal W high. A penalty may arise in the ratio of assimilation to transpiration since forming a canopy of leaves quickly may require a high leaf area-to-mass ratio. This is likely to result in a lower nitrogen content per unit area, a lower assimilation capacity, and therefore a higher Δ. Alternatively, where low Δ is the result of low stomatal conductance, the rate of photosynthesis per unit leaf area may also be lower thereby slowing early canopy growth. In our studies with wheat a positive relationship

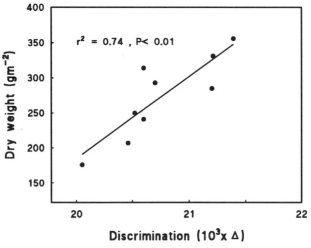

Figure 1. Relationship between carbon isotope discrimination and aboveground dry weight in wheat genotypes 101 days after sowing. Genotypes were grown in 18-m² plots in the absence of water stress and Δ was determined on leaf sheaths before terminal spikelet formation. (Adapted from Richards, 1992.)

between Δ and biomass has been observed in wetter environments, where soil evaporation has been significant, whereas the expected negative relationship has been observed when it is substantially drier and soil evaporation, as a proportion of evapotranspiration, is less (Condon and Richards, Chapter 28, this volume, Fig. 2). Furthermore, in our studies comparing two wheat varieties differing in Δ and grown in large blocks, the low Δ variety consistently had the highest transpiration efficiency (biomass/transpiration) but a lower W_B (biomass/evapotranspiration) in the wetter environments.

There may also be a physiological or a genetic association between Δ and flowering time. A positive association was noted in barley (Crauford *et al.*, 1991), although this may have been due to the sampling procedure as we discussed, and a positive association has also been found in other cereal species (C. López-Castañedo, unpublished data). In this latter case, sampling plant material for Δ was made as early as possible to avoid the effects of drought and vpd (Fig. 2) and although these effects may not have been totally avoided, a link between Δ and flowering time seems likely. We believe that it may have arisen through evolution, or from unconscious selection by breeders for faster growth, in genotypes of short duration, especially in environments where a large leaf area index is seldom achieved. Faster growth may be required to achieve high yields as it compensates for any yield penalty associated with short duration. To achieve a large canopy quickly may only be possible if genotypes have thin leaves (high specific leaf area) or wide open stomata. Such genotypes are likely to have a high Δ. Later flowering lines on the other hand have a longer opportunity to develop their leaf area and would experience less selection pressure for fast

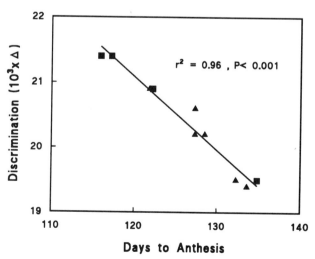

Figure 2. Relationship between time to anthesis and carbon isotope discrimination in wheat (▲) and barley (■) genotypes. Results are averaged over five field sites in southeastern Australia. Δ was determined on stem bases harvested at maturity to represent carbon assimilated early in the crop's life. (C. López-Castañeda, unpublished data.)

canopy development. This suggestion of unconscious selection in short-duration genotypes is supported by Δ values on two unrelated sets of isogenic populations of wheat differing in flowering time. Each population in each set was selected only on its time to anthesis. There were no differences in Δ between any population in any set despite a difference of 20 days in flowering time (Richards and Virgona, unpublished).

V. Use of Δ in Breeding Programs

A final challenge to plant breeders is whether measurement of Δ is sufficiently important to use in plant-breeding programs. At this stage there is no doubt that the use of Δ in breeding shows considerable potential. But it must be put in some context. It will only be one component of any plant-breeding effort and then only part of the armory available to improve W. There are other equally promising ways to improve W that have been discussed recently by Richards (1991). Use of these different traits will depend primarily on the crop species and the climate of the region where the crop is grown. We should not expect Δ to be important in every region and in every water-limited crop. A single crop species may be grown from Alaska to the tropics and crops may vary 10-fold in leaf area and up to 5-fold in crop duration. It is thus little wonder that the relationship between Δ and biomass may vary in different seasons and regions.

If discrimination is used in breeding it will be only one of an array of measurements competing for the limited resources of the breeder. It may not have high priority. Of highest priority in breeding programs are those factors that can readily be seen by the breeder or farmer as limiting to yield or marketing opportunities. Improved resistance to a particular disease, pest, or nutritional limitation are easily visible and these often have high priority and command a large share of the resources in a breeding program. Grain quality also has a high priority because it often determines whether a cultivar will be released. However, improvement in grain yield, let alone W, is less obvious because it is hard to see a 5% increase in yield or biomass and more difficult to select for in early generations. Hence there are often fewer resources put into improving yield than there are in the more visible problems. This implies that strong pressure will not be imposed on breeders to adopt carbon isotope discrimination in breeding programs.

Although the potential use of carbon isotope discrimination in breeding programs is still uncertain we believe it does have considerable promise. Our experience with winter-grown cereals indicates that low discrimination (low Δ) should result in improved W and yield in several water-limited environments: first, where crops are primarily dependent on water stored in the soil from a previous season; second, in vigorous crop species dependent on current rainfall, for less vigorous crop species growing on current rainfall low Δ may need to be combined with plant factors that increase vigor, as the overriding factor in these environments is the amount of leaf area growth achieved during the early part of the season; third, in long-duration crops that maintain a large canopy and full light interception for a long period. Conversely, high Δ may also result in higher yield. This is likely to be the case in high-yielding environments when rainfall or water is not a limiting factor.

VI. Summary and Conclusions

Generations of plant breeders have so far failed to increase W for biomass production in our major grain crops. This alone remains a formidable challenge. Selection for carbon isotope discrimination may provide a worthwhile means to genetically improve W. It fulfills many of the criteria important in selection and it is far simpler to measure than W itself. However, before it is recommended with confidence there are a number of challenges to overcome. Measuring Δ is expensive and simpler, cheaper ways to measure it may be required. Although it is under strong genetic control and G × E is low in the normal field environment or season, G × E is high when variation in Δ is assessed out of season. This may be a substantial handicap in breeding if Δ is assessed in populations several times per year. An important challenge therefore will be to understand what is responsible for the G × E interactions so that they may be reduced. An immediate challenge will be to understand whether there are physiological or genetic associations with Δ which negate any likely improvement in W and, if there are such associations, whether they can be overcome. The final challenge will be to identify the environments where selection based on Δ will result in the biggest on-farm gains and, if Δ is important, to convince breeders of its worth.

References

Condon, A. G., and R. A. Richards. 1992a. Exploiting genetic variation in transpiration efficiency in wheat—an agronomic view. Chapter 28, this volume.

Condon, A. G., and R. A. Richards. 1992b. Broad sense heritability and genotype × environment interaction for carbon isotope discrimination in field-grown wheat. *Aust. J. Agric. Res.* **43:** 921–934.

Condon, A. G., R. A. Richards, and G. D. Farquhar. 1987. Carbon isotope discrimination is positively correlated with grain yield and dry matter production in field grown wheat. *Crop Sci.* **27:** 996–1001.

Condon, A. G., G. D. Farquhar, and R. A. Richards. 1990. Genotypic variation in carbon isotope discrimination and transpiration efficiency in wheat. Leaf gas exchange and whole plant studies. *Aust. J. Plant Physiol.* **17:** 9–22.

Cowan, I. R. 1988. Stomatal physiology and gas exchange in the field, pp 160–172. *In* W. L. Steffen and O. T. Denmead (eds.), Flow and Transport in the Natural Environment: Advances and Applications. Springer-Verlag, Berlin.

Craufurd, P. Q., R. B. Austin, E. Acevedo, and M. A. Hall. 1991. Carbon isotope discrimination and grain yield in barley. *Field Crops Res.* **27:** 301–313.

Ehdaie, B. A., A. E. Hall, G. D. Farquhar, H. T. Nguyen, and J. G. Waines. 1991. Water use efficiency and carbon isotope discrimination in wheat. *Crop Sci.* **31:** 1282–1288.

Ehleringer, J. R., S. Klassen, C. Clayton, D. Sherrill, M. Fuller-Holbrook, Q. A. Fu, and T. A. Cooper. 1991. Carbon isotope discrimination and transpiration efficiency in common bean. *Crop Sci.* **31:** 1611–1615.

Farquhar, G. D., and R. A. Richards. 1984. Isotopic composition of plant carbon correlates with water-use efficiency of wheat genotypes. *Aust. J. Plant Physiol.* **11:** 539–552.

Farquhar, G. D., M. H. O'Leary, and J. A. Berry, 1982. On the relationship between carbon isotope discrimination and intercellular carbon dioxide concentration in leaves. *Aust. J. Plant Physiol.* **9:** 121–137.

Gifford, R. M. 1986. Partitioning of photoassimilate in the development of crop yield, pp. 535–549. *In* J. Cronshaw, W. J. Lucas and T. T. Giaquinta (eds.), Phloem Transport. A. R. Liss, New York.

Hall, A. E., R. G. Mutters, K. T. Hubick, and G. D. Farquhar. 1990. Genotypic differences in carbon isotope discrimination by cowpeas under wet and dry field conditions. *Crop Sci.* **30:** 300–305.

Hanson, P. R., T. J. Riggs, S. J. Klose, and R. B. Austin. 1985. High biomass genotypes in spring barley. *J. Agric. Sci.* (Cambridge) **105:** 73–78.

Hubick, K. T., and G. D. Farquhar 1987. Carbon isotope discrimination—selecting for water-use efficiency. *Aust. Cottongrower* **8:** 66–68.

Hubick, K. T., and G. D. Farquhar. 1989. Genetic variation in carbon isotope discrimination and the ratio of carbon gained to water lost in barley. *Plant Cell Environ.* **12:** 795–804.

Hubick, K. T., G. D. Farquhar, and R. Shorter. 1986. Correlation between water-use efficiency and carbon isotope discrimination in diverse peanut (*Arachis*) germplasm. *Aust. J. Plant Physiol.* **13:** 803–816.

Hubick, K. T., R. Shorter, and G. D. Farquhar. 1988. Heritability and genotype × environment interactions of carbon isotope discrimination and transpiration efficiency in peanut. *Aust. J. Plant Physiol.* **15:** 799–813.

Ismail, A. M., and A. E. Hall. 1992. Correlation between water-use efficiency and carbon isotope discrimination in diverse cowpea genotypes and isogenic lines. *Crop Sci.* **32:** 7–12.

Johnson, R. C., and L. M. Bassett. 1991. Carbon isotope discrimination and water use efficiency in four cool-season grasses. *Crop Sci.* **31:** 157–162.

Johnson, D. A., K. H. Asay, L. T. Tieszen, J. R. Ehleringer, and P. G. Jefferson. 1990. Carbon isotope discrimination: Potential in screening cool-season grasses for water limited environments. *Crop Sci.* **30:** 338–343.

Martin, B., and Y. R. Thorstenson. 1988. Stable carbon isotope composition (δ^{13}C), water use efficiency and biomass productivity of *Lycopersicum esculentum*, *Lycopersicum pennellii*, and the F1 hybrid. *Plant Physiol.* **88:** 213–217.

Martin, B., J. Nienhuis, G. King, and A. Schaefer. 1989. Restriction fragment length polymorphisms associated with water-use efficiency in tomato. *Science* **243:** 1725–1728.

Masle, J., G. D. Farquhar, and S. C. Wong 1992. Transpiration ratio and plant mineral content are related among genotypes of a range of species. *Aust. J. Plant Physiol.* **19:** 709–721.

Read, J. J., D. A. Johnson, K. H. Asay, and L. T. Tieszen. 1991. Carbon isotope discrimination, gas exchange, and water-use efficiency in crested wheatgrass clones. *Crop Sci.* **31:** 1203–1208.

Richards, R. A. 1991. Crop improvement for temperate Australia: Future opportunities. *Field Crops Res.* **26:** 141–169.

Richards, R. A. 1992. The effect of dwarfing genes in spring wheat in dry environments. II. Growth, water use and water-use efficiency. *Aust. J. Agric. Res.* **43:** 529–539.

Siddique, K. H. M., D. Tennant, M. W. Perry, and R. K. Belford. 1990. Water use and water-use efficiency of old and modern wheat cultivars in a Mediterranean environment. *Aust. J. Agric. Res.* **41:** 431–447.

Virgona, J. M., K. T. Hubick, H. M. Rawson, G. D. Farquhar, and R. W. Downes. 1990. Genotypic variation in transpiration efficiency, carbon-isotope discrimination, and carbon allocation during early growth in sunflower. *Aust. J. Plant Physiol.* **17:** 207–214.

Vos, J., and J. Groenwold. 1989. Genetic differences in water-use efficiency, stomatal conductance and carbon isotope fractionation in potato. *Potato Res.* **32:** 113–121.

White, J. W., J. A. Castillo, and J. R. Ehleringer. 1990. Associations between productivity, root growth and carbon isotope discrimination in *Phaseolus vulgaris* under water deficit. *Aust. J. Plant Physiol.* **17:** 189–198.

Wright, G. C., K. T. Hubick, and G. D. Farquhar. 1988. Discrimination in carbon isotopes of leaves correlates with water-use efficiency of field grown peanut cultivars. *Aust. J. Plant Physiol.* **15:** 815–825.

V

Water Relations and Isotopic Composition

Hydrogen and oxygen isotopes in stem waters are revealing new insights into the specific soil layers from which plants extract moisture. As such, analyses of xylem sap provide a nondestructive means of assessing soil–plant water relations and effective rooting zones. In this section, Dawson provides a synthesis of that topic and an examination of how these observations relate to ecological patterning. In a related chapter, Thorburn and Walker examine the specific water uptake patterns for Australian eucalyptus, showing that these trees differentiate between ground and surface waters with counter-intuitive results. Lin and Sternberg provide evidence for a contrasting pattern (salt-stressed mangroves) to that described by Dawson and by Thorburn and Walker, indicating the utility of analyzing both hydrogen and oxygen isotopes in water. In the final chapter, Yakir *et al.* provide experimental evidence for the linkages between the oxygen isotope composition of leaf water and carbon dioxide and of the utility of such information for analyzing gas exchange processes.

30

Water Sources of Plants as Determined from Xylem-Water Isotopic Composition: Perspectives on Plant Competition, Distribution, and Water Relations

Todd E. Dawson

I. Introduction

Sixty years ago Urey *et al.* (1932) reported the discovery of a stable isotope of hydrogen with a mass twice that found in all other naturally occurring stable hydrogen (H) isotopes. Though it is relatively rare (only 0.015% of all stable hydrogen isotopes), the discovery of ^2H, or deuterium (D) as it is now called, and its abundance relative to H, the so-called D/H ratio or δD, has revolutionized research in physics, the atmospheric sciences, geology and geochemistry, and, most recently, the biological sciences. In this chapter, I discuss some recent advances using stable hydrogen isotope analyses in the biological sciences (see also chapters in Rundel *et al.*, 1988). In particular, I focus my discussion on recent research in the plant sciences that has used variation in δD from source and plant water to study plant-water relations especially as they effect plant distribution, plant–plant interactions, and the history of water use.

A. Background

Because of such large differences in the mass between D and H, the ratio, D/H in meteoric water, biological, and geological samples varies by as much as 700‰; more than with any other naturally occurring stable isotope ratio (Taylor, 1974; Hoefs, 1980). In the environment, fractionation of hydrogen isotopes is caused principally by transport processes (Gat, 1982) and phase transitions (e.g., liquid to vapor; Taylor, 1974) through both the atmosphere and lithosphere. In biological systems, fractionation is caused by phase transitions, metabolism, chemical reactions, and the differential incorporation of D and H from H_2O, DHO, and D_2O into tissues due to their different masses and vapor pressures (Craig, 1961; Yapp and Epstein, 1982a). In meteoric waters, δD varies by more than 400‰ (see Taylor, 1974), caused by evaporative fractionation during cloud formation and condensation when precipitation forms and is removed from clouds as either rain or snow (Fig. 1; Craig, 1961; Dansgaard, 1964). When water changes from liquid to vapor it becomes depleted in D (more negative δD). The opposite is true when water condenses out of vapor. Overall there is a decrease in D within meteoric water by 1–5‰/100 m along the thermal gradients from low to high elevation and by 3 to 150‰ from equatorial regions toward the poles. The latitudinal fractionation effect in meteoric water is driven by adiabatic processes which cause fractional losses of D within the initial water vapor as the cloud mass moves inland, over the drier continents, along its storm trajectory (3 to >45‰ per 100 km of movement). When precipitation is formed D is removed from the vapor (see Fig. 1; Ingraham and Taylor, 1991), leading to an overall depletion of D in the atmospheric water (more negative δD). Most of this variation arises because the isotopic composition of water is strongly influenced by atmospheric temperature (Fig. 2; Dansgaard, 1964; Taylor, 1974), which is of course correlated with altitude and latitude (Craig, 1961; Dansgaard, 1964; Yapp and Epstein, 1982b).

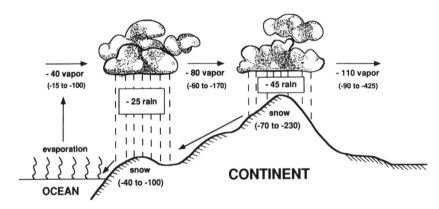

Figure 1. Variation in the hydrogen isotopic ratio of water in the hydrologic cycle. Evaporation causes fractionation against the heavier isotope of hydrogen (deuterium, D), while condensation as rain or snow will be relatively more enriched in D. As an air mass moves inland, the water vapor becomes depleted in D from these processes acting together. The values that appear are what can be expected at mid-latitudes; values in parentheses are the range of values reported in the literature and depend upon the latitude and time of year the samples were collected (from Friedman *et al.*, 1964; White, 1988; and Ingraham and Taylor, 1991).

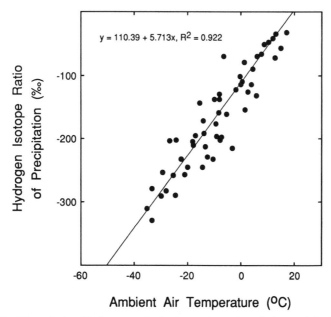

y = 110.39 + 5.713x, R² = 0.922

Figure 2. The relationship between the hydrogen isotopic ratio in precipitation and the ambient air temperature at the time of the rainfall (after Craig, 1961: Dansgaard, 1964; Dawson and Ehleringer, 1991; Dawson, unpublished data). Values are reported in per mil, ‰, notation.

The D/H ratio found in the tissues of organisms, especially plants, reflects the isotopic composition of the water in the environment where they live and function. As shown above, the δD of environmental water is highly variable. Fractionation associated with metabolic processes can further influence tissue δD (Ziegler *et al.*, 1976; Ziegler, 1988; Sternberg, 1988; Nagy, 1988; DeNiro and Cooper, 1989). Therefore, for plant biologists who intend to use stable hydrogen isotopes in their research, the most important ramification they should be aware of is that the marked variation seen in δD makes it very difficult to compare hydrogen (or oxygen) isotopic ratios from tissues collected in different regions unless the source-water δD values at these different locations are known.

II. Water Sources: Using δD as a "Tracer"

The primary source of hydrogen in plants is derived from water. Water sources for plants are precipitation, soil water, runoff (including snowmelt), and groundwater. Soil water, runoff, and groundwater ("environmental water") of course all originally come from precipitation but the δD of these waters is significantly influenced by (1) physical processes such as evaporation, (2) the size and elevational extent of the catchment basin that contributes to the total source-water pool, (3) the depth and geological characteristics of the groundwater aquifer, and (4) the dissolution characteristics and velocity of water movement in subsurface strata into the groundwater (Fontes, 1980; Allison and Hughes, 1983; Barnes and Allison, 1983; Nordstrom *et al.*, 1989; Ingraham and Taylor, 1991). Evaporation plays a dominant role in effecting soil δD and, in areas where only slight differences in the δD of source waters are present, is the primary reason soil water and rainfall can be separated (see Thorburn, Chapter 32,

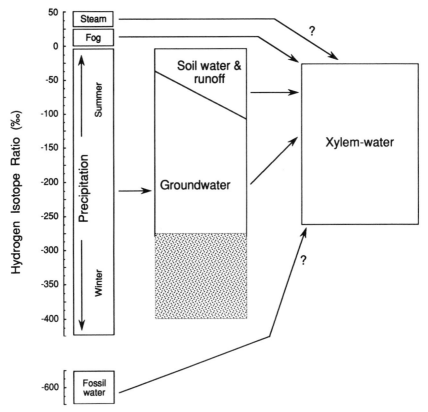

Figure 3. General diagram of the hydrogen isotope ratios of precipitation, groundwater, and xylem water (from extracted plant sap). The range of values that compose each box are those reported from mid to upper latitudinal regions of the world (after Dansgaard, 1964; White, 1988; Ingraham and Taylor, 1991) and are reported in per mil (‰) notation. The diagonal line separating soil water and runoff from groundwater signifies the potential for these two water sources to overlap significantly. The stippled area at the most negative end of the groundwater box signifies a range of possible isotopic values that could occur for groundwater based on precipitation (snow) inputs but are rarely reported in the literature. The data for fog water use are for the California Coastal Redwood, *Sequoia sempervirens* (Dawson, unpublished). The question marks on the lines connecting steam and fossil water to xylem water of plants appear because these sources exist and could potentially be used by plants; however, to date no such reports exist.

this volume). Each of the potential sources of variation in δD cited above is influenced by seasonality, particularly in temperate, continental regions (Fig. 3; Dansgaard, 1964; Lawrence *et al.*, 1982; White, 1988) as well as by the magnitude and nature of biological (microbial) activity within soil surface layers (Schiegel, 1972; M. Magaritz, personal communication).

The water used by plants can come from any one or a combination of the sources mentioned above. With the exception of salt-excluding species (see Lin, Chapter 31, this volume), no fractionation occurs during water uptake by plant roots (Wershaw *et al.*, 1966; Zimmermann *et al.*, 1966; White *et al.*, 1985) so that root water, stem/caudex water, and xylem water that have not been subjected to evaporative or metabolic fractionation (Flanagan and Ehleringer, 1991a; Yakir *et al.*, 1989; DeNiro and Cooper, 1989) will reflect the source or sources of water used by any particular species. In this way, the analysis of δD in water allows one to trace where plants are obtaining their water (White, 1988; Dawson and Ehleringer, 1991; Flanagan *et al.*, 1991; Walker and Richardson, 1991; Thorburn *et al.*, 1992a and this volume). This information may in turn allow us to determine how differential utilization of water sources influences plant distribution, coexistence, and competition for water by plants in both time and space (Dawson and Ehleringer, 1991; Ehleringer *et al.*, 1991; Sternberg *et al.*, 1991; Thorburn *et al.*, 1992; Ehleringer and Dawson, 1992; also see III, B, below).

III. Source- and Plant-Water Determinations and Their Use in Studying Plant Distributions and Interactions

A. An Overview on Methods

To determine which sources of water are used by individual plants, it is essential to determine the stable isotopic composition of all possible sources in the environment—precipitation, fog, soil water, runoff, and groundwater. Comparison of root-water (herbs and grasses), stem/caudex-water (perennial herbs and forbs), or xylem-water (woody taxa) isotopic composition with the isotopic composition of the environmental water allows the water source(s) that plants are using to be determined. Partitioning the contribution of different water sources to the plant's water can be calculated using specific mixing models such as those used by White and coworkers (1985; see equations below), Dawson (1993), or with δD-^{18}O methods (see Thorburn, Chapter 32, this volume).

Source-water samples can be collected in glass vials that can be completely sealed to prevent any water loss that will isotopically alter the sample. Soil water samples should be collected with a soil corer at the depth where water absorption by roots is thought to be most active. For woody plants this depth is often 30–150+ cm below the soil surface (Kramer, 1983). For herbs, forbs, or grasses, it is often shallower, between 15 and 80 cm below the soil surface and below the zone where evaporative enrichment is most pronounced (upper 10 cm; Zimmermann *et al.*, 1966). Soil should be immediately placed in an airtight container and sealed until the water can be extracted (see below).

For woody plants, mature, fully suberized stems are excised and placed immediately into glass (test)tubes, stoppered, and either wrapped in Parafilm or dipped in wax. For herbaceous species, grasses, or forbs, the roots, stem/caudex (the "woody" suberized zone where roots and aboveground parts meet; at the soil surface), or nonphotosynthetic basal rosette can be collected as you would stems. It is extremely important, however, that no leaves or other soft tissues be included (Dawson, 1993). The δD of water in soft tissues, such as leaves, can show significant evaporative isotopic enrichment (White, 1988; Ziegler, 1988; Flanagan *et al.*, 1992a,b; Thorburn *et al.*, 1992a). This would cause significant problems in tracing the water source(s) a particular plant is using (but see discussion on the application of a mixing model below). In fact, even for "green stem" samples, those near actively growing branch apices, where bark has not formed, significant evaporative isotopic enrichment can occur (Fig. 4). Stem samples of this type should not be used (Dawson and Ehleringer, 1993).

To prevent any water loss, all samples should be enclosed in airtight plastic bags and frozen until the water can be extracted. Water is extracted from tissue or soil samples using a cryogenic vacuum distillation apparatus (Ehleringer and Osmond, 1989). The final extract is sealed in glass under a slight vacuum. These samples are prepared for hydrogen isotopic composition by reacting a subsample of approximately 5 μl (in a capillary) of water with zinc in an evacuated Vycor glass tube at 500°C for 30–60 min. (after Coleman *et al.*, 1982). The hydrogen gas resulting from the combustion is analyzed for its isotopic composition on a gas isotope ratio mass spectrometer. A second method for determining the δD of source and plant water is to pass 2–5 μl of water over uranium within a hot (700–750°C) furnace (see Bigeleisen *et al.*, 1952; Friedman and Hardcastle, 1970 for detailed methods). The hydrogen gas that is released from the reaction is collected with a Toepler pump and introduced into the mass spectrometer as before using these methods. Measurement precision of $\pm 1.5\%o$ is achievable.

All δD values are expressed in delta notation (‰) relative to an accepted standard (below) as

$$\delta D = \left[\frac{(D/H)_{sample}}{(D/H)_{standard}} - 1 \right] \times 1000. \tag{1}$$

Currently accepted standards are SMOW (standard mean ocean water), V-SMOW (Vienna-SMOW), with D/H ratios of 155.76×10^6 (± 0.10) or SLAP (standard light antarctic precipitation) with a D/H ratio of $89.01 \pm 0.05 \times 10^6$ (Gonfiantini, 1978; see also Ehleringer and Rundel, 1988, for a discussion of units and standards).

It is important to note that water extracted from trees or other plants that have the capacity to store water is likely to be a time-averaged δD (White, 1988). Time-averaged δD can arise because some plants often have roots distributed throughout the soil and have access to more than one

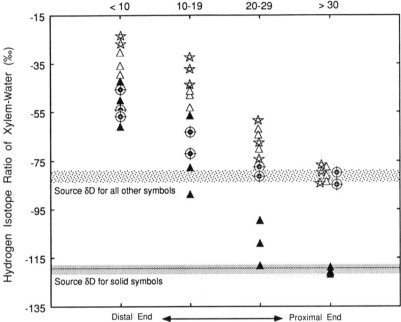

Age of the Stem Tissue (days)

Figure 4. The hydrogen isotope ratio of xylem water from four different North American tree species as a function of the stem age and position. Distal stem samples were the youngest tissues (often "green"). Stem samples toward the proximal end are the samples where the first complete bark layer was formed. The closed triangles are for *Acer negundo*. The open stars are for *Fraxinus pennsylvanica*, open triangles for *Acer saccharum*, and open circles with crosshairs for *Tsuga canadensis*. The source water for *A. negundo* was −121‰ and for the other species −81‰. The data show that significant isotopic enrichment can occur in stems that are not fully suberized, such as those without a well-developed bark layer (from Dawson and Ehleringer, 1993). Variation in source water is shown by the bars across the figure.

water source or have the capacity to store very different types of environmental waters for long periods (e.g., Cacti). Where water storage within the plant is possible, an intensive sampling procedure of both source and plant water is necessary to quantify the range of variation and to minimize errors in estimating which source(s) of environmental water a plant uses (for trees see White *et al.*, 1985, and Eqs. (2)–(4), below). When plants use or store more than one water source it is possible to determine the fractional input of each source by applying a simple two-ended linear mixing model (after White *et al.*, 1985; also see Dawson, 1993). This model can incorporate the fractional input of various water sources—rain, runoff (from snowmelt, streamwater, or other lateral flows), and groundwater (if plants have access to it), and thus determine the contribution of each source to the δD of the extractable plant water. For example, if a plant is believed to derive its water (stored in xylem sap) from both groundwater and summer precipitation that comes as a single monsoonal rain storm, the model can be written as

$$\delta D_{sap} = [X \cdot \delta D_{GW} + (1 - X) \cdot \delta D_R] \cdot (1 - t/d) + t/d \cdot \delta D_{GW}, \qquad (2)$$

where X = initial fractional input of groundwater relative to rain, δD_{GW} = the δD value of the groundwater, δD_R = the δD value of the rain, d = the decay time, in days, for rain in the plant water, t = time, in days after the rain storm event. The quantity that appears in the brackets, $X \cdot \delta D_{GW} + (1 - X) \cdot \delta D_R$, is the isotopic signal obtained from the xylem sap 1 day after the rainfall (the initial mixture of groundwater and rain). The δD of the initial mixture of rain and groundwater is equivalent to the δD of the plant water at time $t = 0$. At time $t = 1$, the δD of the plant water is the same as that of the groundwater. If plants do not have access to groundwater, δD_{GW} may be replaced by the δD of runoff, stream, or soil water. The parameter X is a function of the site hydrology and is determined by sampling the plant's water at daily intervals from the time the rain falls to the time when the plant water comes back to the δD value of the groundwater, runoff, or soil water; X is the slope of this relationship (see White *et al.*, 1985, p. 242). The value d can be determined by knowing the number of days it takes for the δD of the plant water containing the rain signal to decay back to a δD value prior to the storm. The first appearance of the rain signal takes 15–36 h for trees, less time for other woody plants (White *et al.*, 1985; Dawson, 1993), and between 40 min and 8 h for herbaceous monocots and dicots (Table I; Dawson, 1993; see also III, B below).

The model shown in Eq. (2) can be expanded to accommodate two or more rainfall events that would influence the δD of the plant water,

$$\delta D_{sap} = [X_2 \cdot \delta D_{GW} + (1 - X_2) \cdot \delta D_R(\text{rain } 2)] \cdot (1 - t_2/d)$$
$$+ \delta D_{sap}(\text{rain } 1) \cdot t_2/d, \qquad (3)$$

where $\delta D_{sap}(\text{rain } 1)$ is the sap δD value from the first rainfall (from Eq. (1)), X_2 is the X value from the second rainfall, t_2 is the time, in days, after the second rainfall event, or Day 0 after rainfall event number 2 (White *et al.*, 1985). Additional rainfall events can be incorporated in the same way.

The average growing season sap δD ($\delta D'_{sap}$) can be calculated by integrating source-water inputs for an entire growing season as

$$\delta D'_{sap} = f \cdot 1/2 \cdot (1 - X') \cdot (\delta D'_R - \delta D'_{GW}) - \delta D'_{GW}, \qquad (4)$$

where f = the fraction of the growing season, in days, when rainfall contributes to the tree sap δD, $\delta D'_{GW}$ = the average δD value of the groundwater used in that growing season, $\delta D'_R$ = the average δD value of the rain used in that growing season, X' = the initial fractional input of groundwater. Equation (4) is essentially an integration of Eq. (1) over the entire water-use season (White *et al.*, 1985). It does not require that the timing of

Table I The Time It Takes Water of a
Known Hydrogen Isotopic Composition
to Appear in the Xylem Water[a]

Growth form	Time
Trees	
Hardwoods (5)[b]	15–30 h
Conifers (2)[c]	18–36 h
Herbs (5)[d]	1–8 h
Shrubs (2)[e]	5–11 h
Monocots (3)[f]	40 min to 2.5 h

[a]Plants differing in growth form were used. Numbers in parentheses are the number of species sampled.

[b]*Acer negundo, A. grandidentatum, A. saccharum, Betula lutea, Tilia heterophylla* (after Dawson, 1993 and unpublished data).

[c]*Pinus strobus, Taxodium distichum* (White *et al.*, 1985).

[d]*Podophyllum peltatum, Asarum canadense, Fragaria virginiana, Thalictrum dioicum, Solidago flexicaulis* (Dawson, 1993).

[e]*Vaccinium vaccillans, Lindera benzoin* (Dawson, 1993).

[f]*Smilacina racemosa, Trillium grandiflorum, Holcus lanatus* (Dawson, 1993).

each rainfall event be known, such as in Eqs. (2) or (3), but does require that $\delta D'_{GW}$, $\delta D'_R$, and X' are well known. A much simpler mixing model than the formulation shown above has recently been derived and applied to the case of water movement between deep and shallow soil layers and between deep-rooted plants and their shallow-rooted neighbors (see Dawson, 1993). Results from the application of this model are discussed in Section III,B,5. Last, it is important to note that the mixing model (Eq. (4)) outlined above will be limited to cases where two distinct end members (δD signatures) exist. In situations where this is rare or does not occur, the use of soil water potential and δD–$\delta^{18}O$ ratios can be useful in partitioning the input of different water sources used by plants (Thorburn *et al.*, 1992, and Chapter 32, this volume).

B. Case Studies

Several recent investigations have ascertained the patterns of water use in plants and their influence on local distribution and community coexistence using the type of D/H analyses discussed above. I briefly summarize some of these results below and then end this section by showing the results from a recent study that have provided some unexpected insights relevant to studies of plant–plant interactions.

1. Forest Communities In one of the very first studies of differential utilization of water sources by trees, White *et al.* (1985) were able to demonstrate that an analysis of δD in sapwood xylem water showed that trees (*Taxodium distichum*) that inhabited a swamp site in Arkansas did not respond to summer precipitation because their roots were below the shallow water table and the amount of precipitation input was too small to affect the δD value of the groundwater they were using exclusively (Fig. 5). However, in upland dry sites in New York State, white pine (*Pinus strobus*) used almost entirely rainfall for 5 days after a storm; on Day 6 these trees began to draw on water from heartwood that differed significantly in hydrogen isotopic composition from sapwood (Fig. 5). The authors conjectured that the δD of heartwood water was a time-averaged signal from numerous precipitation inputs and that there was no evidence that the dry site trees used any groundwater (see also White, 1988). White pines growing on wet sites had xylem-sap δD values between rainwater and groundwater δD, indicating that trees were using both water sources. Again, after 5 or 6 days

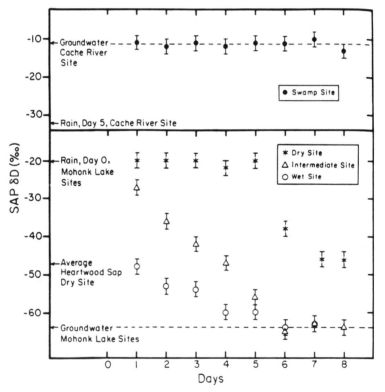

Figure 5. The D/H ratio (δD) of xylem water (sap) extracted from sapwood of two coniferous trees as a function of time after a rainfall of known δD. In the upper panel are data for bald cypress from a swamp site in Arkansas; the rainfall δD at this site was −32‰. In the lower panel are data for Eastern white pine from three sites in New York State; the rainfall δD at these sites was −20‰. The data show that bald cypress used only groundwater, while white pine in dry sites uses only precipitation. White pine growing in the intermediate and wet sites used a combination of water sources (from White *et al.*, 1985; reproduced with permission).

the xylem-sap δD value for wet site trees was nearly the same as the groundwater, providing an estimate of the period over which rain was used by these trees. Using the mixing models outlined above, white pine in dry upland forest sites used between 20 and 32% of the rainfall in dry and wet summers, respectively; only 10% of precipitation inputs were used in dry summers and 16% in wet summers for trees growing in wet sites (White, 1988).

2. Riparian Communities   In the semiarid Intermountain West of North America, the distribution of plant taxa is commonly correlated with steep soil moisture gradients. From the riparian (streamside plant communities) zones within deep canyons to the adjacent ridgecrests there is a marked change in the dominant vegetation cover (West, 1988). The broad-leaved deciduous trees (*Acer, Populus, Betula, Faxinus*) that characterize riparian communities quickly give way to perennial xeromorphic shrubs and trees (*Quercus*) that cooccur with annual and perennial herbs and forbs of xeric site grasslands (see Smedley *et al.,* 1991). Plants that inhabit riparian zones have access to three potential water sources: soil water and stream water, supplied by precipitation and runoff, and groundwater. The hydrogen isotopic composition of stream and soil water was monitored for a period of 2 years (Fig. 6; Dawson and Ehleringer, 1991) and varied between −200‰ (winter snow) and −20‰ (summer rainfall) whereas stream water was constant at ca −121‰. When trees of different size classes were sampled to

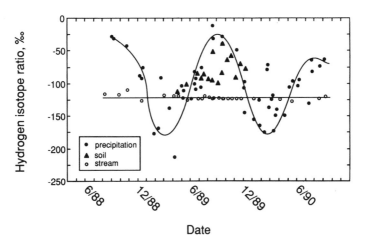

Figure 6.   The hydrogen isotopic composition (‰) of precipitation (●), soil water (at −50 cm; ▲), and surface stream water (○) measured over a 2-year period in Red Butte Canyon Research Natural Area, east of Salt Lake City, Utah. Precipitation values vary with the seasonal temperature regime. Soil water values reflect precipitation inputs. Values for the stream water were essentially unaffected by seasonal variation due to the fact that the catchment basin that contributes to stream runoff is very large, spanning over 350 m in elevation (from Dawson and Ehleringer, 1991; reprinted by permission from *Nature,* **350:** 335–337, copyright © 1991 Macmillan Magazines Ltd.).

determine which water source(s) they were using, Dawson and Ehleringer (1991) discovered that the small trees (<20 cm in circumference) growing in adjacent nonstreamside habitats were using water with a δD most like the soil water in that site (*Quercus gambelli*; Fig. 7). Small trees adjacent to the stream used stream water in (*Acer negundo*), or near (some *A. negundo* and *A. grandidentatum* > 8 cm in circumference; Fig. 7), to where they were growing. Unexpectedly, Dawson and Ehleringer (1991) found that all larger trees, regardless of whether they were growing in a streamside or nonstreamside location, had a δD identical to groundwater (−132‰) and not like the δD of the stream water which they were either growing in or near (Fig. 7). These data indicated that for larger trees inhabiting these streamside habitats, the most active sites of water absorption were limited to deeper strata (Fig. 8). Why might this occur? Long-term weather records for this region show that flow in these canyon streams can be quite low in

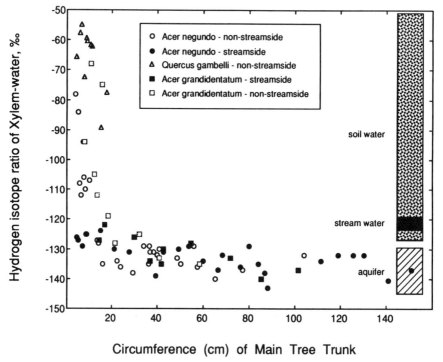

Figure 7. The hydrogen isotope ratio (‰) obtained from xylem water (sap) of three common streamside (closed symbols) and adjacent nonstreamside (open symbols) tree species in the Red Butte Canyon Research Natural Area in 1989 as a function of tree size (circumference in centimeters of the main trunk). Mean δD values of stream water (from Fig. 6) and local well (ground) water were −121.4 ± 0.7‰ and −132.3 ± 2.6‰ respectively; the range of values for each of three water sources is shown by the bars on the right hand side of the figure (from Dawson and Ehleringer, 1991; reprinted by permission from *Nature* 350: 335–337, copyright © 1991 Macmillan Magazines Ltd.).

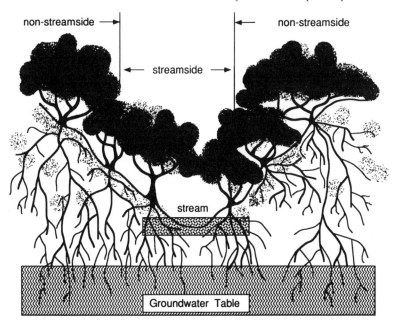

Figure 8. Diagram of the soil profile, tree-rooting patterns, and water sources one would expect to see for the data shown in Fig. 7. As shown, some large trees in both streamside and nonstreamside sites use the groundwater (after Dawson and Ehleringer, 1991; Flanagan *et al.*, 1992).

some years (Dawson and Ehleringer, unpublished) and stream channels can change with major flood events. Thus, the most stable water source in the region is the groundwater aquifer. If trees are to survive intermittent drought, it would be selectively advantageous to use groundwater. Thus, the finding that streamside trees were not using stream water is not surprising given a longer-term view of site hydrology.

These findings have several important implications. First, they demonstrate that the assumption of proximity to a particular water source implies this is where plants derive their water can be false. Second, they show that if groundwater is available, it can enable drought-intolerant species to potentially extend their range. Third, they have direct implications for stream management strategies. For example, a recent study conducted by Smith and co-workers (1991) showed that stream flow diversion can significantly impact plant water use and water relations leading to reduced leaf area and overall productivity of the vegetation. The dominant trees (*Betula, Populus, Salix*) in the Eastern Sierra riparian community they studied demonstrated a progressive shift during the growing season from using soil water to using groundwater (Smith *et al.*, 1991). However, all trees experienced greater water stress, and juvenile plant mortality was greatest in the diverted stream reaches. Though subsurface water utilized by the riparian trees can provide a buffer against severe water stress it may not be sufficient to meet peak transpirational demands. Stream flow diversion will not permit acquifer recharge and eventually not only cause greater juvenile plant mortality but dieback of mature trees (Smith *et al.*, 1991). Application of the δD method like that used by Smith *et al.* (1991) could provide watershed managers with an important way to assess riparian plant-water sources so they can determine if stream flow diversion will influence the ecophysiology of these species and potentially impact the health and stability of a riparian community.

3. Desert Communities   In all deserts plant survival, productivity, and fecundity are limited first and foremost by water (Ehleringer, 1985; Smith and Nowak, 1990). Because water is limited in deserts, the phenology of growth and reproduction of desert plants are inexorably linked to the annual hydrologic cycle (MacMahon and Schimpf, 1981). Soil water is recharged by precipitation from both winter storms and summertime monsoonal fronts (Caldwell, 1985), each having a distinct hydrogen isotopic composition (Winograd *et al.*, 1985; Milne *et al.*, 1987; Ingraham and Taylor, 1991). Several studies recently conducted in Utah have used the variation in hydrogen isotopic composition of precipitation, deep soil water, or groundwater to investigate source-water utilization by a variety of desert plants.

Within a desert scrub community near the Utah–Arizona border, Ehleringer *et al.* (1991) found that the dominant plants used a water source derived from winter storm recharge for their early spring growth (Ehleringer *et al.*, 1991). As the growing season progressed however and this water source was lost or depleted from the upper soil layers, only the deep-rooted species appeared to rely upon this source entirely, while herbaceous and woody perennial taxa utilized a mixture of both summer precipitation and the deep soil water remaining from winter recharge. In contrast, the authors found that annual plants and the succulent perennial, *Yucca angustissima*, used only summer rains (Fig. 9; Ehleringer *et al.*, 1991). Similarly, in a desert Pinyon–Juniper woodland community, Flanagan and Ehleringer (1991b) investigated water-use patterns in four dominant species, two shrubs and two trees. They demonstrated that the species with greater lateral root distributions (*Pinus edulis* and *Artemisia tridentata*, a tree and shrub, respectively) had greater uptake of recent summer precipitation than the two species with deeper roots (*Juniperus osteosperma*, a tree, and *Chrysothamnus nauseosus*, a shrub) that appeared to rely almost entirely on groundwater (Flanagan and Ehleringer, 1991b). Using the mixing model outlined above, these authors estimated that 23 and 30% of the stem water in *P. edulis* and *A. tridentata*, respectively, came from precipitation. Desert annuals and *Yucca* used precipitation exclusively, while woody species and herbaceous perennials used 57 and 91% of summer precipitation, respectively (Ehleringer *et al.*, 1991; Flanagan and Ehleringer, 1991b).

Both the above investigations and studies conducted with juvenile *A. negundo* (in the mountain canyons of Utah) and mature and juvenile *Acer*

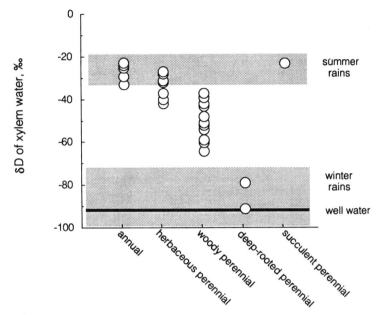

Figure 9.   Hydrogen isotope ratios (δD) of xylem water (‰) during the summer from an array of plants in a desert scrub community in southern Utah. The mean winter precipitation δD value was −88.3‰, well water was −91‰, summer precipitation values ranged between −22 and −80‰ (from Ehleringer *et al.*, 1991; reproduced with permission).

saccharum (sugar maple; in central New York) have extended the results from the source-water and plant-water δD analyses to show a strong correlation between the water source of plant uses and the degree of both plant water stress and water-use efficiency as measured by leaf carbon isotope composition (Fig. 10; Flanagan *et al.*, 1992). For all these studies, plants that relied more heavily on summer precipitation had significantly lower leaf water potentials and experienced greater overall water stress (Fig. 10). Greater utilization of (and perhaps dependence on) summer precipitation correlates with a more conservative water-use pattern (e.g., greater water-use efficiency) for all these species. This is apparent from both leaf gas exchange and leaf carbon isotope discrimination values (Fig. 10). Plants that appeared to use groundwater (*C. nauseosus* and mature *A. saccharum*) were much less water stressed and less water-use efficient relative to the other plants in the same community, presumably because water is rarely ever limiting for these deeply rooted plants (Fig. 10).

Results from these studies have at least two important ramifications. First, differential utilization of source waters will significantly impact coexistence, competition, and community composition. While woody perennials of the desert scrub community and *J. osteosperma* and *C. nauseosus* in the Pinyon–Juniper woodland may have a distinct competitive advantage during prolonged drought because they are deeply rooted and use ground-

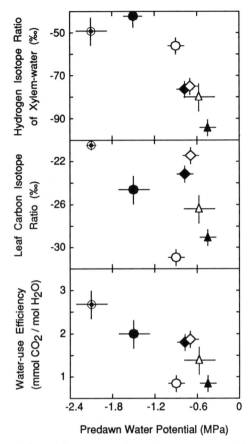

Figure 10. The relationship between predawn plant-water potential, xylem-water δD, leaf carbon isotope composition, and water-use efficiency. The plants, from Flanagan *et al* (1992), are *Chrysothamnus nauseosus* (▲), *Artemisia tridentata* (△), *Juniperous osteosperma* (◆), *Pinus edulus* (○), with the precipitation δD value at −68‰ and the groundwater δD value at −96‰. The plants, from Dawson (unpublished), are *Acer negundo*, juvenile (◇), with a soil water δD value at −48‰, and *A. saccharum*, juvenile (●) and adult (⊕), with the precipitation δD value at −39‰ and groundwater at −56‰. Values are means (*n* = 4 or 5) ± 95% confidence intervals. All plants were significantly different from each other (*P* < 0.05).

water, during rainy summer periods (e.g., monsoons or El Niño) herbaceous perennials will gain a competitive advantage because they have a large proportion of their root biomass near the soil surface (Fonteyn and Mahall, 1978; Robberecht *et al.*, 1983) and can utilize this rainwater source more effectively (Ehleringer *et al.*, 1991). Second, and on a larger scale, global circulation models predict that summer precipitation in desert regions will increase as atmospheric CO_2 increases and global mean temperatures rise (Schlesinger and Mitchell, 1987). A shift toward more summertime precipitation will favor perennial species with roots distributed throughout the entire soil profile over the more deeply rooted woody perennials (Eissenstat and Caldwell, 1988; Ehleringer *et al.*, 1991). Furthermore, a seasonal redistribution of precipitation and use by plants will influence competitive interactions, not only among coexisting taxa that already inhabit these desert communities (Fowler, 1986; Ehleringer *et al.*, 1991), but perhaps also from plants that migrate into the region from other communities that also possess the ability to effectively utilize summertime precipitation (Geber and Dawson, 1993).

4. Coastal Communities Coastal plant communities throughout the world provide ample opportunity to study plant distribution in relation to salinity gradients (Ranwell, 1972). Changes in salinity occur along seawater-to-freshwater gradients at the convergence zone between the intertidal and terrestrial habitats. In the Florida Keys, the interaction between saline waters and runoff derived from precipitation creates a mixture of source waters which influence the development of a vegetation mosaic (Sternberg *et al.*, 1991). Within this mosaic, plants are distributed according to their salinity tolerance and water-use patterns (Sternberg and Swart, 1987; Sternberg *et al.*, 1991). Fresh water is depleted of D relative to seawater (Sternberg and Swart, 1987) and recent studies have shown that different plant species use these water sources differentially (Sternberg and Swart, 1987; Sternberg *et al.*, 1991; Fig. 11). Tolerance to low water potentials (high salinity) and the use of different source waters clearly influence where particular plant species can grow within a mosaic of seawater and fresh water. Tropical and subtropical hardwood species (e.g., *Coccoloba, Bursera, Eugenia, Ficus, Psychotria*) are confined to woodland and hammock habitats (furthest from the ocean) where they use water derived largely from precipitation runoff (Fig. 11). In contrast, salt-tolerant species (e.g., *Salicornia, Cenchrus, Conocarpus*) use mostly seawater, while mangroves (e.g., *Avicennia germinatus* and *Rhizophora mangle*), which are found growing throughout the entire vegetation mosaic, use all water sources (Fig. 11; Sternberg and Swart, 1987; Sternberg *et al.*, 1991). These results that species are partitioned within the vegetation along salinity and D/H gradients (but see Lin, Chapter 31, this volume). The relative use of ocean or fresh water as well as the very different physiological tolerances of taxa to salinity-induced water stress leads to the diverse vegetation mosaic. Because plant distributional patterns are so closely linked to water sources and plant-water relations, changes in the ratio of seawater to fresh water caused by natural disasters such as drought or by runoff diversions and human consumption could have a marked and significant impact on coastal plant community structure and diversity.

There is evidence that fog that forms along coastal shores has a hydrogen istopic ratio that is quite different from precipitation that falls in these same regions (Ingraham and Matthews, 1988, 1990). If the number of foggy days is high, as is the case in many coastal regions, and the interception of fog by vegetation or other surfaces allows this water source to enter the soil or groundwater, then it is likely that fog could constitute an impor-

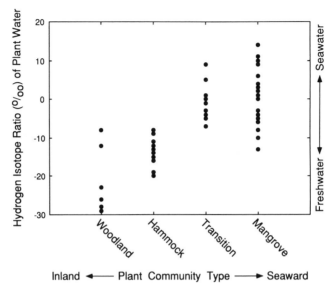

Figure 11. The hydrogen isotope ratio of plant water from four different coastal plant communities in southern Florida as a function of their proximity to seawater (see text for the types of species; drawn from data in Sternberg *et al.*, 1991, and Sternberg and Swart, 1987).

tant source of water for plants (Griffiths, 1991). In fact, ecological studies of the Coast Redwood (*Sequoia sempervirens*) in California have suggested that as much as 25 cm of water is added to the soil as a result of fog-drip off the foliage of this tree (~30% of the annual water input; Ornduff, 1974; Byers, 1953) and that fog is an important factor in limiting this species distribution. Other studies have shown that "fog-collectors" can catch up to 70% more water out of the environment that a rain gauge (Oberlander, 1956; Parsons, 1960; Kerfoot, 1968; Vogelman *et al.*, 1968; Azevedo and Morgan, 1974); many of these studies were done in coastal regions within an otherwise arid or semiarid climate. Future studies in coastal or other moisture-laden environments should account for fog as an important water source influencing plant-water balance and distribution. In fact, Grubb and Whitmore (1966) suggested that the distribution of tropical montane "cloud-forests" in Ecuador may be limited by the presence of "coastal fogs" that move inland and hang against the slopes. It seems likely that if fog-drip can constitute a large fraction of the seasonal water input, then plants, and especially shallow-rooted species, would use a significant proportion of this water source.

5. Plant–Plant Interactions The results discussed above all suggest that across a variety of plant community types the utilization of very different water sources in time and space can have an important influence on competitive interactions among cooccurring species. Plants that are limited to using the same water sources may at some point be competing for that resource (though there is very little direct evidence indicating that plants do compete for water; see Keddy, 1989). As discussed by Ehleringer *et al.* (1991; Fig. 9), those species that can use both precipitation and groundwater may have greater ecological amplitudes with respect to water and as such fair quite well when competing with a variety of coexisting taxa. From analyses of source- and plant-water hydrogen isotopic composition (indices of rooting patterns), standard leaf gas exchange measurements, and leaf carbon isotope discrimination (indices of water use; see Fig. 10), we can assess the potential competitive ability of a plant. Yet to understand the degree of competitive plasticity among all the members of a particular plant community, an analysis of all water sources and their proportional use by each species must be assessed. To accomplish this, the mixing models such as those outlined above can be used. However, these models must be refined so that we can assess the dynamics of water movement belowground. Then the extent and dynamics to which different water sources are used by different species can be assessed (see also Ehleringer and Dawson, 1992).

In a recent set of investigations with the shrub *A. tridentata,* some important insights on the dynamics of water movement below ground have been gained (Richards and Caldwell, 1987; Caldwell and Richards, 1989). Using deuterated water, these authors demonstrated that soil water absorbed at night by this deeply rooted shrub is released into the upper soil layers; the process is believed to be driven by water potential gradients and has been termed hydraulic lift (Richards and Caldwell, 1987; Caldwell and Richards, 1989). Water hydraulically lifted from deep in the soil strata and deposited in the upper soil layers appears to have two functions for *Artemisia.* It aids the shrub in meeting its daily transpirational demands and prolongs root activities (e.g., nutrient absorption) into the long summer drough where surface water is limiting. This occurs however at a cost; some of this water is lost to evaporation or parasitized by neighboring plants such as the grass, *Agropyron deseretorum* (Caldwell and Richards, 1989). The data gathered in these studies provided some exciting new information about the dynamics of water movement in soils as facilitated by deeply rooted plants. Moreover, these studies suggested a potential way to improve our understanding about competition for water in plants using isotopes. However, the δD method I believe is more powerful (and not radioactive) and can be applied so that one can calculate the fraction of hydraulically lifted water used (parasitized) by shallow-rooted neighbors (see also Griffiths, 1991).

For example, during the prolonged summer drought of 1991 in central New York State, one could observe marked differences in the degree of midday wilting that many shallow-rooted herbs, grasses, shrubs, and young (2- to 4-year-old) trees showed. The degree of wilting seemed also to be associated with a plant's distance from large, open-grown trees. Could it be that these plants were using different proportions of water hydraulically lifted by these large trees? To address this question, I used δD analyses of groundwater, soil water, and plant water in conjunction with measurements of plant and soil water potentials and leaf gas exchange. The proportion of water, hydraulically lifted by single, open-grown trees, that was used by other species growing at progressively greater distances from it was also calculated (Dawson, 1993; Fig. 12). Results from an analysis of soil water potentials and δD, as well as from groundwater and xylem-water analyses show that groundwater was in fact hydraulically lifted to upper soil layers by sugar maple, *A. saccharum* (7–40 cm below the surface). More importantly though, there was a marked gradient in the proportion of water in the soil samples that came from the groundwater source; beyond a distance of 2.5 m the δD of the soil water was nearly identical to the δD of precipitation that fell 3 weeks earlier (Fig. 13). An analysis of δD in plant water from several cooccurring species revealed that different species used different proportions of hydraulically lifted water (Dawson, 1993); a subset of these data are shown in Table II. Herbaceous species, represented here by *Fragaria virginiana* (wild strawberry), obtained greater than 50% (46–61%; Dawson, 1993) of their water by using that hydraulically lifted by *A. saccharum*. The shrub, *Vaccinium vacillans* (low-bush blueberry), the perennial herb, *Solidago flexicaulis* (goldenrod), and the grass *Holcus lanatus* obtained between 19 and 25% of their water in this same way. Small beech trees (*Fagus grandifolia*) used the least amount of hydraulically lifted water; only about 7% (Dawson, 1993; Table II). Herbaceous species that used hydraulically lifted water generally had higher leaf-water potentials and leaf conductances than either those that did not have access to this water source (plants beyond 2.5 m from the source tree) or used only a very small proportion of groundwater (Table II). Those species that appeared to be

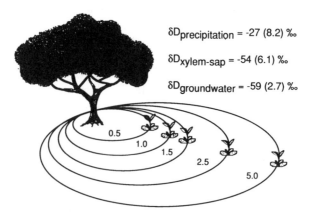

$\delta D_{precipitation}$ = -27 (8.2) ‰

$\delta D_{xylem-sap}$ = -54 (6.1) ‰

$\delta D_{groundwater}$ = -59 (2.7) ‰

Figure 12.   Diagram of sampling zones used at various distances from the source tree (*Acer saccharum*) in a study of hydraulic lift (see text). The δD values shown are mean (SD) for $n = 5$.

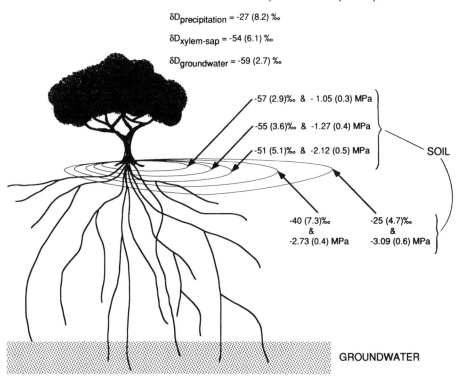

$\delta D_{precipitation} = -27\ (8.2)\ ‰$

$\delta D_{xylem\text{-}sap} = -54\ (6.1)\ ‰$

$\delta D_{groundwater} = -59\ (2.7)\ ‰$

-57 (2.9)‰ & - 1.05 (0.3) MPa

-55 (3.6)‰ & -1.27 (0.4) MPa

-51 (5.1)‰ & -2.12 (0.5) MPa

SOIL

-40 (7.3)‰
&
-2.73 (0.4) MPa

-25 (4.7)‰
&
-3.09 (0.6) MPa

GROUNDWATER

Figure 13.   The hydrogen isotopic composition of soil water (at -35 cm, ‰) and the soil water potential (MPa) at various distances away from the source tree (*Acer saccharum*; see Fig. 12). The δD values shown are means (SD) for $n = 5$ (modified from Dawson, 1993).

using hydraulically lifted water from *Acer* trees possessed the most favorable water balances as well (Table II), suggesting that they may be the best competitors of water (see Dawson, 1993).

These data have several important implications. During extended drought periods, the use of the hydraulically lifted water appears to be especially beneficial to plants in close proximity to the source tree. The data from these same species at 5 m from the source tree suggest that they were water stressed and did not have access to hydraulically lifted water. For several herbaceous species, hydraulically lifted water can compose a very high proportion of its total xylem water. Having access to this water source clearly improves the plant's water status relative to its closest neighbors or to plants more distant from a source tree. Close proximity to neighbors is usually thought to be detrimental because competition for water or other belowground resources is more intense (Richards and Caldwell, 1987). While this may generally be true, during periodic or chronic drought, the above results suggest that herbaceous plants that grow in close proximity to plants that can hydraulically lift groundwater to the upper soil layers directly benefit from it (Dawson, 1993).

Table II   Leaf Water Potential (ψ; MPa), Stomatal Conductance to Water Vapor (g; mol m^{-2} s^{-1}), Hydrogen Isotopic Composition of Xylem Water (δD; ‰), and the Proportion of Water Derived from a Source Tree (%) Measured in Plants at Five Distances from That Tree[a]

		Distance from source tree (*Acer saccharum*)				
Species		0.5 m	1.0 m	1.5 m	2.5 m	5.0 m
Fragaria virgi-	ψ	−1.0(.4)	−0.7(.6)	−1.7(.7)	−1.6(.6)	−2.0(.3)
niana straw-	g	.66(.2)	.59(.1)	.52(.2)	.31(.09)	.17(.06)
berry	δD	−47(1)	−43(3)	−41(4)	−32(4)	−28(2)
	%	58(3)	54(6)	50(5)	13(2)	1(1)
Solidago flexicaulis	ψ	−0.8(.5)	−1.1(.6)	−1.7(.6)	−2.3(.8)	−2.2(.5)
goldenrod	g	.57(.2)	.56(.1)	.58(.1)	.24(.1)	.20(.2)
	δD	−34(4)	−34(2)	−29(3)	−25(1)	−21(3)
	%	20(3)	19(3)	6(2)	0	0
Vaccinium vacil-	ψ	−0.9(.4)	−1.3(.5)	−1.9(.6)	−2.6(.6)	−2.7(.4)
lans shrub	g	.41(.06)	.45(.1)	.31(.2)	.12(.09)	.09(.04)
	δD	−33(3)	−30(1)	−28(4)	−27(2)	−20(3)
	%	19(4)	10(3)	5(2)	1(1)	0
Holcus lanatus	ψ	−1.1(.3)	−1.5(.4)	−1.3(.6)	−1.9(.5)	−1.9(.8)
grass	g	.37(.08)	.30(.1)	.30(.2)	.12(.07)	.09(.08)
	δD	−34(2)	−29(5)	−26(3)	−26(1)	−22(4)
	%	21(3)	7(3)	0	0	0
Fagus grandifolia	ψ	−1.7(.5)	−1.4(.9)	−1.9(.3)	−2.2(.6)	−2.0(.4)
(2 years old)	g	.25(.07)	.26(.9)	.26(.03)	.20(.09)	.14(.06)
beech	δD	−29(3)	−28(2)	−27(1)	−24(2)	−22(2)
	%	1(3)	0	0	0	0

[a]Values are means (±SD) for $n = 3$ per species. δD for precipitation, xylem sap from the source tree, *Acer saccharum*, and groundwater were −33 (8.2), −54 (6.1), and −59 (2.7)‰ respectively.
Source: data from Dawson (1993).

Last, these results also have implications for (1) studies concerned with the factors that limit plant distributions, (2) investigations of drought adaptations and hydraulic properties in plants, and (3) studies that intend to link or model plant water use and water relations to landscape level evapotranspirational and hydrologic processes (Jarvis and McNaughton, 1986; Richards and Caldwell, 1987; Caldwell and Richards, 1989; Whitehead and Hinckley, 1991; Dawson, 1993).

IV. Long-Term Studies of δD Variation in Source and Plant Waters

A. Background and Methods

The above discussion is relevant to determining variation in environmental and plant-water sources for any single point in time or any single growing season. However, it may be desirable to obtain a longer-term record of the patterns of source water variation and plant-water use (Ehleringer and

Dawson, 1992; Dawson and Ehleringer, manuscript). The analysis of δD and δ¹⁸O in tree rings as determined from cellulose nitrate (see below) can provide such a record.

Linking the isotopic composition of complex carbohydrates to the water sources plants are using poses one of the greatest challenges in isotopic studies involving plants and water. It is a challenge because it requires that we understand, first, the fractionation processes during photosynthetic carbohydrate synthesis, processes which involve hydrogen and oxygen but which we know moderately well (Yakir, 1992). Second, and perhaps most important, it requires that we understand the possible post-photosynthetic exchange of hydrogen and/or oxygen isotopes that can and do occur between the carbohydrates and other water pools in the plant (e.g., stems, storage tissues). It is this second requirement that is poorly understood.

Early research demonstrated that there was a linear relationship between the δD in the nonexchangeable hydrogens of cellulose in tree rings and the δD of the principle water source taken up by roots (Fig. 14; Epstein *et al.*, 1976; Yapp and Epstein, 1982a,b; Ramesh *et al.*, 1986). The mechanism underlying this relationship, however, remains unknown (Sternberg, 1988; White, 1988). In fact, it is surprising that the relationship is upheld in all of the studies that have been done to date on linking water sources to tree rings when we know that the hydrogen isotopic signature of carbohydrates can vary 150+‰ either heavier or lighter from when they are first synthesized in the leaf to when they finally become the cellulose of tree rings

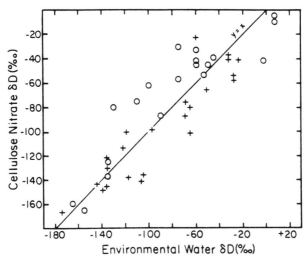

Figure 14. The relationship between the δD value of cellulose nitrate from plants and the δD value of the associated environmental water. Most plants were trees growing throughout North America and the cellulose nitrate values were derived from tree rings. The open circles represent values where the environmental water was a literature value (from Epstein *et al.*, 1976, and Yapp and Epstein, 1982a, from White, 1988; reproduced with permission).

(Yakir, 1992; Dawson and Ehleringer, manuscript). Yakir (1992) has suggested that the very high degree of isotopic exchange during post-photosynthetic metabolism may in fact be the very reason that the good correlation (Yapp and Epstein, 1982a,b) between the δD of tree ring cellulose (more precisely, cellulose nitrate—see below) and the δD of the principle water source a tree is using is upheld. That is, water in the tree trunk is constantly bathing the carbohydrates made in the leaves and isotopic exchange with this water source is what is finally synthesized into the wood. Therefore, despite the complexities of isotopic exchange that occur during carbohydrate metabolism, we should see, that as trees grow and incorporate hydrogen from different water sources into their xylem cellulose, any variation in D/H due to water source(s) should be recorded in the δD of tree rings. This is not true for oxygen isotopes since the O in cellulose can come from either water or carbon dioxide (DeNiro and Cooper, 1989) and the O derived from CO_2 shows isotopic exchange with water prior to cellulose formation (DeNiro and Epstein, 1979; Dawson, unpublished data). However, recent research has shown that the $^{18}O/^{16}O$ ($\delta^{18}O$) in plant organic matter is much more conservative than hydrogen with an isotope effect of +27 (± 5)‰ between the water source and the oxygens found in wood cellulose (Sternberg, 1988; Dawson and Ehleringer, 1993; Dawson, unpublished data; reviewed by Yakir, 1992).

In the past, δD values in tree rings have been used extensively to reconstruct past climates, including precipitation and temperature regimes, and there is an extensive literature on this (for example see White, 1988). Here, I would like to suggest that information on variation in δD found in tree rings can be used in studying more than past climates. They could also be used to study historical plant-water relations that ultimately have influenced patterns of growth and distribution.

For example, we can use the relationship shown in Fig. 14 to reconstruct the history of variation in source-water δD. Furthermore, if the sources of water that a tree used in the past can be determined (e.g., groundwater, precipitation, runoff), we can reconstruct a water-use history for a particular plant as well as past growing conditions (temperatures; see Chapters 33 and 6, respectively, by Yakir and Flanagan, this volume). To obtain a long-term record of water use, trees are cored at a standardized height (often at 1.4 m above the base) at the four cardinal compass directions. Each core must not intersect with any other. Each core must be aged and cross-dated (see Fritts, 1966) with the other cores from the same tree to ensure accurate age estimates and for subsequent isotopic analyses. Once this is accomplished, individual rings or groups of rings (2–5 years) are excised, put together in bulk, and ground to a fine mesh in preparation for the cellulose purification procedure. Before purification, if enough sample exists, a subsample should also be analyzed for $\delta^{13}C$ to test if changes in water sources coincided with changes in the seasonally integrated c_i/c_a (Farquhar *et al.*, 1989). To remove all but the cellulose-bound (nonexchangeable) hydrogens, the ground sample is first extracted in azeotropic mixtures of benzene and methanol in a soxhlet apparatus. This first extraction removes terpenes, fatty acids, alcohol-soluble lignin, and a small fraction of carbohydrate deposited in the wood; each of these fractions have exchangeable hydrogens that must be removed prior to isotopic analysis (Epstein *et al.,* 1976). Dissolved lignin is removed with repeated chlorine bleach washings. The resulting holocellulose is then treated with caustic base washings to separate the hemicellulose and α-cellulose. The purified α-cellulose is then nitrated (see Epstein *et al.,* 1976, and Yapp and Epstein, 1982a, for detailed methods) to remove the exchangeable hydroxyl groups and to form cellulose nitrate. The cellulose nitrate samples are repeatedly washed with cold water, methanol, and then hot water, dried, dissolved in acetone, filtered, and finally precipitated out of the solution with rapid cold water to acetone washings. The final product is then combusted in evacuated glass ampules (as above), the water removed, and then converted to diatomic hydrogen for analysis in a mass spectrometer using the zinc reduction technique described earlier.

B. A Case Study

As woody plants develop, there are often opportunities to exploit new resources as above- and belowground plant parts increase in size and gain access to new microsites. We might imagine that for many tree species, for the first few years of life, when plants are small, there is a relatively narrow zone from which soil resources can be obtained. Water, provided by precipitation as surface runoff, infiltrated soil water, or both are the primary water sources for young (<5-year-old) seedlings and saplings. In arid and semiarid environments as well as in xeric upland habitats, the abundance of these water sources are most important in limiting establishment and survival of plants. As the sapling develops and the root system grows, other sources of water, such as groundwater or other subsurface flows (e.g., springs) may become available. The availability of perennial water sources like groundwater provide opportunities for plants to switch their water use to these more stable sources of water. Shifts in source waters may be recorded in the δD of the nonexchangeable hydrogens in the cellulose of tree rings (White *et al.,* 1985).

We (Dawson and Ehleringer, manuscript) conducted an investigation where both young and mature trees of boxelder, *A. negundo* were cored and these cores analyzed for both ring widths (radial growth increment) and cellulose nitrate δD (CNδD). Our analyses demonstrated that radial growth during the first 20–25 years of life increased erratically (Fig. 15). However, as trees aged (>20 years old), growth appeared to stablize (Fig. 15). The analysis of CNδD in the tree rings showed values in the first 20–25 years of life that resembled δD values of summertime precipitation for this same study site (Dawson and Ehleringer, 1991; see also Fig. 6. *Note:* rings for every 3 years were bulked to provide sufficient cellulose for the analysis). Growth rings formed after trees had reached 25 years of age were larger and in general the CNδD was more similar to the δD of

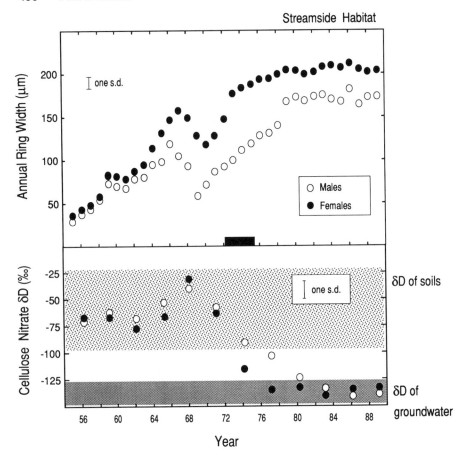

Figure 15. The annual growth ring width (μm) and the cellulose nitrate δD (‰) value measured from tree cores extracted from *Acer negundo* growing adjacent to a stream in Red Butte Canyon Research Natural Area, east of Salt Lake City, Utah. Trees sampled were growing within 3 m of a stream. Growing season precipitation δD values for this site ranged between −23‰ and −98‰, noted next to the stippled areas labeled "δD of soils." The groundwater δD value was −132 (±3.7)‰, noted next to the stippled area labeled "δD of groundwater." The short black bar on the x-axis of the upper panel notes the time when trees at this site reached and began using groundwater. Values are means (s.d.) for $n = 5$ trees per sex (Dawson and Ehleringer, in review). Note: female trees grow more and thus reach the groundwater sooner that do male trees.

groundwater (Dawson and Ehleringer, manuscript; Fig. 15). These findings suggest that when boxelder trees are young, they are restricted to using only surface water sources (e.g., precipitation or streamwater, depending upon which is the dominant source at that site). Years of little radial growth were correlated with years of lower rainfall and stream water runoff. After trees reach a particular size, their radial growth increases and this appears to be the point when they also reached groundwater (Dawson and Ehleringer, 1991; Dawson and Ehleringer, manuscript). These find-

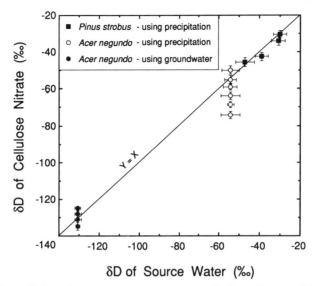

Figure 16. Cellulose nitrate δD versus source-water δD for white pine (*Pinus strobus;* ■, from White, 1983) using precipitation and boxelder (*Acer negundo*, from Dawson and Ehleringer, manuscript) using precipitation (○), and groundwater (●). Values are means (*n* = 5) ± 95% confidence intervals.

ings are similar to those reported by White (1983, pp. 200–258) for *P. strobus* and demonstrate that the cellulose nitrate δD in tree rings can indicate what water source a tree was using at the time radial growth was taking place (Fig. 16).

The information collected from *A. negundo* demonstrates that growth and water-use histories can be reconstructed using an analysis of δD and widths of tree rings. Oxygen isotopic composition is providing identical information (Dawson, unpublished data). Further analyses of this same cellulose for its carbon isotopic composition should provide information about the "physiological history" (e.g., long-term c_i/c_a) of these same trees (Dawson and Ehleringer, manuscript). The use of stable isotope analyses like this could also be applied in studies of landscape changes along watercourses that have meandered across large floodplains. Where rivers may have changed their course over time, this would be recorded in the δD, the carbon isotopic ratio, and the widths of tree rings in species that may have inhabited the floodplain for many years.

V. Conclusions

In this chapter I have shown where the application of hydrogen isotopic composition of source water, plant water, and the cellulose of tree rings can be used to analyze patterns of plant-water use from both a short-term and a long-term perspective and across a broad spectrum of community types and environments. In addition, some of the data shown demonstrate how the analysis of δD can be used in studies of plant–plant interactions. The application of these types of analyses in conjunction with other ecological and physiological research tools I believe holds tremendous promise for research in the future. Furthermore, using δD analyses of plant water and environmental water has great potential for assessing how environmental variation can impact plant-water use and water relations and could play an important role in impact assessment and management strategies of plant communities intimately linked to regional or local landscape hydrology. Last, the application of δD and other stable isotope analyses, reviewed in this volume, may ultimately allow information gathered at different scales to be linked and integrated so that information obtained at the leaf level or whole-plant level might be used to make meaningful predictions at the ecosystem level and vice versa.

Acknowledgments

This chapter grew out of research supported by the National Science Foundation, the United States Department of Agriculture, and Cornell University. Foremost, I thank my friend and colleague Jim Ehleringer for introducing me to the use of stable isotopes in ecophysiological research and for encouraging me to explore their application in new ways. For their comments on an earlier draft of this chapter, I also thank Jim, Graham Farquhar, Tony Hall, and Peter Thorburn; any errors that remain are mine. Lastly, I thank Jim, Graham, Craig Cook, Larry Flanagan, Myles Fisher, Monica Geber, Jill Gregg, Howard Griffiths, Peter Marks, John Marshall, Hester Parker, Jim Richards, and Peter Thorburn for sharing their thoughts and insights with me about many of the topics presented here.

References

Allison, G. B., and M. W. Hughes. 1983. The use of natural tracers as indicators of soil-water movement in a temperate semi-arid region. *J. Hydrol.* **60:** 157–173.

Azevedo, J., and D. L. Morgan. 1974. Fog precipitation in coastal California forests. *J. Ecol.* **55:** 1135–1141.

Barnes, C. J., and G. B. Allison. 1983. The distribution of deuterium and O^{18} in dry soils. *J. Hydrol.* **60:** 141–156.

Bigeleisen, J., M. L. Perlman, and H. C. Prosser. 1952. Conversion of hydrogen materials to hydrogen for isotopic analysis. *Anal. Chem.* **24:** 1356–1357.

Brunel, J. P., G. R. Walker, C. D. Walker, J. C. Dighton, and A. Kennett-Smith. 1991. Using stable isotopes of water to trace plant water uptake, pp. 543–551. *In* Stable Isotopes in Plant Nutrition, Soil Fertility and Environmental Studies: Proceedings of an International Symposium, October 1990. IAEA and FAO, Vienna.

Byers, H. R. 1953. Coastal redwoods and fog drip. *Ecology* **34:** 192–193.

Caldwell, M. M. 1985. Cold deserts, pp. 198–212. *In* B. F. Chabot and H. A. Mooney (eds.), Physiological Ecology of North American Plant Communities. Chapman & Hall, New York.

Caldwell, M. M., and J. R. Richards. 1989. Hydraulic lift: Water efflux from upper roots improves effectiveness of water uptake by deep roots. *Oecologia* **79:** 1–5.

Coleman, M. L., T. J. Shepard, J. J. Durham, J. E. Rouse, and G. R. Moore. 1982. Reduction of water with zinc for hydrogen isotope analysis. *Anal. Chem.* **54**: 993–995.

Craig, H. 1961. Isotopic variation in meteoric waters. *Science* **133**: 1702–1703.

Dansgaard, W. 1964. Stable isotopes in precipitation. *Tellus* **16**: 436–468.

Dawson, T. E. 1993. Hydraulic lift and water use by plants: Implications for performance, water balance and plant–plant interactions. *Oecologia*, in press.

Dawson, T. E., and J. R. Ehleringer. 1991. Streamside trees that do not use stream water. *Nature* **350**: 335–337.

Dawson, T. E., and J. R. Ehleringer. 1993. Isotopic enrichment of water in the "woody" tissues of plants: Implications for plant water source, water uptake, and other studies which use stable isotopic composition of cellulose. *Geochim. Cosmochim. Acta*, in press.

Dawson, T. E., and J. R. Ehleringer. A long-term record of riparian-tree water use and performance. Submitted for publication.

DeNiro, M. J., and S. Epstein. 1979. Relationship between the oxygen isotope ratios of terrestrial plant cellulose, carbon dioxide and water. *Science* **204**: 51–53.

DeNiro, M. J., and L. W. Cooper. 1989. Post-photosynthetic modification of oxygen isotope ratios of carbohydrates in the potato: Implications for paleoclimatic reconstruction based upon isotopic analysis of wood cellulose. *Geochim. Cosmochim. Acta* **53**: 2573–2580.

Ehleringer, J. R. 1985. Annuals and perennials of warm deserts, pp. 162–180. *In* B. F. Chabot and H. A. Mooney (eds.), Physiological Ecology of North American Plant Communities. Chapman & Hall, New York.

Ehleringer, J. R., and T. E. Dawson. 1992. Water uptake by plants: Perspectives from stable isotopes. *Plant Cell Environ.* **15**: 1073–1082.

Ehleringer, J. R., and C. B. Osmond. 1989. Stable Isotopes, pp. 281–300. *In* R. W. Pearcy, J. R. Ehleringer, H. A. Mooney, and P. W. Rundel (eds.), Plant Physiological Ecology: Field Methods and Instrumentation. Chapman & Hall, London.

Ehleringer, J. R., and P. W. Rundel. 1988. Stable isotopes: History, units, and instrumentation, pp. 1–15. *In* P. W. Rundel, J. R. Ehleringer, and K. A. Nagy (eds.), Ecological Studies. Vol. 68. Stable Isotopes in Ecological Research. Springer-Verlag, Heidelberg.

Ehleringer, J. R., S. L. Phillips, W. S. F. Schuster, and D. R. Sandquist. 1991. Differential utilization of summer rains by desert plants. *Oecologia* **88**: 43–434.

Eissenstat, D. M., and M. M. Caldwell. 1988. Competitive ability is linked to rates of water extraction: A field study of two aridland tussock grasses. *Oecologia* **75**: 1–7.

Epstein, S., C. J. Yapp, and J. H. Hall. 1976. The determination of the D/H ratio of non-exchangeable hydrogen in cellulose extracted from aquatic and land plants. *Earth Planet. Sci. Lett.* **30**: 241–251.

Farquhar, G. D., J. R. Ehleringer, and K. T. Hubick. 1989. Carbon isotope discrimination and photosynthesis. *Annu. Rev. Plant Physiol. Plant Mol. Biol.* **40**: 503–537.

Flanagan, L. B., and J. R. Ehleringer. 1991a. Effects of mild water stress and diurnal changes in temperature and humidity on the stable oxygen and hydrogen isotopic composition of leaf water in *Cornus stolonifera*. *Plant Physiol.* **97**: 298–305.

Flanagan, L. B., and J. R. Ehleringer. 1991b. Stable isotopic composition of stem and leaf water: Applications to the study of plant water use. *Funct. Ecol.* **5**: 270–277.

Flanagan, L. B., J. P. Comstock, and J. R. Ehleringer, 1991. Comparison of modeled and observed environmental influences in stable oxygen and hydrogen isotope composition of leaf water in *Phaseolus vulgaris*. *Plant Physiol.* **96**: 588–596.

Flanagan, L. B., J. R. Ehleringer, and T. E. Dawson. 1992. Water sources of plants growing in woodland, desert, and riparian communities: Evidence from stable isotopes, pp. 43–47. *In* Ecology and Management of Riparian Shrub Communities. USDA, Forest Service Symposium, General Report INT–289.

Friedman, I., and K. Harcastle. 1970. A new technique for pumping hydrogen gas. *Geochim. Cosmochim. Acta* **34**: 125–126.

Friedman, I., A. C. Redfield, A. Schoen, and J. Harris. 1964. The variation of deuterium content of natural waters in the hydrologic cycle. *Rev. Geophys.* **2**: 177–224.

Fontes, J-Ch. 1980. Environmental isotopes in groundwater hydrology, pp. 411–440. *In* P.

Fritz and J-Ch. Fontes (eds.), Handbook of Environmental Isotope Geochemistry. Elsevier, Amsterdam.

Fonteyn, P. J., and B. E. Mahall. 1978. Competition among desert perennials. *Nature* **275:** 544–545.

Fowler, N. 1986. The role of competition in plant communities in arid and semi-arid regions. *Annu. Rev. Ecol. Syst.* **17:** 89–110.

Fritts, H. C. 1966. Tree Rings and Climate. Academic Press, New York.

Gat, J. R. 1982. The isotopes of hydrogen and oxygen in precipitation, pp. 21–47. *In* J. Hoefs (ed.), Stable Isotope Geochemistry. Springer-Verlag, Heidelberg, Germany.

Geber, M. A., and T. E. Dawson. 1993. Evolutionary responses of plants to global climate change, pp. 179–197. *In* P. M. Karieva, J. G. Kingsolver, and R. B. Huey (eds.), Biotic Interactions and Global Change. Sinauer Associates, Sunderland, MA.

Gonfiantini, R. 1978. Standards for stable isotope measurements in natural compounds. *Nature* **271:** 534–536.

Griffiths, H. 1991. Application of stable isotope technology in physiological ecology. *Funct. Ecol.* **5:** 254–269.

Grubb, P. J., and T. C. Whitmore. 1966. A comparison of montane and lowland rain forests in Ecuador. *J. Ecol.* **54:** 303–333.

Hoefs, J. 1980. Stable Isotope Geochemistry. Springer-Verlag, Berlin.

Ingraham, N. L., and R. A. Matthews. 1988. Fog drip as a source of groundwater recharge in Northern Kenya. *Water Resources Res.* **24:** 1406–1410.

Ingraham, N. L., and R. A. Matthews. 1990. A stable isotopic study of fog: The Point Reyes Peninsula, California, U.S.A. *Chem. Geol.* (Isotope Geosci. Sect.) **80:** 281–290.

Ingraham, N. L., and B. E. Taylor. 1991. Light stable isotope systematics of large-scale hydrologic regimes in Calfornia and Nevada. *Water Resources Res.* **27:** 77–90.

Jarvis, P. G., and K. G. McNaughton. 1986. Stomatal control of transpiration: Scaling up from leaf to region. *Adv. Ecol. Res.* **15:** 1–49.

Keddy, P. A. 1989. Competition. Chapman & Hall, New York.

Kerfoot, O. 1968. Mist precipitation on vegetation. *For. Abstr.* **29:** 8–20.

Kramer, P. J. 1983. Water Relations of Plants. Academic Press, New York.

Lawrence, J. R., S. D. Gedzelman, J. W. C. White, D. Smiley, and P. Lazov. 1982. Storm trajectories in eastern US D/H isotopic composition of precipitation. *Nature* **296:** 638–640.

MacMahon, J. A., and D. J. Schimpf. 1981. Water as a factor in the biology of North American desert plants, pp. 114–171. *In* D. D. Evans, and J. L. Thames (eds.), Water in Desert Ecosystems. Dowden, Hutchinson & Ross, Stroudsberg, PA.

Milne, W. K., L. V. Benson, and P. W. McKinley. 1987. Isotope content and temperature of precipitation in southern Nevada, August 1983–August 1986, pp. 1–38. *In* United States Geological Survey, Open-File Report 87–463, Washington, DC.

Nagy, K. A. 1988. Doubly-labeled water studies of vertebrate physiological ecology, pp. 270–287. *In* P. W. Rundel, J. R. Ehleringer, and K. A. Nagy (eds.), Ecological Studies. Vol. 68. Stable Isotopes in Ecological Research. Springer-Verlag, Heidelberg.

Nordstrom, D. K., J. W. Ball, R. J. Donahoe, and D. Wittemore. 1989. Groundwater chemistry and water-rock interactions of Stripa. *Geochim. Cosmochim. Acta* **53:** 1727–1740.

Oberlander, G. T. 1956. Summer fog precipitation on the San Franciso peninsula. *Ecology* **37:** 851–852.

Ornduff, R. 1974. Introduction to California Plant Life. Univ. of California Press, Berkeley, CA.

Parsons, J. J. 1960. "Fog-drip" from coastal stratus, with special reference to California. *Weather* **15:** 58–62.

Ramesh, R., S. K. Bhattacharya, and K. Gopalan. 1986. Climatic correlations in the stable isotope records of silver fir (*Abies pindrow*) trees from Kashmir, India. *Earth Planet. Sci. Lett.* **79:** 66–74.

Ranwell, D. S. 1972. Ecology of Salt Marshes and Sand Dunes. Chapman & Hall, London.

Richards, J. H., and M. M. Caldwell. 1987. Hydraulic lift: Substantial nocturnal water transport between soil layers by *Artemisia tridentata* roots. *Oecologia* **73:** 486–489.

Robberecht, R., B. E. Mahall, and P. S. Nobel. 1983. Experimental removal of intraspecific competitors—effects on water relations and productivity of a desert bunchgrass, *Hilaria rigida. Oecologia* **60:** 21–24.

Rundel, P. W., J. R. Ehleringer, and K. A. Nagy (eds.), 1988. Ecological Studies. Vol. 68. Stable Isotopes in Ecological Research. Springer-Verlag, Heidelberg.

Schiegel, W. E. 1972. Deuterium content of peat as a paleoclimatic recorder. *Science* **175:** 512–513.

Schlesinger, M. E., and J. F. B. Mitchell. 1987. Climate model simulations of the equilibrium climatic response to increased carbon dioxide. *Rev. Geophys.* **25:** 760–798.

Smedley, M. P., T. E. Dawson, J. P. Comstock, L. A. Donovan, D. E. Sherrill, C. S. Cook, and J. R. Ehleringer. 1991. Seasonal carbon isotope discrimination in a grassland community. *Oecologia* **85:** 314–320.

Smith, S. D., and R. S. Nowak. 1990. Ecophysiology of plants in the intermountain lowlands, pp. 179–241. *In* C. B. Osmond, L. F. Pitelka, and G. M. Hidy (eds.), Ecological Studies. Vol. 80. Plant Biology of the Basin and Range. Springer-Verlag, New York.

Smith, S. D., A. B. Wellington, J. L. Nachlinger, and C. A. Fox. 1991. Functional responses of riparian vegetation to streamflow diversion in the eastern Sierra Nevada. *Ecol. Appl.* **1:** 89–97.

Sternberg, L.d. S. L. 1988. Oxygen and hydrogen isotope ratios in plant cellulose: Mechanisms and applications, pp 124–141. *In* P. W. Rundel, J. R. Ehleringer, and K. A. Nagy, (eds.), Ecological Studies. Vol. 68. Stable Isotopes in Ecological Research. Springer-Verlag, Heidelberg.

Sternberg, L.d. S. L., and P. K. Swart. 1987. Utilization of freshwater and ocean water by coastal plants of southern Florida. *Ecology* **68:** 1898–1905.

Sternberg, L.d. S. L., N. Ish-Shalom-Gordon, M. Ross, and J. O'Brien. 1991. Water relations of coastal plant communities near the ocean/freshwater boundary. *Oecologia* **88:** 305–310.

Taylor, H. P. 1974. The application of oxygen and hydrogen isotope studies to problems of hydrothermal alterations and ore deposition. *Econ. Geol.* **69:** 843–883.

Thorburn, P. J., G. R. Walker, and T. J. Hatton. 1992. Are River Red Gums taking water from soil, groundwater, or streams? pp 37–42. *In* Catchments of Green—The Proceedings of the National Conference on Vegetation and Water Management. Adelaide, Australia.

Thorburn, P. J., G. R. Walker, and J-P. Brunel. 1993. Extraction of water from *Eucalyptus* trees for analysis of deuterium and oxygen-18: Laboratory and field techniques. *Plant Cell Environ.,* in press.

Urey, H. C., I. G. Brickwedde, and G. M. Murphy. 1932. A hydrogen isotope of mass 2 and its concentration. *Phys. Res.* **39:** 1–15.

Vogelmann, H. W., T. Siccama, D. Leedy, and D. C. Ovitt. 1968. Precipitation from fog moisture in the Green Mountains of Vermont. *J. Ecol.* **49:** 1205–1207.

Walker, C. D., and S. B. Richardson. 1991. The use of stable isotopes of water in characterizing the sources of water in vegetation. *Chem. Geol.* (Isotope Geosci. Sect.) **94:** 145–158.

Wershaw, R. L., I. Friedman, and S. J. Heller. 1966. Hydrogen isotope fractionation of water passing through trees, pp. 55–67. *In* F. Hobson and M. Speers (eds.), Advances in Organic Geochemistry. Pergamon, New York.

West, N. E. 1988. Intermountain deserts, shrub steppes, and woodlands, pp. 209–230. *In* M. G. Barbour and W. D. Billings (eds.), North American Terrestrial Vegetation. Cambridge Univ. Press, Cambridge, MA.

White, J. W. C. 1983. The climatic significance of D/H ratios in white pine in the Northeastern United States. Dissertation, Columbia Univ., New York.

White, J. W. C. 1988. Stable hydrogen isotope ratios in plants: A review of current theory and some potential applications, pp. 142–162. *In* P. W. Rundel, J. R. Ehleringer, and K. A. Nagy (eds.), Ecological Studies. Vol. 68. Stable Isotopes in Ecological Research. Springer-Verlag, Heidelberg.

White, J. W. C., E. R. Cook, J. R. Lawrence, and W. S. Broecker. 1985. The D/H ratios of sap in trees: Implications of water sources and tree ring D/H ratios. *Geochim. Cosmochim. Acta* **49:** 237–246.

Whitehead, D., and T. M. Hinckley. 1991. Models of water flux through forest stands: Critical leaf and stand parameters. *Tree Physiol.* **9:** 35–57.

Winograd, I. J., B. J. Szabo, T. B. Coplen, A. C. Riggs, and P. T. Kolesar. 1985. Two-million-year record of deuterium depletion in Great Basin ground waters. *Science* **227:** 519–522.

Yakir, D. 1992. Variations in the natural abundance of oxygen-18 and deuterium in plant carbohydrates. *Plant, Cell and Environ.* **15:** 1005–1020.

Yakir, D., M. J. DeNiro, and P. W. Rundel. 1989. Isotopic inhomogeneity of leaf water: Evidence and implications for use of isotopic signals transduced by plants. *Geochim. Cosmochim. Acta* **53:** 2769–2773.

Yapp, C. J., and S. Epstein. 1982a. A reexamination of cellulose carbon-bound hydrogen δD measurements and some factors affecting plant-water D/H relationships. *Geochim. Cosmochim. Acta* **46:** 955–965.

Yapp, C. J., and S. Epstein. 1982b. Climatic significance of the hydrogen isotope ratios in tree cellulose. *Nature* **297:** 636–639.

Ziegler, H. 1988. Hydrogen isotope fractionation in plant tissues, pp 105–123. *In* P. W. Rundel, J. R. Ehleringer, and K. A. Nagy (eds.), Ecological Studies. Vol. 68. Stable Isotopes in Ecological Research. Springer-Verlag, Heidelberg.

Ziegler, H., C. B. Osmond, W. Stichler, and P. Trimborn. 1976. Hydrogen isotope discrimination in higher plants: Correlations with photosynthetic pathway and environment. *Planta* **128:** 85–92.

Zimmermann, V., D. Ehhalt, and K. O. Munnich. 1966. Soil-water movement and evapotranspiration: Changes in the isotopic composition of water, pp. 567–585. *In* Proceedings of the Symposium of Isotopes in Hydrology. Intern. At. Energy Assoc. Pub.

31

Hydrogen Isotopic Fractionation by Plant Roots during Water Uptake in Coastal Wetland Plants

Guanghui Lin and Leonel da S. L. Sternberg

I. Introduction

Hydrogen and oxygen isotope analyses of plant stem water have been used to quantitatively determine the use of different water sources by various plants in different environments, such as white pines (White *et al.*, 1985), desert plants (Ehleringer *et al.*, 1991), streamside trees (Dawson and Ehleringer, 1991), and coastal plants (Sternberg and Swart, 1987; Sternberg *et al.*, 1991; Lin and Sternberg, 1992a). These studies underscore the importance of stable isotope analysis in understanding physiological responses of various plants to environmental factors. In coastal plants, for example, results from the study of Sternberg and Swart (1987) indicate that, in Florida Keys, hardwood hammock species growing inland are using fresh water and succulent species growing in the margin of mangrove forests are using a mixture of ocean and fresh water, while mangrove species have a spectrum of water usage, ranging from fresh water to ocean water along a tidal inundation gradient from the inland to the seaward margins. Their observations supported the hypothesis that mangrove species are fully capable of growing on fresh water, but are limited to ocean margins because of competitive exclusion (Sternberg and Swart, 1987). Lin and Sternberg (1992a) also observed that a frequent shift of water source between ocean water and fresh water, as reflected by seasonal variation in δD and $\delta^{18}O$ values of stem water, may be responsible for the occurrence of scrub or dwarf mangroves in southern Florida.

The approach used in these studies is based on two basic premises. First, there are significant differences in isotopic composition between possible water sources. Second, there is no isotopic fractionation by plant roots during water uptake. Previous studies have demonstrated that there are significant differences in isotope ratios of possible water sources such as between precipitation and ground water in continental habitats (White *et al.*, 1985; Dawson and Ehleringer, 1991) or between fresh water and ocean water in coastal habitats of southern Florida (Sternberg and Swart, 1987). Although there is evidence indicating no isotopic fractionation during water uptake in terrestrial plants (Gonfiantini *et al.*, 1965; Wershaw *et al.*, 1966; Ziegler, 1988; Dawson and Ehleringer, 1991), the second premise has not been tested for coastal wetland plants. In the studies by Sternberg and Swart (1987), Sternberg *et al.* (1991), and Lin and Sternberg (1992a), both hydrogen and oxygen isotope ratios of stem water were analyzed to determine the relative use of ocean and fresh water. However, interpretation of water utilization using δD values frequently differs from the interpretation using $\delta^{18}O$ values. Sternberg and Swart (1987), for example, observed that $\delta^{18}O$ values of stem water from mangrove trees whose surface roots were bathed in ocean water matched those of ocean water, suggesting utilization of typical ocean water, while the δD values of water from the same stems were about 10‰ more negative relative to those of ocean water, suggesting uptake of both ocean water and fresh water (see Fig. 2 in the reference). Similarly, Lin and Sternberg (1992a) found that some scrub mangroves during the wet season were probably using pure fresh water if δD values of stem water were used to interpret water use, but they might be taking up both fresh water and ocean water if $\delta^{18}O$ values were used instead (see Fig. 2 in the reference).

These disagreements challenged us to question the reliability of the second premise mentioned above for coastal wetland plants such as mangroves. In this study, we compared δD and $\delta^{18}O$ values between stem and source water from several plant species growing naturally in coastal wetland habitats. Both hydrogen and oxygen isotope ratios of stem water will match those of source water under both field and greenhouse conditions if there is no isotopic fractionation during water uptake. On the other hand, differences in isotopic compositions between stem and source water indicate isotopic fractionation during water uptake. Our results indicate that there is a significant hydrogen isotopic fractionation during water uptake in coastal wetland plants. Oxygen isotope ratios of stem water from coastal wetland plants, however, match those of source water.

II. Materials and Methods

A. Sampling of Stem and Source Water in the Field

During the period from June 1990 to December 1991, stem and source water were collected from various species of coastal wetland plants having their roots submerged in water along the coast of southern Florida. These species were divided into four groups according to their salt tolerance, which were salt-secreting halophytes (*Avicennia germinans* (L.) L.), salt-excluding halophytes (*Conocarpus erecta* L., *Laguncularia racemosa* Gaert., *Rhizophora mangle* L.), slightly salt-tolerant glycophytes (*Acrostichum aureum* L., *Annona glabra* L., *Ilex cassine* L., *Myrica cerifera* L., *Schinus terebinthifolius* Raddi.), and salt-sensitive glycophytes (*Magnolia virginiana* L., *Plucea foetida* (L.) DC., *Salix caroliniana* Michx.) (Long and Lakela, 1971; Odum *et al.*, 1982; Myers and Ewel, 1990). Plant stem samples were collected from sections of stems at least 50 cm away from transpiring leaf surfaces of all species at noon on sunny dates and were cryogenically distilled by a vacuum system in the laboratory (Sternberg and Swart, 1987). Source waters were collected from bodies of water surrounding the whole-plant root systems.

In order to test for possible isotopic fractionations associated with the water distillation process of stem samples, we collected stem water by squeezing a small section of stem from 15 plants grown in various salinities and using the remainder of the stem for water distillation. Water samples were analyzed for hydrogen isotope ratios only since the amount of water gained by squeezing was too small for oxygen isotope analysis.

B. Greenhouse Studies

1. Experiment I In the first experiment, we tested whether there is isotopic fractionation during water uptake in both halophytes and glycophytes under greenhouse conditions. Seedlings of four halophyte species (*C. erecta*, *L. racemosa*, *R. mangle*, and *A. germinans*) as well as two glycophyte species (*Pinus elliottii* var. *densa* Little & Dorman, and *Lycopersicon esculentum* Mill.) were hydroponically cultivated in regular nutrient solutions with no salt added in well-sealed foam boxes. The hydroponic solutions were aerated by aquarium pumps and water levels were maintained by the addition of water every day. After seedlings were grown for 6 weeks or longer, stem water and culture solution samples were taken with three to six replicates for hydrogen and oxygen isotopic analyses. δD and $\delta^{18}O$ values of stem water were compared with those of the culture solutions in each box. A previous experiment indicated that isotopic composition of water in such foam boxes remains constant over the study period (Ish-Shalom-Gordon *et al.*, 1992).

2. Experiment II Forty-eight seedlings of *R. mangle* with similar size dimensions were grown under 12 different conditions combining three salinities (100, 250, 500 mM NaCl), two nutrient levels (10 and 100% strength nutrient solution), and two sulfide concentrations (0, 2.0 mM Na_2S) in the greenhouse for 6 months. At the end of the experiment, stem water from these seedlings and the culture solution (three samples per container) were collected for hydrogen isotopic analyses. Deuterium depletion in stem water relative to source water, or the difference in hydrogen isotope ratio between source and stem water ($\delta D_{source} - \delta D_{stem}$), was calculated for each seedling. In addition, dry weight was determined and used to calculate plant growth rate, or increase in dry biomass per unit time per plant. The mean values of plant growth rates for the different treatments have been reported previously (Lin and Sternberg, 1992b). Transpiration rates were measured with a porometer (LI-1600, LICOR, Lincoln, NE) and total leaf area was measured with a portable area meter (LI-3000, LICOR, Lincoln, NE) for each seedling. Water-loss rate via leaf transpiration of each seedling was estimated by the product of mean transpiration rate and total leaf area. The relationship between δD_{source}-δD_{stem} and plant growth rate was examined by regression analysis according to the least-square method.

3. Experiment III   In this experiment, we further investigated the relationship between δD_{source}-δD_{stem} and water-loss rate by directly controlling the leaf area of each plant. Six to eight seedlings of three salt-excluding halophytes (*C. erecta, L. racemosa,* and *R. mangle*) were grown hydroponically in culture boxes in a greenhouse. After these seedlings acclimated to greenhouse conditions, half of the seedlings were defoliated in such a way that only a single leaf per seedling was maintained throughout the experiment. The seedlings were then grown in the greenhouse for 6 weeks, at which time stem water and solution water were collected for hydrogen isotopic analysis. δD_{source}-δD_{stem} values were compared between the defoliated seedlings and the control seedlings.

C. Hydrogen and Oxygen Isotopic Analysis

Hydrogen isotope ratios of water samples were determined by passing 2 to 4 μl of water through a hot uranium furnace (750°C) as described by Bigeleisen *et al.* (1952). Hydrogen released from this reaction was collected with a Toepler pump and used for isotopic analysis. Oxygen isotope ratios were determined by equilibration with carbon dioxide as described by Epstein and Mayeda (1953). All isotopic analyses were performed in a mass spectrometer (VG PRISM). Isotope ratios are reported in δ units,

$$\delta\%o = [(R_{sample}/R_{standard}) - 1]1000,$$

where R is the D/H or $^{18}O/^{16}O$ of sample and standard (SMOW), respectively.

III. Results

A. Field Studies

In salt-excluding halophytes, stem water showed about 10 (8 to 13)‰ deuterium depletion relative to source water when grown in both fresh and saline water (Fig. 1A, Table I). Stem water from a salt-secreting halophyte demonstrated a similar extent of deuterium depletion (7–9‰) when grown in saline water, but less deuterium depletion (2‰) when grown in fresh water (Fig. 1B, Table I). Stem water from glycophytes, however, showed considerable variation in deuterium depletion relative to source water

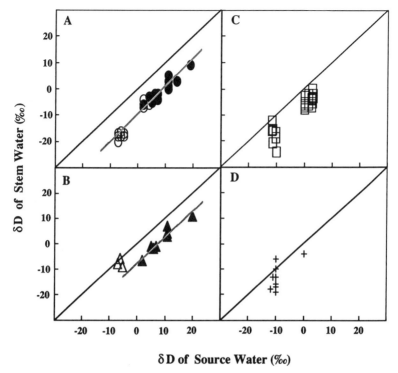

δD of Source Water (‰)

Figure 1. Comparison of δD values between stem and source water from salt-excluding halophytes (A), a salt-secreting halophyte (B), salt-tolerant glycophytes (C), and salt-sensitive glycophytes (D) grown in coastal regions of southern Florida. Open symbols in A and B represent δD values from halophytes growing in fresh water, while closed ones represent δD values from those growing in saline water. The solid lines show 1:1 relationship, while the gray line in A and B is the best fit linear regression line for all salt-excluding halophytes ($Y = -10 + X$, $r = 0.97$, $P < 0.001$) and for salt-secreting plants when grown in saline water ($Y = -8 + X$, $r = 0.95$, $P < 0.001$).

(Figs. 1C and 1D), with higher deuterium depletion in slightly salt-tolerant species (5 to 8‰) than in salt-sensitive species (2 to 5‰) (Table I). In contrast, oxygen isotope ratios of stem water were not significantly different from those of source water in all of these coastal wetland plants (Fig. 2, Table I). δD value of stem water by distillation was not significantly different from that obtained by squeezing ($P < 0.01$, Fig. 3), suggesting that the observed deuterium depletion in stem water of these coastal wetland plants cannot be attributed to an isotopic fractionation occurring during the process of water distillation of stem samples.

B. Greenhouse Studies

Stem water from salt-excluding halophytes grown hydroponically in the greenhouse with fresh water (Experiment I) showed 10 to 11‰ depletion in deuterium relative to source water, while the salt-secreting halophyte (*A.*

Table I Differences in δD and $\delta^{18}O$ Values (Mean $\pm$ SE) between Source and Stem Water (δ_{source}-δ_{stem}) from Various Coastal Wetland Plants Grown in Different Salinities along the Coast of Southern Florida

Plant species and type	Salinity (ppt)	N	δ_{source}-δ_{stem} (‰) Hydrogen	Oxygen
Salt-excluding halophytes				
Conocarpus erecta	28	3	13 ± 0.3	0.5 ± 0.1
	0	3	12 ± 0.7	0.0 ± 0.1
Laguncularia racemosa	35	4	9 ± 1.1	0.0 ± 0.1
	28	3	8 ± 0.6	0.4 ± 0.1
	0	3	11 ± 0.6	0.3 ± 0.1
Rhizophora mangle	93	1	10	0.4
	40	1	11	0.2
	35	3	9 ± 0.7	0.0 ± 0.1
	28	4	10 ± 0.6	0.0 ± 0.2
	0	7	10 ± 0.8	−0.2 ± 0.1
Salt-secreting halophyte				
Avicennia germinans	93	1	9	0.5
	35	4	7 ± 0.9	0.3 ± 0.1
	28	4	8 ± 0.6	0.3 ± 0.1
	0	3	2 ± 0.9	−0.2 ± 0.1
Salt-tolerant glycophytes				
Acrostichum aureum	0	3	5 ± 1.5	0.1 ± 0.1
Annona glabra	0	5	6 ± 1.0	0.6 ± 0.1
Ilex cassine	0	5	6 ± 1.6	−0.3 ± 0.1
Myrica cerifera	0	4	8 ± 1.9	0.3 ± 0.1
Schinus terebinthifolius	0	4	7 ± 0.4	0.3 ± 0.1
Salt-sensitive glycophytes				
Magnolia virginiana	0	2	5	0.0
Plucea foetida	0	2	2	0.2
Salix caroliniana	0	5	4 ± 1.3	−0.3 ± 0.4

germinans) showed significantly less deuterium depletion (3‰) in stem water when grown with fresh water under the same greenhouse conditions (Fig. 4A). Two terrestrial glycophytes, however, exhibited insignificant deuterium depletion (Fig. 4A). As in our field study, oxygen isotope ratios showed no significant differences between stem and source water in all plants tested (Fig. 4B).

In *R. mangle*, deuterium depletion in stem water relative to source water, or δD_{source}-δD_{stem}, was highly correlated with plant growth rate resulting from different environmental conditions (Experiment II) (Fig. 5A). The higher the plant growth rate, the higher the δD_{source}-δD_{stem}. Similarly, there was a correlation between the δD_{source}-δD_{stem} and water-loss rate (Fig. 5B). δD_{source}-δD_{stem} increased with increasing water-loss rate until a maximum value was reached, whereupon δD_{source}-δD_{stem} remained relatively constant with further increase of water-loss rate. In all three salt-excluding halo-

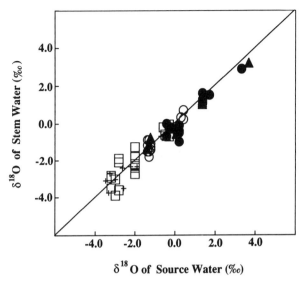

Figure 2.   Comparison of $\delta^{18}O$ values between stem and source water from salt-excluding halophytes, a salt-secreting halophyte, salt-tolerant glycophytes, and salt-sensitive glycophytes grown in coastal regions of southern Florida. The lines show the 1:1 relationship. Symbols are the same as in Fig. 1.

phytes, defoliation treatments (Experiment III) significantly reduced the δD_{source}-δD_{stem} values (Fig. 6). When leaf area was reduced from about 200 to 20 cm^2/plant, δD_{source}-δD_{stem} values changed significantly from 8–12 to 2–7‰ in seedlings of these three species.

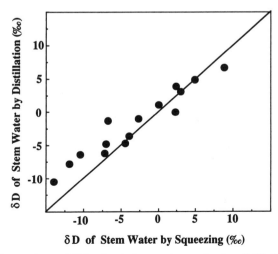

Figure 3.   Comparison of δD value between stem water by distillation and by direct squeezing from plants grown in different salinities. The line shows the 1:1 relationship.

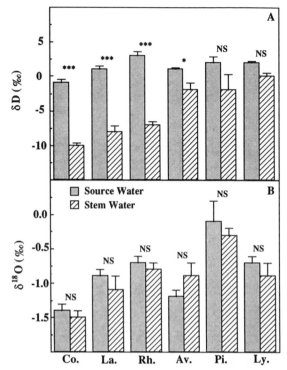

Figure 4. Comparison of δD and δ^{18}O values between stem and source water from three salt-excluding halophytes (Co., *Conocarpus erecta;* La., *Laguncularia racemosa*; Rh., *Rhizophora mangle*), one salt-secreting halophyte (Av., *Avicennia germinans*), and two glycophytes (Pi., *Pinus elliottii* var. *densa*; Ly., *Lycopersicon esculentum*) grown hydroponically with fresh water in the greenhouse. NS, not significant at $P > 0.05$; *significant at $P < 0.05$; ***highly significant at $P < 0.001$.

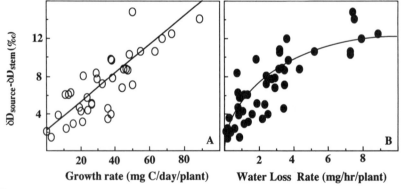

Figure 5. Relation between deuterium depletion in stem water (δD_{source}-δD_{stem}) and plant growth rate (A) and water-loss rate through leaf transpiration (B) of *R. mangle* seedlings grown under different conditions. The line shown in Fig. 5A is the regression line for the equation, δD_{source}-$\delta D_{stem} = 2.2049 + 0.1396 \cdot$ (growth rate) ($r = 0.816$, $P < 0.001$), while the line in Fig. 5B was drawn by eye.

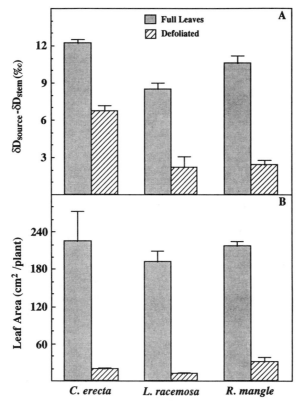

Figure 6. Comparison of deuterium depletion in stem water (δD_{source}-δD_{stem}) and total leaf area for the defoliated seedlings and the seedlings with full leaves of three salt-excluding halophytes grown hydroponically with fresh water in the greenhouse. The differences in both δD_{source}-δD_{stem} and leaf area between treatment and control are highly significant ($P < 0.001$) for all three species.

IV. Discussion

A. Deuterium Depletion in Stem Water from Coastal Wetland Plants

Most coastal wetland plants, especially halophytes, showed significant deuterium depletion (about 10‰) in stem water relative to source water under field (Fig. 1, Table I) and greenhouse conditions (Fig. 4). In contrast, all previous observations on terrestrial plants indicate that there are no significant differences in δD values between stem and source water (Gonfiantini *et al.*, 1965; Wershaw *et al.*, 1966; Ziegler, 1988; White *et al.*, 1985; Dawson and Ehleringer, 1991). Such deuterium depletion in stem water from coastal wetland plants cannot be attributed to an isotopic fractionation associated with our stem-water distillation process, since there were no significant differences between δD values of distilled and squeezed stem water (Fig. 3). Wershaw *et al.* (1966) have also demonstrated that the distillation process for extracting stem water does not cause any isotopic fractionation as long as the distillation is carried to completion, which is the case in our distillation procedures.

Deuterium depletion in stem water from seedlings of a salt-excluding halophyte (*R. mangle*) was highly correlated with plant growth rate and water-loss rate (Fig. 5). This relation cannot be the result of changes in root metabolism caused by treatments of salinity, nutrient, and sulfide, since the reduction in total leaf area (or water-loss rate) by direct defoliation also significantly reduced deuterium depletion in stem water from all three salt-excluding halophytes, as observed in Experiment III (Fig. 6). Thus, deuterium depletion in stem water from these coastal wetland plants probably results from deuterium discrimination occurring in roots, which is not related to root metabolism, but to plant growth rate or water-loss rate via leaf transpiration.

In contrast to the differences in hydrogen isotope ratios between stem and source water, there were no significant differences between $\delta^{18}O$ values of stem and source water in all coastal wetland plants tested in this study (Fig. 2, Table I), indicating that there is no significant oxygen isotopic fractionation during water uptake in these plants. Our greenhouse experiment (Experiment I) also showed that oxygen isotope ratios of stem water were not significantly different from those of source water in all species tested, although $\delta^{18}O$ values of stem water were slightly lower than those of source water in five of the six species tested (Fig. 4). Therefore, oxygen analysis is a more effective approach for determining the relative use of different water sources by coastal wetland plants. Consequently, oxygen isotope ratios of cellulose from stems will be a more effective indicator of plant utilization of different water sources in coastal regions over a long-term period than hydrogen isotope ratios.

B. Hypothesis Concerning Deuterium Depletion in Stem Water

We hypothesize that the deuterium depletion in stem water from these plants is caused by isotopic fractionations occurring in roots during water uptake. Water entering the xylem of most plant roots passes through the root cortex via either the apoplastic or the symplastic pathways, but must cross the cell membrane of the endoderm because of the Casparian strip (Weatherley, 1982). It has been observed that water can enter root xylem freely without crossing a membrane via a "pure" apoplastic pathway located near the root tip where the Casparian band is not mature (Perry and Greeway, 1973) or by leakage in the region of developing laterals where the Casparian band is discontinuous (Dumbroff and Peirson, 1971; Peterson *et al.*, 1981). Apoplastic uptake by these pure pathways, however, was observed only in some terrestrial glycophytes such as corn (*Zea mays*), broad bean (*Vicia faba*), and tomato (*L. esculentum*) (Dumbroff and Peirson, 1971; Peterson *et al.*, 1981). In halophytes such as mangroves, there is no evidence for such pure apoplastic uptake by roots. Moon *et al.* (1986) demonstrated in a mangrove species (*Avicennia marina*) that there was no major pure apoplastic pathway at the root tips and at mature lateral/parent root junctions due to the continuity of the periderm. They found that the uptake of ions and water in intact roots of this mangrove species occurred mainly via the symplast of younger regions of the third-order and fourth-order roots because of the formation of the Casparian band at both the periderm and exoderm, which caused most of the root system to be isolated from seawater. Therefore, it is reasonable to conclude that most water absorbed by roots of coastal halophytes, such as mangroves, passes through membranes at both the periderm and the exoderm (Moon *et al.*, 1986).

Membranes in coastal halophytes are probably less porous than those in terrestrial glycophytes because of their ultrafiltration function associated with salt exclusion (Scholander, 1968; Campbell and Thomson, 1976; Moon *et al.*, 1986). Thus large aggregates of water molecules must be dissociated by breakage of hydrogen bonds in order to pass through the root membranes according to the model of Schönherr and Schmidt (1979). Because of the twofold difference in mass between hydrogen and deuterium, water uptake by such roots may result in significant discrimination against DHO molecules (as opposed to H_2O), which have stronger hydrogen bonds with other water molecules. According to Barrow (1966), the strength of a hydrogen bond is proportional to the square root of $m_1 \cdot m_2/(m_1 + m_2)$, where m_1 and m_2 are the masses of hydrogen and oxygen atoms forming the bond, respectively. Thus, if a hydrogen atom of a hydrogen bond is replaced by deuterium, the strength of the bond will increase by about 41.4%. In contrast, replacement of ^{16}O by ^{18}O will only increase the strength of the bond by about 3.3%. Therefore, any hypothetical membrane effect involving breakage of hydrogen bonds will discriminate significantly against deuterium, but to a much less degree against ^{18}O. Our results are consistent with this hypothetical membrane effect on isotopic compositions of stem water, since only deuterium depletion was observed in stem water of the coastal plants (Figs. 1–3, Table I).

Although salt secretion by salt glands is the major salt-tolerance mechanism in salt-secreting halophytes such as *A. germinans*, their roots also have a salt-exclusion or ultrafiltration function (Ball, 1988). It has been shown that root membrane resistance for another salt-secreting mangrove (*A. marina*) is much higher than that of other terrestrial glycophytes and increases with salinity (Field, 1984). This characteristic of roots of salt-secreting halophytes may explain why *A. germinans* shows less deuterium depletion in stem water when grown in fresh water than in saline water (Figs. 1B and 4A, Table I).

In terrestrial glycophytes where ultrafiltration is not necessary, however, pores in the membranes are probably large enough to let large aggregates of water molecules pass through. Thus no isotopic fractionation during water uptake should be observed (Fig. 3, also see Gonfiantini *et al.*, 1965; Wershaw *et al.*, 1966; White *et al.*, 1985; Dawson and Ehleringer, 1991). Most coastal glycophytes also show some salt tolerance and can survive in areas with occasional saltwater intrusion (Long and Lakela, 1971); their root membranes may also discriminate against deuterium during water uptake. These plants, however, show considerable variation in deuterium discrimination, probably due to their differences in salt tolerance (Fig. 1, Table I).

C. Relation between Transpiration and Deuterium Depletion in Stem Water

We have observed that discrimination against deuterium during water uptake in *R. mangle* is related to transpiration (Figs. 5 and 6). The higher the transpiration, the greater the discrimination. In order to explain this relationship, we consider that the flow of stem water is controlled by two major processes. The first is the reversible diffusion of water through the ultrafiltration membrane from the external medium to the root xylem. This process discriminates against deuterium as previously discussed. The second process is the mass flow of water from the root xylem through the stem to the leaf. This process is controlled by transpiration, does not discriminate against deuterium, and most likely is irreversible. We hypothesize that when stem flow is limited by transpiration, (i.e., during low transpiration), isotopic discrimination by the membrane is not evident. However, when membrane permeability limits stem flow, (i.e., during high transpiration), the isotopic discrimination by the ultrafiltration membrane becomes evident.

V. Summary

Our results suggest that there are significant hydrogen isotopic fractionations during water uptake in coastal wetland plants, resulting from discrimination against deuterium by their roots in association with salt ultrafiltration. The discrimination at its highest was about 10‰ for salt-excluding halophytes, about 2–6‰ for salt-secreting halophytes depending on salinity conditions, about 6‰ for slightly salt-tolerant glycophytes in coastal habitats, and only about 3‰ for salt-sensitive glycophytes in coastal areas. Deuterium depletion in stem water relative to source water ($\delta D_{source}-\delta D_{stem}$) increased with increasing water-loss rate via leaf transpiration. In contrast to hydrogen isotopic fractionation, there was no significant oxygen isotopic fractionation during water uptake in these coastal wetlands plants. Therefore, oxygen isotopic analysis of xylem water will be a more effective approach to determine the relative use of different water sources by coastal halophytes. Since some desert halophytes have an ultrafiltration process similar to that in coastal halophytes (Sharma, 1982), their roots may also discriminate against deuterium during water uptake. Thus, oxygen isotope analysis of their xylem water may also be more appropriate for the determination of their water sources. In addition, measurement of deuterium depletion in stem water relative to source water ($\delta D_{source}-\delta D_{stem}$) may be an effective method for determining the extent of the pure apoplastic pathway and of the ultrafiltration functions in the root membranes of halophytes.

Acknowledgments

This study was supported by NSF Grant Number BSR 8908240. We thank Mr. Tim Banks for his technical support on stable isotopic analysis. Senior author is grateful to Drs. James Ehleringer, Anthony Hall, and Graham Farquhar for their encouragement and critical comments on the manuscript. This is publication No. 397 from the program in Ecology, Behavior and Evolution of the Department of Biology, University of Miami, Coral Gables, Florida.

References

Ball, M. C. 1988. Ecophysiology of mangroves. *Trees* **2**: 129–142.

Barrow, G. M. 1966. Introduction to Molecular Spectroscopy. McGraw–Hill, New York.

Bigeleisen, J., M. L. Perlman, and H. C. Prosser. 1952. Conversion of hydrogenic materials to hydrogen for isotopic analysis. *Anal. Chem.* **24**: 1356–1357.

Campbell, N., and W. W. Thomson. 1976. The ultrastructural basis of chloride tolerance in the leaf of *Frankenia*. *Ann. Bot.* **40**: 687–693.

Dawson, T. E., and J. R. Ehleringer. 1991. Streamside trees that do not use stream water: Evidence from hydrogen isotope ratios. *Nature* **350**: 335–337.

Dumbroff, E. B., and D R. Peirson. 1971. Probable sites for passive movement of ions across the endodermis. *Can. J. Bot.* **49**: 35–38.

Ehleringer, J. R., S. L. Phillips, W. S. F. Schuster, and D. R. Sandquist. 1991. Differential utilization of summer rains by desert plants: Implications for competition and climate change. *Oecologia* (Berl.) **88**: 430–434.

Epstein, S., and T. Mayeda. 1953. Variation of ^{18}O content of water from natural sources. *Geochim. Cosmochim. Acta* **42**: 241–245.

Field, C. D. 1984. Ions in mangroves, pp. 43–48. *In* H. J. Teas, (ed.), Tasks for Vegetation Science. 9. Physiology and Management of Mangroves. Junk, The Hague/Boston.

Gonfiantini, R., S. Gratziu, and S. Tongiori. 1965. Oxygen isotopic composition of water in leaves, pp. 405–410. *In* Isotope and radiation in soil-plant-nutrition studies. Intern. At. Energy Agency, Vienna.

Ish-Shalom-Gordon, N., G. Lin, and L. da S. L. Sternberg. 1992. Isotopic fractionation during cellulose synthesis in two mangrove species: Salinity effects. *Phytochemistry* **31**: 2623–2626.

Lin, G., and L. da S. L. Sternberg. 1992a. Comparative study of water uptake and photosynthetic gas exchange between scrub and fringe red mangroves, *Rhizophora mangle* L. *Oecologia* (Ber.) **90**: 399–403.

Lin, G., and L. da S. L. Sternberg. 1992b. Effect of growth form, salinity, nutrient, and sulfide on photosynthesis, carbon isotope discrimination, and growth of red mangrove (*Rhizophora mangle* L.). *Aust. J. Plant Physiol.* **19**: 509–517.

Long, R. W., and O. Lakela. 1971. A Flora of Tropical Florida. Univ. of Miami Press, Coral Gables, FL.

Moon, G. J., B. F. Clough, C. A. Peterson, and W. G. Allaway. 1986. Apoplastic and symplastic pathways in *Avicennia marina* (Forsk.) Vierh. roots revealed by fluorescent tracer dyes. *Aust. J. Plant Physiol.* **13**: 637–648.

Myers, R. L., and J. J. Ewel. 1990. Ecosystems of Florida. Univ. of Central Florida, Orlando, FL.

Odum, W. E., C. C. McIor, and T. J. Smith III. 1982. The Ecology of the Mangroves of South Florida: A Community Profile. U.S. Fish and Wildlife Service, Office of Biological Services, Washington, D.C. FWS/OBS-81/24.

Perry, M. W., and H. Greenway. 1973. Permeation of uncharged organic molecules and water through tomato roots. *Ann. Bot.* **37:** 225–232.

Peterson, C. A., M. E. Emanuel, and G. B. Humphreys. 1981. Pathway of movement of apoplastic fluorescent dye tracers through the endodermis at the site of secondary root formation in corn (*Zea mays*) and broad bean (*Vicia faba*) roots. *Can. J. Bot.* **60:** 1529–1535.

Scholander, P. F. 1968. How mangrove desalinate seawater? *Physiol. Plant.* **21:** 251–261.

Schönherr, J., and H. W. Schmidt. 1979. Water permeability of plant cuticles: Dependence of permeability coefficients of cuticular transpiration on vapor pressure saturation deficit. *Planta* **144:** 391–400.

Sharma, M. L. 1982. Aspects of salinity and water relations of Australian chenopods, pp. 155–172. *In* D. N. Sen and K. S. Rajpurohit (eds.), Task for Vegetation Science. 2. Contribution to the Biology of Halophytes. Junk, The Hague/Boston.

Sternberg, L. da S. L., and P. K. Swart. 1987. Utilization of freshwater and ocean water by coastal plants of southern Florida. *Ecology* **68:** 1898–1905.

Sternberg, L. da S. L., I. Ish-Shalom-Gordon, M. Ross, and J. O'Brein. 1991. Water relations of coastal plant communities near the ocean/freshwater boundary. *Oecologia* (Ber.) **88:** 305–310.

Weatherley, P. E. 1982. Water uptake and flow in roots, pp. 79–109. *In* O. L. Lange, P. S. Nobel, C. B. Osmond, and H. Ziegler (eds.), Physiological Plant Ecology. II. Water Relations and Carbon Assimilation. Springer-Verlag, New York.

Wershaw, R. L., I. Friedman, and S. J. Heller. 1966. Hydrogen isotopic fractionation of water passing through trees, pp. 55–67. *In* G. D. Hobson and G. C. Speers (eds.), Advances in Organic Geochemistry. Pergamon, New York.

White, J. W. C., E. R. Cook, J. R. Lawerence, and W. S. Broecker. 1985. The D/H ratios of sap in trees: Implications for water sources and tree ring D/H ratios. *Geochim. Cosmochim. Acta* **49:** 237–246.

Ziegler, H. 1988. Hydrogen isotope fractionation in plant tissues, pp. 105–123. *In* P. W. Rundel, J. R. Ehleringer, and K. A. Nagy (eds.), Ecological Studies. 68. Stable Isotopes in Ecological Research. Springer-Verlag, New York.

32

The Source of Water Transpired by *Eucalyptus camaldulensis:* Soil, Groundwater, or Streams?

Peter J. Thorburn and Glen R. Walker

I. Introduction

There is great interest in the water relations of *Eucalyptus camaldulensis* Dehnh. (river red gums) on the floodplains of Australia's longest river—the River Murray. *E. camaldulensis* forests and woodlands are prized for recreation, agriculture, and forestry, and more recently conservation. However, manipulation of river flows during this century has altered the hydrology of the forests (Bren, 1988), which in turn is reducing the vigour of *E. camaldulensis* and altering forest composition (Chesterfield, 1986).

This chapter describes an investigation into the relative importance of various possible sources of water available to *E. camaldulensis* (i.e., from streams, soil, or groundwaters). The main technique used in this study was monitoring of stable isotope composition of water from tree sap and different possible water sources, as described in Chapter 30 by Todd Dawson. However, detailed analysis of isotope data (that is comparison of deuterium and oxygen-18 concentrations) and measurement of soil water status were required to provide insights into the water relations of this species. The study thus not only addresses an important ecological problem, it provides an example of the power of isotopic methods for studying plant-water sources.

II. Water Relations of *E. camaldulensis* on the River Murray Floodplain

E. camaldulensis forests and woodlands occur along the banks of the River Murray and creeks of the floodplain, in areas subjected to relatively frequent flooding (Bren, 1988; Roberts and Ludwig, 1990). Flow of the river is highly controlled by a series of weirs and managed lakes, to reduce flooding and provide irrigation water supplies. However, flooding is important in the ecology of the plant communities on the floodplains, notably the *E. camaldulensis* forests and woodlands. *E. camaldulensis* are relatively tolerant of flooding (Blake and Reid, 1981; Van der Moezel *et al.*, 1989). Floodwaters normally provide suitable conditions for seed germination and establishment of *E. camaldulensis* and suppress growth of competing species and pathogens (Dexter, 1978). Reduction of the flood frequency has thus inhibited the natural regeneration of *E. camaldulensis* (Chesterfield, 1986; Bren, 1988). It is possible that further alterations in flow of creeks on the floodplain may occur under possible salinity mitigation schemes.

To understand the effects of changing local hydrology on the health of *E. camaldulensis* communities it is necessary to understand their current water relations. Dexter (1978) found that groundwater was an important water source of *E. camaldulensis* in the Barmah forests on the River Murray floodplain in northern Victoria and that floods were important for recharging soil water deficits. However, the Barmah forest is underlain by low–salinity groundwater. Other areas of the floodplain are underlain by shallow, saline groundwaters, for example, the Chowilla anabranch system of eastern South Australia (Jolly *et al.*, 1991). It would be expected that *E. camaldulensis* forests in this region depend more heavily on creek water for their survival. It logically follows that altering the flow (or salinity) in the creeks will possibly be more detrimental to these communities than those elsewhere on the floodplain. Therefore it was important that the sources of water exploited by the *E. camaldulensis* in areas of saline groundwater were defined.

III. Methods

A. Field Site

An experimental site was established in an *E. camaldulensis* forest on Monoman Island (Fig. 1), adjacent to Chowilla Creek in the Chowilla anabranch system on the River Murray floodplain. The area has an average rainfall of ~250 mm year^{-1} and potential evaporation of ~2000 mm year^{-1}. The River Murray and many of the creeks in the anabranch system are perennial. However, their water level varies considerably, both seasonally and due to the effects of management of river flows by the local water authori-

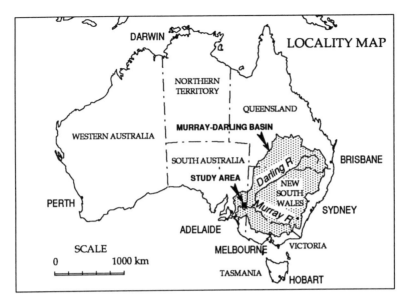

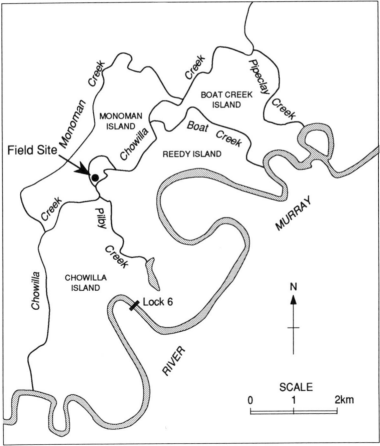

Figure 1. Location of the study area, with the position of the field site indicated.

ties. This variation in water level gives rise to many ephemeral creeks leading off the main perennial creeks.

The study area was underlain by saline shallow groundwater (Fig. 2). The creeks generally have a poor hydraulic connection with the adjacent groundwater, due to the medium to heavy clay texture of the surface soil over most of the floodplain (including the study site). This results in large differences in creek and groundwater salinity (e.g., Fig. 2).

The site consisted of two groups of trees. The first was on the bank of an ephemeral creek (running directly off Chowilla Creek) and is referred to as the "creekside" group. The creek bank at this site had a slope of ~13%, so water was at variable distances from the trees depending on the amount of creek flow. The creekside trees would be inundated in winter and/or spring of most years.

The second group of trees was 40 m from the creek and is referred to as the "inland" group. This site was ~2 m higher in elevation than the creekside site and would be inundated approximately every 5 years. A piezometer was installed at the inland site to provide access to the groundwater.

Trees in both groups were 0.4 to 1.2 m diameter and 10 to 30 m tall.

B. Sourcing Water in Trees Using Stable Isotopes of Water

Naturally occurring stable isotopes of water were used to determine the source of water transpired by the trees. Water was extracted from both groups of trees at various times, and its deuterium (D) and oxygen-18 (^{18}O) composition compared with water sampled from the possible sources in the root zone of the trees. The method is based on two main assumptions: that the isotopic composition of water is not altered when (1) water is taken up

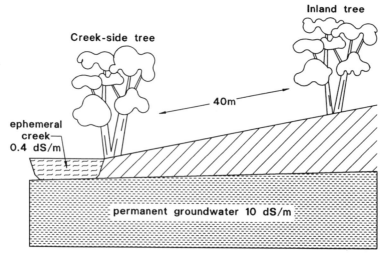

Figure 2. Schematic cross section of the field site.

Table I   Mean δD Values (‰) of Water Extracted (by Azeotropic Distillation Using Kerosene) from Soil, Roots, and Stems of *Eucalyptus* Seedlings Grown in Pots[a]

Species	Soil	Root	Stem
E. camaldulensis	−22.8	−21.8	−21.3
E. largiflorens	−22.0	−21.4	−21.9

Source: from Thorburn *et al.* (1993).
[a]None of the values are significantly different ($P < 0.05$; LSD = 1.5%).

by roots, nor (2) when it moves from roots to leaves. These assumptions were previously tested in glasshouse studies for *E. camaldulensis* (and *E. largiflorens*), where it was found that there was no difference in D concentration between waters extracted from soil, roots, or stems (Table I).

For the method to be successful water must also be extracted from trees and water sources in the root zone without any significant alteration of isotopic composition, and representative samples must be obtained from the trees. The similarity of the δD[1] values in Table I indicates the accuracy of the extraction techniques for soils (also see Revesz and Woods, 1990) and *Eucalyptus* stems and roots. Sampling different parts of two *Eucalyptus* trees also indicated that there was no change in δD values of sap between the trunks and small twigs (Table II). This result was further confirmed when statistical analysis of water extracted from 18 paired twig (3 to 10 mm in diameter) and trunk samples showed no significant difference in δD values (Thorburn *et al.*, 1993). Twigs were chosen to be sampled because repeated long-term sampling of sapwood from the trunk may be harmful to trees.

C. Field Sampling

Water sources of the trees were studied during winter to early spring (July to September), 1991. Tree sap and water samples were taken five times during this period, and rainfall and creek levels recorded. Soil profiles were also sampled to the water table, twice.

1. Tree Samples   Two or three trees in each group, covering an area of ~250 m², were examined. Three to ten twigs, approximately 1 cm diameter and 10 cm long, were cut from the trees at each sampling time. Outer bark was lightly scraped from the twigs, which were then cut to ~1 cm length, placed in a 180-ml glass jar containing kerosene (Jet A-1 aviation fuel), and sealed airtight.

2. Soil Samples   Soil profiles were hand augered to ~30 cm below the water table (which was ~2.5 to 3 m deep) at the inland site on July 26th and

---

[1]Isotope concentrations are expressed as delta notation: $\delta(‰) = (R_i/R_s - 1)1000$, where R is the ratio of heavy : light isotope, i indicates the isotope sample, and s the standard. δ^2H refers to ^2H, while $\delta^{18}O$ refers to ^{18}O.

Table II   Variation in δD Values (‰) in Water Extracted from Different Parts of a *Eucalyptus camaldulensis* and a *E. largiflorens* in a Native Forest[a]

	Tree part			
	Small twig	Large twig	Branch sapwood	Trunk sapwood
Diameter of tree part	~3 mm	~10 mm	80–120 mm	0.4–1.2 m
E. camaldulensis	−29.2	−31.8	−28.6	−27.5
E. largiflorens	−20.6	−20.9	−22.3	−19.8

Source: after Thorburn *et al.* (1993).
[a]Precision of the analyses was 1.3‰.

September 3rd. Soil profiles were not augered at the creekside site, because the soil profile was too wet due to the proximity of the creek water. Soil samples were immediately placed in 500-ml glass jars and sealed airtight. Soil matric potential was determined on these samples and water was extracted for isotopic analysis. Smaller soil samples (~50 g) were also taken in the field for determination of soil water content and chloride concentration.

3. Creek and Groundwater Samples   Creek water samples were taken at the five sampling times. Groundwater samples were obtained from the augered holes when soil samples were taken at the inland site and from the piezometer at other times.

4. Creek Water Levels and Rainfall   Creek levels fluctuated during the course of the study. Water was near to, or over, the base of the creekside trees at these sampling times (Table III), and was within 4 m of the base of the trees for a week prior to the study.

Table III   Distance of Water from the Base of the Creekside Trees, the Estimated Proportion of Groundwater Contained in the Sap of Inland Trees at Each Sampling Time, and the Estimated Proportion of Groundwater or Creek Water Contained in the Sap of Creekside Trees at Each Sampling Time

	Date in 1991				
	21 July	26 July	6 Aug	21 Aug	3 Sept
Distance of creek water from creekside trees (m)[a]	−1	0	+3	−1.5	−1
Proportion of groundwater in inland tree sap (%)[b]	40	47	57	63	50
Proportion of groundwater in Creekside tree sap (%)[b]	37	0	42	52	60
Proportion of creek water in Creekside tree sap (%)[b]	36	0	31	39	45

[a]Negative values indicate distance downslope from the tree bases, while positive values are upslope with water covering the base of the trees.
[b]Estimated from the linear mixing model described in the Appendix.

During the study, 24 mm of rain fell in the area, with a similar amount falling in the preceding 2 weeks. The largest rainfall events during the study were 10 mm, which fell 2 days before the second sampling, and 4 mm, which fell 4 days prior to the final sampling. The remaining rain fell in amounts ≤ 1 mm day^{-1}. The δD value of rainfall was -35 to $-40‰$.

D. Analyses

δD and δ^{18}O values were determined in all creek water and groundwater samples and water extracted from all twigs and selected soil samples. Water was extracted from twigs within a week of field sampling by azeotropic distillation with kerosene (Thorburn *et al.*, 1993). Water was also extracted from soils using this method (Revesz and Woods, 1990). Isotopic analyses for δD were performed by reduction of 25 μl of water to H$_2$ over uranium at 800°C. δ^{18}O was determined by a modification of the Epstein and Mayeda (1953) technique, using 1 ml of water. Isotopic concentrations were expressed relative to the standard V-SMOW. Precision of the analyses (including both sampling and analytical errors) was estimated to be $\pm 1.3‰$ for δD and $\pm 0.3‰$ for δ^{18}O for soil (Revesz and Woods, 1990) and plant (Thorburn *et al.*, 1993) samples.

Chloride concentration and matric potential were also measured on soil samples. Matric potential was determined using the "filter paper" technique (Greacen *et al.*, 1989). Chloride concentration was used to estimate osmotic potential. Creek and groundwater salinity were also measured.

IV. Results

A. Soil, Groundwater, and Creek δ Values

The δD and δ^{18}O soil profiles from the two sampling times were similar, so only the July profile is shown (Fig. 3). The profiles all showed a distinct minimum at ~ 1 m depth (e.g., Fig. 3), rather than the more expected monotonic decline with depth. The observed profile shape could be due to either infiltration of rain or flood water to this depth at some past time or to water movement in the vapor phase (possibly caused by temperature gradients in the soil profile; Barnes and Allison, 1984; Barnes *et al.*, 1989).

The creek δD and δ^{18}D values decreased between the first and second sampling (Fig. 4), then remained constant. They were significantly lower than the groundwater δ values at all but the first sampling time. The groundwater δD and δ^{18}O values did not vary significantly during the sampling period (Fig. 4).

B. Tree Sap δ Values

The inland tree sap δD and δ^{18}O values showed little variation through the study period (Fig. 4). Tree sap δD values matched soil values at two regions in the profile (at ~ 0.2 m and below 1.4 m; Fig. 3a), while the tree sap δ^{18}O values corresponded to soil at only one depth (~ 0.4 m; Fig. 3b). The

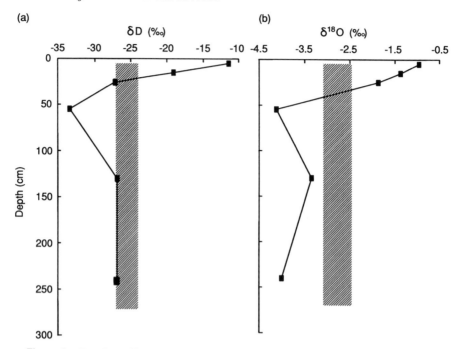

Figure 3.   Depth profiles of soil water δD (a) and $\delta^{18}O$ (b) at the inland site in winter (July 26), 1991. Hatched area shows the range of tree sap δ values for five sampling times between July 24 and September 3.

correspondence between inland tree and soil water δ values was similar at both the July and September sampling times (September soil data not shown).

The creekside tree sap δD and $\delta^{18}O$ values were similar to the inland tree values at four of the five sampling times (Fig. 4). All the tree sap $\delta^{18}O$ values were greater than those of both the groundwater and the creek water at all times (Fig. 4b). For the δD results, however, tree values were similar to the groundwater values later in the study period (Fig. 4a).

C. $\delta D–\delta^{18}O$ Plot

The δD and $\delta^{18}O$ data from all the tree sap, creek water, and groundwater samples (Fig. 5) lie to the right of the meteoric water line, indicating that these samples are evaporatively enriched relative to rainfall. The four groups of samples, i.e., soil, creek water, groundwater, and tree sap, tend to plot in separate locations on Fig. 5. However, the tree sap, creek water, and groundwater data form a straight line. Regression of these data gave the relationship

$$\delta D = 3.1 \; \delta^{18}O - 16.1 \qquad (r^2 = 0.87). \qquad (1)$$

The soil samples from 0.2 and 1.4 m depth lie on the regression line (Eq. (1)), even though these points were not included in the regression. The

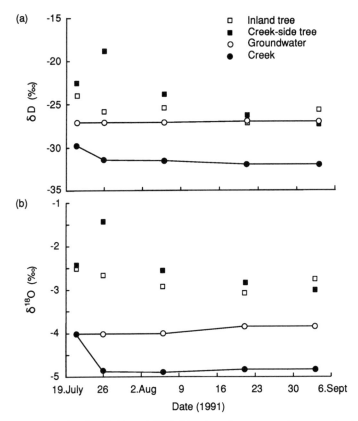

Figure 4. Variation in δD (a) and δ¹⁸O (b) values from groundwater at the inland site, creek water, and sap from trees at the inland and creekside sites, during winter and early spring, 1991.

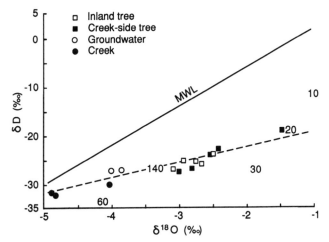

Figure 5. δD–δ¹⁸O plot for all tree sap and water source samples collected during winter and early spring 1991. (The numbers indicate the depth in cm of the soil water sample.) The meteoric water line and the regression line (dashed line) for all samples (excluding the soil samples) are also shown.

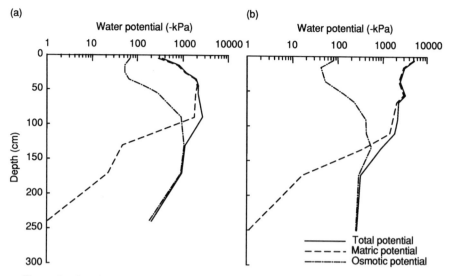

Figure 6.   Depth profiles of soil matric, osmotic, and total potential at the inland site on (a) July 26 and (b) September 3, 1991.

isotopic values of soils samples below 1.4 m will also lie on the regression line since the groundwater value, which corresponds to the bottom of the soil profile, does.

D. Soil Water Potential

Matric potentials were high (>-500 kPa) in the surface soil on July 26 (Fig. 6a), in response to recent rain. By September the surface soil had dried and matric potentials were <-2 MPa (Fig. 6b). Osmotic potential varied from ~ -60 kPa at the soil surface to -1 MPa at depth, and profiles showed little variation between each sampling time (Fig. 6).

Total water potential was calculated at each depth by summing the matric, osmotic, and gravitational potentials. At all times total potentials were -300 to -500 kPa near the water table (Fig. 6), due to the effects of subsoil osmotic potential. In the upper profile the total potential mainly reflected the matric potential because of the low surface soil salinities. Thus, in July the total potentials showed the effect of the winter rain (Fig. 6a), while in September the total potentials of the surface soil were low (Fig. 6b).

V. Discussion

A. Sources of Water Transpired by the Trees

There are several ways to determine the source of water being transpired by trees from sap δD and δ^{18}O data. The simplest of these is to compare the sap δ value of one isotope (either D or ^{18}O) with those of the various possible water sources (e.g., White *et al.*, 1985; Dawson and Ehleringer, 1991). However, sap δ values will not always match only one of the sources, possibly because of simultaneous uptake from two or more of the possible sources. Determination of both δD and $\delta^{18}O$ values of all possible water sources provides the greatest likelihood of determining the source of water transpired by plants (e.g., Walker *et al.*, 1992). The various comparisons within the data from this study, and the interpretations made from them, are discussed below.

1. Inland Trees   Tree sap δD values (Fig. 3a) match those of more than one possible water source. The sources are water at ~0.2 m soil depth and water below 1.4 m soil depth (including groundwater). The similarity of sap δD values to these two sources suggests uptake from either source, or from both. The $\delta^{18}O$ data (Fig. 3b) can be similarly analyzed, but indicate only one source of water; from the soil at ~0.4 m depth. This source is different from those indicated by the δD values. For the $\delta^{18}O$ data to be consistent with the δD data trees must have taken up water from two sources, which had similar δD values but different $\delta^{18}O$ values. The tree sap isotope composition would be an uptake-weighted average between these sources.

The high correlation between the δD and $\delta^{18}O$ data of the creek water, groundwater, tree sap (Eq. (1)), and some of the soil samples also suggests that the isotopic composition of the tree sap is explained by uptake from two water sources, those being groundwater and soil water from 0.2 m depth. This can be understood by considering that if there was significant uptake of soil water from any depths other than 0.2 m and below 1.4 m depth the tree sap δD-$\delta^{18}O$ data would have been displaced from the regression line. Soil water from 1.4 m depth is not considered a source of the sap water, because water at this soil depth would itself be a product of groundwater and surface soil water mixing.

The proportion of the "end-member" sources (i.e., groundwater or soil water) in each sample can be estimated from the δD and $\delta^{18}O$ data using a linear mixing model (described in the Appendix). The proportions of groundwater in the tree sap ranged from 40 to 63% over the five sampling times (Table III). There was an increase in the amount of groundwater in the sap samples through the first four sampling times, as expected from the drying of the surface soil over the sampling period (as indicated by the matric potential profiles, Fig. 6). However, there was a reduction in the proportion of groundwater in the tree sap at the fifth sampling time, from 63 to 50% (Table III), presumably due to the 4 mm of rain that fell 4 days prior to the final sampling.

2. Creekside Trees   The difference between tree sap and creek and groundwater δD and $\delta^{18}O$ values (Fig. 4) shows clearly that the creekside trees were not relying *solely* on creek water. The higher δD and $\delta^{18}O$ values of the tree sap compared with the groundwater and creek water values suggests that trees were also extracting water from the surface soil, as this was the only water source more isotopically enriched than the groundwater and creek water at the site (e.g., Fig. 3). The similarity between creekside and inland tree δD and $\delta^{18}O$ values at four of the five sampling times also suggests that the creekside trees were taking up shallow soil water, even though creek water covered some or all of the soil beneath the creekside trees at all times. However, even though these results show that shallow soil water was being used by the creekside trees, it is not possible to differentiate between the contribution of groundwater and creek water to these trees from the δD and $\delta^{18}O$ data of Fig. 4.

The contribution of different water sources to the creekside trees can be calculated from the δD-$\delta^{18}O$ data (Fig. 5) using the linear mixing model, as done with the inland trees. From the above discussion, water from 0.2 m soil depth can be taken as the isotopically enriched end-member in the mixing calculations. However, there is a problem in choosing an isotopically depleted end-member for the analysis. Since both creek water and groundwater lie on the regression line (Eq. (1)), it is not possible to distinguish between these waters for mixing calculations. Examination of the mixing model calculations will provide insight into which water source was used by the creekside trees.

If it is assumed that the creekside trees were using creek water and shallow soil water, the mixing model calculations indicate that there was 31 to 45% creek water in the tree sap at four sampling times and no creek water in the sap at the second sampling time (Table III). These results appear inconsistent with the creekside trees taking up creek water for two reasons: First, there was no relationship between the proportion of creek water in the tree sap and the advance of the creek water up the creek bank; and second, there is an absence of creek water in the tree sap at the second sampling time. Creek water covered an increasingly greater area under the trees during the first three sampling times, for example, almost completely covering the soil at the third sampling time. If the trees were exploiting this source of water, the amount of creek water in the tree sap would be expected to increase through time. More importantly, if significant amounts of creek water were used by the trees, the flux of creek water into the soil would have leached away the isotopically enriched water previously resident in the soil. Thus, there would eventually be no isotopic "signature" of the shallow soil water, and creek water and groundwater would be the end-members of the mixing model for the sap δ values.

From the above discussion, and given the strong similarity between inland and creekside tree sap isotopic compositions (Figs. 4 and 5), we suggest that the creekside trees were using soil water and groundwater, rather than creek water. In this case, groundwater would be the isotopically depleted end-member in the mixing calculations. The proportion of groundwater in the creekside tree sap increased through time, from 37 to 60%, excluding the second sampling time when it was 0% (Table III). The increase through time may be due to a reduction in shallow root activity, caused by anoxic conditions in the flooded soil.

The absence of groundwater (or creek water) at the second sampling time is not easily explained, unless it indicates that as the creek water wetted a large area of the soil it mobilized the resident soil water which was then absorbed by the trees' roots. However, this must not have been a significant process over the whole study period. If it was the isotopic signature of the soil water would have been leached away (as discussed above).

3. Rainfall Isotopic Signature   In past studies rainfall has been assumed as a possible source of water for trees. Paradoxically, the stable isotope data do not indicate that rainfall was directly used by the trees in this study, even though the matric potential data show the surface soil was wet by rain. Soil water contents were >15% prior to rainfall, and this water was isotopically enriched due to evaporation (Fig. 3). Rainfall would have mixed with this enriched soil water, and thus not been isotopically distinct. By the time of soil sampling some evaporation would have taken place further obscuring the isotopic signature of the rain. Thus rain was probably a source of water for the trees, even though there was no clear isotopic signature of rain in the tree sap data.

B. Ecological Implications of the Results

1. Activity of Surface Roots of the Inland Trees   Uptake of water from near-surface soil depths by the inland trees was expected in the early part of the study period, because matric (and total) potentials were high in this region (Fig. 6a). However, at the end of the study period surface soil matric potentials were very low (−2.5 MPa, Fig. 6b), yet around half the water transpired by the trees was derived from these soil depths (Table III). There may be some error in matric potentials and mixing model results due to factors such as spatial variability in soil water matric potentials or δ values, or the travel time of sap from roots to twigs. However, it is unlikely that these errors would have been large enough to have caused substantial changes in the range of surface soil matric potentials or the mixing model results.

The is other evidence that the trees were extracting water from shallow soil layers to low potentials at this site. Thorburn *et al.* (1993) sampled water from roots of these trees (and nearby *E. largiflorens*) in January, 1991, and compared its isotopic composition with that of soil around the roots. The roots were 10–20 mm in diameter and were taken from the top 0.2 m of soil which had very low matric potentials (−2.5 to −5.0 MPa). In 10 of the 14 roots sampled, δD values of the root waters were similar to those of the soil around the roots, but different from twig sap δD values (twig sap δD values were the same as groundwater δD values). A further two roots had δD values intermediate between those of the groundwater and surrounding soil. The similarity between the soil and root δD values showed that these large shallow roots were generally not participating in groundwater uptake, but were in communication with water in the surrounding dry soil.

Shallow roots of these trees may be active in dry soils to facilitate uptake of nutrients. Additionally, high salinity levels in the lower profile may reduce root activity, therefore making the surface roots relatively more important in water uptake.

2. Response of the Creekside Trees to Flooding   In this arid and saline environment, it is unexpected that the trees studied did not respond to being flooded with fresh creek water, given that inundation occurred comparatively slowly as creek water advanced up the creek bank. There are several possible reasons why the trees did not respond to flooding: (1) there were too few roots in the surface soil and creek bank to allow significant amounts of creek water to be absorbed by the creekside trees; (2) the hydraulic conductivity of the soil lining the creek was too low to allow significant amounts of water to flow to the roots; or (3) flooding of the soil by creek water reduced the supply of oxygen to the roots below that necessary for the root to function. Given the activity of shallow roots in dry soils (with consequently very low unsaturated soil hydraulic conductivities) at the inland site, it seems unlikely that the first two reasons account for the behavior of the creekside trees.

The ephemeral creek water does not provide a constant source of water for the trees, so exploitation of this resource would require annual (or biannual) changes in resource allocation (e.g., to grow adventitious roots) by the trees. In contrast, the groundwater is a permanent water supply. If this water was not saline the reason why trees would not divert resources to utilize the seasonal flooding would be clearer (e.g., Dawson and Ehleringer, 1991). However, our results indicate that the saline groundwater is a preferable water source to the trees. Thus, this level of salinity poses a lower level of stress on the trees than does the "work" required to utilize the seasonally available creek water.

If the microenvironmental conditions were altered, however, the tree response may be different. Possible factors that may be altered are the permanence of the creek water supply or the salinity of the groundwater. To address these factors we are investigating the source of water taken up by *E. camaldulensis* on the banks (i.e., within 1.5 m) of two permanent creeks and in an area where the groundwater is three to four times saline as at this site (Mensforth *et al.* 1993).

C. Practical Implications of the Results

The results of this study have implications for the management of *E. camaldulensis* forests in the region. For example, the water relations of trees near ephemeral creeks would not be as greatly affected by changes in river or creek flow and salinity as previously thought, as creek water does not appear to be an important source of water for the trees. Also, the importance of creek water inundation in recharging soil water deficits for maintenance of *E. camaldulensis* health must be questioned in these forests. However, tree health may still be linked with river hydrology. Since groundwater is an important water source for the trees, changes in groundwater levels or salinity (which may be caused by variations in river or creek flow) may affect the health of the trees.

VI. Summary

Naturally occurring differences in the concentration of the stable isotopes D and ^{18}O were used to determine the sources of water transpired by *E. camaldulensis*. The water relations of these trees may be complex because the *E. camaldulensis* forests in the region studied occur near perennial and ephemeral streams and are underlain by saline groundwater (electrical conductivity of 10 dS m^{-1} at our site). Simple comparison of tree sap and creek water δD and δ^{18}O data showed that creek water was not the dominant source of water for trees flooded by creek water. Also, the isotopic composition of the sap of flooded trees was similar to that of nearby *E. camaldulensis* which were not flooded. This level of data analysis was not able to provide further insights into the water source of these trees, however, as tree sap δ values could result from simultaneous water uptake at more than one source.

The combination of the δD and δ^{18}O data allowed application of a simple two end-member mixing model to estimate the proportion of groundwater and shallow soil water used by the trees. The proportion of groundwater used by trees distant from a creek increased from 40 to 63% as surface soil dried out. However, water was still being extracted from the surface soil even when soil water potentials were low (< -2.0 MPa). Previous studies at the site had also found that large surface roots were withdrawing water from soils at these matric potentials.

Flooded creekside trees did not respond to the flood waters, but took up water from similar sources and in similar proportions to the inland trees. They used an increasing amount of groundwater as the flooded area under the trees increased. This suggests that flooding was reducing surface root activity, presumably due to water logging. Although the groundwater underlying the trees at this site was saline, it provided a preferable water resource for the trees in comparison with the ephemeral creek water.

Appendix: Calculating the Proportion of Groundwater in Tree Sap from δD and δ^{18}O Data

When the chemical composition of different waters plot on a straight line, it indicates that they may have been derived by mixing of two waters of significantly different chemical composition (these waters are called end-members of the mixing process). These mixing concepts are commonly used in interpreting hydrochemistry data (e.g., Mazor, 1991). The conditions under which mixing could be assumed to occur are that there are no sinks or other sources of the chemicals involved and that it is physically plausible for mixing to occur. The proportion of each of the end-members in any sample on the mixing line can be readily calculated from a two end-member linear mixing model. In this section we will derive this model, as it is applied to the isotopic composition of waters from creeks, soil, underground sources, and tree sap in this study.

If the possible end members for the isotopic composition of tree sap are the δD and $\delta^{18}O$ values of the groundwater and soil water, the proportion of groundwater in tree sap samples is given by the ratio of the distance between groundwater and soil water to the distance between tree sap and soil water on the δD-$\delta^{18}O$ graph. The distance between groundwater and soil water data (D_{s-g}) on the δD-$\delta^{18}O$ graph can be calculated from

$$D_{s-g} = [(X_s - X_g)^2 + (Y_s - Y_g)^2],\qquad\text{(A1)}$$

where X is the $\delta^{18}O$ value of the sample, Y is the δD value of the sample (see below), and the subscripts s and g refer to soil water and groundwater. Similarly, the distance between the tree sap sample and the soil water data (D_{s-t}) is calculated from

$$D_{s-t} = [(X_s - X_t)^2 + (Y_s - Y_t)^2],\qquad\text{(A2)}$$

where the subscript t refers to tree sap.

The proportion of groundwater (P_g) in the tree sap is then given by,

$$P_g = D_{s-t}/D_{s-g}.\qquad\text{(A3)}$$

The mixing line for the tree sap samples is defined by the regression line of the creek, groundwater, and tree sap data (Eq. (1)). The isotopically enriched end-member of the mixing line is taken as the δD-$\delta^{18}O$ value of the 0.2-m depth soil sample, while the isotopically depleted groundwater end-member is the δD-$\delta^{18}O$ value of the groundwater or creek water (Fig. 5). The δD-$\delta^{18}O$ data of all the samples involved in the calculations were adjusted to fit the regression equation. This adjustment was done by calculating δD values from the measured $\delta^{18}O$ using Eq. (1). Thus the Y values referred in Eqs. A1 and A2 are the calculated δD values.

It should be noted that the proportion of groundwater in tree sap can also be calculated from concentrations of only one isotope (e.g., D; White *et al.*, 1985). However, the combination of both D and ^{18}O described above provides greater accuracy for cases where the range in δ values is small compared with the analytical precision of the values. Such was the case in this study.

Acknowledgments

The senior author acknowledges the support of an Australian Postgraduate Priority Research Award and a Centre for Groundwater Studies Bursary during this work, which was conducted while on leave from the Queensland Department of Primary Industries. This work was supported by research grants from the Australian Water Research Advisory Council. The assistance of Kerryn McEwan with the isotopic analysis is greatfully acknowledged.

References

Barnes, C. J., and G. B. Allison. 1984. The distribution of deuterium and ^{18}O in dry soils. 3. Theory for non-isothermal water movement. *J. Hydrol.* **74:** 119–135.

Barnes, C. J., G. B. Allison, and M. W. Hughes. 1989. Temperature gradient effects on stable isotope and chloride profiles in dry soils. *J. Hydrol.* **112:** 69–87.

Blake, T. J., and D. M. Reid. 1981. Ethylene, water relations and tolerance to waterlogging of three *Eucalyptus* species. *Aust. J. Plant Physiol.* **8:** 497–505.

Bren, L. J. 1988. Effects of river regulation on flooding of a riparian red gum forest on the River Murray, Australia. *Reg Rivers Res Manage.* **2:** 65–77.

Chesterfield, E. A. 1986. Changes in the vegetation of the river red gum forest at Barmah, Victoria. *Aust. For.* **49:** 4–15.

Dawson, T. E., and J. R. Ehleringer. 1991. Streamside trees that do not use stream water. *Nature* **350:** 335–337.

Dexter, B. D. 1978. Silviculture of the river red gum forests of the central Murray floodplain. *Proc. R. Soc. Victoria* **90:** 175–191.

Epstein, S., and T. Mayeda. 1953. Variations of ^{18}O content of water from natural sources. *Geochim. Cosmochim. Acta* **42:** 213–224.

Greacen, E. L., G. R. Walker, and P. G. Cook. 1989. Procedure for the Filter Paper Method of Measuring Soil Water Suction. Divisional Rep. No. 108, CSIRO Div. of Soils, CSIRO, Australia.

Jolly, I. D., G. R. Walker, K. R., McEwan, P. J. Thorburn, T. J. Hatton, J. Bennett, and J. M. Hacker. 1991. The role of floodplain vegetation in salt accession to the River Murray in South Australia, pp. 53–59. *In* International Hydrology and Water Resources Symposium 1991, Perth 2–4 October. National Conf. Publ. No. 91/22, Preprints Vol. 1, The Institution of Engineers, Australia.

Mazor, E. 1991. *In* Applied Chemical and Isotopic Groundwater Hydrology, pp. 85–91. Open University Press, Buckingham, Great Britain.

Mensforth L. J., Thorburn P. J., Tyerman S. D., and Walker G. R. 1993. Sources of water used by riparian *Eucalyptus camaldulensis* overlying highly saline groundwater. *Oecologia,* submitted.

Revesz, K., and P. H. Woods. 1990. A method to extract soil water for stable isotope analysis. *J. Hydrol.* **115:** 397–406.

Roberts, J., and J. A. Ludwig. 1990. Riparian habitats on the Chowilla floodplain of the River Murray, South Australia. *Wetlands (Aust.)* **9:** 13–19.

Thorburn, P. J., G. R. Walker, and J. P. Brunel. 1993. Extraction of water from *Eucalyptus* trees for analysis of deuterium and oxygen-18: Laboratory and field techniques. *Plant Cell Environ.* **16:** 269–277.

Van der Moezel, P. G., L. E. Watson, and D. T. Bell. 1989. Gas exchange responses of two *Eucalyptus* species to salinity and waterlogging. *Tree Physiol.* **5:** 251–257.

Walker, G. R., J. P. Brunel, and A. K. Kennett-Smith. 1992. Using stable isotopes of water to trace plant water uptake, pp. 81–100. *In* Kennett-Smith, A. K., and G. R. Walker (eds.), Point Recharge and Diffuse Discharge in the Murray Geological Basin. CSIRO Division of Water Resources, Water Resource Series, No. 7, CSIRO, Australia.

White, J. W. C., E. R. Cook, J. R. Lawrence, and W. S. Broecker. 1985. The D/H ratios of sap in trees: Implications for water sources and tree ring D/H ratios. *Geochim. Cosmochim. Acta* **49:** 237–246.

33

The ^{18}O of Water in the Metabolic Compartment of Transpiring Leaves

Dan Yakir, Joseph A. Berry, Larry Giles, and C. Barry Osmond

I. Introduction

Photosynthesis in terrestrial plants is invariably associated with transpiration of water with associated isotopic fractionation. Consequently, water at the site of photosynthetic metabolism is enriched in ^{18}O relative to the water that is transpired (the source water). It is important to know the extent of this enrichment to be able to decipher the isotopic signature in plant organic matter and to interpret the role of the terrestrial biosphere in regional and seasonal changes in the $\delta^{18}O$ of atmospheric CO_2 (Mook *et al.*, 1983; Francey and Tans, 1987; Friedli *et al.*, 1987).

Craig and Gordon (1965) developed an equation which can be used to predict the isotopic enrichment of a well-mixed body of water undergoing steady-state evaporation. Following the notation of Farquhar and Lloyd (Chapter 5, this volume), this may be written as

$$\Delta_E = h(\Delta_V - \varepsilon_K) + \varepsilon_K + \varepsilon^*, \tag{1}$$

where Δ_E, the enrichment of the evaporating water body relative to source water, is defined as $\Delta_E = R_E/R_S - 1$ (where R represents $^{18}O/^{16}O$), and is approximately equal to $\delta_E - \delta_S$; Δ_V is the corresponding difference of atmospheric water vapor and source water, $\Delta_V = R_V/R_S - 1$; h, is the relative humidity of the air at leaf temperatures; ε^* is the isotopic fractionation for equilibrium evaporation (ca. 9‰); and ε_K is the kinetic fractionation in diffusion of water vapor in air (ca. 28‰) in the stomatal pore.

Application of this theory to leaves is difficult because water in leaves is not well mixed, isotopically. The theory should, however, accurately predict the average composition of water at the site of evaporation. Diffusion of water is the principal mechanism of mixing operating in a leaf and this relatively slow process is opposed by a large net flux of liquid water to the evaporating surfaces. Under conditions where there is rapid evaporation, mixing of source water entering the leaf with evaporating water may be slow. As a result bulk water extracted from transpiring leaves is generally less enriched than that predicted by Eq. (1) (Dongmann and Nurnberg, 1974; Förstel, 1978; Leaney *et al.*, 1985; Walker *et al.*, 1989; Flanagan *et al.*, 1991; Farquhar *et al.*, 1992).

The heterogeneous nature of leaf water has been clearly demonstrated by analyzing water expressed from the petiole of a detached leaf in a pressure chamber (Yakir *et al.*, 1989), a process which essentially reverses the normal direction of flow. The first water expressed was similar to source water, and subsequently expressed water became progressively more enriched. The significant question before us here is what is the isotopic composition of the water in which leaf metabolism occurs along this gradient.

In the experiments reported here we make use of the well-documented exchange of the oxygens of CO_2 with those of H_2O. The effect of CO_2 exchange on oxygen isotope balance of leaf water should be insignificant (1:100–1000) in comparison to that of transpiration. Thus, CO_2 can be used to "probe" leaf water, without significantly perturbing its isotopic composition. In leaves the equilibration of CO_2 with H_2O is catalyzed by carbonic anhydrase, an enzyme that is predominantly localized in the chloroplasts. Other compartments lacking carbonic anhydrase should not contribute significantly to the exchange in leaves because the uncatalyzed rate is many orders of magnitude slower (cf. Yoshida and Mitzutani, 1986).

As discussed by Farquhar and Lloyd (Chapter 5, this volume), the isotopic exchange of CO_2 generally occurs under nonequilibrium conditions, and factors such as incomplete isotopic equilibrium of CO_2 (because of limited enzyme activity or mass flow), and kinetic isotope effects associated with diffusion across the stomata and leaf boundary layer, need to be considered. The measurement protocol developed here is designed to minimize these nonequilibrium effects. We have used the $\delta^{18}O$ of CO_2 at the CO_2 compensation point (Γ in a closed recirculating system (which equilibrates with chloroplast water due to the action of carbonic anhydrase), as well as the $\delta^{18}O$ of the O_2 from photosynthetic water oxidation (Guy *et al.*, 1987, 1993), to report the $\delta^{18}O$ value of water in the metabolic compartment of leaf cells.

The experiments described here were done in the phytotron, Duke University in 1990–1991, and preliminary reports of data have been presented in the proceedings of several symposia (Yakir *et al.*, 1992; Yakir, 1992; Berry, 1991). This paper embraces a series of corrections to earlier data sets which have been suggested in the course of widespread discussion among colleagues. The corrections do not alter the conclusion that leaf metabolism in transpiring leaves of sunflower occurs in a water fraction the oxygen isotopic composition of which is distinctly less positive than that predicted for water at the sites of evaporation.

II. Material and Methods

A. Plant Material and Growth Conditions

Sunflower plants (*Helianthus annuus* cv. giant mammoth) were germinated in vermiculite. After 1 week sprouts were transferred to a hydroponic system containing half-strength Hoagland solution inside a growth chamber (Environmental Growth Chambers, Model M-13). Conditions in the growth chamber were kept at 26°C and 1000 μE m^{-2} s^{-1} for a 16-h day and at 20°C for the 8-h dark. Plants were used for gas exchange measurements after about 3 weeks growth.

B. Gas Exchange and Sample Collection

Attached leaves were enclosed in a leak-tight, fan-stirred gas exchange cuvette, connected to a gas flow system which could be configured for open flow or a recirculating flow via a loop containing a cold trap for drying the air (Fig. 1). The relative humidity in the chamber was monitored with a

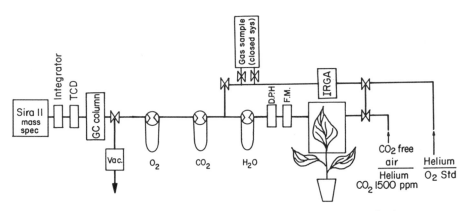

Figure 1. A schematic representation of the on-line gas exchange system for the measurement of $^{18}O/^{16}O$ ratios in CO_2 and O_2 released by leaves. An intact leaf was enclosed in a gas-tight, fan-stirred cuvette fitted with a glass top and placed under a light source (150 μE m^{-2} s^{-1}). Relative humidity in the cuvette was measured using a dew-point hygrometer (D.P.H.) in the exit flow of the chamber and was controlled by adjusting the flow rate (F.M.). The system could be used in three modes: (1) CO_2-free air flow was used to drive all respired CO_2 from the leaf into the CO_2 trap (liquid nitrogen) after drying the gas stream in the H_2O trap (-80°C); (2) the system loop was closed and air at 45 ppm CO_2 (as measured by infrared gas analyzer, IRGA) was circulated through the leaf cuvette, the H_2O trap, and the 1-liter gas-sampling volume; after equilibrium was attained the sample tube could be isolated and the CO_2 separated and used for isotopic analysis; (3) a flowing mixture of helium/CO_2 (1500 ppm) was used to purge O_2 from the system and to suppress respiration by the leaf. Photosynthetically produced O_2 was trapped by the O_2 trap (molecular sieve at liquid nitrogen temperature) after drying (H_2O trap) and CO_2 removal (CO_2 trap). Standard samples of O_2 from a cylinder were injected into the line and trapped as above. CO_2 or O_2 released by warming the respective traps was fed into the mass spectrometer (Sira II) by a helium flow. Resultant peaks were integrated for quantitative analysis. Transpired water trapped by the H_2O trap was also recovered and used for isotopic analysis by conventional methods.

dew-point sensor and controlled by adjusting the flow of dry air through the chamber. Dry CO_2-free air was fed into the system in the open flow model while the CO_2 concentration in the closed system was allowed to reach the CO_2 compensation point. Water vapor transpired by the leaf was trapped for analysis of $\delta^{18}O$, and Fig. 2 shows that, in sunflower, approximately 2 h were required to establish isotopic steady state, when the $\delta^{18}O$ of transpired water becomes the same as that of source water.

Once the leaf attained steady state at the chosen humidity and CO_2 concentration, one of two protocols was used to estimate the $\delta^{18}O$ of chloroplast water.

1. A flask containing 1 liter of air at CO_2 compensation concentration (45 ppm CO_2) was switched into the line (Fig. 1) while flow through a loop containing a water trap was adjusted to keep the relative humidity constant.
2. The line was left in flow-through mode with flowing CO_2-free air. The photorespired CO_2 from the leaf was trapped cryogenically and analyzed (Fig. 1).

In the first case there was no net flux of CO_2 across the stomata and the same population of CO_2 molecules could be exposed to the leaf indefinitely. Given sufficient time for exchange, the CO_2 in the sample flask should come to a complete isotopic equilibrium with chloroplast water. The kinetics of the approach to equilibrium will be considered later. This method was theoretically the more pleasing since complete equilibrium

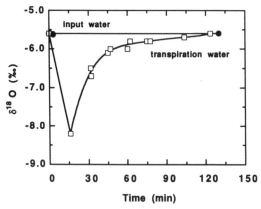

Figure 2. Establishment of isotopic steady-state in leaf water. An intact leaf of a sunflower was enclosed in a leaf gas exchange chamber (Fig. 1) at 80% relative humidity for 4 hours before the relative humidity in the chamber was changed to 60% (Time 0). Transpiration water was trapped at the times indicated (□) and its $\delta^{18}O$ value determined by mass spectrometer. Equality of the isotopic composition of source water (●) with that of transpiration water indicates the attainment of isotopic steady state in leaf water. The experiment was then repeated and the two data sets were combined.

should be established, while the second protocol was much easier and quicker. We used the first method primarily to control for possible artifacts in the second.

In the second protocol, we could be sure that all of the CO_2 trapped had contacted the leaf, since it was photorespired principally by mitochondria closely associated with the chloroplasts of the mesophyll cells in the leaf. Nearly complete isotopic equilibrium of CO_2 with chloroplast water within the leaf seemed likely, but since there was a net outward flux of CO_2, a kinetic effect due to diffusion was to be expected. This correction was calculated from measurements of the flow rate through the stirred cuvette and the stomatal conductance of the leaves, as discussed later.

We also used an independent method for assessing the $\delta^{18}O$ of chloroplast water. Since O_2 is produced irreversibly without discrimination from chloroplast water (Guy *et al.*, 1987, 1993), the $\delta^{18}O$ of the O_2 should be identical to that of the chloroplast water. This provided a check on the assumptions that CO_2 exchanged predominantly with chloroplast water. A leaf in the cuvette was provided with a flow of O_2-free helium containing CO_2 (1500 ppm). O_2 produced by chloroplasts was trapped at liquid nitrogen temperatures on a trap of molecular sieve 5A as that gas left the cuvette.

C. Isotope Ratio Measurements

At the end of each experiment the leaf was cut and placed in a septum-sealed 15-ml test tube. The tube was then attached to a vacuum line and leaf water was quantitatively distilled at 80°C and collected cryogenically. Subsamples of 0.5 ml from leaf-water samples, and of water from the hydroponic system (source water), were equilibrated with CO_2 at 25°C for 36 h. An aliquot of the CO_2 was cryogenically purified under vacuum before mass spectrometric analysis. In several cases stem water was also distilled and analyzed to confirm the lack of isotopic fractionation during uptake.

Samples of CO_2 from the gas exchange system were dried and trapped in the double trap inlet system of the mass spectrometer. Noncondensable air was pumped out and the CO_2 samples were analyzed in the dual-inlet mode.

Oxygen samples were dried and concentrated with molecular sieve (Mallinckrodt 5 A) on line and fed with a helium flow into the mass spectrometer via a direct inlet (Fig. 1). In this mode a working standard O_2 was injected into the line between samples. The $\delta^{18}O$ of the oxygen standard was determined by conversion to CO_2 on hot graphite (Yakir and DeNiro, 1990). It was analyzed in the dual-inlet mode with the same standard used in the CO_2 measurements described above.

All mass spectrometer measurements were carried out on a VG-Sira II mass spectrometer and expressed in the δ notation where

$$\delta^{18}O = \{[(^{18}O/^{16}O)_{sample}/(^{18}O/^{16}O)_{standard}] - 1\} \times 1000\%o$$

and the standard is standard mean ocean water (SMOW). Internal precision for the isotopic determinations were better than 0.1‰ in the dual-inlet mode and better than 0.3‰ in the direct-inlet mode.

D. Theory

In the first protocol, CO_2 recirculating in the closed system at Γ should eventually reach isotopic equilibrium with chloroplast water. We can estimate the kinetics of approach to equilibrium from measurements of the flow rate of air circulating through the chamber and the stomatal conductance of the leaf. The system may be viewed as three linked CO_2 pools,

$$c_i \underset{g_s}{\overset{\longleftarrow}{\longrightarrow}} c_a \underset{g_f}{\overset{\longleftarrow}{\longrightarrow}} c_v, \tag{2}$$

where c_i is the CO_2 within the leaf, c_a within the chamber, and c_v within the sample tube. These pools are linked by stomatal conductance (g_s) and the "effective conductance" (flow/leaf area) of the forced flow through the sample tube (g_f). At the start of the experiment $c_i = c_a = c_v$, but these pools are not necessarily at isotopic equilibrium, δ_v will approach δ_i with a relaxation time; $\tau \cong M_V/F_E$, where M_V is the quantity of CO_2 in the enclosed air and F_E is the gross exchange flux (mol s^{-1}) of CO_2 across the conductances g_s and g_f; $F_E = m \, c_v[g_s g_f/1.6(g_s + g_f)]$, where m is leaf area. In a typical experiment $M_V = 1.8 \, \mu$mol; $g_s = 0.623$ mol m^{-2}s^{-1}; $m = 6.27 \times 10^{-3}$ m^2; and the air flow $g_f = 1.64$ l min^{-1} or 0.172 mol m^{-2}s^{-1}, giving $\tau \cong 85$ s. A sample that exchanged 5 min (about 3.5 turnovers) gave the same $\delta^{18}O$ value as samples that exchanged 10, 15, or 30 min with the same leaf (± 0.5‰).

In the second protocol there was a net flux of CO_2 from photorespiration in the leaf which was flushed out of the chamber by the mass flow of air. This may be viewed as

$$c_i \underset{F_E}{\overset{F_{E'}}{\longleftrightarrow}} c_a \overset{F_f}{\longrightarrow} \text{trap}. \tag{3}$$

At the steady state c_a is smaller than c_i and F_E greater than $F_{E'}$. Consequently the diffusional fractionation will be partly expressed with δ_i will be greater than δ_t, where subscripts i and t refer to intercellular air (in equilibrium with chloroplast water) and trapped air, respectively. This system is essentially analogous to normal CO_2 fixation, except that the flux of CO_2 is reversed. Following Farquhar *et al.* (1982) we may write that

$$\Delta = [(c_i - c_a)/c_i]a + [c_a/c_i]b, \tag{4}$$

where a is the diffusional fractionation, 8.8‰ (calculated from the reduced masses; cf. Craig and Gordon, 1965), and b, the discrimination associated with F_f, is zero. The second term is, thus, always zero. We need to know the CO_2 concentrations, and these can be estimated from the measured conductances. By noting that the net flux

$$F_N = g_s(c_i - c_a) = g_f(c_a) = g_t(c_i), \tag{5}$$

where g_t is the series conductance, we may write

$$(c_i - c_a)/c_i = g_t/g_s = g_f/(g_f + g_s) \quad (6)$$

or

$$\Delta = g_f/(g_f + g_s)a \quad (7)$$

and

$$\delta_E \cong \delta_T + \Delta, \quad (8)$$

where subscripts E and T represent the CO_2 at equilibrium with chloroplast water and CO_2 trapped, respectively. More CO_2 could be obtained by this protocol and it was easier to keep the system at steady state. The overall precision was about 0.3‰.

III. Results

Figure 3 shows the $\delta^{18}O$ values of water in the leaf which equilibrates with photorespiratory CO_2 during photosynthesis under a range of different relative humidities and different transpiration conditions. The $\delta^{18}O$ of chloroplast water were calculated from the $\delta^{18}O$ of CO_2 based on the well-characterized equilibrium isotope effects (Brenninkmeijer *et al.*, 1983).

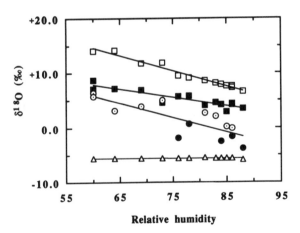

Figure 3.   Water compartmentation in sunflower leaves, the CO_2 approach. An attached sunflower leaf was enclosed in a gas exchange chamber (Fig. 1) at each of the indicated relative humidities and constant conditions were maintained until isotopic steady state was attained (as shown in Fig. 2). CO_2 was then collected in a sample tube (closed system) or a CO_2 trap (open system) and its $\delta^{18}O$ value was used to reconstruct the $\delta^{18}O$ value of water in the chloroplasts ($\bigcirc$, $\bullet$) by correcting for CO_2–H_2O equilibrium fractionation. Data collected in the open, flow-through, mode (see Fig. 1) were also corrected for diffusional fractionation as described under II,D, Theory. Total leaf water ($\blacksquare$) and stem water ($\triangle$) were then extracted for determination of their $\delta^{18}O$ values. The isotopic composition of the water at the evaporating surfaces ($\square$) was calculated using Eq. (1).

These $\delta^{18}O$ values were corrected for diffusional isotopic effects, where appropriate, as discussed in the theory above. No correction is needed for the measurements in the closed system, and for the open system, corrections range from 1 to 4‰.

The values obtained from CO_2 compensation point experiments at high RH were slightly but consistently less positive than those obtained by flow-through technique. These values, which require no assumptions to be made with respect to diffusional isotope fractionations, indicate metabolic water is about 6‰ less positive than the bulk leaf water and is much closer in isotopic composition to source water than to water at the evaporating sites. In eight of nine experiments the values obtained by the flow-through method are also less positive than bulk leaf water. The line fitted to both flow-through and closed-system data sets is about parallel to the Craig line for water at the sites of evaporation, but almost 10‰ less positive.

These observations were confirmed by direct measurement of $\delta^{18}O$ in O_2 evolved from water oxidation in photosynthesis (Fig. 4). The line fitted to these data is again about parallel to the Craig line for water at the sites of evaporation, but some 6‰ less positive. The calculated $\delta^{18}O$ values for the water at the evaporating site have been corrected for the effects of the helium atmosphere on fractionation during evaporation according to Craig and Gordon (1965).

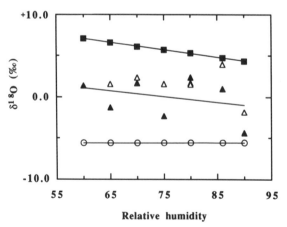

Figure 4. Water compartmentation in sunflower leaves, the O_2 approach. Leaves were treated as described for Fig. 3 but under helium/CO_2 mixture (1500 ppm CO_2). Photosynthetic O_2 evolved from water in chloroplasts was trapped, concentrated, and fed directly into a mass spectrometer for $\delta^{18}O$ determination (Fig. 1). The results (▲) are compared with those obtained for bulk leaf water (△) and source water (○) extracted at the end of each experiment. The $\delta^{18}O$ values of water at the evaporating sites, under the helium atmosphere, were calculated using Eq. (1) (■).

IV. Discussion

The observations presented here confirm that leaf water appears to show significant compartmentation with respect to isotopic composition. The apparent compartmentation persists at high transpirational fluxes and appears to be of a similar magnitude to that which is observed using the pressure expression method (Yakir *et al.*, 1989). The processes which might maintain this apparent compartmentation have been discussed, but no satisfactory mechanistic explanation is currently available (Yakir, 1992). One interpretation discussed by White (1989) with acknowledgment to Farquhar, and by Farquhar and Lloyd (Chapter 5, this volume), assumes that leaf water is a continuum of isotopic composition, ranging between that at the evaporating sites in leaves (defined by the Craig model) and those of source water in veins. Farquhar *et al.* (1993) report measurements of the $\delta^{18}O$ of CO_2 exchanged under nonequilibrium conditions, from which they conclude that metabolic water has an isotopic composition indistinguishable from that at the evaporating sites. This is consistent with the continuum model since the chloroplasts should be physically, and hence isotopically, very "close" to the evaporating sites. Our results showing that the isotopic composition of bulk leaf water is intermediate between the source water and that at the evaporating sites (Figs. 3 and 4) are also consistent with this model. However, our experiments using independent methods based on isotopic exchange of CO_2 with leaf water and direct photosynthetic production of O_2 yield results that are difficult to reconcile with this simple model. The water that is the source of photosynthetic O_2 is apparently substantially less enriched than the water at the evaporating sites and could not be distinguished from the bulk leaf water (Fig. 4). Furthermore, CO_2 exchanges with a pool of leaf water that is also less enriched than evaporating water, and in this case, this pool appears to be "closer" to the source water than the bulk leaf water (Fig. 3). Both methods should probe the isotopic composition of chloroplast water, which appears to be very different from that at the evaporating sites.

At this point we can not explain the difference between our studies and those of Farquhar *et al.* (1993). It is possible that we each have chosen species or environmental conditions that lead to a different relationship between the water of chloroplasts and that at the evaporating sites. Nevertheless, our experiments provide useful confirmation that, despite the isotopic gradients within a leaf, the isotopic composition of both CO_2 and O_2 are uniquely determined by chloroplast water.

We note that assumptions regarding the mechanisms controlling the isotopic composition of chloroplast water have implications for the ¹⁸O balance of atmospheric O_2 (Berry, 1991) and CO_2 (Francey and Tans, 1987). Farquhar *et al.* (1993) have developed a model of the ¹⁸O balance of atmospheric CO_2 that takes into account the global distribution of local fluxes of CO_2 between the atmosphere, ocean, and biosphere and the isotopic composition of waters that exchange with CO_2. On a global scale, the largest of these exchange fluxes is with chloroplast water. Therefore, efforts to understand the isotope balance of the atmosphere depend critically on theoretical or empirical models that can be used to predict the isotopic composition of chloroplast water. The Farquhar *et al.* (1993) model uses meteorological data to predict the isotopic composition of water at steady-state evaporation and assumes that this is identical to the isotopic composition of chloroplast water.

Our limited experiments do not negate the theoretical approach taken by Farquhar *et al.* (1993). However, further empirical confirmation of the theoretical approach should be sought. A gas exchange method based on the studies described here provides one possible way to assess the isotopic composition of chloroplast water in plants in natural environments.

We are attracted to the notion that plant organic matter might report an integrated value for the $\delta^{18}O$ of leaf water in which metabolism takes place, serving as an integrator of the isotopic signature conveyed from the plant to atmospheric CO_2 over the time that it grew. Furthermore, if this information could be extracted, useful information about past climatic or hydrologic regimens could be obtained from analysis of plant samples (White, 1989). Studies with submerged aquatic plants and with model systems show a remarkably constant oxygen isotopic discrimination of about 27‰ between carbon-bound oxygens in compounds such as cellulose and the water (Epstein *et al.*, 1977; DeNiro and Epstein, 1979; Yakir and DeNiro, 1990). However, there are large diurnal fluctuations in the $\delta^{18}O$ of the water in leaves transpiring in air (Förstel, 1978; Leaney *et al.*, 1985). Superimposed on these are isotopic gradients within the transpiring tissues. These uncertainties concerning the isotopic composition of the relevant water complicate the interpretation of the isotopic signature in carbon-bound oxygen of leaf carbohydrates. When first formed in photosynthesis, the oxygens of carbohydrates should reflect the composition of chloroplast water. However, in subsequent metabolic steps leading to cellulose deposition up to 50% of these oxygens re-equilibrate with water. This would tend to obscure the signature of chloroplast water if this occurs in a different compartment of the leaf or even at a different time in the day. The technical approaches explored here should make it possible to evaluate the impact of "postphotosynthetic metabolism" on the $\delta^{18}O$ of various chemical fractions of the leaf, and it seems plausible that a useful record of chloroplast water may be preserved in the $\delta^{18}O$ of some leaf constituents.

The isotopic exchange of CO_2 between the leaves and the atmosphere, as we presently understand it (Farquhar *et al.*, 1993; Osmond *et al.*, 1993), is exceptionally dynamic and probably beset with species-dependent, capacitance-related effects on isotopic gradients within leaves (e.g., Tissue *et al.*, 1991). However, it is impossible to reproduce such detailed processes in global scale models of atmospheric isotope balance. Studies that combine modeling with direct empirical measurements in nature are clearly needed to resolve this dilemma.

Acknowledgments

This research was supported by grants from the North Carolina Biotechnology Center, from Duke University, by NSF Grant BSR87-06429 to the Duke University Phytotron, and by U.S. Department of Energy DE-FG05-89ER14005. We are grateful to Dr. John Hays, Biogeochemistry Laboratories, Indiana University, Bloomington, for advice on handling of O_2 which made possible the direct, on-line method described here. Contribution No. 57, Department of Environmental Sciences and Energy Research, The Weizmann Institute of Sciences.

References

Berry, J. A. 1991. Biosphere, atmosphere, ocean interactions, a plant physiologist's perspective, pp. 441–456. *In* P. G. Falkowski and A. O. Woodhead (eds.), Primary Productivity and Biogeochemical Cycles in the Sea. Plenum, New York.

Brenninkmeijer, C. A. M., P. Craft, and W. G. Mook. 1983. Oxygen isotope fractionation between CO_2 and H_2O. *Isot. Geochem.* **1:** 181–190.

Craig, H., and L. Gordon. 1965. Deuterium and oxygen-18 variation in the ocean and marine atmosphere, pp. 9–130. *In* Proceeding of the Conference on Stable Isotopes in Oceanography Studies and Paleotemperatures. Laboratory of Geology and Nuclear Science, Pisa.

DeNiro, M. J., and S. Epstein. 1979. Relationships between the oxygen isotope ratios of terrestrial plant cellulose, carbon dioxide and water. *Science* **204:** 51–53.

Dongmann, G., and H. W. Nurnberg. 1974. On the enrichment of H_2 ^{18}O in leaves of transpiring plants. *Radiat. Environ. Biophys.* **11:** 41–52.

Epstein, S., P. Thompson, and C. J. Yapp. 1977. Oxygen and hydrogen isotopic ratios in plant cellulose. *Science* **198:** 1209–1215.

Farquhar, G. D., M. H. O'Leary, and J. A. Berry. 1982. On the relationship between carbon isotope discrimination and intercellular carbon dioxide concentration in leaves. *Aust. J. Plant Physiol.* **9:** 121–137.

Farquhar, G. D., J. Lloyd, J. A. Taylor, L. B. Flanagan, J. P. Syvertsen, K. T. Hubick, S. C. Wong, and J. R. Ehleringer. 1993. Vegetation effects on the isotope composition of oxygen in atmospheric CO_2, in press.

Flanagan, L. B., J. P. Comstock, and J. R. Ehleringer. 1991. Comparison of modeled and observed environmental influences on the stable oxygen and hydrogen isotope composition of leaf water in *Phaseolus vulgaris* L. *Plant Physiol* **96:** 588–596.

Förstel, H. 1978. The enrichment of ^{18}O under field and under laboratory conditions. *Radiat. Environ. Biophys.* **15:** 323–344.

Francey, R. J., and P. P. Tans. 1987. Latitudinal variation in oxygen-18 of atmospheric CO_2. *Nature* **327:** 495–497.

Friedli, H., U. Siegenthaler, D. Raudber, and H. Oeschger. 1987. Measurements of concentration, $^{13}C/^{12}C$ and $^{18}O/^{16}O$ ratios of tropospheric carbon dioxide over Switzerland. *Tellus* **39B:** 80–88.

Guy, R. D., M. L. Fogel, J. A. Berry, and T. C. Hoering. 1987. Isotopic fractionation during oxygen production and consumption by plants, pp. 597–600. *In* J. Biggins (ed.), Progress in Photosynthesis Research III. Nijhoff, Dordrecht.

Guy, R. D., M. L. Fogel, J. A. Berry. 1993. Photosynthetic fractionation of the stable isotopes of oxygen and carbon. *Plant Physiol* **101:** 37–47.

Leaney, F. W., C. B. Osmond, G. B. Allison, and H. Ziegler. 1985. Hydrogen-isotope composition of leaf water in C_3 and C_4 plants: Its relationship to the hydrogen-isotope composition of dry matter. *Planta* **164:** 215–220.

Mook, W. G., M. Koopmans, A. F. Carter, and C. D. Keeling. 1983. Seasonal, latitudinal and secular variations in the abundance and isotopic ratios of atmospheric carbon dioxide. *J. Geophys. Res.* **88:** 10915–10933.

Osmond, C. B., D. Yakir, L. Giles, and J. Morrison. 1993. From corn shucks to global green-house: Stable isotopes as integrators of photosynthetic metabolism. *In* N. E. Tolbert and J. Preiss (eds.), Photosynthetic Carbon Metabolism and Regulation of Atmospheric CO_2 and O_2. Academic Press, San Diego.

Sternberg, L. S. L. 1989. Oxygen and hydrogen isotope ratios in plant cellulose: Mechanisms and applications, vol. 68, pp. 124–143. *In* P. W. Rundel, J. R. Ehleringer, and K. A. Nagy (eds.), Stable Isotopes in Ecological Research: Ecological Studies. Springer-Verlag, Berlin/Heidelberg.

Tissue, D. T., D. Yakir, and P. S. Nobel. 1991. Diel water movement between parenchyma and chlorenchyma of two desert CAM plants under dry and wet conditions. *Plant Cell Environ.* **14**: 407–413.

Walker, C. D., F. W. Leaney, J. C. Dighton, and G. B. Allison. 1989. The influence of transpiration on the equilibration of leaf water with atmospheric water vapor. *Plant Cell Environ.* **12**: 221–234.

White, J. W. C. 1989. Stable hydrogen isotopic ratios in plants: A review of current theory and some potential applications, pp. 142–162. *In* P. W. Rundel, J. R. Ehleringer, and K. A. Nagy (eds.), Stable Isotopes in Ecological Research. Springer-Verlag, New York.

Yakir, D. 1992. Water compartmentation in plant tissue: Isotopic evidence, pp. 205–222. *In* S. G. N. Somero, C. B. Osmond and C. L. Bolis, (eds.), Water and Life. Springer-Verlag, Berlin/Heidelberg.

Yakir, D., and M. J. DeNiro. 1990. Oxygen and hydrogen isotope fractionation during cellulose metabolism in *Lemna gibba* L. *Plant Physiol* **93**: 325–332.

Yakir, D., M. J. DeNiro, and P. W. Rundel. 1989. Isotopic inhomogeneity of leaf water: Evidence and implications for the use of isotopic signals transduced by plants. *Geochim. Cosmochim. Acta* **53**: 2769–2773.

Yakir, D., J. A. Berry, L. Giles, C. B. Osmond, and R. Thomas. 1992. Applications of stable isotopes to scaling biospheric photosynthetic activity. *In* J. R. Ehleringer and C. Field (eds.), Scaling Processes between Leaf and Landscape Levels. Academic Press, San Diego.

Yoshida, N., and Y. Mizutani. 1986. Preparation of carbon dioxide for oxygen-18 determination of water by use of a plastic syringe. *Anal. Chem.* **58**: 1273–1275.

Index

Accumulated transpiration, tree ring
carbon isotope fractionation, 145,
149–150

Acer saccharum, soil water potential, 484

Acremonium coenophialum, temperate grass,
283, 293–295

Algae, carbon isotope fractionation, 41

Altitude
gradients, conifer carbon isotope discrimination, 194–195
lichen carbon isotope discrimination,
205
variation, conifer carbon isotope discrimination, 191–194

Arabidopsis thaliana
carbon isotope discrimination
carbon dioxide exchange, 380
carbon isotopic composition, 378–379
DNA polymorphisms, 376–377
gas exchange, 379–380
gene isolation, 384–385
genetic analysis of associated traits,
383–384
genetic and nongenetic interactions,
378–383
genetic variation, 375
humidity effects, 379–380
light effects, 379–380
mineral concentration, 384
seed effects, 380–383
genetic features, 373
RFLP linkage map, 378
RFLP prospects, 383–385

Artemisia tridentata, water movement, 483

Atmosphere
carbon isotope exchange between plants
and, 50–58, 110–127, 134–139
isotopic composition, 50, 110–115,
134–139

Bald cypress, deuterium/hydrogen ratio in
xylem-water, 474

Banksia, carbon isotope discrimination and
transpiration efficiency, 320–321

Barley
carbon isotope discrimination
biomass relationship, 456–458
distribution, 317–318
flowering time relationship, 459–460
genotype, 351
genotypic ranking, 352
variation, 314–416
water-use efficiency relationship, 453
water variation, 322
transpiration efficiency, water variation
and, 322

Barley breeding
carbon isotope discrimination
broad-sense heritability, 412–414
drought resistance, 407–410
drought-stress interaction, 401–402
genotype × environment interaction,
414
grain yield, 402, 412–414
ICARDA experimental work, 401–
402
screening, 410–414
spaced versus dense plantings, 411–
412
temporal variation, 407
transpiration efficiency, 400–401,
405–407, 412–414
water variation, 405–406
yield stability, 407–410

Basal area increment index, tree ring,
144–145, 147, 151

Bean, *see also Phaseolus*
carbon isotope discrimination
biomass relationship, 456–458
water-use efficiency relationship, 453

Biomass, genetic intransigency, 451–452

Boundary layer conductance, isotope
fractionation, 74–76

Boundary layer discrimination, CO_2 exchange, 51–52

Boxelder, cellulose nitrate deuterium/
hydrogen ratio, 489–490

Bread wheat breeding
  broad-sense heritabilities, 429–430
  carbon isotope discrimination
    broad-sense heritability, 429–430
    coefficient of variation, 423–426
    environment effects, 428–429
    gene effects, 428–429
    generation means, 423–426
    implications, 432
    interaction effects, 428–429
    test of significance, 423–426, 427–428
    transpiration efficiency correlation
      with, 430–431
  genetic analyses, 426–428
  genetic models and analyses, 422
  genotypes, 420–421
  growth characters
    broad-sense heritability, 429–430
    carbon isotope discrimination correlation, 430–431
    coefficient of variation, 423–426
    environment effects, 428–429
    gene effects, 426–428
    generation means, 423–426
    interaction effects, 428–429
    test of significance, 423–426
  implications, 431–432
  indirect selection, 431
  response to selection, 422
  transpiration efficiency
    broad-sense heritability, 429–430
    carbon isotope discrimination correlation, 430–431
    coefficient of variation, 423–426
    gene effects, 426–428
    generation means, 423–426
    implications, 432
    studies, 419–420
    test of significance, 423–426
  water treatments, 421,423–426
Broad-sense heritability
  barley, carbon isotope discrimination, 412–414
  bean, 391–393
  bread wheat, 429–430
  cowpea, 355–359
  wheat, 451–460

Canopy effect, isotope variation, 13, 110–111, 134–137
Carbon-13
  carbon dioxide, measurement, 19–20
  content
    Aloe, 31

  forest canopy, 110–111, 134–137
  Poaceae, 29–30
  tree ring, water status, 126–127, 146–149
  leaf, applications, 12–13
  photosynthetic pathway and content, 12
  plant depletion, history, 10
  variation, basis of plant, 12–13
Carbon-13/carbon-12 ratio, *see also* Carbon
    isotope ratio
  C_3 and C_4 plant, 29–30
  fractionation, terms applied, 19–20
  kinetic effects, 48
  plant surveys, history, 10–11
  standards, 47
Carbon dioxide
  assimilation
    mathematics, 32
    water content, 102–104
    water stress, 98–100
  bicarbonate interconversion, 22
  $^{13}C/^{12}C$ content, measurement, 19–20
  carbon isotope composition, 50
    desert plant, 159–160
  carboxylation, isotope fractionation, 22–23
  compensation point, 35
    lichen, 203–204
  diffusion
    carbon isotope effect on atmospheric, 48
    conifer carbon isotope discrimination, 195–197
    isotope fractionation, 21–22
    separation, 34–35
  diffusive conductance to, transpiration rate, 349
  discrimination, leaf-water $^{18}O/^{16}O$ ratio, 86–88
  discrimination between bicarbonate, 38
  hydration, carbon isotope effect, 49
  intercellular, conifer photosynthetic rates, 196
  leakage, C_4 plant discrimination, 38–39
Carbon dioxide exchange
  carbon isotope effects
    boundary layer discrimination, 51–52
    C_3 photosynthesis, 19–25, 56–58
    C_4 photosynthesis, 22–23, 25–26, 58
    carboxylation fractionations, 55
    decarboxylations, 55–56
    intercellular air space diffusion, 53–54
    leaf cell wall resistance, 54–55
    liquid phase pathway, 54–55

physical chemistry, 47–50
processes affecting, 50–56
stomata diffusion, 52–55
Jeffrey pine, 216, 218
leaves, oxygen isotope fractionation,
59–66, 72–86, 535–537
lichen, 205–206
loblolly pine, 236
oxygen isotope effects
leaves, 62–66
$^{18}O/^{16}O$ in leaf water, 59–60
photosynthetic characteristics, 57–58
shortleaf pine, 236
tall fescue, 288–293
tropical forest, 135–137
Carbon dioxide source
coniferous canopy, carbon isotope
discrimination, 119–122
forest canopy, carbon isotope discrimination, 110–115
lichen, carbon isotope discrimination,
203
tropical forest, 132
carbon isotope discrimination and,
116–117
Carbonic anhydrase
CO_2 exchange and, 54
CO_2 and HCO_3^- interconversion, 22
Carbon isotope
carbon dioxide exchange and effects of,
see Carbon dioxide exchange, carbon isotope effects
composition, 47
photosynthetic pathway distribution, 12
ratio, physical chemistry, 47–50
Carbon isotope composition
Arabidopsis thaliana, carbon isotope
discrimination and, 378–379
coniferous canopy
CO_2 source, 120–122
diurnal changes, 121
effect of height, 119–120
desert plants
gas exchange activity, 157–158, 159
intercellular CO_2 concentration, 158,
159–160
water-use efficiency, 159–160
factors affecting, 110
forest canopy, 110–115
changes in, 114–115
lichen, 202
loblolly pine, *see* Loblolly pine, carbon
isotope composition
plant breeding, tissue type and, 350–
352

shortleaf pine, *see* Shortleaf pine, carbon isotope composition
tropical forest
CO_2 concentration, 134
CO_2 flux, 117–119
CO_2 source, 116–117
gas exchange, 135–137
height effect, 115–116
light effects, 134–135
materials and methods, 133–134
principles, 131–133
water-use efficiency, 135
variations in, canopy CO_2 and water
fluxes, 125–127
water-use efficiency relationship to,
desert community, 479
Carbon isotope discrimination
*Arabidopsis thaliana, see Arabidopsis
thaliana*, carbon isotope discrimination
barley, *see* Barley, carbon isotope discrimination
bean germplasm, variation, 388–390
biomass relationship, 456–458
bread wheat, *see* Bread wheat breeding
breeding
biomass relationship, 456–458
canopy growth relationship, 458
canopy temperature as substitute, 455
cost, 454
flowering time relationship, 459–460
genetic challenges, 455–456
measurement, 453–455
mineral content as substitute, 454
sampling factor, 455
specific leaf area as substitute, 454
use of, 460–461
water-use efficiency, 452–453, 456–
460
C_3 plant
against $^{13}CO_2$, 20–25
components, 131
canopy growth relationship, 458
coffee, *see* Coffee, carbon isotope discrimination
conifer
altitudinal gradients and plasticity,
194–195
altitudinal variation, 191
genetic variation, 191–194
leaves, 188–189
plasticity, 195–197
water-use efficiency, 187, 188–191
coniferous canopy, 119–124
CO_2 responses, 123–124

light variations, 122
water stress, 122–123
cowpea, *see* Cowpea, carbon isotope
discrimination
desert plant
CO$_2$ concentration as indicator, 159–
160
growth rate and neighbor removal,
168–170
interpopulation variation, 163–164
life expectancy, 161, 168
natural selection, 166–170
wash microhabitats, 161–162
water competition, 164–166
water-use efficiency, 159–160, 162–
163
drought resistance, 407–410
forest canopies, studies with, 109–110
genetic variation for, nature of, 456
genotype × environment interaction,
414
genotypic differences in
date of sampling, 351
drought, 353–354
environmental conditions, 351–352
gas exchange, 354
physiological basis, 353–355
tissue type, 350–351
water-use efficiency, 354–355
genotypic ranking, consistency, 352–
353
grain yield and, water stress, 363, 402–
405
harvest index, 364
humidity effects, 379–380
inheritance, 355–360, 391–393, 412–
414, 451–460
breeding strategies, 361–367
environmental influence, 359–360
flowering effect, 360
harvest index effect, 360–361
sampling time effect, 357–359
intrinsic gas exchange and, equation, 4–6
Jeffrey pine, *see* Jeffrey pine, carbon
isotope discrimination
Larrea tridentata
carbon isotope ratios, 176
field sites, 175
physiological ecology, 174–175
leaf soluble sugars
and insoluble fraction, 104–106
materials and methods, 94–96
water-use efficiency, 102–104
lichen
altitude effects, 205
CO$_2$ source, 203

gas exchange techniques, 202–204,
208–210
photobiont association, 204–208
physiology, 208–210
source, 202
thallus respiration, 206–208
light effects, 379–380
mineral nutrition effects, 304
peanut, *see* Peanut, carbon isotope
discrimination
Phaseolus, *see* Phaseolus, carbon isotope
discrimination
plant dry matter, water-use efficiency
and, 100–102
as predictor of transpiration efficiency,
316–318
screening, 410–414
as selection trait in breeding, 400
shoot biomass and, water stress, 363
spaced versus dense plantings, 411–
412
tall fescue accessions, 282–283
temperate grass, *see* Temperate grass,
carbon isotope discrimination
transpiration efficiency and
Banksia, 320–321
C$_3$ species, 318–323
carbon discrimination, 314–316
Eucalyptus camaldulensis, 318–320
factors affecting, 312–316
genetic correlation between, 316–
318
herbaceous crop species, 321–323
nutrition effect, 313–314
peanut, 312–314
relationship between, 312
studies, 247
temporal variation, 316
tropical forest, 115–119
calculation, 134
gas exchange, 135–137, 138
stomatal conductance and assimilation, 137
water-use efficiency, 135, 138
water-use efficiency and, plant breeding, 452–453, 456–460
wheat, *see* Wheat, carbon isotope discrimination
wheat grass, *see* Wheat grass, carbon
isotope discrimination
yield stability, 407–410
Carbon isotope effects
CO$_2$ exchange and, 47–58
C$_3$ metabolism, 56–58
C$_4$ metabolism, 58
kinetic effects, 48

processes affecting, 50–56
thermodynamic effects, 49
Carbon isotope fractionation
algae, 41
biochemical basis, 19–26
C₃ plant, 22, 25, 35–37, 43
C₄ plant, 23, 25–26, 37–40, 43–44
carboxylation resistance, 40
δ values, defined, 29
history, 10–11
maize, 42
process contributing to
carbon dioxide diffusion, 21–22
carboxylation, 22–23
environmental, 23
enzymatic, 22–23
kinetic, 21
physical, 21–22
species effects, 23
thermodynamic, 21
qualitative approach, 24–25
rye, 42
separation of diffusional steps, 34–35
theory, 24–26
tree ring
accumulated transpiration, 145, 149–
150
basal area increment index, 144–145,
147, 151
growth analysis, 144–145
isotope analysis, 145
isotope ratios, 148–150
methods, 142–145
site description, 143–144
source-water variation, 488
studies, 141–142
transpiration deficit, 145, 146–147
tree growth and, 147–148
water balance analysis, 145
water balance model, 142–143
two stage, 31–34
mathematics, 32–34
variation, 40–43
verification, 40
Carbon isotope ratio, *see also* Carbon-13/
carbon-12 ratio
Larrea tridentata, 176–182
ozone, 214, 223
physical chemistry, 47–50
tree ring carbon isotope fractionation,
148–150
Carboxylation, carbon isotope fractionations associated with, 55
Cellulose
deuterium/hydrogen ratio, 14
hydrogen removal, 488–489

Cellulose nitrate, water source versus
deuterium/hydrogen ratio, 487–488
Chloroplast water
CO₂ exchange and, 62–66
¹⁸O concentration in, 532–533, 538
Coastal wetland plants
deuterium depletion
growth rate comparison, 502
stem water and leaf area comparison,
503
stem water, mechanism, 505–508
water-loss rate comparison, 502
hydrogen isotope analysis, 500
isotopic fractionation, 499
greenhouse studies, 499–500
membrane characteristics, 507
oxygen isotope analysis, 500
stem and source water
deuterium/hydrogen ratio, 500–501
¹⁸O/¹⁶O ratio, 500
sampling, 498–499
stem water
deuterium depletion and transpiration, 508
Coffea, see Coffee
Coffee
carbon isotope discrimination
bean yield, 343
drought effects, 330–332, 341–343
gas exchange, 328, 334–335
effects, 332–333
genotypic variation, 329
hydraulic efficiency, 338
intrinsic water-use efficiency and,
332–333
leaf dry weight relationship, 330–332
as predictor of performance, 341–
343
root hydraulic conductance, 339
soil moisture alterations, 333–335
spatial scales affecting, 340–341
temporal scales affecting, 339–340
CO₂ assimilation and stomatal conductance, 334–335
gas exchange
drought effects on greenhouse, 337–
338
spatial scales affecting, 340–341
temporal scales affecting, 339–340
intrinsic water-use efficiency, leaf absolute symplastic volume, 336–337
leaf-water potential
drought effects on field, 341–343
drought effects on greenhouse, 337–
338
water variation, 329–330

soil drying response, 329–333
water availability, sensing, 335–339
Conifer
    altitudinal gradients, 194–195
    altitudinal variation, 191
    carbon isotope discrimination, 188–191
    photosynthetic rates in, CO_2 concentrations, 196
    water-use efficiency
    variables, 187
Coniferous canopy
    carbon isotope composition, 119
    carbon isotope discrimination, controlled conditions, 122–124
    CO_2 source, 119–122
Convection, oxygen isotope fractionation and, 60–62
Cotton
    carbon isotope discrimination
        variation, 314–316
        water-use efficiency relationship, 453
        water variation, 322
    transpiration efficiency, water variation, 322–323
Cowpea
    carbon isotope discrimination
        breeding strategies, 361–363
        drought effects, 353–354
        flowering effect, 360
        gas exchange, 354
        genotype, 351
        grain yield, 363
        harvest index, 360–361, 362, 364
        inheritance, 355–360
        shoot biomass, 363
        variation, 314–316
        water-use efficiency, 354–355, 453
        water variation, 322–323
    transpiration efficiency, water variation, 322
    water-use efficiency
        inheritance, 355–360
Craig–Gordon model, isotope fractionation, 71
Crassulacean acid metabolism (CAM), carbon isotope composition, 30–31
Creosote bush, *see Larrea tridentata*
Crop water-use efficiency, defined, 7

Decarboxylation
    carbon dioxide exchange and, 55–56
    glycine, photorespiration and, 55–56
Desert plants
    carbon isotope composition, *see* Carbon isotope composition, desert plants

carbon isotope discrimination, *see* Carbon isotope discrimination, desert plants
    climate and variation, 155–157
Desert precipitation, 156
Deuterium, source water and plant, 14–15
Deuterium/hydrogen ratio
    boxelder, tree ring growth, 489–490
    cellulose nitrate versus water source, 487–488
    coastal wetland plant, source and stem water, 499–501
    groundwater, 469
        in riparian communities, 475
    precipitation, 469
        in riparian communities, 475
    soil water, 484
    standards, 470
    as tracer, water sources and, 467–469
    transpiration, 74
    tree ring, water source and, 487–488
    water
        in coastal communities, 481–482
        in forest communities, 473–475
        plant, 470
        source, 470
        surface stream, 475
        variation, 465–466, 467–468, 488
    white pine, cellulose nitrate versus source-water, 491
    xylem-water, 469
        as function of stem age and position, 470
        linear mixing model, 471–472
        time of appearance, 472–473
Diffusion, oxygen isotope fractionation and, 60–62
Discrimination, *see also specific isotopes*
    carbon isotope
        defined, 34
        terms applied, 19–20
    CO_2, C_4 plants, 58
DNA polymorphism analysis
    Arabidopsis thaliana, carbon isotope discrimination, 376–377
    mutagenesis versus, 372–373
    principles, 373–374
Douglas fir, *see* Conifer

Eucalyptus camaldulensis
    carbon isotope discrimination and transpiration efficiency, 318
    creekside trees, flooding response, 524

deuterium concentration
  soil, root, and stem water, 514–515
  variations, 515
deuterium/hydrogen ratio
  soil, groundwater, and creek, 517
  tree sap, 517–518
inland trees, water uptake, 523–524
^{18}O concentration
  proportion of groundwater in tree
    sap, 525–526
  tree sap, creek water, and ground-
    water, 518
soil water potential, 520
water relations, 512
  analyses, 517
  ecological implications, 523–524
  field sampling, 515–517
  field site, 512–514
  practical implications, 524–525
  source water, 514–515
water source transpiration
  creekside trees, 521–523
  inland trees, 521
  rainfall, 523
  technique for determining, 520–
    521
Eucalyptus largiflorens, deuterium concentration, 514–515
Evaporation, oxygen isotope fractionations, 59–62
Evaporative-enrichment model, *see* Transpiration, isotope fractionation model
Evapotranspiration, tree ring carbon
  isotope fractionation, 143

Forest canopy
  carbon isotope composition, 110–115,
    131–138, 141–152, 187–197
Fractionation factor, 20–25, 32, 47–55,
  72–76

General combining ability (GCA), *Phaseolus vulgaris,* carbon isotope discrimination, 392
Genetic variation
  Arabidopsis thaliana, carbon isotope
    discrimination, 375
  conifer, carbon isotope discrimination,
    191–194
  wheat, carbon isotope discrimination,
    439–441
  wheatgrass, carbon isotope discrimina-
    tion, 276
Grass, *see specific type of grass*

Helianthus annuus
  carbon isotope discrimination
    insoluble fraction, 104–106
    leaf soluble sugars, 102–104
    leaves, stems, and roots, 100
    materials and methods, 94–96
  growth
    biomass and leaf area relationship, 96
    CO_2 assimilation, 98–100
    water availability, 96–97
  leaf water ^{18}O, *see* Leaf water, ^{18}O
Hydrogen isotope, xylem sap, desert
  plant, 165
Hydrogen isotope composition, *see also*
  Deuterium/hydrogen ratio
  leaf water
    boundary layer conductance, 75–76
    canopy position, 84–85
    diurnal variations, 77–78
    environmental and biological effects,
      77–80
    interspecies variation, 82–84
    intraspecies variation, 84
    leaf position, 85–86
    stomatal conductance, 85
    water stress effects, 81–82
    within-plant variation, 84–86
Hydrogen isotope fractionation
  coastal wetland plant, stem water, 499–
    500
  plant root, basis for study, 497–498
Hydrogen/oxygen isotope, history, 13–14
Hydrologic cycle, water deuterium/hydro-
  gen ratio, 465–466

Instantaneous water-use efficiency, defined, 5
Intrinsic water-use efficiency, defined, 6
Isotope discrimination, defined, 49–50
Isotope enrichment
  evaporating water body, equation for,
    529
  leaf water, 529–530
Isotope fractionation
  coastal wetland plants, *see* Coastal wet-
    land plants, isotope fractionation
  model, 71
  transpiration
    isotope effects, 72–73
    model, 73–74

Jeffrey pine
  carbon isotope discrimination
    factors affecting, 214–215
    measurements, 216–217
    needle age, 218, 222–223
    ozone sensitivity, 218, 221–223

ozone-sensitive
  CO_2 response measurements, 216
  factors, 213–215
  gas exchange measurements, 216
  intercellular CO_2 concentration in
    resistant and, 218
  ozone levels, 215
  photosynthetic rates in resistant and,
    217–218, 220, 221–222
  plant material, 215–216
  stomatal conductance in resistant and,
    218, 220, 222
  study site, 215–216
  water-use efficiency, 222–223

Kinetic fractionation factor, transpiration,
  72
Knudsen diffusion, CO_2 exchange, 52–53

Larrea tridentata
  carbon isotope discrimination in, *see*
    Carbon isotope discrimination,
    Larrea tridentata
  carbon isotope ratio
    assimilation rate, 181–182
    distribution ranges, 183
    photosynthetic capacity, 179
    vapor pressure experiments, 182
    water availability, 176–177
    water-use efficiency, 177, 179, 182
  conductance, assimilation relationship,
    180–181
  gas exchange data, 180–182
  photosynthetic capacity, nitrogen avail-
    ability, 179–180
  physiological ecology, 174–175
  water-use efficiency, field and con-
    trolled conditions, 180–181
Leaf
  CO_2 diffusion, 34–35
  CO_2 released by, $^{18}O/^{16}O$ ratio measure-
    ment, 531
  CO_2 transport, carbon isotope effects,
    52–55
  O_2 released by, $^{18}O/^{16}O$ ratio measure-
    ment, 531
  water, $^{18}O/^{16}O$, 59–62
  water-vapor concentrations, 5
Leaf absolute symplastic volume, water-
    use efficiency, 336–337
  characteristics, 328
Leaf soluble sugars
  carbon isotope discrimination
    CO_2 assimilation, 105
    materials and methods, 94–96
    water availability effect, 96–97

Leaf water, *see also* Transpiration
  compartmentation
    CO_2 approach, 535–536
    interpretation, 537–538
    isotopic composition, 537–538
    O_2 approach, 536
  discrimination, transpiration effects,
    80–81
  hydrogen isotope composition, 76–
    80
  isotope composition
    canopy position, 84–85
    diurnal patterns, 77–78
    environmental and biological effects,
      76–81
    interspecies variation, 82–84
    intraspecies variation, 84
    leaf position, 85–86
    observed and modeled, 76–81
    stomatal conductance, 85
    water stress effects, 81–82
    within-plant variation, 84–86
  isotope ratio measurements, 533–534
  isotopic steady state, 532
  $^{18}O/^{16}O$ ratio
    CO_2 approach, 535–536
    gas exchange, 531–533
    materials and methods, 531–535
    O_2 approach, 536
    sample collection, 531–533
    theory, 534–535
  $C^{18}O^{16}O$ discrimination and, 86–88
  oxygen isotope composition, 76–80
Leaf-water potential
  coffee, *see* Coffee, leaf-water potential
  Fagus gradifolia, 485–486
  Fragaria virginiana, 485–486
  Holcus lanatus, 485–486
  Solidago flexicaulis, 485–486
  temperate grass, carbon isotope discrim-
    ination, 284–286
  Vaccinium vacillans, 485–486
Lichen
  carbon isotope composition, 202
  carbon isotope discrimination, *see* Car-
    bon isotope discrimination, lichen
  characteristics, 201–202
  CO_2 compensation point, 203
  CO_2 concentrating mechanisms, 208–
    210
  CO_2 exchange measurements, 203,
    208–210
  photobiont association
    carbon isotope discrimination, 204–
      205
    gas exchange, 205–206, 208–210

photosynthetic rate, 206
thallus compensation point, 206
thallus respiration, 206–208
photosynthetic characteristics, 206–208
physiology, carbon isotope discrimination, 208–210
Loblolly pine
carbon isotope composition
measurements, 230
ozone exposure, 232–236
phosphoenolpyruvate carboxylase, 239
water stress, 232–236
gas exchange measurements, 230
ozone affected
carbon isotope composition, 232–236
experimental design, 229–230
exposure chamber, 229
internal CO_2 concentration, 236, 238
net photosynthesis, 236
ozone exposure and meteorology, 231
phosphoenolpyruvate carboxylase, 239
sensitivity factors, 240
stomatal conductance, 236, 239–240
study area, 228–229
treatment regimes, 229–230
water stressed
carbon isotope composition, 232–236
experimental design, 229–230
internal CO_2 concentration, 236
net photosynthesis, 236
stomatal conductance, 236, 239–240
treatment regimes, 229–230
Lod score, RFLP and, 373

Maize, carbon isotope fractionation, 42
Mutagenesis
DNA polymorphism analysis versus, 372–373
transpiration efficiency, 372

Natural selection, desert plant, carbon isotope discrimination, 166–170
Nitrogen, effect on transpiration efficiency, 313–314
Nothofagus species, carbon isotope discrimination and altitude, 194

Orchardgrass, *see also* Temperate grass
drought resistance, 284
water-use efficiency, 288
Oxygen
evaporative enrichment, 13–14
evolution, water stress, 98–100

Oxygen-18
content, studies, 14
concentration
see Eucalyptus camaldulensis, 518, 525–526
proportion of groundwater in tree sap, 525–526
tree sap, creek water, and groundwater, 518
Oxygen-18/oxygen-16 ratio
boundary layer effects, 75–76
coastal wetland plants, stem and source water, 501
leaf water
$C^{18}O^{16}O$ discrimination, 86–88
convection and diffusion factors, 60–62
evaporation factor, 59–60
leaf CO_2 exchange, 62–66
water, CO_2 exchange, 62–66
Oxygen isotope, atmospheric CO_2, 58
Oxygen isotope composition
leaf water
boundary layer conductance, 74–76
canopy position, 84–85
diurnal variations, 77–78
environmental and biological effects, 77–80
interspecies variation, 82–84
intraspecies variation, 84
leaf position, 85–86
stomatal conductance, 85
water stress effects, 81–82
within-plant variation, 84–86
Oxygen isotope effects
CO_2 exchange and, 58–66
convection and diffusion as factors, 60–62
evaporation fractionations, 59–60
leaf, 62–66
Oxygen isotope fractionation
evaporation, 59–60
transpiration, 72–73
Ozone
Jeffrey pine injury from, 213–215
loblolly pine and, *see* Loblolly pine, ozone affected
phosphoenolpyruvate carboxylase and, 239
phytotoxicity, mechanism, 227–228
shortleaf pine and, *see* Short leaf pine, ozone affected
uptake
carbon isotope ratio, 214
stomatal conductance, 214, 228, 239–240

Peanut
   carbon isotope discrimination
      biomass relationship, 456–458
      carbon discrimination, 314–316
      distribution, 316–317
      genotype × environment, 257–258
      genotypic ranking, 352
      harvest index, 262–264
      leaf area versus, 258–261, 265
      nutrition effect, 313–314
      temporal change, 264
      total dry matter, 255–257
      transpiration efficiency versus, 249,
         255–257, 312–313
      water-use efficiency relationship, 453
   leaf area, water regimes, 260
   transpiration efficiency
      carbon isotope discrimination versus,
         249, 255–257, 312–313
      field canopy measurements, 252–253
      harvest index association, 261–264
      selection, 264–265
      water stress, 253–254, 261–262
      whole plant, 248–252
Péclet number, oxygen isotope fractionation, 60–62
Pee Dee limestone, 20, 47
Phaseolus, see also Bean
   carbon isotope discrimination
      developmental stage effect, 304–305
      environmental and developmental
         effects, 299–301, 307
      interspecific variation, 301
      leaf age effect, 305–307
      materials and methods, 298–299
      photosynthetic photon flux density,
         304–305
      soil moisture effects, 301–303
      soil nitrogen effect, 303–304
   P. acutifolius
      characteristics, 298
      germplasm, carbon isotope discrimi-
         nation, 388
      stomatal conductance, 303
   P. lunatus germplasm, carbon isotope
      discrimination, 388
   P. vulgaris
      breeding
         characteristics, 387
         implications, 396–397
      carbon isotope discrimination
         breeding and implications, 396–
            397
         crop dry weight, 388
         crop growth association, 393–394

flowering, 388
         general and specific combining
            ability, 392
         inheritance, 391–392
         physiological traits association, 393
         root length density, 395–396
         seed yield, 388
         seed yield association, 393–394
         water deficits, 388, 394–396
      characteristics, 298
      germplasm, carbon isotope discrimi-
         nation variation, 388–390
      stomatal conductance, 303
Phosphoenolpyruvate carboxylase, 37–
   38
   C_4 plant carboxylation, 23
   CO_2 exchange, 55
   ozone exposure, 239
Photorespiration, glycine decarboxylation,
   55–56
Photosynthesis
   C_3
      carbon dioxide exchange, 56–58
      carbon isotope effects, 56–58
      carbon isotope fractionation, 22, 25,
         29–30, 35–37, 43
      CO_2 fixation, 22
      daylight respiration, 35–37
      isotope fractionation, 25
      transpiration, 158
   C_4
      carbon isotope discrimination, 37–40
      carbon isotope effects, 58
      carbon isotope fractionation, 23, 25–
         26, 29–30, 37–40, 43–44
      CO_2 exchange, 58
      CO_2 fixation, 23
      discovery, 11
      isotope fractionation, 25–26
   carbon and water flux concepts, 5
   CO_2 exchange in, heterogeneity effects,
      57–58
   leaf water $^{18}O/^{16}O$ ratio, $C^{18}O^{16}O$ dis-
      crimination, 86–88
   loblolly pine, 236
   pathways, carbon isotopes, 12
   shortleaf pine, 236
Photosynthetic photon flux density, coniferous canopy, 121, 122–123
Photosynthetic water-use efficiency, *see*
   Water-use efficiency
Pine
   Jeffrey, *see* Jeffrey pine
   loblolly, *see* Loblolly pine
   shortleaf, *see* Shortleaf pine

Plant breeding
  barley, carbon isotope discrimination,
    400–414
  biomass genetic intransigency, 451–452
  bread wheat, *see* Bread wheat breeding
  carbon isotope discrimination as crite-
    rion, 316–318, 355–359, 375–383,
    391–395, 410–414, 431–432, 453–
    460
  genetic challenges, 455–456
  genotypic ranking, 162–164, 352–353
  physiological selection criteria, 399–
    400
  steps, 399
  strategies, 361–367
  water-use efficiency improving, 451–
    453
Plant canopy, isotopic composition of air,
  50–51, 76–86, 109–126, 131–139
Plant roots, hydrogen isotope fractionation, basis for study, 497–498
Plants, *see also specific plants*
  C_3
    CO_2 fixation, 22
    daylight respiration, 35–37
    isotope fractionation, 25
  C_3 and C_4, carbon isotope fractionation,
    29–30
  C_4
    carbon isotope discrimination, 37–40
    CO_2 fixation, 23
    discovery, 11
    isotope fractionation, 25–26
  CAM
    carbon isotope composition, 30–31
    carbon isotope fractionation, 44
  carbon isotope discrimination, *see specific
    species*
  carbon isotope exchange, air isotope
    effect, 50–51
  carbon isotope fractionation, history,
    10–11
  coastal wetland, *see* Coastal wetland
    plants
  diffusion, isotope fractionation, 21–22
  growth
    water-use concepts, 4–7
    water-use efficiency, 3–4
  isotope composition, defined, 49–50
  standard, 20
  uptake scheme, isotope fractionation,
    24–25
Plant-water sources
  deuterium/hydrogen ratio
    in coastal communities, 481–482

  in riparian communities, 475–478
  as tracer, 467–469
  methods for determining, 469–473
  sampling, 469–470
Polyethylene glycol, temperate grass
  stress, 289
Potato, carbon isotope discrimination and
  water-use efficiency relationship, 453
Precipitation, coefficients of variation,
  desert, 156

Ranunculus, photosynthesis versus CO_2
  concentration, 196
Respiration, C_3 daylight, 35–37
Respiratory recycling, forest canopy, 114–
  115
Restriction fragment length polymorphism (RFLP) analysis
  Arabidopsis thaliana, carbon isotope
    discrimination and, 376–377
  principles, 373–374
Ribulose 1,5-bisphosphate carboxylase/
  oxygenase
  CO_2 exchange and, 55
  isotope fractionation by, 22–23
Root hydraulic conductance, coffee carbon isotope discrimination, 338–339
Rye, carbon isotope fractionation, 42
Ryegrass, *see also* Temperate grass
  water-use efficiency, 288

Shortleaf pine
  carbon isotope composition
    measurements, 230
    ozone exposure, 232–236
    phosphoenolpyruvate carboxylase,
      239
    water stress, 232–236
  ozone affected
    carbon isotope composition, 232–236,
      240
    data analysis and statistics, 230–231
    experimental design, 229–230
    exposure chamber, 229
    exposures and meteorology, 231
    internal CO_2 concentration, 236, 238
    measurement errors, 238
    net photosynthesis, 236
    phosphoenolpyruvate carboxylase,
      239
    sensitivity factors, 240
    stomatal conductance, 236, 239–240
    study area, 228–229
    treatment regimes, 229–230

water stressed
    carbon isotope composition, 232–236,
        240
    data analysis and statistics, 230–231
    experimental design, 229–230
    internal CO_2 concentration, 236
    measurement errors, 238
    net photosynthesis, 236
    stomatal conductance, 236, 239–240
    treatment regimes, 229–230
    water-use efficiency, 238
Solute potential, temperate grass, carbon
    isotope discrimination, 284–286
Specific combining ability (SGA), *Phaseolus
    vulgaris*, carbon isotope discrimination
    and, 392
Stable isotopes, historical aspects, 9–15
Standard light antarctic precipitation, 470
Standard mean ocean water (SMOW), 470
Stomata
    closure, sunflower growth, 98–100, 104
    CO_2 diffusion, 34–35
        carbon isotope effects, 52
        isotope fractionation, 24–25
        variation, CO2 exchange, 57
Stomatal conductance
    coffee, CO_2 assimilation, 334–335
    Fagus gradifolia, 485–486
    Fragaria virginiana, 485–486
    Holcus lanatus, 485–486
    ozone and, 214, 223, 227–228
        Jeffrey pine, 218, 220
        loblolly pine, 236
        shortleaf pine, 236
    Phaseolus, 303
    Solidago flexicaulis, 485–486
    Vaccinium vacillans, 485–486
    wheat, 440–441
Sunflower, carbon isotope discrimination
    and water-use efficiency relationship,
    453

Tall fescue
    accessions, carbon isotope discrimina-
        tion, 282–283, 284–286
    Acremonium coenophialum effects, 293–
        295
    carbon isotope discrimination, *Acremo-
        nium coenophialum* effects, 293–295
    drought resistance, 284
    gas exchange, 288–293
    germplasm enhancement, 293–295
    seasonal forage yield, dryland condi-
        tions, 286
    solute potential, dryland conditions, 286

turgor pressure, dryland conditions,
    286
    water potential, dryland conditions, 286
    water-use efficiency, 288
Temperate grass
    carbon isotope discrimination
        dryland conditions, 286
        field evaluation, 282–286
        gas exchange, 288–293
        leaf-water potential, 284–286
        materials and methods, 282–283
        solute potential, 284–286
        turgor pressure, 284–286
        water stress, 283–284
    water-use efficiency
        gas exchange, 288–293
        whole plant, 286–288
Thallus CO_2 compensation point, lichen,
    203–204
Tomato, carbon isotope discrimination
    and water-use efficiency relationship,
    453
Transpiration
    isotope effects during, 72–73
    isotope fractionation model, 73–74
        conditions for using, 74
        boundary layer influence, 74–76
    concept, 5
Transpiration deficit, tree ring carbon
    isotope fractionation, 145, 146–147
Transpiration efficiency
    barley
        biomass relationship, 405
        carbon isotope discrimination, 405–
            407, 412–414
    bread wheat, *see* Bread wheat, transpira-
        tion efficiency
    carbon isotope discrimination
        Banksia, 320–321
        C_3 species, 318–323
        carbon discrimination and, 314–316
        Eucalyptus camaldulensis, 318–320
        factors affecting, 312–316
        genetic correlation between, 316–318
        herbaceous crop species, 321–323
        nutrition effect, 313–314
        peanut, 312–314
        relationship between, 312
        studies, 247
        temporal variation, 316
    crop, 435
    crop productivity, 93
    defined, 6, 93–94
    DNA polymorphisms, 372–373
    dry matter gain, 311

intrinsic, enhancement, 328
molecular analysis, 372–373
mutagenesis, 372
ozone, 228
peanut
  field canopies, 252–257
  genotype × environment, 257–258
  harvest index association, 261–264
  leaf area versus, 258–261, 265
  partitioning ratio versus, 261–262
  selection, 264–265
  water stress, 249, 251–252
  whole plant, 248–249
variation, C_3 plants, 248
water deficit, 228
water-use efficiency and crop yield
  relationship, 435
wheat, *see* Wheat, transpiration efficiency
Transpiration ratio, defined, 7
Tree ring
  carbon isotope fractionation
    accumulated transpiration, 145, 149–150
    basal area increment index, 144–145, 147, 151
    growth analysis, 144–145
    isotope analysis, 145
    isotope ratios, 148–150
    methods, 142–145
    site description, 142–143
    source-water variation, 488
    studies, 141–142
    transpiration deficit, 145, 146–147
    tree growth, 147–148
    water balance analysis, 145
    water balance model, 142–143
  cellulose nitrate deuterium/hydrogen
    ratio, boxelder, 489–490
  deuterium/hydrogen ratio, source-water
    variation, 488
Tropical forest
  carbon isotope composition
    CO_2 flux, 117–119
    CO_2 source, 116–117
    effect of height, 115–116
Turgor pressure, temperate grass, carbon
  isotope discrimination, 284–286

Vapor pressure deficit, water-use, 5

Water, *see also* Xylem-water; Plant water
  sources

apoplastic uptake, 506–507
deuterium/hydrogen ratio
  desert community, 478–479
  forest community, 473–475
  methods for determining, 470
  riparian community, 475–478
  soil, 484
  standards, 470
  as tracer, 467–469
  variation, 465–466
  variation in plant, 486–492
evaporative enrichment, 13–14
hydraulic lift, sampling zones, 484
leaf, *see* Leaf water
movement, studies, 15
sources, 467–469
  coastal community, 481
  desert community, 478
  deuterium/hydrogen ratio variation, 486–492
  methods for determining, 469–473
  plant–plant interactions, 482–486
  riparian community, 475
  sampling, 469–470
  water absorption relationship, 476–478
Water availability, coffee, sensing of soil, 335–339
Water balance
  model, tree ring, 142–143
  tree ring carbon isotope fractionation
    and, 146–147
Water stress, *see also* Transpiration efficiency
  barley, carbon isotope discrimination, 402–405
  loblolly pine, *see* Loblolly pine, water
    stressed
  peanut
    carbon isotope discrimination, 253–254
    transpiration efficiency, 249, 251–252
  shortleaf pine, *see* Shortleaf pine, water
    stressed
  temperate grass, carbon isotope discrimination, 283–284
  transpiration efficiency, 228
Water-use efficiency, *see also* Transpiration
  efficiency
  biomass, 451–452
  C_3 transpiration, 158
  carbon isotope discrimination and, 270
  leaf sugars, 102–104
  plant breeding, 452–453, 456–460,
  plant dry matter, 100–102

carbon isotope composition, 6
  desert, 479
CO₂ assimilation, 98–100
coffee
  carbon isotope discrimination, 332–333
  hydraulic efficiency, 338–339
  leaf absolute symplastic volume, 336–337
conifer
  altitudinal variation, 191
  carbon isotope discrimination, 187, 188–191
cowpea, inheritance, 355–360
crop, 435
  defined, 7
defined, 93–94
desert plant, carbon isotope composition, 159–160, 162–163
factors affecting, 100–101, 269
improving, plant breeding, 451–453
inheritance, 355–360
instantaneous, defined, 5
interest in, 3–4
intrinsic, defined, 6
Jeffrey pine, ozone injury, 222–223
Larrea tridentata
  carbon isotope ratio, 177, 179, 182
  field and controlled conditions, 180–181
  phenotypic components, 183–184
  vapor pressure experiments, 182
mineral concentration, 384
plant-breeding programs, 269–270
plant growth and, concepts, 4–7
plant-water potential relationship to, desert, 479
shortleaf pine, water stress, 238
soil surface evaporation, 7
temperate grass, 286–288
transpiration efficiency relationship, 435
tropical forest, carbon isotope composition, 135, 138
water availability, 96–97
wheat, 444–445
wheatgrass, *see* Wheatgrass, water-use efficiency
xylem-water δD relationship, desert, 479
Water vapor, isotope composition, fractionation, 74
Wheat
  carbon isotope discrimination
    biomass relationship, 456–458
    canopy growth relationship, 458
    flowering time relationship, 459–460
    grain yield relationship, 437–439
    photosynthetic capacity change effect, 448
    seasonal variation, 445
    sources of variation, 439–441
    transpiration efficiency, 436–437
    water-use efficiency relationship, 453
  dry matter production, 442–443
  gas exchange, variation, 439–440
  grain yield, carbon isotope discrimination relationship, 437–439
  photosynthetic capacity changes, 447–448
  scale experiments, conclusions, 447
  stomatal conductance, 440–441
  transpiration efficiency
    carbon isotope discrimination, 436–437
    crop, 444–445
    genetic variation, 436–437
    photosynthetic capacity change effect, 448
    scale effects impact, 441–447
    seasonal variation, 445
  water use, 442–443
  water-use efficiency, crop, 444–445
Wheatgrass
  carbon isotope discrimination
    clone stability, 276–277
    environmental effects, 273–275
    forage yield, 277–279
    genetic variation, 276
    genotypic ranking, 352
    water level, 274
    water-use efficiency, 271–273, 453
  characteristics, 270
  drought resistance, 284
  water-use efficiency
    carbon isotope discrimination, 271–273
    early studies, 270
    intrinsic, 272
White pine
  deuterium/hydrogen ratio
    cellulose nitrate versus source-water, 491
    xylem-water, 474
Wildrye, carbon isotope discrimination and water-use efficiency relationship, 453

Xylem water
  deuterium/hydrogen ratio, 469
  Fagus gradifolia, 485–486

forest communities, 473–475
Fragaria virginiana, 485–486
as function of stem age and position, 470
Holcus lanatus, 485–486
linear mixing model for, 471–472
riparian communities, 476

soil profile and, 476–477
Solidago flexicaulis, 485–486
time of appearance, 472–473
tree-rooting patterns and, 476–477
Vaccinium vacillans, 485–486
water sources and, 476–477

Physiological Ecology

A Series of Monographs, Texts, and Treatises

Continued from page ii

T. D. SHARKEY, E. A. HOLLAND, and H. A. MOONEY (Eds.). Trace Gas Emissions by Plants, 1991

U. SEELIGER, (Ed.). Coastal Plant Communities of Latin America, 1992

J. R. EHLERINGER and C. B. FIELD (Eds.). Scaling Physiological Processes: Leaf to Globe, 1993